Industry 5.0

Other related titles:

You may also like

- PBME027 | Leong | ESG Innovation for Sustainable Manufacturing Technology: Applications, designs and standards | 2024
- PBTE108 | Imran | The Role of 6G and Beyond on the Road to Net-Zero Carbon | 2023

We also publish a wide range of books on the following topics:
Computing and Networks
Control, Robotics and Sensors
Electrical Regulations
Electromagnetics and Radar
Energy Engineering
Healthcare Technologies
History and Management of Technology
IET Codes and Guidance
Materials, Circuits and Devices
Model Forms
Nanomaterials and Nanotechnologies
Optics, Photonics and Lasers
Production, Design and Manufacturing
Security
Telecommunications
Transportation

All books are available in print via https://shop.theiet.org or as eBooks via our Digital Library https://digital-library.theiet.org.

IET MANUFACTURING SERIES 26

Industry 5.0

Design, standards, techniques and applications for manufacturing

Edited by
Wai Yie Leong

The Institution of Engineering and Technology

About the IET

This book is published by the Institution of Engineering and Technology (The IET).

We inspire, inform and influence the global engineering community to engineer a better world. As a diverse home across engineering and technology, we share knowledge that helps make better sense of the world, to accelerate innovation and solve the global challenges that matter.

The IET is a not-for-profit organisation. The surplus we make from our books is used to support activities and products for the engineering community and promote the positive role of science, engineering and technology in the world. This includes education resources and outreach, scholarships and awards, events and courses, publications, professional development and mentoring, and advocacy to governments.

To discover more about the IET please visit https://www.theiet.org/.

About IET books

The IET publishes books across many engineering and technology disciplines. Our authors and editors offer fresh perspectives from universities and industry. Within our subject areas, we have several book series steered by editorial boards made up of leading subject experts.

We peer review each book at the proposal stage to ensure the quality and relevance of our publications.

Get involved

If you are interested in becoming an author, editor, series advisor, or peer reviewer please visit https://www.theiet.org/publishing/publishing-with-iet-books/ or contact author_support@theiet.org.

Discovering our electronic content

All of our books are available online via the IET's Digital Library. Our Digital Library is the home of technical documents, eBooks, conference publications, real-life case studies and journal articles. To find out more, please visit https://digital-library.theiet.org.

In collaboration with the United Nations and the International Publishers Association, the IET is a Signatory member of the SDG Publishers Compact. The Compact aims to accelerate progress to achieve the Sustainable Development Goals (SDGs) by 2030. Signatories aspire to develop sustainable practices and act as champions of the SDGs during the Decade of Action (2020-2030), publishing books and journals that will help inform, develop, and inspire action in that direction.

In line with our sustainable goals, our UK printing partner has FSC accreditation, which is reducing our environmental impact to the planet. We use a print-on-demand model to further reduce our carbon footprint.

British Library Cataloguing in Publication Data
A catalogue record for this product is available from the British Library

ISBN 978-1-83724-009-8 (hardback)
ISBN 978-1-83724-010-4 (PDF)

Typeset in India by MPS Limited

Cover image: Shanghai Miji Network Technology Co. Ltd

Original AI intelligent factory assembly line by Peng Guinong, Shanghai Miji Network Technology Co. Ltd

Contents

Preface

The world stands on the brink of another industrial transformation – one that builds upon the digital foundations of Industry 4.0 and reaches new heights with the emergence of Industry Revolution 5.0. Unlike its predecessor, which focused heavily on automation, cyber-physical systems and the Internet of Things, Industry 5.0 shifts the paradigm towards human-centred solutions, blending the power of cutting-edge technologies with human creativity, expertise and ethical considerations. This evolution is not merely technological; it is social and economic, reshaping how we live, work and interact with our environment.

The convergence of advanced artificial intelligence, robotics, 3D printing and the Internet of Everything is now being augmented with a renewed focus on collaboration between humans and machines. This human-machine synergy unlocks new potential in industries ranging from manufacturing to healthcare, from energy to education, as intelligent systems learn to complement, rather than replace, human ingenuity. The aim of Industry 5.0 is to not only enhance productivity but also create a more sustainable, ethical and personalised approach to industrial processes.

This book, *Industry 5.0: Design, standards, techniques and applications for manufacturing*, provides a comprehensive exploration of the key components that define this new industrial era. It is designed to serve as a foundational guide for engineers, designers, researchers and policymakers who seek to understand and contribute to the future of Industry 5.0. Through an examination of its applications, innovative designs, emerging standards and the latest techniques, this work offers a critical resource for those striving to navigate the complex yet exciting landscape of this revolution.

The chapters in this book delve into the core technologies and methodologies driving Industry 5.0. We discuss practical applications across various sectors, providing insights into how human-centric designs are redefining production, how standardised protocols are being adapted to accommodate intelligent systems and how innovative techniques are being implemented to address societal and environmental challenges. Moreover, this book addresses the ethical implications of human-machine collaboration, the need for sustainable practices and the role of regulation and standardisation in ensuring the responsible development of these transformative technologies.

As we embark on this next phase of industrial evolution, the balance between technological advancement and human well-being becomes ever more critical. I hope this book serves not only as an educational tool but also as an inspiration for

future innovations that uphold the values of sustainability, inclusivity and collaboration. Let us embrace Industry 5.0, not just as an advancement in technology but
as a path towards a more equitable and harmonious industrial future.

I am grateful to the contributors, collaborators and experts who have shared
their insights to make this book a reality. Their collective knowledge and vision
illuminate the transformative potential of Industry 5.0, paving the way for a new
chapter in human and technological development.

About the editor

Wai Yie Leong is a senior professor at INTI International University, Malaysia. Wai Yie is the past chairperson of IET Malaysia Local Network, council member of IET UK and vice president of the Institution of Engineers Malaysia (IEM). She received her PhD in Electrical Engineering from The University of Queensland, Australia. She specialises in medical signal processing, industrial revolution 4.0 technology, 5G and telecommunications.

Chapter 1

Introduction to Industry Revolution 5.0— current and future systems, features and benefits

R. Kamalakannan[1], P. Sivakumar[2], Dadapeer Doddamani[2], Tan Koon Tatt[3] and Nasser Mohammed Nasser Al Jahadhmi[2]

With the arrival of Industry Revolution 5.0, industrial practices will undergo a radical change as human ingenuity and cutting-edge technology are combined to build production systems that are efficient, sustainable, and harmonious. This chapter examines the features, systems, and advantages of Industry Revolution 5.0, both present and future. Industry 5.0 is centered on human-centric solutions, machine-human collaboration, and the integration of personalized experiences into manufacturing. This is in contrast to Industry 4.0, which was centered on the automation and interconnectivity of production systems through technologies like Internet of Things (IoT), artificial intelligence (AI), and big data. In-depth discussion of Industry 5.0's key components is provided in this chapter, including the application of cobots (collaborative robots), improved customization options, and environmentally friendly production techniques. It also emphasizes the advantages of this revolution, including enhanced productivity, better-quality products, and the development of more satisfying work conditions. This study attempts to provide a thorough knowledge of how Industry Revolution 5.0 is poised to transform the landscape of industrial operations, promoting innovation and fostering a more resilient and adaptive industrial ecosystem by looking at existing implementations and future projections.

1.1 Introduction

With every industrial revolution, from Industry 1.0's mechanization to Industry 4.0's automation and digitization, the environment of industrial production has

[1]Department of Mechanical Engineering, Thiagarajar College of Engineering, India
[2]Mechancial and Industrial Engineering, University of Technology and Applied Sciences, Oman
[3]School of Technology and Engineering Science, Wawasan Open University, Malaysia

experienced profound changes. As we approach yet another critical turning point, Industry Revolution 5.0 appears, bringing with it a new era that seeks to balance the complementary effects of cutting-edge technologies and human creativity. Industry 5.0 prioritizes a human-centric strategy, in contrast to its predecessor, which was primarily focused on increasing efficiency and productivity through automated processes and networked technologies. The goal of this paradigm shift is to improve customization, sustainability, and the general standard of work settings by incorporating collaborative robots, or cobots, into production lines. Industry 5.0 is distinguished by its emphasis on integrating smart technologies with human creativity and emotional intelligence to create production systems that are resilient, adaptive, and efficient all at the same time. By putting an emphasis on sustainability and customized manufacturing, it overcomes the drawbacks of Industry 4.0 and guarantees that technical breakthroughs result in significant and lasting changes to the workplace and the products being produced [1]. The purpose of this introduction is to examine the present situation and future prospects of Industry 5.0, stressing its salient characteristics and the advantages it seeks to provide. We will learn more about how this next industrial revolution is likely to change manufacturing by encouraging innovation and developing more sustainable and human-centered industrial processes by closely examining these elements.

1.1.1   Strategies of Industry 5.0

Industry 5.0 is about bringing sophisticated technologies and human talents together in a way that is harmonious. It is a paradigm change in manufacturing and industrial processes. The following comprehensive tactics capture the spirit of Industry 5.0.

1.1.1.1   Human centric manufacturing

Including collaboration robots, or cobots, to increase productivity and security while assisting human workers. Cobots can perform dangerous or repetitive duties, freeing up humans to concentrate on more strategic and creative work. Putting money into programs for ongoing education and training will help the workforce become more skilled so they can work together with cutting-edge technology and change with the times.

1.1.1.2   Customization and personalization

Putting in place adaptable manufacturing technologies that facilitate mass customization and can quickly adjust to shifting customer demands. This covers adaptable machinery and modular production lines. creating customized products that cater to each customer's requirements and preferences through data analytics and consumer feedback, increasing customer happiness and loyalty.

1.1.1.3   Sustainability and green manufacturing

Adopting techniques like recycling, energy-efficient procedures, and the use of sustainable materials that reduce waste and maximize resource utilization.

Reducing environmental effect by designing items with their whole lifecycle in mind and encouraging recycling, refurbishing, and reuse.

1.1.1.4   Advanced technologies integration

Using AI and ML to improve decision-making, forecast maintenance requirements, and optimize production processes. AI is capable of analyzing enormous volumes of data to find trends and streamline processes. IoT-enabled device and system connections allow for real-time data gathering, communication, and monitoring throughout the manufacturing line, increasing responsiveness and efficiency.

1.1.1.5   Enhanced connectivity and data utilization

Establishing networked production environments that facilitate seamless communication between humans, machines, and systems. Using digital twins to model and improve industrial processes is one example of this. Applying big data analytics to understand consumer behavior, market trends, and production processes in order to improve strategic planning and decision-making.

1.1.1.6   Resilience and agility

Creating supply networks that can react swiftly to demand fluctuations and interruptions. This entails introducing flexible logistical solutions, expanding transparency, and broadening the pool of providers. Utilizing digital tools and technology to build production environments that are responsive and flexible. This covers the application of cutting-edge cybersecurity techniques, blockchain technology for supply chain transparency, and cloud computing.

1.1.1.7   Ergonomics and safety

Creating work environments that put employee health and safety first by implementing improved safety procedures, automated heavy lifting help, and ergonomic equipment. Putting an emphasis on workers' overall well-being by encouraging work-life balance, offering mental health help, and developing a pleasant workplace culture

1.1.1.8   Innovation and continuous improvement

Making R&D investments to stay abreast of market trends and technical breakthroughs. Fostering innovation by collaborating with startups, academic institutions, and research centers, putting into practice continuous improvement techniques like Six Sigma and lean manufacturing concepts to continuously improve and optimize production operations.

1.1.1.9   Ethical and inclusive practices

Encouraging a workforce that is inclusive and diverse, bringing a range of viewpoints and ideas that foster creativity and innovation. Ensuring that automation and AI are applied morally, taking into account issues of bias, privacy, and employment.

1.2   Industry 5.0—revolutionizing the factory floor with human-centric manufacturing

After more than ten years of benefits from Industry 4.0, the manufacturing sector is ready for Industry 5.0. The Fourth Industrial Revolution, or industry 4.0, has given businesses countless options in the areas of automation in the manufacturing sector, digitization, efficiency, mass production, interconnection, mass data collecting, and customization of any component of the production line [2]. But even with all of these benefits, Industry 4.0 is not without its drawbacks. Workers in manufacturing and other industries frequently feel underappreciated in the competitive modern era, since their duties frequently veer toward managerial responsibilities. They are not directly involved in the production process to satisfy the market's demands for speed, security, and customization as well as the ever-changing preferences of users. The next industrial revolution, known as Industry 5.0, seeks to solve these issues by fusing human innovation with automated efficiency. Robots in Industrial Revolution 5.0 are cognitively aware of humans and cooperate with them rather than competing with them. By utilizing their intuition, abilities, and capacity for creative problem-solving, it increasingly integrates workers into production processes rather than completely eliminating human involvement in the manufacturing process. Because of this, the idea of industry 5.0 manufacturing improves on earlier ideas by placing workers back at the center of the plant and fostering intimate communication between humans and machines. Let's explore the manufacturing 5.0 revolution in more detail, including its advantages, obstacles to its implementation, and prospects for human-centric industry 5.0 in this revolutionary period.

1.2.1   Recognizing Industry 5.0's significance in the manufacturing sector

Staying ahead in the fast-paced corporate environment of today requires embracing the bleeding edge of technology advancement, and this is where Industry 5.0 excels [3]. The goal of this next stage of industrial innovation is to transform the factory floor based on three fundamental ideas: resilience, sustainability, and human-centeredness. It is not simply about machines (Figure 1.1).

Sustainability: Industry 5.0 is dedicated to environmentally friendly methods that support long-term ecological balance and reduce their negative effects on the environment. The manufacturing 5.0 revolution promotes sustainability by optimizing resource use, energy efficiency, and waste reduction, helping to create a future that is more robust and ecologically conscientious.

Resilience: Industry 5.0 places a high priority on resilience in the face of changing problems including the COVID-19 pandemic, global supply shortages, natural disasters, and geopolitical instability. It places a high priority on creating flexible, adaptive production systems that can survive interruptions. Industry 5.0 provides businesses with creative solutions to help them deal with uncertainty by utilizing cutting-edge technologies and flexible manufacturing processes.

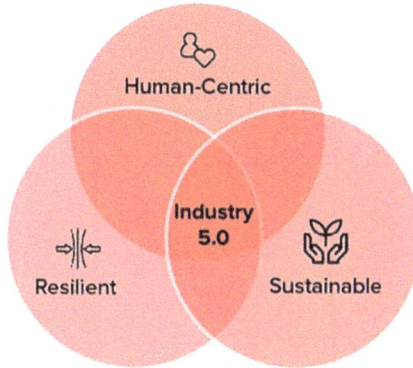

Figure 1.1   Industry 5.0 strategies

Human-centeredness: In addition to the strength of intelligent machinery and state-of-the-art technologies, the Fifth Industrial Revolution highlights the value of human interaction, inventiveness, and teamwork. Industry 5.0 aims to improve worker quality of life, job happiness, and workplace safety by encouraging human-machine collaboration. Numerous studies carried out in the manufacturing industry have revealed that 71% of the occupations that people perform on a daily basis are unsafe, boring, or unclean. Furthermore, more than 80% of these workers have disclosed that they had been injured at work. At this point, Industry 5.0 shows itself as a reliable savior, providing a symbiotic partnership between humans and robots to securely handle physically hard tasks and free up workers to concentrate more on activities that bring value. Consider yourself a manufacturing line worker who must assemble a product that requires multiple hefty components. Therefore, you can make use of robotic hands that will perform all the heavy lifting for you rather than endangering your back and shoulders by carrying a large metal component. Additionally, if you are in an unsafe location, a cobot can safeguard you from harm by miming your motions with extra thought. These instances of Industry 5.0 demonstrate how humans and machines work together and progress instead of replacing one another.

1.3   Benefits of Industry 5.0 in manufacturing

Numerous advantages are provided by Industry 5.0 to workers, production processes, and society at large. Let's examine each of the benefits of Industry 5.0 for the expansion of the manufacturing sector in more detail (Figure 1.2).

1.3.1   Enhanced efficiency and productivity

Businesses may increase productivity, decrease downtime, and improve overall efficiency by utilizing cutting-edge technology like AI, the IoT, and robotics [4]. Human labor may focus on more intricate and valuable tasks because of this optimization, which raises productivity levels. When a system is not producing more

Figure 1.2   Benefits of Industry 5.0 in manufacturing

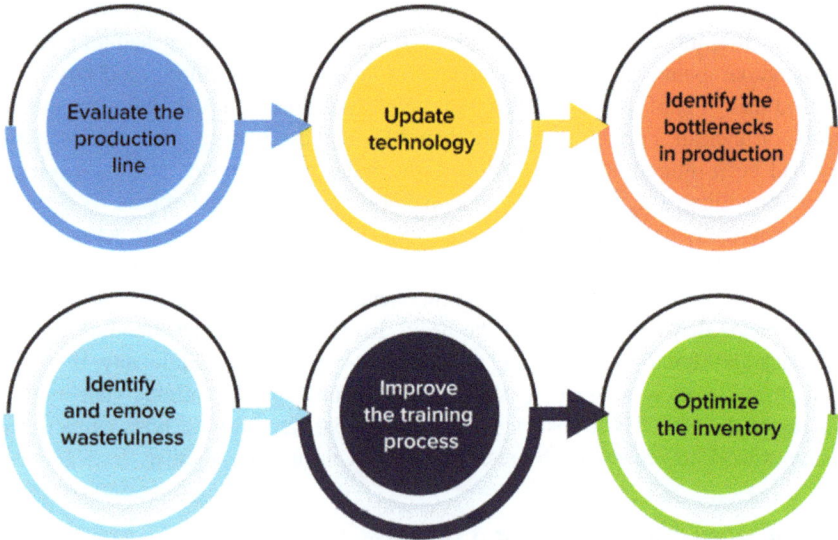

Figure 1.3   Steps to achieve production efficiency

goods without affecting the production of other things, it is said to be operating efficiently. This occurs when the best possible combination of labor and capital is used to produce materials at the lowest feasible cost. Generally, the percentage of scheduled production time that is actually productive is highlighted by the overall equipment effectiveness statistic, which is used to quantify production efficiency. Although a 100% OEE score is ideal, where the manufacturing line produces only high-quality items quickly and with no downtime, most businesses are only able to reach 60%–85% (Figure 1.3).

1.3.1.1   Steps to achieve production efficiency

Evaluate the production line

Gaining insight into current manufacturing conditions is the first step toward improving efficiency. Analysis of manufacturing line efficiency is useful in this situation. Assessing every facet of the manufacturing line, starting with the quantity of units produced by your business in X amount of time, should be your first step. A critical indicator to consider would be evaluating your capacity utilization. It entails calculating each factory's potential for manufacturing output over an X amount of time [5]. It allows you to gauge which factory is operating at maximum capacity.

Update technologies

Working with outdated technologies is one of the main problems that modern organizations have. You won't be able to provide your team with the competitive advantage they need to become more efficient until you have the correct combination of technologies in place, like AI in manufacturing. Even though technologies frequently require significant upfront costs, they might be beneficial once you adopt them. Allow me to illustrate this with an example. If your computer hangs for five minutes, the information would not go to the manufacturing line for five minutes. This could result in workers starting the incorrect procedure.

Identify the bottlenecks in production

Breakdowns in the production line result in bottlenecks. These bottlenecks may result from shortcomings in either of the two components: people or equipment. A manufacturing company must be able to recognize the true cause of the bottleneck and take the appropriate steps to remove it.

Identify and remove wastefulness

Waste occurs at a manufacturing facility in a variety of ways, some of which the staff may not even be aware of. It may be inadequate ventilation, inefficient heating or cooling, or another issue. Furthermore, waste can also be defined as mismanagement or needless labor expenses. Whatever the cause, the end effect is almost always the same: a reduction in the efficiency of manufacturing [6]. Finding the wasteful component of your production process is the first step in doing this. Then, developing standards is the first step toward solving this problem. Each team should have a different set of standards based on its own KRAs. Once those criteria are set, it is the responsibility of a manager to make sure they are being met.

Improve the training process

Contrary to popular belief, people rather than machinery are the primary source of bottlenecks and inefficiencies. Furthermore, it's usually not even their fault. Employees should receive the appropriate training, making sure that it satisfies the requirements of the business. Standardized operating procedures, or SOPs, should be created as soon as the issue of inadequate training has been determined. Employees can measure and enhance their processes by using these SOPs as a guidebook.

Optimize the inventory

Any manufacturing company's inventory has room for improvement, both in terms of raw materials and final goods. Building a manufacturing efficiency software in collaboration with a custom enterprise software development firm is essential. You may generate automated notifications for material billing, inventory management, and invoicing using these technologies. Additionally, it would expedite the inventory process by removing the labor-intensive human component of creating spreadsheets and reports. Additionally, by keeping an eye on inventory counts and preventing product shortages, these ERP solutions can assist manufacturing companies in maintaining ideal stock levels.

1.3.2   Enhanced quality level

Large volumes of data can be quickly and readily analyzed in real time by AI-driven quality control systems, making it easier to find flaws and deviations from requirements. Additionally, IoT sensors offer ongoing production parameter monitoring, making it possible to identify possible quality problems early on and lowering the possibility that faulty goods will be distributed to customers.

1.3.3   Improved safety

In order to guarantee safety in the manufacturing sector, Industry 5.0 is essential. Predictive maintenance solutions driven by AI and collaborative robotics assist in identifying and reducing safety risks in the manufacturing setting. Moreover, virtual simulations and immersive training experiences made possible by AR and VR technologies give employees a secure setting in which to practice risky activities.

1.3.4   Cost reduction

Industry 5.0 helps to cut expenses. Predictive maintenance and AI-driven optimization cut maintenance costs, increase equipment longevity, and minimize downtime. Moreover, the on-demand fabrication of bespoke components and spare parts is made possible by additive manufacturing technologies, which lowers inventory costs and does away with the requirement for large amounts of warehousing.

1.3.5   Sustainable manufacturing

Industry 5.0 promotes resource-saving procedures and environmentally beneficial practices, which helps achieve environmental sustainability objectives. Environmental impact is decreased by optimizing resource use and energy efficiency using AI-driven algorithms and IoT-enabled monitoring systems. Moreover, compared to conventional manufacturing techniques, additive manufacturing procedures produce less waste, resulting in a more sustainable production strategy.

1.4   Industry 5.0 technology applications in the manufacturing industry

Industry 5.0 propels the advancement of manufacturing by the amalgamation of cutting-edge technologies and human-centered methodologies, hence augmenting efficiency, productivity, and safety within the production floor [7]. The following are some of the main technologies and applications related to Industry 5.0 (Figure 1.4).

1.4.1   Collaborative robotics

Cobots are intended to collaborate with people to increase safety and productivity. They free up human workers for more difficult or creative activities because they are effective at handling physically demanding, repetitive, strenuous, or dangerous duties. They are extensively utilized in material handling, assembly, pick-and-place tasks, and quality control [8]. Without the requirement for safety cages, cobots may cooperate with human workers and adapt to different industrial processes. BMW, for instance, has incorporated cobots into its production procedures to increase productivity and efficiency. Collaborating closely with human workers, these cobots support them in physically taxing and repetitious duties.

1.4.2   Artificial intelligence and machine learning

Large-scale data is analyzed by AI and machine learning (ML) algorithms to enhance quality control, forecast equipment failures, and optimize production processes. Predictive maintenance enabled by AI can foresee equipment failures before they happen, saving downtime and maintenance expenses. Moreover, ML algorithms can improve supply chain logistics, inventory control, and production schedules, resulting in more effective resource allocation and lower waste.

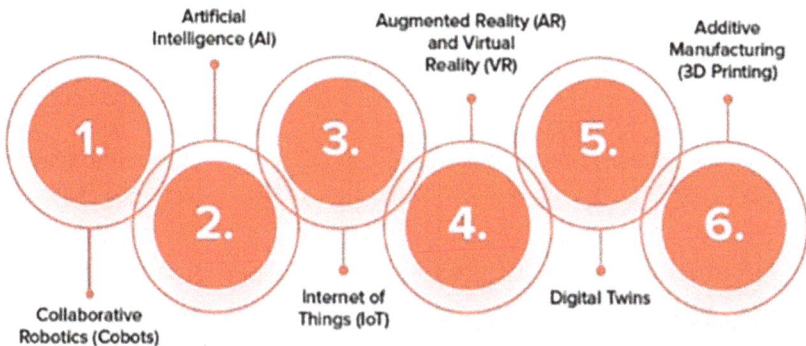

Figure 1.4   Industry 5.0 technology applications

1.4.3   Internet of Things

Real-time data from machines, equipment, and processes is gathered by IoT devices and sensors, offering insightful information for manufacturing operations, remote monitoring, and optimization. Manufacturers can use this information to pinpoint equipment performance inefficiencies and put corrective measures in place. IoT in manufacturing has the ability to identify abnormalities in machine behavior and notify maintenance personnel so they can handle possible problems before they get worse.

1.4.4   Augmented reality and virtual reality

Through immersive, interactive experiences, augmented reality (AR) and virtual reality (VR) technology improve worker training, maintenance, and troubleshooting procedures. By superimposing digital data on the real world, AR helps workers with intricate jobs like quality checks, maintenance schedules, and equipment setup [9]. By providing personnel with a virtual environment to practice operating machines, VR reduces the chance of accidents and improves skill competency.

1.4.5   Digital twins

With the ability to simulate, evaluate, and optimize operations in a virtual environment, digital twins have the potential to completely transform manufacturing processes. Manufacturers may foresee maintenance needs, find areas for improvement, and obtain deep insights into machine performance by building virtual replicas of physical assets, processes, and systems. Businesses may achieve unprecedented levels of productivity, flexibility, cost savings, and creativity by utilizing the potential of digital twins, which will ultimately lead to success in the Industry 5.0 age.

1.4.6   Additive manufacturing (3D printing)

A key element of human-centric manufacturing is 3D printing, which provides unparalleled production process flexibility and makes it possible to create three-dimensional items layer by layer. This ground-breaking technology makes it possible to manufacture complicated components on demand, revolutionizing small-batch production, customization, and prototyping. When compared to conventional subtractive manufacturing techniques, this industry 5.0 manufacturing technology lowers lead times, material waste, and tooling costs. Additionally, it improves human-machine production since workers can incorporate their artistic abilities into product creation, and 3D printing makes these concepts a reality. Industry 5.0 increases productivity, fosters creativity on the manufacturing floor, and empowers people through the seamless integration of these cutting-edge technology.

1.5   Problems and solutions for manufacturing Industry 5.0 implementation

As more and more companies adopt the manufacturing 5.0 revolution, it becomes simpler to ignore the possible difficulties. However, in order for the manufacturing

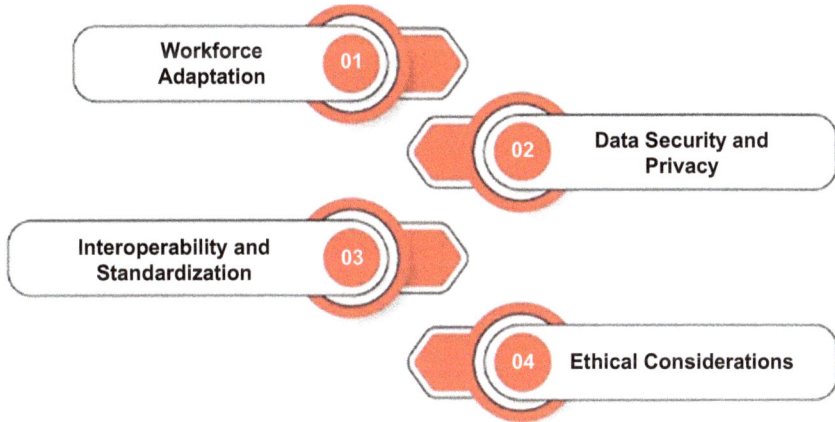

Figure 1.5   Setbacks and solutions for Industry 5.0

sector to expand, organizations need to recognize its obstacles and effectively address them (Figure 1.5).

1.5.1   Workforce adaption

To work with sophisticated robots and intelligent machinery, the workforce needs new skill sets as a result of the adoption of human-centric Industry 5.0 technology. The solution is to upskill current employees and prepare the next generation of workers for future occupations; organizations need to invest in workforce training and education programs.

1.5.2   Data security and privacy

Industry 5.0's increased connection and data interchange give rise to worries about cybersecurity risks and protecting sensitive data. The solution is to protect data security and privacy, businesses need to put strong cybersecurity safeguards, encryption mechanisms, and access controls in place.

1.5.3   Interoperability and standardization

Integration of diverse technologies and software solutions may lead to compatibility issues and interoperability challenges. The solution is manufacturers should develop industry-compliant solutions that guarantee seamless communication and interoperability between devices, machines, and software platforms.

1.5.4   Ethical considerations challenge

The widespread adoption of AI, automation, and robotics raises ethical concerns related to job displacement, algorithmic bias, and data misuse. The solution is to address this challenge; businesses should follow ethical guidelines, regulatory frameworks, and oversight mechanisms to ensure the responsible deployment of Industry 5.0 technologies.

1.6   Future of Industry 5.0 in manufacturing

Manufacturing's transition to Industry 5.0 has enormous potential because of continuing technical developments, shifting customer needs, and shifting global patterns. It will encourage greater human-machine cooperation, improving flexibility, safety, and productivity in the manufacturing process. With flexible and adaptable production processes, Industry 5.0 will allow for mass customization of products in response to changing consumer tastes and market demands. There will be behavioral as well as physical variations between factories in 2034 and those in today's world. The capacities of humans and machines will coexist and codevelop, with humans being able to further hone their special talents in problem-solving, creativity, and lateral thinking. Production tools and manufacturing 5.0 technologies—such as exoskeletons, augmented reality, virtual reality, AI, big data, and digital twins—are anticipated to significantly improve the sector throughout the next ten years. Sustainability will also gain prominence as businesses use manufacturing 5.0 technologies to optimize resource utilization, cut waste, and create more robust supply chains. Furthermore, since AI is becoming more and more prevalent in this new era, ethical and responsible AI practices—which guarantee accountability, transparency, and fairness in decision-making processes—will become increasingly important [10]. To put it briefly, Industry 5.0 is expected to bring about a revolutionary shift toward intelligent, robust, human-connected, and sustainable manufacturing ecosystems that empower employees, please consumers, and spur economic expansion. Thus, as this new era dawns, companies should choose reputable software development services to create ground-breaking solutions that promote the expansion of the manufacturing sector in a way that is both responsible and inclusive.

1.7   Conclusion

Industry Revolution 5.0 is a paradigm shift that combines cutting-edge technology with a fresh focus on the creativity and well-being of people to create more individualized, human-centric industrial systems. Industry 5.0 seeks to establish a synergistic interaction between humans and machines, in contrast to its predecessor, Industry 4.0, which concentrated on automation and data interchange in industrial technology. Increased sustainability, innovation, and productivity are the goals of this partnership. The combination of human creativity and decision-making abilities with robotics, AI, and the IoT characterizes current systems. These systems are made to be adaptive, flexible, and sensitive to the shifting demands of the consumer and the community. More developments in AI, ML, and human-machine interfaces are probably in store for future systems, which will result in more natural and seamless interactions between people and technology. In conclusion, Industry 5.0 marks the beginning of a new phase in industrial development that places an emphasis on innovation, sustainability, and human-centric principles. In order to fully realize the promise of this ground-breaking strategy, governments, corporations, and society as a whole must work together to successfully implement Industry 5.0.

References

[1] Akundi, A., Euresti, D., Luna, S., Ankobiah, W., Lopes, A., and Edinbarough, I. (2022). State of Industry 5.0—analysis and identification of current research trends. *Applied System Innovation*, 5(1), 27. https://doi.org/10.3390/asi5010027.

[2] Alvarez-Aros, E. L., and Bernal-Torres, C. A. (2021). Technological competitiveness and emerging technologies in Industry 4.0 and Industry 5.0. *Anais Da Academia Brasileira De Ciências*, 93(1), e20191290. https://doi.org/10.1590/0001-3765202120191290.

[3] Broo, D. G., Kaynak, O., and Sait, S. M. (2022). Rethinking engineering education at the age of Industry 5.0. *Journal of Industrial Information Integration*, 25, 100311. https://doi.org/10.1016/j.jii.2021.100311.

[4] Iyengar, K. P., Zaw, E., Jalli, J., *et al.* (2022). Industry 5.0 technology capabilities in trauma and orthopaedics. *Journal of Orthopaedics*, 32, 125–132. https://doi.org/10.1016/j.jor.2022.06.001.

[5] Kaasinen, E., Anttila, A. H., Heikkilä, P., Laarni, J., Koskinen, H., and Väätänen, A. (2022). Smooth and resilient human–machine teamwork as an Industry 5.0 design challenge. *Sustainability*, 14(5), 2773. https://doi.org/10.3390/su14052773.

[6] Singh, D. K., and Sobti, R. (2022). Long-range real-time monitoring strategy for precision irrigation in urban and rural farming in society 5.0. *Computers & Industrial Engineering*, 167, 107997. https://doi.org/10.1016/j.cie.2022.107997.

[7] Lu, Y., Zheng, H., Chand, S., *et al.* (2022). Outlook on human-centric manufacturing towards Industry 5.0. *Journal of Manufacturing Systems*, 62, 612–627. https://doi.org/10.1016/j.jmsy.2022.02.001.

[8] Srivastava, S. (2024, June 26). *Industry 5.0 – Revolutionizing the Factory Floor with Human-Centric Manufacturing*. Appinventiv. https://appinventiv.com/blog/industry-5-0-manufacturing/ (accessed 28 March 2024).

[9] Kamalakannan, R., and Pandian, R. S. (2018). A Tabu Search strategy to solve cell formation problem with ratio level data. *International Journal of Business Intelligence and Data Mining*, 13(1/2/3), 209. https://doi.org/10.1504/ijbidm.2018.088431.

[10] Nagarajan, N., and Kamalakannan, R. (2022). Analyze the effect of crater cutting tool wear modeling in the machining of aluminium composite. *Materials Research*, 25, e20220239. https://doi.org/10.1590/1980-5373-mr-2022-0239.

Chapter 2

Challenges and impact of Industry 5.0

Alex Looi Tink Huey[1,2]

Industry 4.0 marks a significant shift in manufacturing and production industries, integrating advanced manufacturing techniques with digital transformation. This revolution is enabled by connected technologies that create intelligent manufacturing systems—systems that are not only interconnected but also capable of communication, analysis, forecasting, and executing intelligent actions based on the information they process. For many companies, Industry 4.0 remains the "next big thing"—a trend they're still actively adapting or planning to focus on. It represents a revolution driven by advancements in information technology (IT), with key elements including automation, robotics, big data analytics, smart systems, virtualization, artificial intelligence (AI), machine learning, and the Internet of Things (IoT) [1]. Industry 4.0 has unlocked new levels of connectivity, innovation, and economic strength, enabling businesses to become more flexible, efficient, and resource-conscious [2]. At the heart of this digital revolution lies data. The ability to efficiently transform raw data into meaningful information is crucial for the success of future businesses. The core pillars of Industry 4.0 include autonomous systems and AI, which are driving the digitization and interconnection of production and manufacturing processes through cyber-physical systems. The industry recognizes that the widespread adoption of AI and automation will fundamentally redefine how businesses and industries work. As such, AI is poised to be one of the most critical enablers and a leading force in the evolution of IT.

Yet, even as companies and entire industry supply chains are still navigating the complexities of Industry 4.0, the next wave, Industry 5.0 is already on the horizon. Hence, it is crucial to understand what Industry 5.0 entails and how it might impact both the current operations and future business strategy. According to the European Commission, Industry 5.0 approach envisions an industry that goes beyond just efficiency and productivity but emphasizing its broader role and contribution to society [3]. This approach prioritizes the well-being of workers within the production process and leverages new technologies to promote prosperity that

[1]Malim Consulting Engineers Sdn Bhd, Selangor, Malaysia
[2]IEC Young Professional, International Electrotechnical Commission (IEC), Busan, South Korea

Table 2.1   Differences between Industry 4.0 and Industry 5.0 [4]

Industry 4.0	Industry 5.0
• Centered around enhanced efficiency through digital connectivity and artificial intelligence. • Technology—centered around the emergence of cyber-physical objectives. • Aligned with optimization of business models within existing capital market dynamics and economic models—i.e. ultimately directed at minimization of costs and maximization of profit for shareholders. • No focus on design and performance dimensions essential for systemic transformation and decoupling of resource and material use from negative environmental, climate, and social impacts.	• Ensures a framework for industry that combines competitiveness and sustainability, allowing industry to realize its potential as one of the pillars of transformation. • Emphasizes impact of alternative modes of (technology) governance for sustainability and resilience. • Empowers workers through the use of digital devices, endorsing a human-centric approach to technology. • Builds transition pathways toward environmentally sustainable uses of technology. • Expands the remit of corporation's responsibility to their whole value chains. • Introduces indicators that show, for each industrial ecosystem, the progress achieved on the path to well-being, resilience, and overall sustainability.

extends beyond mere job creation and economic growth, all while respecting the planet's environmental limits. Industry 5.0 complements the existing Industry 4.0 framework by focusing on research and innovation as essential tools for transitioning to a sustainable, human-centric and resilient industry. It envisions a harmonious collaboration between humans and machines, where advanced technologies augment human capabilities rather than replacing them, and recognizing the unique value that human creativity, empathy, and critical thinking bring to the table. The human-centric approach is a response to the growing concerns about the dehumanization of work in an increasingly automated world. This new approach shifts the focus from technology and economic growth to human progress and well-being.

Like every industrial revolution, Industry 5.0 introduces a range of challenges and opportunities to the industries. This chapter explores the complexities of Industry 5.0, drawing comparisons with Industry 4.0 (Table 2.1), addressing challenges it poses, and assessing its potential impact on both industries and society as a whole.

2.1   The three pillars of Industry 5.0

Industry 5.0 has three key pillars: (1) human-centric, (2) resilient, and (3) sustainable (Figure 2.1), which all have significant implications for business strategy.

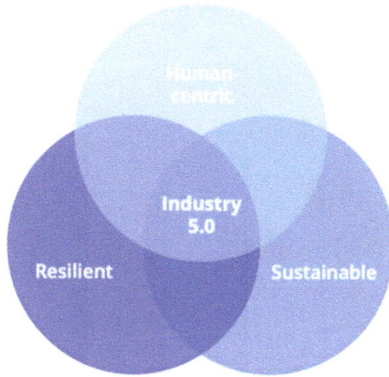

Figure 2.1   The three pillars of Industry 5.0 [1]

A human-centric strategy is one that promotes talent, diversity, and empowerment. The key shift here is moving from seeing people as resources (e.g. human resources) to seeing them as ends in themselves. In other words, it is a paradigm shift from organizations being served by people to organizations serving people. This shift is more transformative than it might initially appear and aligns closely with the current trends in the job market. In many industries and countries, attracting, retaining, and developing talent has become more challenging than acquiring and maintaining customers. If this trend continues, business strategies will need to adapt accordingly, and that is what Industry 5.0 aims to address. Traditionally, strategy focuses on gaining a competitive advantage to create unique added value for customers. However, in a truly human-centric organization, the first strategic priority shifts toward gaining a competitive advantage and using it to create unique added value for employees (Figure 2.2). This approach empowers employees by meeting their evolving skills and training needs, enhancing industry competitiveness and attracting top talent.

A resilient strategy must be agile and adaptable, utilizing flexible technologies. In the wake of COVID-19, global supply shortages, and the war in Ukraine, few would argue against the importance of resilience—today and in the future. While agility and flexibility have been long-standing goals for many organizations, they do not automatically equate to resilience. Businesses today are predominantly driven by efficiency and optimizing profits, not resilience. In fact, the push for efficiency in its "lean" form often prioritizes agility and flexibility at the cost of resilience, potentially making businesses more vulnerable. For us to realize that resilience to truly become a cornerstone of Industry 5.0, it means that strategy must shift away from a sole focus on growth, profit, and efficiency. Instead, the emphases should be on building organizations that are "anti-fragile," meaning organizations are able to anticipate, respond to, and learn from crises in a timely and systematic manner, ensuring the organization's stable and sustainable performance.

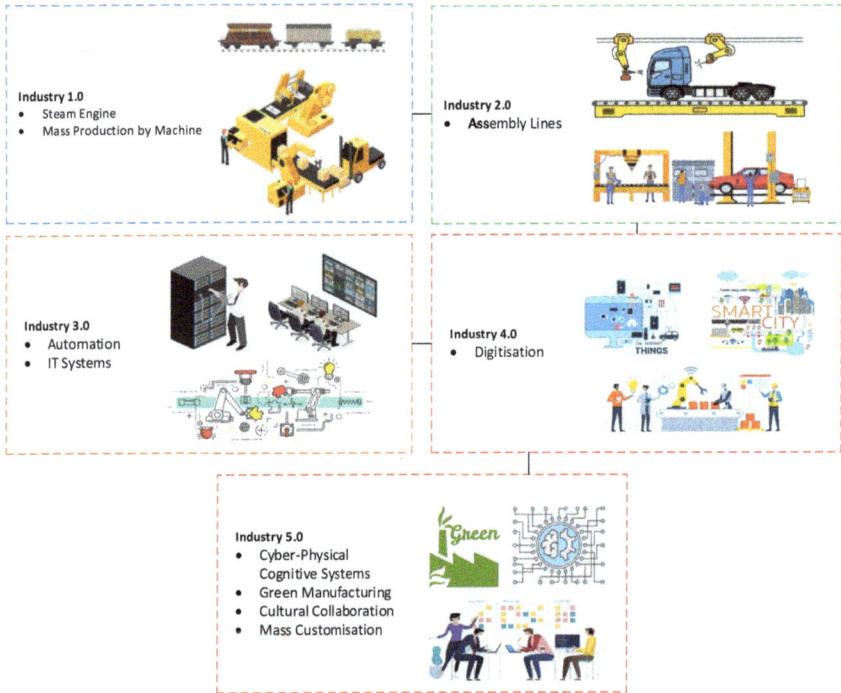

Industry 5.0: Potential Opportunities and Adoption Challenges

Industry 1.0
- Steam Engine
- Mass Production by Machine

Industry 2.0
- Assembly Lines

Industry 3.0
- Automation
- IT Systems

Industry 4.0
- Digitisation

SMART CITY

THINGS

Industry 5.0
- Cyber-Physical Cognitive Systems
- Green Manufacturing
- Cultural Collaboration
- Mass Customisation

Green

Figure 2.2   Industry 5.0 differs from all previous revolutions, as humans are at the present face with the center in production procedures [8]

At the recent United Nations climate summit, COP28 in Dubai, a harsh truth was laid bare: fossil fuels remain the primary culprits behind climate change. More than 100 nations committed to tripling renewable energy capacity and doubling global energy efficiency rates by 2030 [5]. However, the most pressing concern is the significant shortfall in progress, putting the critical 1.5 °C warming threshold established by the Paris Agreement in serious jeopardy. The Paris Agreement calls for substantial reductions in global greenhouse gas (GHG) emissions to keep the long-term global average surface temperature rise well below 2 °C [6]. But why is this limit so vital? Surpassing it could trigger multiple climate tipping points, leading to sudden, irreversible, and catastrophic consequences for humanity (Figure 2.3).

In today's climate-conscious world, the concept of sustainability needs little introduction. A sustainable strategy leads to action on sustainability and respects the planetary boundaries. Organizations need to prioritize all three Ps of the Triple Bottom Line—people, planet, and profit and align with the United Nations (UN) 17 Sustainable Development Goals (SDGs) and Environmental Social Governance

Figure 2.3   Surpassing the 2 °C threshold could trigger multiple climate tipping points

(ESG). Similar to the two previous pillars, this strategy represents a radical change. To date, corporate sustainability efforts have mostly centered on reducing harm, or in some cases, have been superficial attempts at appearing green—greenwashing. However, truly integrating sustainability into an organization's strategy involves much more than simply minimizing negative impacts. Genuinely, sustainable organizations focus on amplifying their positive contributions. This approach describes a business model that benefits society and the environment, transformative changes that aim to make the world a better place. In other words, Industry 5.0 demands that businesses become part of the solution, and not part of the problem. By rethinking the value chains and supply chains, energy consumption practices, implementing renewable energy, and optimizing the use of natural resources, businesses can become more resilient to external shocks. A circular economy focuses on minimizing waste, conserving natural resources, and reducing pollution, all of which contribute to mitigating climate change [7]. Businesses have the potential to be key players in addressing societal challenges, including resource preservation, decarbonization, climate change, and social stability.

2.2   Challenges and impact

Skill gaps and socioeconomic impact: As advanced robots become more integrated into the workforce, human workers must develop competencies to collaborate effectively with smart machines and robot manufacturers [8]. In addition to necessary managerial and soft skills, acquiring technical skills remains a significant challenge. Programming industrial robots and managing transitions to new roles demand high levels of technical expertise. A widening gap in expertise and highly skilled workforce, particularly in areas such as robotics, AI, and big data analytics, presents a significant challenge. While Industry 5.0 holds the promise of enhanced productivity, it may also disrupt traditional employment models [9]. Addressing the potential socioeconomic effects, including job displacement, is vital to maintaining societal stability. According to McKinsey, a complex organizational structure can contribute to inefficiency due to unclear roles and responsibilities [10]. Therefore, it is essential to allocate resources where they are most needed. McKinsey also reported that by 2030, as many as 375 million workers globally may need to switch occupational categories due to automation [10].

Bridging this gap is essential to realizing the full potential of Industry 5.0. Both governments and industry stakeholders must focus on investing in education and training programs that provide the necessary skills to prepare the workforce for Industry 5.0. Governments need to implement safety nets, such as reskilling programs and unemployment support, to lessen the socioeconomic impacts of technological progress [9]. Businesses must also focus on enhancing user experiences across factories and extended enterprises by improving wearables and human–machine interfaces, optimizing collaborative robots (cobots), and leveraging safety technologies. Additionally, they should implement strong diversity, equity, and inclusion initiatives while enhancing and cross-leveraging IT and operational technology skill sets and knowledge [11]. As technology evolves, so do organizational roles. Organizations should either create new positions or adjust existing ones to ensure employees possess the skills necessary to keep pace with Industry 5.0's rapid developments [12]. Emerging roles may include chief robotics officer or senior AI strategist, along with positions such as data analysts or data scientists specializing in predictive analytics or customer experience optimization. Economic growth is vital for generating jobs, as stagnant or slowly growing economies tend to produce few, if any, new positions. Nations with robust economic expansion, productivity gains, and innovation are likely to see a higher demand for labor. Leaders need to effectively assemble teams by understanding the skills present within the organization. This involves creating detailed skill profiles for each employee and ensuring these profiles are accessible across the company [13]. The employee directory should include comprehensive skill-based information, such as certifications, trainings, and recent project experience, which is automatically updated as projects conclude.

Technology adoption: Embracing advanced technologies requires significant time and effort from human workers. In an ideal Industry 5.0 environment, customized software-connected factories, collaborative robotics, AI, real-time data management, and the IoT must be fully adopted. Similar to other initiatives and trends, challenges have arisen from the lack of alignment between business and technology, as well as failure to adapt and customize for different sites [11]. Without cultural shifts, a robust communication framework, and the development of necessary skills, organizations may face poor return-on-investment and negative impacts on their dynamics.

It is crucial to understand and leverage automation and robotics to enhance the organization's production process. Adopting automation technologies free up resources, allowing businesses to concentrate on other crucial areas such as customer service and product innovation—both vital for success [12]. By identifying the tasks that are most time-consuming or repetitive, implementation of automation with robotics or machine learning algorithms can be the solutions. Investing in data and AI technology helps businesses gain a competitive advantage. By investing in cloud computing systems for real-time data insights and leveraging AI algorithms for process automation, businesses can save time and money while optimizing operations with unprecedented accuracy for decision-making. Businesses need to define and establish clear boundaries to guide operations, including company priorities, product strategies, project team missions, and short-term goals (Figure 2.4). Senior management should provide stringent oversight to identify opportunities for resource sharing or collaboration among teams and prevent mission creep [13].

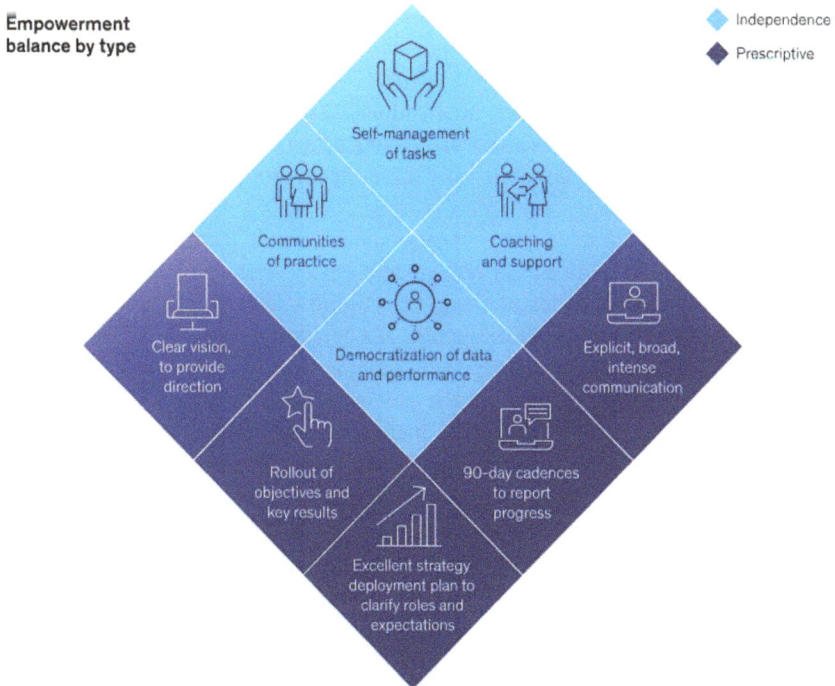

Figure 2.4   Getting empowerment right requires a delicate balance between worker independence and prescriptive management [13]

Investment costs: Implementation of advanced technologies involves considerable upfront capital costs. Cobots are expensive. Additionally, training human workers for new roles incurs extra costs. Businesses often struggle with the financial burden of upgrading production lines for Industry 5.0 as the adoption of Industry 5.0 is costly.

It is essential to ensure these innovations are accessible to everyone, not just those who can afford them.

Data privacy and security concerns: Security remains a critical challenge for Industry 5.0, as establishing trust within ecosystems is paramount. The authentication processes used in the industry must be robust enough to interact with various devices and withstand future quantum computing applications. The reliance on AI and automation in industry presents new threats to businesses, necessitating trusted security measures such as ethical use of AI, data privacy, and the potential for bias in decision-making algorithms.

As connectivity and data sharing between humans and machines increase, safeguarding data privacy and security becomes critical. Since Industry 5.0 applications rely heavily on information and communication technology (ICT) systems, strict cybersecurity protocols and measures are required to prevent potential security breaches. Prioritizing ethical AI development and establishing strong data governance frameworks are key to addressing privacy concerns in Industry 5.0. AI systems need to be transparent, accountable, and free from bias which are crucial for maintaining public trust.

Energy concerns: Industry 5.0 involves management of high data rates required for various AI and IoT applications. As more smart devices become interconnected with the application of AI and data centers, energy management emerges as a critical concern as these technologies require significant computational power.

Industry 5.0 seeks to optimize energy management by not only improving energy consumption practices via implementation of energy management system but also incorporating methods of energy harvesting via development of renewable energy. Sustainability and the development of a "green" ecosystem within extended manufacturing organizations have been gaining significance even before the COVID-19 era. Leaders recognized the powerful synergy between digital transformation and sustainability. Additive manufacturing, already more sustainable than traditional subtractive methods, is further complemented by circular green manufacturing, which emphasizes waste reduction, recycling, energy optimization and auditing, the use of eco-friendly materials, and leveraging data and AI for predictive maintenance, and paperless digital communication [11]. Beyond their internal operations, leadership teams are increasingly promoting and rewarding green initiatives throughout their supply chains, among employees, and with customers. Manufacturers should strive to make technologies more energy-efficient and environmentally friendly [12]. The adoption of people-centric energy transition approach, which emphasizes the importance of individuals and communities, is emerging as a significant driver in transforming energy systems while promoting positive social outcomes in line with Industry 5.0 [14]. This concept is central to achieving a just energy transition. Essentially, a people-centric energy transition is an approach that prioritizes individuals and communities in decision-making processes, ensuring their active participation, empowerment, and fair access to clean energy resources. It acknowledges that moving toward sustainable energy involves not just adopting new technologies but also understanding the broader social, economic, and environmental contexts. When long-term collaborative efforts support inclusive participation, a people-centric energy transition delivers a range of multifaceted benefits. Notably, these include equitable access to clean energy, improved human health and environmental conditions, increased job inclusivity, and poverty reduction.

Regulatory challenges: The face-paced advancement of Industry 5.0 technologies often outpaces existing regulatory frameworks. This creates a gap between innovation and regulation, leading to uncertainties and risks. As industries become more globalized and interconnected, harmonizing standards and regulations across different jurisdictions become increasingly complex.

Governments and regulatory bodies must work closely with industrial stakeholders to establish policies and regulations that strike a balance between fostering innovation and ensuring safety and ethical standards. International cooperation is essential for creating common standards and regulations that ensure the safe and ethical integration of Industry 5.0 technologies [9]. Governments at all levels must collaborate to establish international standards that will lay the foundation for Industry 5.0. Governments need to maintain strong overall demand growth, which is crucial for fostering new job creation, alongside supporting the establishment of new businesses and innovation [10]. Effective fiscal and monetary policies that bolster aggregate demand and encourage business investment and innovation are vital. Additionally, targeted initiatives in specific sectors, such as boosting investments in infrastructure and energy transitions, could further support these goals.

2.3   Keys to the success of Industry 5.0

For businesses to continue driving prosperity across nations, it must evolve into a catalyst for change and innovation. Industry 5.0 plays a crucial role in providing the technological advancements needed for the industry to reaffirm its role as a solution-provider for society and an appealing employer for young professionals seeking meaningful careers [15]. Instead of representing a technological leap forward, Industry 5.0 builds upon the foundations of Industry 4.0, offering a broader perspective on people-centric technological evolution of industrial production. By focusing on people-centric technological development, Industry 5.0 can support and empower workers, enhancing industry resilience and sustainability.

Businesses need to take a leading role in adopting new technologies and reimagining production processes with environmental impacts in mind (Figure 2.5). Industry must set an example in the green transition. A transformed industry will have a profound impact on society, particularly for workers whose roles and skill sets will evolve.

Figure 2.5   Businesses need to take a leading role in adopting new technologies with environmental impacts in mind

Workers of the future will increasingly focus on tasks that machines struggle with, such as managing people, applying specialized knowledge, and engaging in communication. They will spend less time on routine physical tasks and data collection, areas where machines already outperform humans. The demand for skills will evolve, emphasizing social and emotional intelligence as well as advanced cognitive abilities such as logical reasoning and creativity.

References

[1]   Forbes. What is industry 5.0 and how it will radically change your business strategy? https://www.forbes.com/sites/amazon-web-services-asean/2024/05/17/8-ethical-challenges-for-generative-ai/ (accessed 3 May 2024).

[2]   Winterhalter, C. A new revolution in the making, *ISO Focus: The New Industrial Revolution*, 2018, 131, 2–3.

[3]   European Commission. Industry 5.0. https://research-and-innovation.ec.europa.eu/research-area/industrial-research-and-innovation/industry-50_en (accessed 10 July 2024).

[4]   European Commission, Directorate-General for Research and Innovation. Industry 5.0: a transformative vision for Europe – governing systematic transformations towards a sustainable industry. https://op.europa.eu/en/web/eu-law-and-publications/publication-detail/-/publication/38a2fa08-728e-11ec-9136-01aa75ed71a1 (accessed 3 May 2024).

[5]   World Economic Forum. What were the key outcomes of COP28? https://www.weforum.org/agenda/2023/12/cop28-key-outcomes-un-climate-summit/ (accessed 3 May 2024).

[6]   United Nations. 1.5 °C: what it means and why it matters. https://www.un.org/en/climatechange/science/climate-issues/degrees-matter (accessed 3 May 2024).

[7]   Knight, C. What is the linear economy? https://www.eib.org/en/stories/linear-economy-recycling. European Investment Bank. Accessed December 2024.

[8]   Adel, A. Future of Industry 5.0 in society: human-centric solutions, challenges and prospective research areas. *Journal of Cloud Computing* 2022, 11, 40. https://journalofcloudcomputing.springeropen.com/articles/10.1186/s13677-022-00314-5 (accessed 15 June 2024).

[9]   Binus University. Industry 5.0 challenges amidst society 5.0. https://ie.binus.ac.id/2023/09/22/industry-5-0-challenges-amidst-society-5-0/#:~:text=Challenges%20of%20Industry%205.0%20Amidst,artificial%20intelligence%2C%20and%20data%20analytics (accessed 15 June 2024).

[10]   Manyika, J., Lund, S., Chui, M., *et al. Jobs Lost, Jobs Gained: What the Future of Work Will Mean for Jobs, Skills, and Wages.* New York: McKinsey & Company; 2024.

[11]   The Economic Times Business Verticals. Are we on our way to Industry 5.0? https://ciosea.economictimes.indiatimes.com/news/strategy-and-management/are-we-on-our-way-to-industry-5-0/96047166?utm_source=copy&utm_medium=pshare (accessed 15 June 2024).

[12]   AG5. Industry 5.0 manufacturing. https://www.ag5.com/industry-5-0-manufacturing/ (accessed 15 June 2024).

[13]   Misljencevic, D., Houthuys, S., Welchman, T., and Schrader, U. *Today's Industrial Revolution Calls for an Organization to Match.* New York: McKinsey & Company; 2024.

[14]   Energy Asia. From inclusion to empowerment: the social impacts of a people-centric energy transition. https://www.officialenergyasia.com/the_social_impacts_of_a_people_centric_energy_transition/ (accessed 10 July 2024).

[15]   European Commission, Directorate-General for Research and Innovation. Industry 5.0: towards more sustainable, resilient and human-centric industry. https://research-and-innovation.ec.europa.eu/news/all-research-and-innovation-news/industry-50-towards-more-sustainable-resilient-and-human-centric-industry-2021-01-07_en (accessed 3 May 2024).

Chapter 3

Leading the Industry Revolution 5.0 transformation: a technology roadmap for Industry Revolution 5.0

Wai Yie Leong[1]

Industry Revolution 5.0 (IR 5.0) represents the next step in the evolution of industrial practices, focusing on human-centered innovation, sustainability, and the integration of advanced digital technologies with human intelligence. While Industry 4.0 emphasized the automation and connectivity of machines (IoT, AI, big data), IR 5.0 emphasizes synergy between humans and technology, fostering personalized solutions and sustainable growth. The roadmap of IR 5.0 involves several critical aspects: human–machine collaboration, AI-driven innovation, sustainability, ethical concerns, and new industrial standards.

3.1 Introduction

The shift from Industry 4.0 to Industry 5.0 signals an evolution from automation-driven efficiency to human-centric innovation [1,2]. Industry 4.0 integrated technologies such as artificial intelligence (AI), the Internet of Things (IoT), and cyber-physical systems (CPS) to create smart, interconnected manufacturing ecosystems [3,4]. However, Industry 5.0 emphasizes the role of humans in guiding and augmenting these systems to achieve a balance between technological advances and societal well-being.

Industry 5.0 aims to leverage the strengths of both humans and machines by fostering greater collaboration. It promotes sustainability, inclusivity, and resilience while integrating ethical considerations into the use of technology [5]. This chapter presents a technology roadmap to navigate the key milestones in the Industry 5.0 transformation, ensuring that companies and sectors remain competitive in this new era as shown in Figure 3.1.

3.2 Literature review on the technology roadmap for Industry Revolution 5.0

A review on the Technology Roadmap for Industry Revolution 5.0 involves exploring the evolution of industrial revolutions, from early mechanization to

[1]Faculty of Engineering and Quantity Surveying, INTI International University, Malaysia

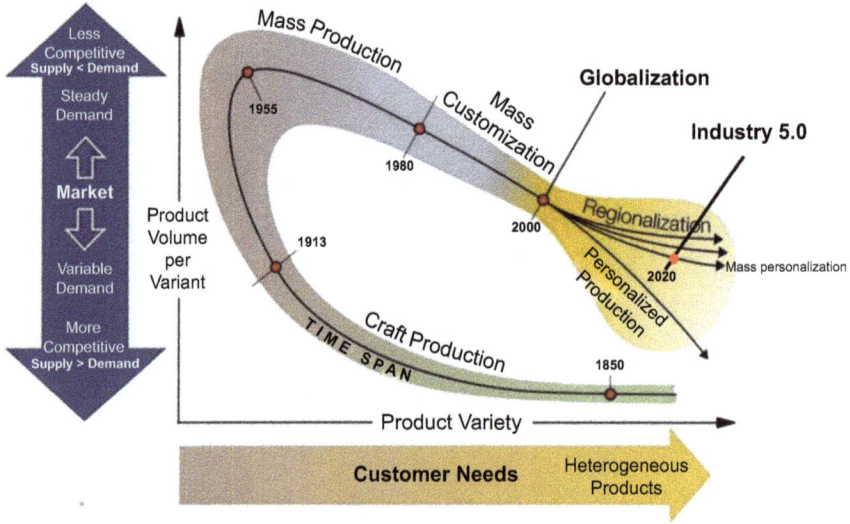

Figure 3.1   Shifting manufacturing paradigm

Table 3.1   Industry Revolution 5.0 roadmap

Stage 1: Research and early development (2024–27)

Focus areas
(a) Development of AI, human-robot collaboration, and sustainable technologies.
(b) Integration of XR into business processes for training and collaboration.
(c) Ethical framework development for responsible AI and human-centric systems.

Stage 2: Adoption and integration (2027–30)

Focus areas
(a) Broad adoption of AI and XR across sectors such as healthcare, manufacturing, and education.
(b) Strengthened focus on sustainability through eco-efficient production methods.
(c) Increasing public-private partnerships for the promotion of IR 5.0 technologies.

Stage 3: Full-scale deployment (2030–35)

Focus areas
(a) Human-centric workplaces powered by AI and cobots.
(b) Sustainable, zero-waste production lines utilizing advanced recycling and renewable energy technologies.
(c) Robust legal frameworks and global standards for the deployment of ethical AI.

---

advanced integration of human-centered technologies, while examining literature focused on the roadmap guiding the development and implementation of Industry 5.0 technologies as shown in Table 3.1 and Figure 3.2.

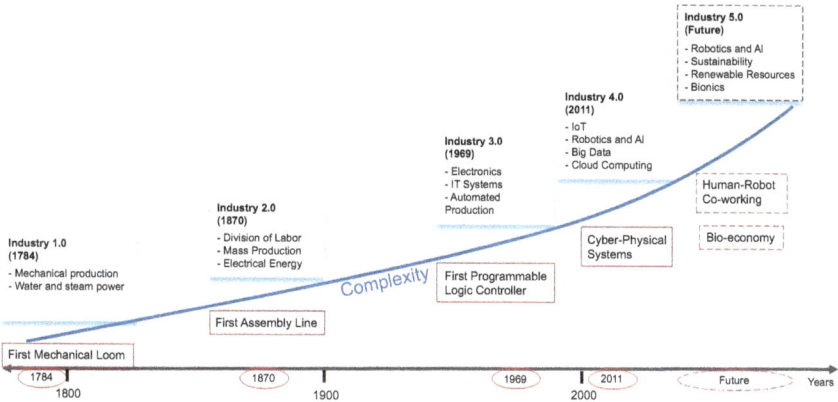

Figure 3.2 Illustration of evolution from Industry 1.0 to Industry 5.0

Industry 1.0 (18th century): The first industrial revolution was marked by the advent of mechanization, driven by water and steam power. It facilitated mass production and efficiency in agriculture, textiles, and transportation.

Industry 2.0 (late 19th–early 20th century): The second revolution introduced electric power, enabling further automation, assembly lines, and increased productivity in factories. Major developments in transportation and communication, including railways and the telegraph, were also significant.

Industry 3.0 (mid-20th century): This phase was characterized by the rise of digital technologies, electronics, and IT systems. The development of computers and automation systems transformed manufacturing processes, allowing for precision and greater control [6].

Industry 4.0 (early 21st century): The fourth industrial revolution brought CPS, the IoT, and big data into manufacturing and other industries. Real-time data exchange between machines and systems has facilitated the rise of smart factories, where self-optimization and decentralized decision-making are prevalent.

Industry 5.0 (late 2010s–present): Unlike Industry 4.0, which focused heavily on automation and efficiency, Industry 5.0 emphasizes human–machine collaboration. It integrates advanced technologies like AI, robotics, cybernetics, and biotechnology with a focus on sustainability, customization, and societal well-being [7].

Industry 5.0 aims to create more synergies between humans and machines, leading to a shift from solely automating processes to augmenting human intelligence and creativity. This approach aligns technology with social and environmental goals, contrasting with Industry 4.0's focus on efficiency and productivity. Articles such as Nahavandi [8] emphasize the role of human–robot collaboration to drive this revolution forward.

Significant advancements in AI, machine learning (ML), robotics, 3D printing, and digital twins are driving the growth of Industry 5.0. These technologies facilitate the creation of customizable solutions and collaborative systems that can cater to individual needs, marking a clear departure from mass production. A review by

Lasi *et al.* [9] focuses on how IoT and automation laid the groundwork for these developments, while Xu *et al.* [10] extend this by discussing Industry 5.0's emphasis on hyper-customization.

Industry 5.0 is not only about advancing technology but also about addressing sustainability challenges. The literature reflects an increasing focus on the circular economy, eco-friendly production methods, and energy-efficient technologies. In this context, works like Bai *et al.* [1] explore how technologies such as AI and 3D printing can reduce resource consumption and waste.

As Industry 5.0 reshapes the manufacturing and industrial landscape, ethical concerns arise around the future of work, data privacy, and technology governance. Kusiak [11] explores how human-centric design in Industry 5.0 requires balancing automation with meaningful employment, raising questions about retraining and upskilling the workforce.

The roadmap for Industry 5.0 involves the transition from pure automation to hybrid systems where AI augments human decision-making. Major industrial players and governmental bodies are crafting guidelines to shape this future. Reports such as European Commission's Industry 5.0 [12] highlight the importance of fostering collaboration between policymakers, industry leaders, and academic institutions to ensure technological advances are aligned with ethical and sustainability goals.

3.3   Key drivers of Industry 5.0

Industry Revolution 5.0 is driven by several critical factors that form the foundation of the transformation: human-centric design, ethical AI and automation, sustainability.

Human-centered approaches focus on enhancing the synergy between workers and intelligent machines. The emphasis is on designing systems that improve human well-being, creativity, and productivity. Human–machine collaboration is key, wherein machines handle repetitive tasks while humans focus on problem-solving and innovation.

Unlike Industry 4.0, which focused heavily on automation, Industry 5.0 integrates ethical AI. The new framework ensures AI systems are transparent, fair, and aligned with societal values. By integrating ethics into AI systems, Industry 5.0 seeks to prevent unintended consequences and promote trust in intelligent systems.

Sustainability is a central tenet of Industry 5.0. Technologies such as smart materials, green manufacturing processes, and circular economy principles are designed to reduce environmental impact. Industry 5.0 embraces renewable energy, waste reduction, and energy-efficient technologies to achieve net-zero emissions.

3.4   Technologies enabling Industry 5.0

Industry Revolution 5.0 (IR 5.0) represents the next step in the evolution of industrial practices, focusing on human-centered innovation, sustainability, and the integration of advanced digital technologies with human intelligence [13] as shown in Figure 3.3.

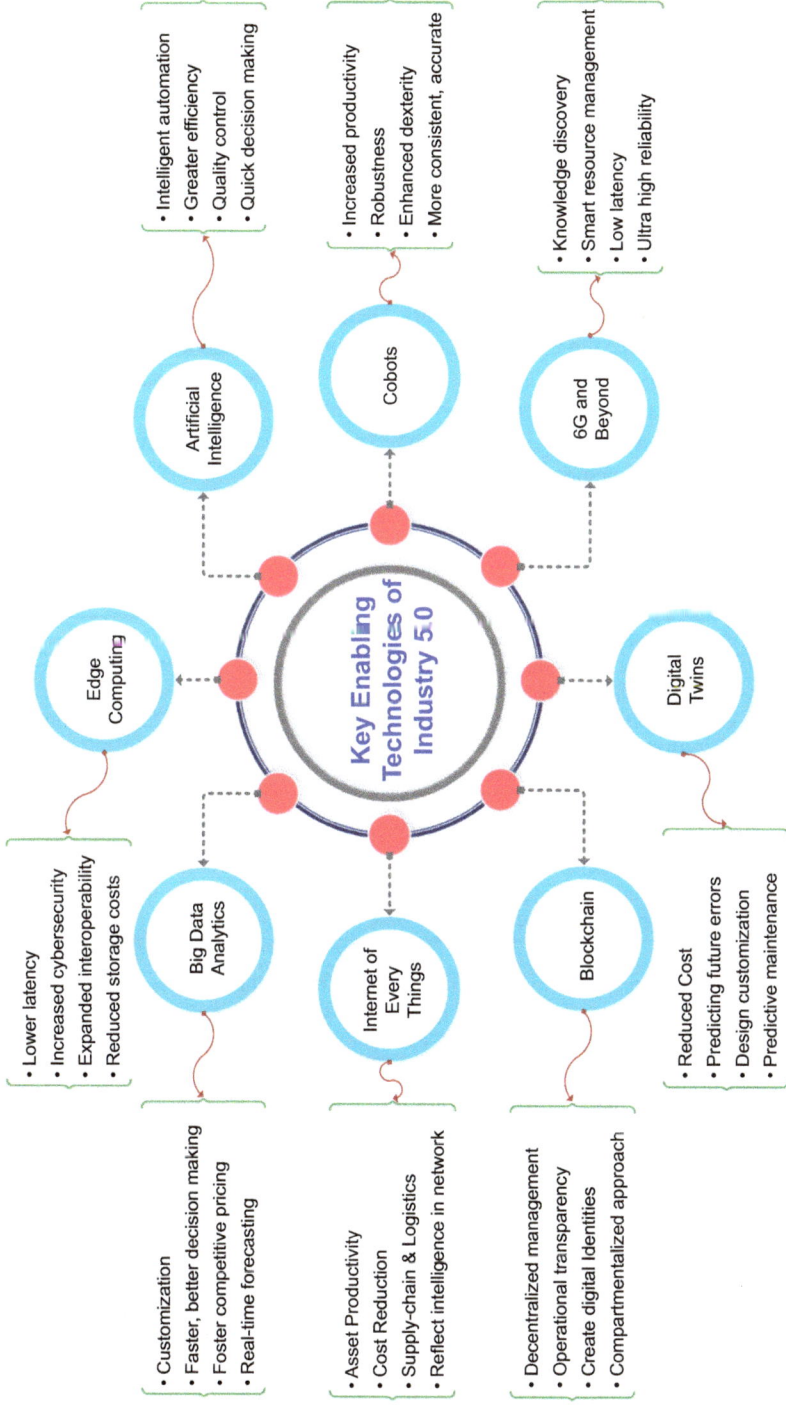

Figure 3.3 *Key enabling technologies of Industry 5.0*

Artificial Intelligence
- Intelligent automation
- Greater efficiency
- Quality control
- Quick decision making

Cobots
- Increased productivity
- Robustness
- Enhanced dexterity
- More consistent, accurate

6G and Beyond
- Knowledge discovery
- Smart resource management
- Low latency
- Ultra high reliability

Edge Computing
- Lower latency
- Increased cybersecurity
- Expanded interoperability
- Reduced storage costs

Big Data Analytics
- Customization
- Faster, better decision making
- Foster competitive pricing
- Real-time forecasting

Internet of Every Things
- Asset Productivity
- Cost Reduction
- Supply-chain & Logistics
- Reflect intelligence in network

Blockchain
- Decentralized management
- Operational transparency
- Create digital Identities
- Compartmentalized approach

Digital Twins
- Reduced Cost
- Predicting future errors
- Design customization
- Predictive maintenance

Key Enabling Technologies of Industry 5.0

While Industry 4.0 emphasized the automation and connectivity of machines (IoT, AI, big data), IR 5.0 emphasizes synergy between humans and technology, fostering personalized solutions and sustainable growth. The roadmap of IR 5.0 involves several critical aspects: human–machine collaboration, AI-driven innovation, sustainability, ethical concerns, and new industrial standards as shown in Table 3.2.

Collaborative robots, or cobots, are designed to work alongside human operators. These robots are equipped with sensors and AI to adapt to human behavior, enabling safe and efficient collaboration. Cobots can handle repetitive or dangerous tasks, allowing human workers to focus on creative, decision-making activities.

AI and ML drive Industry 5.0 by optimizing processes, enabling predictive maintenance, and providing decision-making support. These systems learn from human input, enhancing their performance in environments where human intelligence is irreplaceable.

The Internet of Everything expands on IoT by integrating people, processes, data, and things into a seamless ecosystem. IoE connects not only devices but also human interactions, creating a fully integrated network that improves decision-making, efficiency, and responsiveness [14].

Table 3.2   Key technologies in Industry Revolution 5.0

Human–robot collaboration (cobots)	
Role	Robots and humans will work side by side, augmenting human capabilities in various sectors.
Key technology	Collaborative robots (cobots) equipped with AI-driven decision-making algorithms.
Applications	Healthcare, precision manufacturing, and logistics.
Artificial intelligence and machine learning	
Role	AI will drive decision-making, automation, and enhanced productivity across industries, while ensuring more human oversight.
Applications	Predictive maintenance, personalized customer experiences, smart energy management.
Extended reality (XR)	
Role	Virtual, augmented, and mixed reality technologies to provide immersive experiences for training, product design, and customer interaction.
Applications	Design and prototyping, education, remote work.
Sustainable technologies	
Role	Focus on eco-friendly practices through resource efficiency, recycling, and renewable energy sources.
Key technologies	3D printing for circular economy, solar and wind energy integration, and water recycling technologies
Applications	Energy-efficient smart grids, waste management, and eco-friendly materials in manufacturing
Blockchain technology	
Role	Secure and transparent systems for tracking supply chains, ensuring data integrity, and supporting decentralized transactions.
Applications	Supply chain management, digital identity verification, and financial transactions.

3D printing continues to revolutionize manufacturing in Industry 5.0. With advancements in materials science, additive manufacturing allows for more personalized, sustainable production. It reduces waste, enhances customization, and shortens supply chains, aligning with the sustainability goals of Industry 5.0.

3.5   Strategies for transitioning to Industry 5.0

The transition to Industry 5.0 requires a comprehensive and well-planned approach that integrates human-centric technologies, sustainable practices, and ethical AI deployment. Table 3.3 shows the key strategies to facilitate a smooth shift to this new industrial paradigm.

Table 3.3   Strategies for transitioning to Industry 5.0

Prioritizing human–machine collaboration	
Strategy	Industry 5.0 emphasizes collaboration between humans and machines (cobots). Companies should focus on enhancing the synergy between human creativity and technological efficiency.
Actions	• Invest in collaborative robots (cobots): Deploy robots that can work safely alongside human workers.
	• Develop augmented workspaces: Leverage technologies like AI, AR, and VR to create environments where humans and machines can interact intuitively.
	• Upskill the workforce: Train workers to collaborate effectively with advanced machines, focusing on tasks that complement human abilities (creativity, problem-solving).
Example	Automotive manufacturing companies are incorporating cobots on assembly lines to support human workers in tasks like precision welding, where human oversight ensures quality while machines provide speed.
Integrating sustainable practices	
Strategy	Sustainability is a core pillar of Industry 5.0, requiring industries to transition toward green and circular economies.
Actions	• Adopt circular economy models: Reduce waste and increase the recycling of materials by integrating additive manufacturing (3D printing) and eco-friendly production methods.
	• Embrace renewable energy sources: Integrate solar, wind, and hydropower into factory operations.
	• Implement smart energy management systems: Use AI-driven systems to optimize energy consumption and improve efficiency.
Example	A 3D printing company can use recycled materials for manufacturing products, reducing the need for new raw materials and minimizing waste in the production process.
Ethical AI and responsible data use	
Strategy	AI technologies in Industry 5.0 should be developed and deployed in a manner that aligns with ethical standards and ensures transparency, fairness, and accountability.

(Continues)

Table 3.3   (Continued)

Actions	• Develop an AI ethics framework: Define principles for the ethical use of AI, focusing on issues like bias, data privacy, and accountability.
	• Ensure responsible data management: Implement secure data protocols to protect sensitive information and ensure the ethical use of data in decision-making.
	• Collaborate with governments and organizations to set global standards for AI ethics and ensure regulatory compliance.
Example	Healthcare companies can implement AI in diagnostic systems that are regularly audited for bias, ensuring fairness in treatment recommendations across demographics.

Emphasizing personalization and customization

Strategy	Industry 5.0 focuses on providing personalized products and services through AI-driven mass customization techniques.
Actions	• Implement AI-driven customization platforms: Use AI to tailor products and services based on individual customer preferences.
	• Utilize advanced manufacturing techniques: Invest in smart manufacturing systems that can rapidly shift between different production lines to meet customer demands.
	• Leverage extended reality (XR): Use augmented reality (AR) and virtual reality (VR) to enhance customer experiences and enable real-time customization.
Example	In the fashion industry, companies are using AI to offer personalized clothing designs based on consumer preferences, fitting data, and lifestyle patterns, leading to unique, customer-specific products.

Workforce upskilling and education

Strategy	The transition to Industry 5.0 requires a skilled workforce capable of managing new technologies and collaborating with AI and machines.
Actions	• Launch continuous learning programs: Develop training initiatives that focus on the skills required for Industry 5.0 such as AI literacy, robotics, and data analytics.
	• Promote lifelong learning: Encourage employees to engage in lifelong learning by offering certification programs and workshops in emerging technologies.
	• Collaboration with educational institutions: Form partnerships with universities and training institutes to develop specialized courses tailored to Industry 5.0 roles.
Example	Manufacturing companies are establishing digital apprenticeships where workers learn advanced manufacturing techniques alongside robots, preparing them for future roles in smart factories.

Cross-industry collaboration and partnerships

Strategy	The complexity of Industry 5.0 technologies requires collaboration between multiple industries, governments, and academic institutions.
Actions	• Form public-private partnerships: Collaborate with governments, research institutions, and private firms to develop and adopt Industry 5.0 technologies.
	• Foster industry alliances: Work with other industries to develop shared standards, technologies, and best practices that benefit all stakeholders.
	• Support open innovation ecosystems: Encourage collaboration with startups, universities, and tech companies to foster innovation and accelerate technological development.

Table 3.3   (Continued)

Example	Automotive and electronics industries can collaborate to create standardized technologies for smart cities, leveraging cross-industry innovation to enhance sustainability and automation in urban infrastructure.

Phased adoption and pilot testing

Strategy	Companies should take a phased approach to adopting Industry 5.0 technologies, beginning with pilot projects before scaling up.
Actions	• Start with pilot projects: Implement small-scale pilot projects to test the feasibility and impact of new technologies. • Use data-driven decision-making: Analyze the results of pilot projects to understand key challenges, benefits, and ROI before scaling up. • Adopt a flexible technology roadmap: Keep the adoption plan adaptable to integrate new technological advances as they emerge.
Example	Manufacturers in high-tech industries often start by introducing cobots in select areas of the production line and gradually scale their use based on success metrics and feedback from human operators.

Enhancing cybersecurity and resilience

Strategy	With increased reliance on AI, cobots, and connected systems, cybersecurity and system resilience become essential.
Actions	• Strengthen cybersecurity protocols: Adopt advanced security measures such as AI-driven threat detection and blockchain for secure data transactions. • Build resilient systems: Design industrial systems that are resilient to cyber-attacks and can recover quickly from potential disruptions. • Invest in data privacy measures: Ensure data privacy compliance with global regulations like GDPR and CCPA, and protect customer data in all AI and customization processes.
Example	Smart factories that rely on connected devices (IoT) are increasingly using blockchain-based security for supply chain management, ensuring that data integrity is maintained and minimizing the risk of cyberattacks.

3.5.1   *Workforce upskilling and reskilling*

As Industry 5.0 relies heavily on human expertise, upskilling and reskilling the workforce are essential. Organizations must invest in continuous learning programs to enable employees to work alongside AI, robotics, and other advanced technologies. Training programs should focus on critical thinking, creativity, and ethical decision-making.

3.5.2   *Adoption of ethical standards*

Ethical standards in AI and automation are paramount in Industry 5.0. Companies must integrate ethical principles in their development of AI systems, ensuring fairness, transparency, and accountability. Regulatory bodies will play a critical role in defining and enforcing these standards.

3.5.3   Sustainability initiatives

Organizations must adopt sustainability initiatives, aligning with the principles of Industry 5.0. This includes transitioning to renewable energy, implementing energy-efficient technologies, and reducing waste through the circular economy [15]. The push for sustainable practices will drive innovation in energy storage, recycling technologies, and resource management.

3.6   Implications for various sectors

In manufacturing, Industry 5.0 will lead to the widespread adoption of cobots, smart factories, and personalized production as shown in Figure 3.4. It will create a shift from mass production to bespoke, customer-driven solutions, enabled by advanced robotics and AI-driven optimization.

3.6.1   Healthcare

In healthcare, Industry 5.0 promises to improve patient care through precision medicine, personalized treatment plans, and AI-assisted diagnostics. Human-machine collaboration in medical procedures and diagnostics will enhance the accuracy and efficiency of healthcare delivery [16].

3.6.2   Energy

The energy sector will benefit from Industry 5.0 through the integration of renewable energy sources and smart grids. AI-driven energy management systems will optimize energy distribution, reduce wastage, and improve grid resilience (Table 3.4).

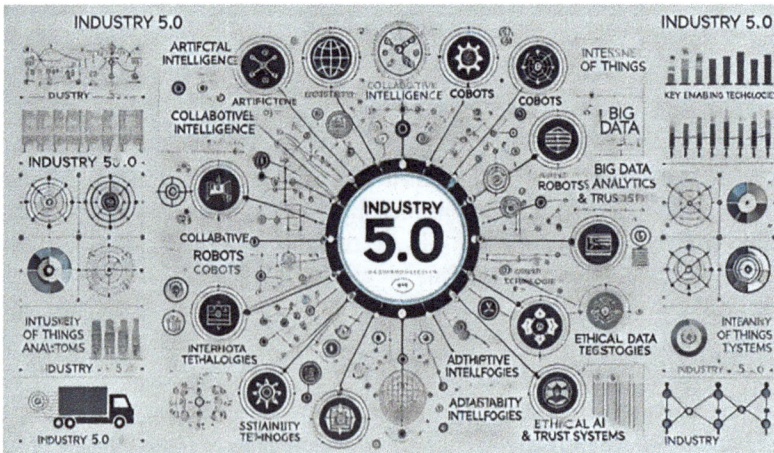

Figure 3.4   Key technology

Table 3.4   Impacts of the technology roadmap for Industry Revolution 5.0

Enhanced human–machine collaboration

Impact   Industry 5.0 will reshape the relationship between humans and machines by emphasizing collaboration over automation. Cobots (collaborative robots) will assist human workers, enhancing their capabilities rather than replacing them. This human–machine symbiosis will lead to more customized products, innovative solutions, and efficient decision-making processes.

Example   In manufacturing, robots working alongside human operators will allow for mass customization, producing personalized products efficiently. Xu *et al.* [10] highlight how human creativity combined with machine precision can significantly improve productivity and innovation.

Job creation and workforce transformation

Impact   Contrary to fears of mass unemployment due to automation, Industry 5.0 is expected to create new job roles centered around supervising, programming, and interacting with AI and robots. It will also drive the need for reskilling and upskilling, focusing on STEM education and lifelong learning.

Example   AI specialists, data scientists, and cobot coordinators are expected to become essential roles in industries like manufacturing, healthcare, and logistics. Tvenge and Martinsen [17] argue that while some jobs may be displaced, Industry 5.0 will offer more complex, high-skill roles requiring human oversight and creative problem-solving.

Economic growth and new business models

Impact   The integration of AI, 3D printing, IoT, and big data will drive economic growth by enabling more efficient production, lower costs, and the rise of new business models like on-demand manufacturing and hyper-personalization. Additionally, Industry-as-a-Service (IaaS) models will emerge, allowing businesses to adopt flexible and scalable industrial services.

Example   Businesses will be able to adopt just-in-time production practices, reducing waste and inventory costs while delivering customized products. According to Nahavandi [8], businesses can leverage AI to dynamically adjust production lines based on real-time demand, driving agility and profitability.

Sustainability and green technology adoption

Impact   Industry 5.0 aligns with global sustainability goals, with a strong emphasis on reducing waste, conserving resources, and minimizing environmental impact. 3D printing and AI-driven energy management systems will play a key role in resource efficiency, contributing to the circular economy.

Example   Companies can leverage AI to optimize resource usage and implement predictive maintenance to reduce energy consumption and extend equipment lifespan. Bai *et al.* [1] stress that Industry 5.0 technologies, especially AI, will help industrial systems transition to greener and more sustainable practices, supporting environmental goals like carbon neutrality.

Improved quality of life and social impact

Impact   Industry 5.0 promotes technologies that directly improve the quality of life by addressing societal challenges such as healthcare, education, and social equity. The use of AI and advanced robotics in healthcare, for example, will revolutionize patient care, leading to earlier diagnoses, personalized treatment plans, and improved healthcare accessibility.

Example   AI-powered diagnostics, robot-assisted surgeries, and telemedicine platforms will enhance the healthcare experience for patients, especially in remote areas. Floridi *et al.* [18] that Industry 5.0 technologies will have profound implications for social welfare, creating opportunities to use AI for good and foster more inclusive innovation.

(Continues)

Table 3.4   (Continued)

Technological innovation and customization

Impact   Industry 5.0 will spur innovation across sectors by leveraging AI, machine learning, blockchain, and 5G to create more customized and adaptive solutions. It will also facilitate real-time data analytics and AI-driven innovation, allowing companies to make informed decisions more quickly.

Example   Real-time analytics and predictive algorithms will allow industries to optimize production and supply chains dynamically, improving operational efficiency. According to Kusiak [11], this capability will enable industries to respond quickly to market changes, ensuring resilience and competitiveness.

Ethical AI and data privacy

Impact   As AI becomes more deeply integrated into industries, ethical considerations will be at the forefront. Industry 5.0 places significant emphasis on ethical AI practices, ensuring that AI systems are transparent, fair, and free of bias. Additionally, data privacy will be a key focus, with strict guidelines on data usage and protection.

Example   Ethical AI frameworks and data governance protocols will be established to ensure AI is used responsibly in decision-making processes. Floridi *et al.* [18] emphasize the importance of designing AI systems that prioritize transparency, accountability, and fairness, ensuring trust in human–machine collaboration.

Resilience through advanced cyber-physical systems

Impact   The integration of cyber-physical systems (CPS) and digital twins will improve the resilience of industrial systems by enabling real-time monitoring, control, and optimization. These technologies will help industries become more agile and responsive to disruptions, ensuring continuity in critical operations.

Example   Digital twins will allow businesses to simulate production environments and test changes before implementing them in the real world, reducing downtime and improving decision-making. Lasi *et al.* [9] argue that CPS will enable industries to optimize resource allocation and adapt more quickly to changing market conditions.

Global connectivity and interoperability

Impact   Industry 5.0 will foster global connectivity and collaboration through the increased use of IoT and advanced communication technologies like 5G. Interoperability between systems and devices across industries and regions will become essential for real-time data exchange and coordinated production efforts.

Example   A connected global network of smart factories will enable companies to collaborate across borders, improving supply chain efficiency and enabling global responsiveness. Xu *et al.* [19] emphasize the importance of developing unified standards to ensure interoperability and seamless integration of technologies worldwide.

3.7   Challenges and barriers of the technology roadmap for Industry Revolution 5.0

Industry Revolution 5.0 focuses on enhancing human–machine collaboration, sustainability, and addressing societal goals. However, this shift from Industry 4.0 to Industry 5.0 comes with a set of challenges and barriers that must be overcome to fully realize its potential. These challenges can be categorized into technological, organizational, workforce, societal, ethical, and regulatory dimensions as shown in Table 3.5.

*Table 3.5   Challenges and barriers of the technology roadmap for Industry
Revolution 5.0*

Technological challenges	
Integration of advanced technologies	Industry 5.0 requires the seamless integration of AI, robotics, big data, cyber-physical systems, biotechnology, and IoT. These technologies often operate in silos, creating difficulties in achieving synergy. Xu *et al.* [10] emphasize the complexity of creating a cohesive framework where various technologies function in a unified manner.
Cybersecurity risks	With increased connectivity and human–machine collaboration, the potential for cyber-attacks also rises. The integration of AI and IoT into manufacturing creates more vulnerability points. Protecting sensitive industrial and personal data remains a significant challenge, as highlighted by Sari *et al.* [20].
Data management and interoperability	Managing the massive amount of data generated by AI, IoT, and other technologies is challenging. Interoperability among different systems and platforms is critical but often lacking, making it difficult for organizations to derive meaningful insights from data. Borgia [21] points to issues in standardization as a major hurdle in ensuring systems can communicate efficiently.
Organizational challenges	
Lack of a clear vision	While the concept of Industry 5.0 is gaining momentum, many organizations lack a clear roadmap or strategy for integrating these technologies with a human-centered focus. Without alignment between technological advancements and corporate goals, the transformation can become fragmented.
Investment and cost barriers	Industry 5.0 involves substantial upfront costs in technology development, infrastructure, and workforce training. For small and medium-sized enterprises (SMEs), this poses a significant challenge, as noted by Mittal *et al.* [22]. The return on investment (ROI) may take longer to materialize, causing hesitation in adopting these technologies.
Workforce challenges	
Skill gaps and workforce readiness	One of the most significant barriers is the mismatch between the existing workforce's skills and the demands of Industry 5.0 technologies. Tvenge and Martinsen [17] stress the need for upskilling and reskilling workers to handle advanced AI systems, robotics, and human–machine interfaces. The current educational systems and corporate training programs are often inadequate in addressing these evolving needs.
Resistance to change	Employees and management may resist the shift towards Industry 5.0 due to fear of job displacement, as automation and robotics become more integrated into the workplace. Overcoming this resistance requires transparency, education, and creating an organizational culture that embraces change.
Societal and ethical challenges	
Job displacement and social inequality	While Industry 5.0 is designed to be more human-centric, there is still a risk that automation and robotics will lead to job displacement. Certain industries, particularly in manufacturing, may experience labor shifts, disproportionately affecting low-skilled workers. Kusiak [11] highlights the need to find a balance between automation and meaningful human employment.

(Continues)

Table 3.5   (Continued)

Ethical AI usage	Ensuring ethical AI practices is a key barrier. Questions surrounding privacy, data usage, AI bias, and decision-making accountability arise. Floridi *et al.* [18] emphasize the importance of designing AI systems that are transparent, fair, and accountable. As AI becomes more integrated into production processes, it becomes essential to create frameworks for ethical decision-making.
Regulatory and standardization challenges	
Lack of regulatory frameworks	One of the key barriers to Industry 5.0 adoption is the lack of clear and consistent global regulatory frameworks governing AI, data privacy, and robotics. Many countries are still working on regulations for Industry 4.0, and Industry 5.0 adds complexity due to its focus on human interaction with machines and ethical AI usage. Schwab [23] points out that the legal frameworks often lag behind technological advancements.
Standardization issues	Developing common standards for AI, robotics, IoT, and cybersecurity is critical for enabling interoperability and collaboration. The lack of universally accepted standards hampers cross-industry and cross-border collaboration, as noted by Xu *et al.* [10].
Environmental and sustainability challenges	
Sustainable technology development	The transition to Industry 5.0 demands the development of technologies that are not only efficient but also sustainable. Ensuring that advanced manufacturing systems are energy-efficient and environmentally friendly requires significant research and innovation. Bai *et al.* [1] explore how sustainability must be built into the design of industrial systems from the outset.
Resource availability	The materials required for building advanced AI systems, robotics, and smart factories are not always sustainably sourced. There is growing concern over the availability of rare earth metals that are crucial for advanced technologies. Addressing these resource constraints is vital for the sustainable implementation of Industry 5.0.

3.8   Conclusions

The Technology Roadmap for Industry 5.0 reflects the integration of advanced technologies with human-centered design, sustainability, and ethical considerations. Unlike its predecessor, Industry 4.0, which primarily focused on automation, Industry 5.0 places a strong emphasis on human–machine collaboration and leveraging technology to benefit society. The roadmap outlines the steps required to achieve a more balanced and forward-looking industrial ecosystem.

Human-centric systems will dominate future industries: Industry 5.0 shifts the focus from pure automation to collaboration between humans and intelligent machines. The technology roadmap highlights how AI, robotics, and cobots (collaborative robots) will work alongside humans to enhance creativity and decision-making. Unlike the fully autonomous systems of Industry 4.0, these technologies will augment human capabilities in design, problem-solving, and customization of products and services. Xu *et al.* [10] emphasize the necessity of designing systems that enhance the cognitive and physical tasks humans perform, rather than replacing them.

Hyper-personalization as a key driver: Industry 5.0 technologies enable hyper-personalization in manufacturing and service delivery. This is driven by advances in AI, 3D printing, and data analytics, which allows companies to offer products tailored to individual customer preferences on a mass scale. This capability shifts production paradigms from mass production to mass customization. Nahavandi [8] points out that this move toward personalization will create more value for customers while reducing waste through on-demand production.

Sustainability and green technology integration: Sustainability is a central pillar of the Industry 5.0 roadmap. The integration of technologies such as AI for resource optimization, 3D printing for reduced material waste, and renewable energy systems is crucial for minimizing the environmental footprint of industrial processes. Bai *et al.* [1] emphasize that the roadmap for Industry 5.0 goes beyond technological advancement; it aims to align industrial practices with global environmental goals, including the United Nations Sustainable Development Goals (SDGs).

Hybrid workplaces with humans and AI collaboration: The roadmap suggests a future where humans work closely with AI systems, machine learning, and robotics to achieve higher efficiency, innovation, and safety. While AI will handle repetitive and data-heavy tasks, humans will engage in more complex problem-solving and decision-making roles. This hybrid workforce model will lead to more flexible work environments, which are crucial for future industries. Kusiak [11] notes that the key challenge will be to develop new workplace structures and management systems that optimize this synergy.

Advanced cyber-physical systems for enhanced automation: The roadmap shows that CPS that connect physical processes with digital networks will play a fundamental role in Industry 5.0. These systems will enable real-time monitoring, control, and decision-making across industries. In particular, digital twins—virtual representations of physical assets—will facilitate better planning, simulation, and operational efficiency. Lasi *et al.* [9] identified these technologies as a key element in realizing the smart factories of Industry 4.0, and now in Industry 5.0, their applications are extended to human–machine cooperation.

Ethical AI and regulatory frameworks: AI ethics will be central to Industry 5.0's success, requiring strong governance frameworks to manage AI decision-making, data privacy, and fairness. Ensuring that AI systems operate transparently and without bias will be crucial in building trust between humans and intelligent machines. Floridi *et al.* [18] emphasize that the roadmap must include robust ethical guidelines to govern AI and its interaction with humans in industries, ensuring that technology serves human needs without infringing on rights [24].

Skilled workforce and lifelong learning: Industry 5.0 will require a highly skilled workforce capable of interacting with advanced AI systems, robots, and other smart technologies. The roadmap calls for a significant investment in education, training, and re-skilling to meet this demand. Developing educational programs that emphasize STEM fields (Science, Technology, Engineering, Mathematics) along with soft skills like creativity, critical thinking, and problem-solving will be critical. Tvenge and Martinsen [17] suggest that continuous learning and adaptability will be essential in enabling workers to thrive in this evolving landscape.

Challenges in interoperability and standardization: The complexity of integrating various technologies—ranging from AI and IoT to 3D printing and advanced robotics—presents significant challenges. Interoperability among different systems and devices is critical to ensure smooth communication and data flow. The lack of unified standards remains a barrier to large-scale adoption, as highlighted by Schwab [23]. The roadmap outlines the need for international collaboration to establish common protocols and standards that facilitate seamless technology integration.

Cybersecurity as a critical component: With increased reliance on connected devices and AI systems, cybersecurity becomes a critical concern in Industry 5.0. The roadmap identifies the need for robust cybersecurity frameworks that can protect industrial operations from cyber-attacks, data breaches, and system failures. Sari *et al.* [20] highlight that without proper cybersecurity measures, the benefits of Industry 5.0 technologies could be undermined by vulnerabilities in digital infrastructure.

References

[1]   Bai, C., Dallasega, P., Orzes, G., and Sarkis, J. (2020). Industry 4.0 technologies assessment: A sustainability perspective. *International Journal of Production Economics*, 229, 107776.

[2]   Leong, W. Y. (2024). Fostering creative thinking through immersive virtual reality environments in education. *Educational Innovations and Emerging Technologies*. 4(3), 8–25, https://doi.org/10.35745/eiet2024v04.03.0002.

[3]   Brettel, M., Friederichsen, N., Keller, M., and Rosenberg, M. (2014). How virtualization, decentralization, and network building change the manufacturing landscape: An Industry 4.0 perspective. *International Journal of Mechanical, Aerospace, Industrial, Mechatronic and Manufacturing Engineering*, 8(1), 37–44.

[4]   Leong, W. Y., Leong, Y. Z., and San Leong, W. (2024). Human–machine interaction in the electric vehicle battery industry. In *2024 10th International Conference on Applied System Innovation (ICASI)*, pp. 69–71. Piscataway, NJ: IEEE.

[5]   Leong, W. Y., Chuah, J. H., and Tee, B. T. (2020). *The Nine Pillars of Technologies for Industry 4.0*. Stevenage: The Institution of Engineering and Technology.

[6] Leong, W. Y., Leong, Y. Z. and Leong, W. S. (2024). Strengthening security in computing, *2024 IEEE Symposium on Wireless Technology & Applications (ISWTA)*, Kuala Lumpur, Malaysia, 2024, pp. 113–116, doi: 10.1109/ISWTA62130.2024.10651781.

[7] Romero, D., Stahre, J., Wuest, T., and Noran, O. (2018). Towards a human-centred reference architecture for next-generation balanced automation systems: Industry 5.0. *Computers in Industry*, 99, 20–33.

[8] Nahavandi, S. (2019). Industry 5.0—A human-centric solution. *Sustainability*, 11(16), 4371.

[9] Lasi, H., Fettke, P., Kemper, H.-G., Feld, T., and Hoffmann, M. (2014). Industry 4.0. *Business & Information Systems Engineering*, 6(4), 239–242.

[10] Xu, L. D., Xu, E. L., and Li, L. (2021). Industry 4.0 and Industry 5.0—Inception, conception, and perception. *Journal of Manufacturing Systems*, 61, 254–258.

[11] Kusiak, A. (2018). Smart manufacturing. *International Journal of Production Research*, 56(1–2), 508–517.

[12] European Commission. (2021). *Industry 5.0: Towards a Sustainable, Human-Centric, and Resilient European Industry*. European Commission, Brussels.

[13] Shrouf, F., Ordieres, J., and Miragliotta, G. (2014). Smart factories in Industry 4.0: A review of the concept and of energy management approaches in production based on the Internet of Things paradigm. *Proceedings of the 2014 IEEE International Conference on Industrial Engineering and Engineering Management*, pp. 697–701.

[14] Leong, W. Y. (2022). *Human Machine Collaboration and Interaction for Smart Manufacturing: Automation, Robotics, Sensing, Artificial Intelligence, 5G, IoTs and Blockchain*. Stevenage: The Institution of Engineering and Technology.

[15] European Commission: Directorate-General for Research and Innovation. (2024). ERA industrial technologies roadmap on human-centric research and innovation for the manufacturing sector. Publications Office of the European Union. Available from: https://data.europa.eu/doi/10.2777/0266

[16] Zawadzki, P., and Żywicki, K. (2016). Smart product design and production control for effective mass customization in the Industry 4.0 concept. *Management and Production Engineering Review*, 7(3), 105–112.

[17] Leong, W. Y. (2024). *ESG Innovation for Sustainable Manufacturing Technology: Applications, designs and standards*. Stevenage: The Institution of Engineering and Technology.

[18] L. Floridi, J. Cowls, M. Beltrametti, *et al.* (2018). AI4People—An ethical framework for a good AI society: Opportunities, risks, principles, and recommendations. *Minds and Machines*, 28(4), 689–707.

[19] Xu, L. D., Xu, E. L., and Li, L. (2018). Industry 4.0: State of the art and future trends. *International Journal of Production Research*, 56(8), 2941–2962.

[20] A. Sari, C. Acarturk and E. Aydogan (2020). Cybersecurity risks of Industry 4.0 and roadmaps of cybersecurity in smart manufacturing. *Future Internet*, 12(11), 185.

[21] E. Borgia (2014). The Internet of Things vision: Key features, applications, and open issues. *Computer Communications*, 54, 1–31.

[22] S. Mittal, M. A. Khan, D. Romero, and T. Wuest (2018). A critical review of smart manufacturing & Industry 4.0 maturity models: Implications for small and medium–sized enterprises (SMEs). *Journal of Manufacturing Systems*, 49, 194–214.

[23] K. Schwab (2017). *The Fourth Industrial Revolution*. New York: Crown Publishing Group.

[24] Longo, F., Padovano, A., and Umbrello, S. (2020). Value-oriented and ethical technology engineering in Industry 5.0: A human-centric perspective for the design of the factory of the future. *Applied Sciences*, 10(12), 4182.

Chapter 4

Industry 5.0 and digital transformation for Malaysia construction industry

Chee Fui Wong[1]

The Malaysian government has acknowledged the importance of Industry 4.0 towards the digital transformation of the construction industry through national policies such as National Construction Policy 2023 and Construction 4.0 Strategic Plan 2021–2025. In 2021, the European Commission has introduced Industry 5.0 as an industry that recognises the power of the industry to achieve societal goals beyond jobs and growth to become a provider of prosperity, by making production respect the boundaries of our planet and placing the well-being of the worker at the centre of the production process. It is of paramount importance that Malaysia should start to incorporate Industry 5.0 into the digital transformation of the construction industry in Malaysia. Industry 5.0 aims to transform the technological driven digital transformation to a value driven mass personalisation industry. Industry Revolution 5.0 (IR5.0) complements Industry 4.0 by adopting all the enabling technology of Industry 4.0, while putting focus on human–centric, sustainability and resilient. Digital transformation in the Malaysia construction industry incorporates both Industry 4.0 and Industry 5.0 by adopting the emerging technologies in the construction industry such as building information modelling, industrialised building system, drone technology and other emerging technologies. It is paramount that the construction industry moves towards digital transformation through the adoption of these technologies.

4.1 Construction 4.0

The "Fourth Industrial Revolution" or "Industry 4.0" is the era of "Cyber Physical Systems (CPS)" which comprises smart machines, storage systems and production facilities capable of autonomously exchange information, triggering actions and control each other independently [1]. The "Fourth Industrial Revolution" is the disruptive transformation of industries through the application of emerging technology

Industry 4.0 transforms the processes of product design, fabrication, usage, operation, maintenance and servicing. Furthermore, it will revolutionise the

[1]Department of Civil Engineering, Lee Kong Chien Faculty of Engineering, Universiti Tunku Abdul Rahman, Selangor, Malaysia

operation, procedures, logistical management and carbon footprint of the production. Industry 4.0 revolutionises global industrial competition, diminishing the comparative edge of low-cost nations that depend on inexpensive workforce [2]. In simple terms, Industry 4.0 integrate the human, machine and processes.

Globally, the term Industry 4.0 refers to the technologies such as autonomous robot; simulation and artificial intelligence (AI), system integration, Internet of Things (IoT), cyber security, cloud computing technologies; additive manufacturing (3D printing), augmented reality and big data analytics [3].

However, in Malaysian context, we have re-defined the main technologies for Construction 4.0 under the Malaysian Construction Industry Development Board (CIDB) – Construction 4.0 Strategic Plan 2021–2025 [4] to encompass also

1. Building information modelling (BIM)
2. Prefabrication and modular construction
3. Advance building material

Construction 4.0 Strategic Plan is a roadmap for the Malaysian construction sector to embrace the Industry 4.0 in ways that would revolutionise the construction industry, productivity and competitiveness. The Construction 4.0 strategic framework is as shown in Figure 4.1. CIDB defined Construction 4.0 as the transformation process for the construction industry to implement the emerging technologies in Industry 4.0 towards the digitalisation of the construction industry [4].

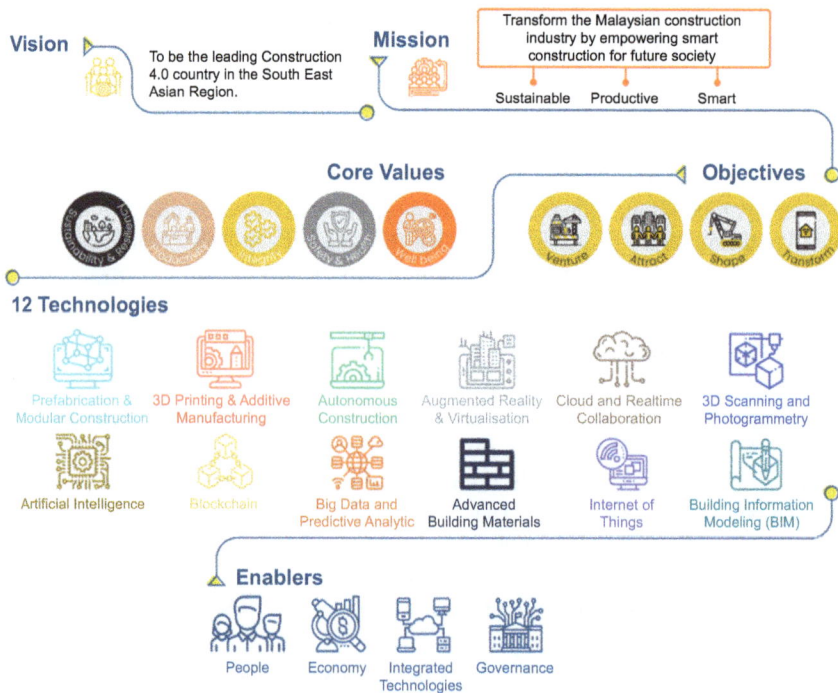

Figure 4.1   Construction 4.0 strategic framework [4]

4.2　Industry 5.0

The European Commission in the policy paper published in 2021 has introduced the concept of Industry 5.0 with the following definition:

> *The Industry 5.0 recognized the power of industry to achieve societal goals beyond jobs and growth to become a resilient provider of prosperity, by making production respect the boundaries of our planet and placing the well being of the industry worker at the center of the production process* [5].

The Industry 5.0 focused on the three pillars (Figure 4.2), namely

1. Human-centric approach
2. Sustainable
3. Resilience.

This means that Industry 5.0 is value driven towards a sustainable, human-centric and resilient industry with the aim to bring human back into the production in a sustainable and resilient way.

The Fifth Industrial Revolution or Industry 5.0 will be focused on the collaboration between human and machine, as human intelligence works in harmony with cognitive artificial intelligence. The workers will be upskilled to provide value-added tasks in production and manufacturing by putting human back into industry production with collaborative machine and robots. In simple terms, the Fifth Industry Revolution focused to personalisation and customisation to meet the customer need and expectation while adopting a sustainable and resilience approach.

The Malaysian government has realised the paramount importance of the digital transformation in improving the productivity and increasing the competitiveness of the industry. The government has launched various policies to promote the transformation of the industry towards embracing the digital transformation and productivity-enhancing innovations. These include

1. National Policy on Industry 4.0 (Industry4WRD) – launched 31 October 2018
2. Malaysia Digital Economy Blueprint (MyDIGITAL) – launched 19 February 2021

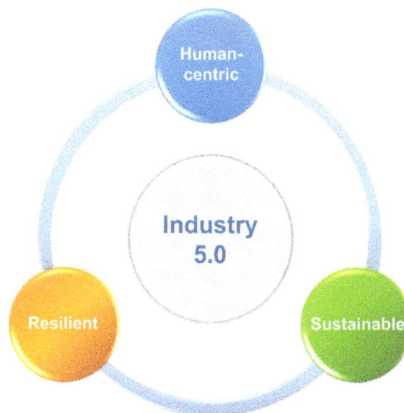

Figure 4.2　The three pillars of Industry 5.0 [5]

3. National Fourth Industrial Revolution (4IR) Policy – launched 1 July 2021
4. CIDB Construction Strategic Plan 4.0 (CR4.0) (2021–2025) – launched November 2021

These policies set the framework for the potential of digital transformation through the adoption of emerging and disruptive technologies to enhance the productivity and competitiveness in Malaysia. It is of paramount importance to embrace these emerging technologies towards Industry 4.0 that are technological driven to enhance the productivity through automation, mass customisation, intelligent supply chain that removes the manpower from the industries. However, these current policies have overlooked the human-centric, sustainability and resilience factors.

The concept of Industry 5.0 plays an important role to complement Industry 4.0 in embracing all the enabling technologies of Industry 4.0, while putting the focus back into human-centric, sustainability and resilience. The Industry 5.0 focus on bringing back the manpower into the industry productivity through mass personalisation and customisation. The human factors will have a more significant role to play in the delivery of customer experience, hyper customisation, developing intelligent responsive and distributed supply chain as well as interactive activated experience product.

In simple terms, Industry 5.0 aims to transform the technological driven digital transformation in a value driven mass personalisation industry. It is important to understand that Industry 5.0 is not a chronological continuation or the next stage of Industry 4.0. The digital transformation of the construction industry must not wait until they have fully implemented Industry 4.0 before embarking on Industry 5.0. The Industry 5.0 transformation complements Industry 4.0 through the adoption of all the enabling technologies of Industry 4.0 while putting focus on human-centric, sustainability and resilient (Figure 4.3).

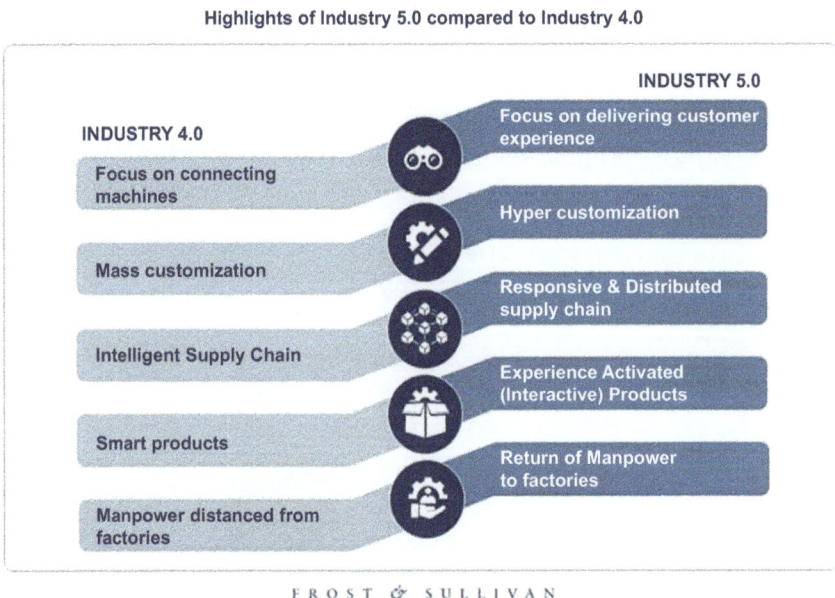

Figure 4.3   Highlight of Industry 5.0 compared to Industry 4.0 [6]

4.3 Digital transformation in the construction industry

Construction sector is one of the less digitalised sectors. According to a study by the McKinsey Global Institute Industry digitisation index in 2016, the construction industry is among the least digitalised and only behind "Agricultural and hunting" in terms of digitalisation as shown in Figure 4.4 [7].

However, the construction industry is beginning to realise the impact of digital transformation in terms of the technologies to be used and building specifications. The construction industry is embarking on a transformation with innovative new technologies pushing for industry penetration as well as updating and upgrading current the technologies.

Some of the emerging technologies in construction industry includes BIM, industrialised building system (IBS), drone technologies, green building and sustainable construction, 3D printing and other new emerging technologies

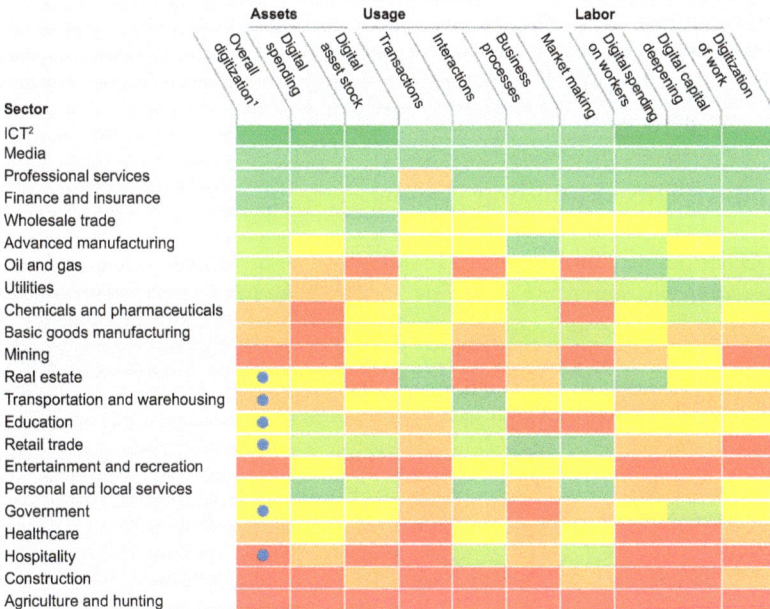

Figure 4.4   McKinsey Global Institute Industry digitisation index [7]

4.4   Building information modelling

BIM technology is a disruptive technologies that will transform the global construction industry [8]. BIM technology provides digital information modelling environments that allow the construction data information to be managed in a collaborative process among all the stakeholders that are involved through the whole project life cycle.

In Malaysian context, the CIDB has defined BIM as a "*modelling technology and associated set of process to produce, communicate, analyse and the use of digital information models throughout the construction project life-cycle*" [9].

BIM technology not only provides accurate visualisation (BIM 3D) but also incorporates the time element or scheduling into the design (BIM 4D) as well as cost element (BIM 5D), which in BIM 6D and BIM 7D, we incorporate as-built information and even O&M element into it for facility management [10]. The benefits of BIM include reduced disciplinary conflicts among stakeholders, reduced crashes during construction and provide resolution; improved construction efficiency and productivity, reduced construction reworks as well as accurate visualisation.

The European Commission (2020) report has identified real-time-based digital twins and simulation as an enabling technology for Industry 5.0 [11]. The BIM can be considered as digital twin, if the both the physical building and digital model are interlinked, used throughout the whole project life cycles which includes the facilities management and even the demolition stage.

4.5   Industrialised building system

The CIDB has defined "*Industrialised building system or IBS means the technique of construction whereby components are manufactured in a controlled environment, either at site or off site, and subsequently transported, positioned and assembled into construction works*" [12]. Globally, IBS is also known as modular construction, prefabricated construction, off-site construction as well as prefinished prefabricated volumetric construction (PPVC).

In Malaysian context, IBS systems include six types of systems including pre-cast concrete framing; steel frameworks system; pre-fabricated timber framing systems; steel frame systems; and blockworks systems and innovative systems.

The CIDB Construction Industry Standard CIS18: 2023 Manual for IBS content scoring system (IBS score) has categorised the IBS score calculation into three major components, namely structural system (50 points), wall system (20 points) and other simplified construction (30 points) [12].

With industrialised building systems (IBS), the construction material can be produced through mass customisation and standardisation where we have Malaysian Standard MS1064 to specified the standard size of material so that it will be more feasible for the industry to mass produce these construction materials.

One of the famous examples of building construction that utilised IBS or modular construction is the Avenue South Residence condominium in Singapore which is the world's tallest modular building project using PPVC [13].

4.6 Drone technology

Drone technology is a disruptive and emerging technology that transforms the construction industry by revolutionising the processes of data collecting, analysis and decision-making. Drone sensor technology employs unmanned aerial vehicles (UAVs) to understand their environment and collect important data through the installed sensors that allow the UAV to record images, estimate distances, sense changes in the environment and navigate through complicated terrains without human intervention.

With drone mapping survey technology, ground survey works that takes weeks or months to complete can be completed within days. Drone mapping survey technology utilises "Light Detection and Ranging (LIDAR)" that uses sensor to send out pulses of laser light and measures the exact time it takes for these pulses to return as they bounce from the ground to produce 3D-Digital Terrain Model with GIS and aerial mapping [14].

Drone technology can also be used for site supervision and construction progress monitoring, where drone can be used to view all activity, with daily automated reporting of equipment, sub/direct personnel and materials. Drone technology applied in construction monitoring can detect safety hazards before they become safety incidents and flag violating behaviour as it occurs to reduce accidents.

4.7 3D construction

Additive manufacturing (AM) is the process of building physical objects by layering materials like metal, plastic or concrete using dedicated software and equipment. The AM technology is also known as 3D-printing that produces components directly from a digital design file and prints the special raw material layer by layer to create specially tailored material for producers. The paradigm shift of AM technology research is ongoing in the construction industry for the process called 3D construction printing (3DCP) [15].

The CIDB, CIDB IBS, CIDB Construction Research Institute of Malaysia (CREAM) and KA Bina have collaborated to develop the first 3D-printed home in ASEAN. The first mock-up unit was completed in April 2022 at CIDB headquarters in Jalan Chan Sow Lin (Figure 4.5).

In Sarawak, Malaysia, a sample 3D printed house is constructed at Akademi Binaan Malaysia (ABM) Kuching Sarawak. The 3D-printed house consisting of a

Figure 4.5   3D printed house in CIDB Headquarter, KL

Figure 4.6   3D printed house in Kuching, Sarawak

built-up area of 1000 sq ft (90 m^2) was completed in 46 hours using the 3D con-struction printer where the total length of the print was over 9 km that was extruded layer by layer on top of each other in a total of 145 layers each of 2 cm height (Figure 4.6) [16].

With emerging technologies such as additive manufacturing and 3D printing, there is improvement in efficiency and productivity of construction projects towards the digital transformation of the construction industry.

4.8   Digitalisation of construction documentation

In the transformation towards Construction 4.0 and digital transformation of the construction industry, it is important to embrace the digitalisation of the construction record management. The construction record management includes the site diary, inspection forms; project specification, contract documents, project quality plan, construction schedule; project safety and health record can be digitalised, analysed and incorporated into the BIM for information integration.

The benefits of digitalisation of construction record management include ease of access; enhance database management; improve data collaboration; enable information sharing; enhance safety and security; efficient information storage and data safekeeping.

In the future, for Industry 5.0 and digital transformation of the construction industry, real-time digital construction record can be implemented where all construction records and databases can be shared among all stakeholders in real time through digital platforms.

4.9   Transformation towards Industry 5.0 and digital transformation

The construction industry still lacks awareness of the concept of Industry 5.0 and emerging technologies for digital transformation particularly for the Construction 4.0 technology. Generally, there is still a misperception that the adoption of new construction technologies will have cost implications. However, construction stakeholders need to consider the improvement in efficiency and productivity as well as manpower reduction and time saving that will eventually add value and improve the construction projects.

The construction industry should also leverage in government policies and incentives to improve their capabilities and company readiness to adopt Industry 5.0 and digital transformation technologies in their process and construction methods.

More training and human capacity building need to be developed and trained in order to reduce the mismatch in skillsets and lack of right human capital to implement Industry 5.0 and digital transformation technologies.

4.10   Conclusion

Overall, construction industry players must thoroughly consider the evolving needs of the construction industry in end-to-end project management to draw on the Construction 4.0 and Industry 5.0 emerging technologies. The only way to achieve this is to embrace technology and productivity-enhancing innovations to improve decision-making and work procedures.

It is important that the construction industry to adopt emerging technologies in Industry 4.0 as part of the Malaysia Construction 4.0 Strategic Plan. However, the construction industry must also be ready to embark Industry Revolution 5.0.

It is of paramount importance for stakeholders in the construction industry to understand that Industry Revolution 5.0 is not a chronological continuation or the next stage of Industry 4.0 and the construction players must not wait until they have fully implemented Industry 4.0 before embarking on Industry 5.0. The Industry 5.0 complements Industry 4.0 and digital transformation of the construction industry through the adoption of all the emerging technologies, while emphasising on human-centric, sustainability and resilient. The aim of Industry 5.0 is to revolutionise the technological driven digital transformation into a value driven mass personalisation industry.

References

[1] Wong, CF. (2022), "Human–machine interaction (HMI) technology – Malaysia national technology roadmap Industry4WRD leading the human intelligence transformation in smart manufacturing", in Leong WY (eds.) *Human Machine Collaboration and Interaction for Smart Manufacturing. Automation, Robotics, Sensing, Artificial Intelligence, 5G, IoTs and Blockchain*, Institution of Engineering and Technology (IET), London, pp. 9–21.

[2] Economic Planning Unit, Malaysia (EPU). (2021), National 4IR Policy. Putrajaya: Economic Planning Unit, Prime Minister's Department.

[3] Chuah, JH. (2020), "The nine pillars of technologies for Industry 4.0", in Leong WY, Chuah JH, and Tee BT (eds.) *The Nine Pillars of Technologies for Industry 4.0*. Institution of Engineering and Technology (IET), London, 2020, pp. 1–21.

[4] CIDB. (2020), *Construction 4.0 Strategic Plan (2021–2025)*. Construction Industry Development Board (CIDB), Malaysia.

[5] European Commission, Directorate-General for Research and Innovation. (2021), *Industry 5.0: Human-Centric, Sustainable and Resilient*. Publications Office. Available at: https://data.europa.eu/doi/10.2777/073781.

[6] Frost and Sullivan. (2019), Industry 5.0—bringing empowered humans back to the shop floor. Available at: https://www.frost.com/frost-perspectives/industry-5-0-bringing-empowered-humans-back-to-the-shop-floor/ (accessed 1 July 2024).

[7] Agarwat, R., Chadrasekaran, S. and Sridhar, M. (2016), *Imagining Construction's Digital Future*. Singapore: McKinsey Productivity Sciences Center.

[8] WEF. (2018), *The Fourth Industrial Revolution is About to Hit the Construction Industry. Here's How It Can Thrive*. World Economic Forum. Available at: https://www.weforum.org/agenda/2018/06/construction-industry-future-scenarios-labour-technology/.

[9]   CIDB. (2016), *BIM Guide 1: Awareness, Construction Industry Development Board Malaysia*. Construction Industry Development Board (CIDB). Available at: https://mybim.cidb.gov.my/download/bim-guide-book-1/#.

[10]   Charef, R., Alaka, H. and Emmitt, S. (2018), "Beyond the third dimension of BIM: A systematic review of literature and assessment of professional views", *Journal of Building Engineering*, 19(1), 242–257.

[11]   European Commission, Directorate-General for Research and Innovation. (2020), *Enabling Technologies for Industry 5.0*. European Commission. (doi: 10.2777/082634). Available at: https://data.europa.eu/doi/10.2777/073781.

[12]   CIDB. (2023), *Construction Industry Standard CIS18:2023 Manual for IBS Content Scoring System (IBS Score)*. Construction Industry Development Board (CIDB), Malaysia.

[13]   Building and Construction Authority (BCA). (2022), *How United Tec Build the World's Tallest Modular Building Project*. Building and Construction Authority (BCA), Singapore. Available at: https://www1.bca.gov.sg/buildsg-emag/articles/how-united-tec-built-the-world-s-tallest-modular-building-project.

[14]   Emimi M., Khaleel M., and Alkrash A. (2023), "The current opportunities and challenges in drone technology", *International Journal of Electrical Engineering and Sustainability (IJEES)*, 1(3), 74–89.

[15]   MIDA. (2022), "Additive manufacturing in Industrial building system modular construction", *MIDA E-Newsletter* 6. Available at: https://www.mida.gov.my/wp-content/uploads/2022/02/1-JANUARY-MIDA-INVEST-MALAYSIA.pdf.

[16]   COBOD. (2022), "Even Borneo now has it first 3D printed house", COVID Press Release, 27 October. Available at: https://cobod.com/even-borneo-now-has-its-first-3d-printed-house/#:~:text=The%20new%203D%20printed%20house,in%20collaboration%20with%20the%20agency (accessed 1 July 2024).

Chapter 5

Insights from Industry 5.0 on the cultivation of engineering talent based on industrial needs

Zeng Cheng[1] and Wai Yie Leong[2]

In April 2021, the European Union announced the notion of "Industry 5.0: Towards a Sustainable, Human-centric, and Resilient European Industry." Industry 5.0 expands on the current Industry 4.0 framework by emphasizing the significance of research and innovation in facilitating a shift toward a sustainable, human-centered, and resilient European industry. The objective of Industry 5.0 is to facilitate the seamless integration of technology and society into the process of economic development.

5.1 Industry and education

In the age of Industry 5.0, highly trained technicians have become indispensable catalysts for economic progress. It is well recognized that educational institutions play a crucial role in improving students' skills to fulfill the requirements of emerging sectors. Educational institutions must meticulously evaluate how to adapt their educational focus, revise their curriculum, and enhance their teaching techniques to guarantee that the students they educate develop future-oriented abilities that align with industry demands and are globally relevant. The analysis of the functions and strategies that educational institutions should implement in this particular setting is of utmost importance.

Industry 5.0 characterizes substantial transformations in the progress of industrial production and manufacturing. It signifies significant changes in the operational methodology, technological integration, and response to market and societal demands of industries. Undoubtedly, talent is the essential resource upon which the industry relies to accomplish its goals. Scholarly studies on human capital theory consistently highlight a direct correlation between increased knowledge, high-quality education, and increased productivity in the labor market [1]. Drawing on human capital theory, education augments an individual's abilities and knowledge, therefore boosting their ability to participate in industrial

[1]Faculty of Education and Liberal Arts, INTI International University, Malaysia
[2]Faculty of Engineering and Quantity Surveying, INTI International University, Malaysia

operations. Such analysis can provide significant perspectives on the cultivation of engineering aptitude.

5.1.1   Globe employment situation

According to UNESCO, the global gross enrollment ratio for upper secondary schools has consistently increased from 64.87% in 2013 to 69.72% in 2022, despite the temporary impact of the COVID-19 pandemic [2]. Furthermore, according to data presented by the International Labour Organization (ILO), the COVID-19 pandemic had a substantial adverse effect on the global economy in 2020, resulting in widespread business closures and an increase in unemployment [3]. During this period, the worldwide unemployment rate for those aged 15 and above rose to 6.574%, hitting its highest point in a decade. The global unemployment rate decreased to 5.276% between 2021 and 2022 as a result of the widespread distribution of vaccines and the restart of economic activities (Figures 5.1 and 5.2). However, there are still substantial structural barriers that need to be addressed, particularly given that the unemployment rate in low-income countries was 5.326% in 2024. Subsequent to the pandemic, the impact of automation and artificial intelligence (AI) on the labor force has been more apparent, significantly influencing employment in industries that primarily depend on manual labor. The situation has also underscored the necessity of technical and creative labor.

Furthermore, with the emergence of Industry 5.0, there is an increasing demand for changes across all disciplines and industries. This is especially noteworthy given the emergence of new enterprises and economies that are linked to AI, cloud computing, big data, and genetic engineering. These advancements are resulting in substantial alterations in the skill and knowledge sets that are necessary for the labor market. Some educational institutions lack long-term planning and clear goals in their program offerings, resulting in a lack of interaction between

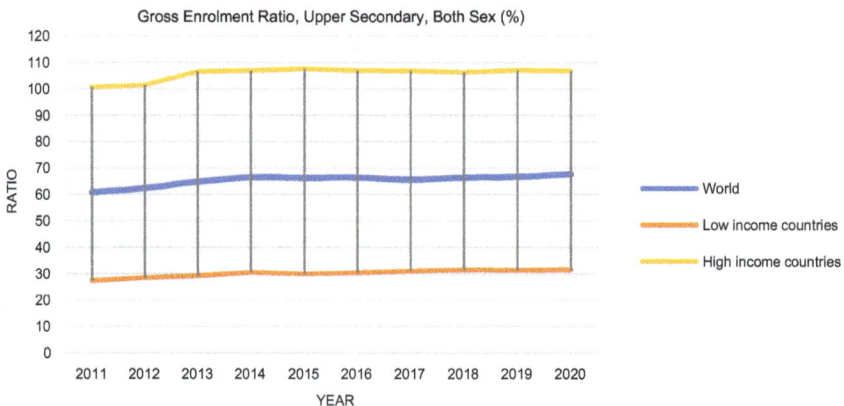

Figure 5.1   The gross enrollment ratio by annual

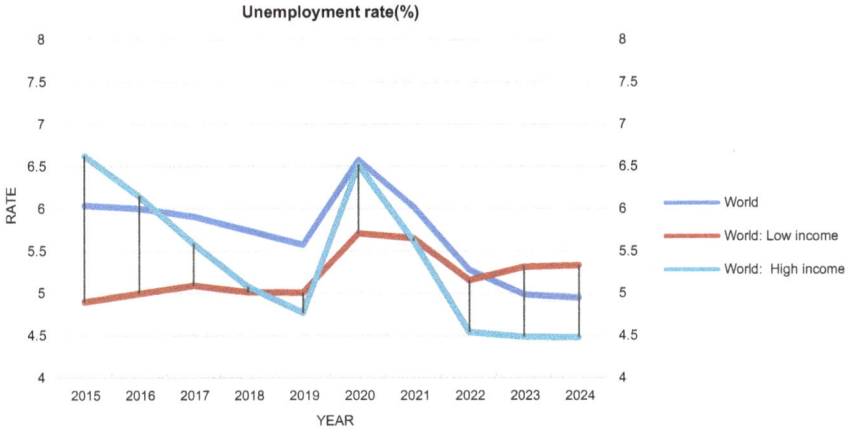

Figure 5.2 The unemployment rate by annual

talent cultivation programs and society. As a result, there is a lack of coordination between these programs and the demands of the job market. Talented persons in the job market are encountering significant challenges as a result of a combination of a growing number of graduates and structural imbalances between talent and the labor market. Educational institutions ought to carefully consider how to align themselves with the demands of Industry 5.0 by cultivating the necessary workforce.

5.1.2 Evolution from Industry 4.0 to Industry 5.0

Industry 4.0 and Industry 5.0 are important shifts in the development of industrial production and manufacturing. Gaining a comprehensive awareness of the definitions, connections, and distinctions between Industry 4.0 and Industry 5.0 is essential for comprehending the future direction of industrial processes and acquiring a more profound insight into the mechanisms involved in the evolution of Education 4.0 and Education 5.0.

5.1.2.1 Supporting technologies of Industry 4.0

Industry 4.0 refers to the present movement toward automation and the interchange of data in manufacturing technology. The integration of cyber-physical systems (CPS), the Internet of Things (IoT), cloud computing, and cognitive computing enables the creation of a smart manufacturing environment [4]. In this environment, machines are interconnected, data is continuously analyzed, and systems have the ability to autonomously make decisions. The link between Industry 4.0 and its supporting technologies is fundamental, as these technologies facilitate the essential functions of Industry 4.0 by supplying the required infrastructure, data processing, and communication capabilities (Figure 5.3).

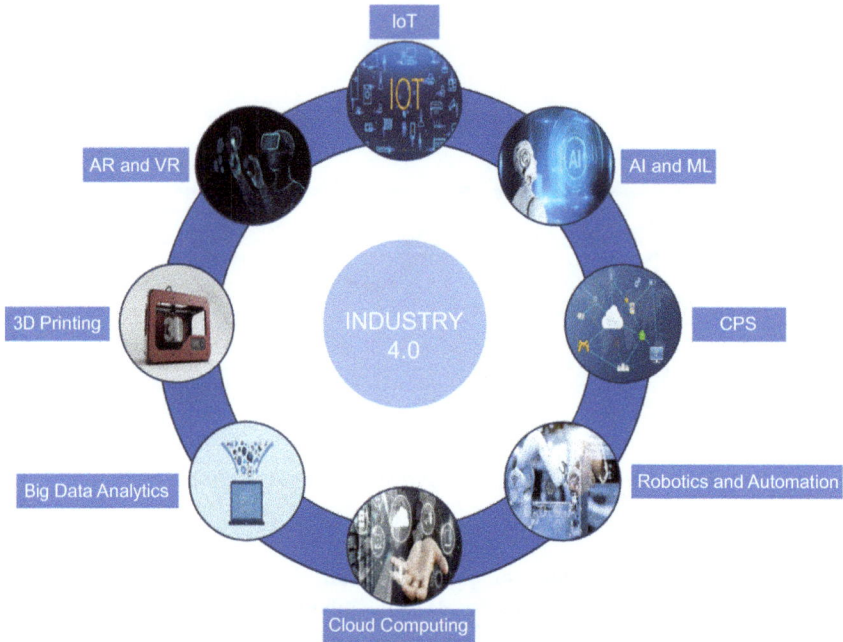

Figure 5.3   Key enabling technologies in Industry 4.0

- IoT: IoT is fundamental to Industry 4.0 by enabling seamless communication between devices, machines, and central systems. It facilitates the collection of vast amounts of real-time data from various sources, which is critical for informed decision-making and optimizing production processes. This interconnected network of devices enhances operational visibility and control, leading to more efficient and responsive manufacturing environments [5].
- AI and machine learning (ML): AI and ML are essential in Industry 4.0 for analyzing the massive volumes of data generated by IoT devices. These technologies optimize operations by improving decision-making processes and automating complex tasks, from production scheduling to quality control. AI and ML drive greater efficiency, reduce errors, and enable predictive maintenance, ensuring smoother and more adaptive industrial processes.
- CPS: CPS is the backbone of Industry 4.0, integrating computational algorithms, networking, and physical processes. CPS bridges the gap between the physical world—comprising machines and sensors—and digital platforms, enabling real-time data collection, monitoring, and control of manufacturing processes. This integration enhances the precision, adaptability, and efficiency of production systems, making them more intelligent and responsive to changes.
- Robotics and automation: Robotics and automation are key enablers of Industry 4.0, with robots increasingly equipped with advanced sensors and AI

capabilities [6]. These robots perform complex tasks either independently or in collaboration with human workers, enhancing productivity and safety. In hazardous environments, robotic automation minimizes human exposure to danger, while in production settings, it improves consistency and output quality.

- Cloud computing and big data analytics: Cloud computing provides the scalable infrastructure necessary to store and process the enormous amounts of data generated in Industry 4.0 environments [7]. Big data analytics processes this data to uncover actionable insights, which are crucial for strategic planning and operational optimization. Together, these technologies enable real-time analytics, driving smarter, data-driven decisions across the industrial value chain [8].

- Additive manufacturing (3D printing): Additive manufacturing, commonly known as 3D printing, plays a significant role in Industry 4.0 by enabling rapid prototyping, on-demand production, and the creation of complex structures with minimal waste [9]. This technology supports the customization of products at scale, allowing manufacturers to meet specific customer needs more efficiently while reducing material usage and production time.

- Augmented reality (AR) and virtual reality (VR): AR and VR are transformative tools in Industry 4.0, providing immersive and interactive training for employees, enhancing maintenance procedures, and visualizing complex machine systems [10]. These technologies overlay digital information onto the real world, offering real-time, hands-on assistance for troubleshooting and operational tasks, thereby improving workforce efficiency and reducing downtime.

5.1.2.2 Supporting technologies of Industry 5.0

Industry 5.0 expands upon the progress made in Industry 4.0 by prioritizing a focus on humans, sustainability, and resilience. It achieves these objectives by incorporating modern technologies. The technologies that support Industry 5.0 have a vital role in facilitating these aims (Figure 5.4). Industry 5.0 aims to integrate these technical breakthroughs with human needs and environmental concerns.

- Blockchain: Blockchain is a decentralized technology that leverages a distributed ledger to enhance transparency and reduce reliance on centralized authorities [11]. This is pivotal for securing and streamlining supply chain management in Industry 5.0. Its cryptographic features ensure secure transactions and data exchanges, critical in sectors where data integrity is paramount, such as pharmaceuticals and luxury goods. Blockchain provides an immutable, transparent record of every transaction, ensuring product authenticity and trustworthiness. By utilizing consensus mechanisms, it validates transactions, allowing all participants to reach an agreement on the ledger's current state without needing a trusted intermediary, fostering a more decentralized and secure industrial ecosystem.

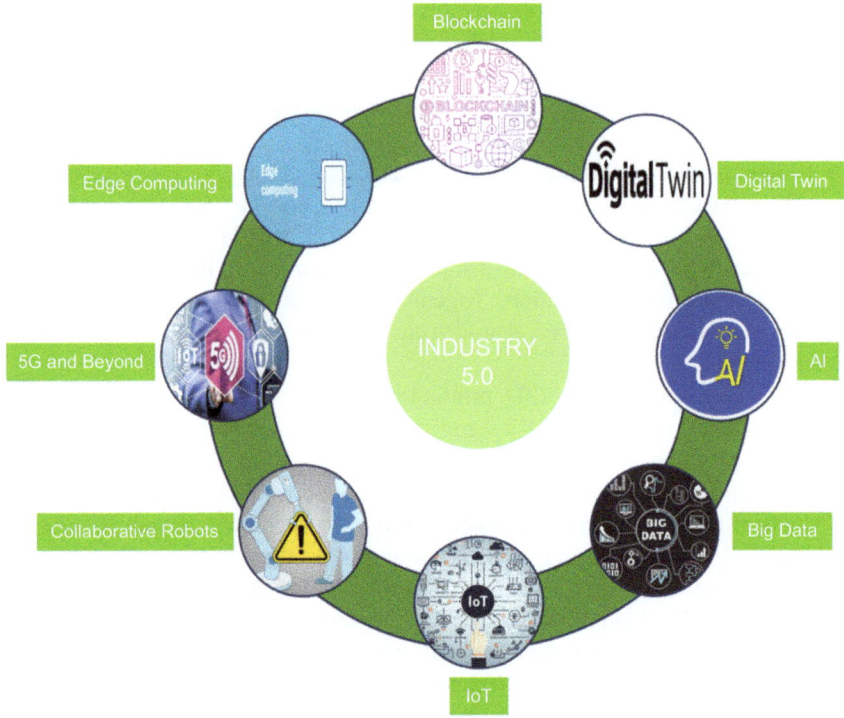

Figure 5.4   Key enabling technologies in Industry 5.0

- Digital twin: Digital twins are virtual replicas of physical assets, enabling companies to simulate and predict equipment behavior under varying conditions, thereby optimizing performance [12]. By integrating data from multiple sources, digital twins enhance system operations and facilitate informed decision-making, aligning with Industry 5.0's goals of efficiency and customization. The secure handling of sensitive information within digital twins is crucial for critical infrastructures like energy and utilities.

- AI: AI drives substantial advancements across multiple domains in Industry 5.0 by personalizing products and services to meet individual customer needs, thus boosting satisfaction and loyalty [13]. AI's predictive analytics capabilities help anticipate market trends and customer behaviors, enabling proactive strategic planning. In manufacturing, AI reduces errors by enhancing accuracy and consistency in operations.

- Big data: Big data is fundamental in managing and processing vast amounts of information, thereby improving operational efficiency and providing deep strategic insights. It employs advanced statistical tools and predictive algorithms to forecast outcomes based on historical data, which enhances accuracy in forecasting and risk management.

- IoT: IoT is essential for ensuring seamless connectivity, allowing continuous data exchange between devices and central systems, which is critical for real-time monitoring and control in Industry 5.0. IoT incorporates advanced security protocols to safeguard interconnected systems against breaches and cyber threats, facilitating proactive maintenance and improving operational uptime.
- Collaborative Robots: Collaborative robots play a crucial role in Industry 5.0 by enhancing human–robot collaboration. Unlike traditional robots, cobots are designed to work alongside humans, combining human creativity and problem-solving abilities with the precision and endurance of robots, leading to more flexible and personalized manufacturing processes.
- 5G and beyond: 5G and next-generation communication technologies are foundational to Industry 5.0, offering ultra-low latency that is crucial for real-time applications, such as the remote control of machinery and autonomous vehicle operations [14]. These technologies are more energy-efficient, significantly lowering the overall energy consumption of connected devices, which aligns with Industry 5.0's sustainability goals.
- Edge computing: Edge computing plays a critical role in Industry 5.0 by enabling cost-effective data management through localized processing, thereby reducing the reliance on extensive cloud storage and associated costs [15]. By processing data close to the source, edge computing minimizes latency and accelerates response times for critical applications, making real-time decision-making more efficient.

5.1.2.3 Relationship

Both Industry 4.0 and Industry 5.0 signify crucial phases in the ongoing evolution of industrial processes. The transition from Industry 4.0 to Industry 5.0 represents a significant shift in focus from the optimization of production processes through advanced technologies to a more human-centric approach that integrates technology with human values. This paradigm shift is not just about adopting new technologies but about changing the way businesses operate and value their human resources and environmental impact. This evolution builds on the technological foundation laid by Industry 4.0 while introducing new dimensions that address the social and ethical implications of industrial advancements. The future of the industry lies in balancing these elements, using technology not just to replace human efforts but to enhance human capabilities and achieve a sustainable development model. The following points outline the relationship between Industry 4.0 and Industry 5.0:

Coexistence and transition

Industry 4.0 is a transformative stage in which cutting-edge digital technologies are incorporated into manufacturing and industrial operations. The main objective was to improve effectiveness, streamline processes, and make decisions based on real-time data. These advancements led to the development of intelligent manufacturing facilities, where interconnected machines, sensors, and human operators collaborate to enhance efficiency and reduce inefficiencies. The focus is on technological advancements and economic benefits, which include improving productivity and cost-effectiveness through economies of scale and high levels of automation.

Industry 5.0, on the other hand, expands upon this base while shifting the emphasis toward a more human-centered approach. Industry 4.0 focuses on automating processes and enhancing efficiency, whereas Industry 5.0 brings back the human element into the industrial equation. It emphasizes a human-centric approach, aiming to create a symbiosis between humans and machines. This shift is not about replacing Industry 4.0 but rather complementing it by addressing its limitations, particularly regarding sustainability, social impact, and human involvement. The objective is to establish an equilibrium between human ingenuity and machine effectiveness, guaranteeing that technology is utilized to amplify human abilities and welfare rather than supplant them [16]. Industry 5.0 focuses on the customization and personalization of products and services, aiming to meet individual customer needs while also ensuring worker well-being and promoting sustainable practices. This method is in line with wider social objectives of sustainability, ethical production, and the establishment of significant employment opportunities.

The two paradigms coexist, with Industry 5.0 providing a pathway for industries to evolve beyond automation. The transition from Industry 4.0 to Industry 5.0 does not imply that Industry 4.0 becomes obsolete. Instead, the two paradigms coexist, with Industry 5.0 building upon the advancements of Industry 4.0. This coexistence allows for a more gradual and integrated evolution of industrial practices, where the technological innovations of Industry 4.0 are enhanced by the human-centric focus of Industry 5.0.

Technological continuity and innovation
Both Industry 4.0 and Industry 5.0 primarily depend on technological progress, yet they have distinct focuses and applications. Industry 4.0 technologies, including the IoT, big data, and AI, continue to be essential in Industry 5.0. Industry 5.0 incorporates advancements such as collaborative robots (cobots) that are specifically built to operate in conjunction with people. These collaborative robots, or cobots, serve as more than mere automation tools. They function as collaborators in both the creative and production processes, augmenting human abilities instead of displacing them [17].

Furthermore, Industry 5.0 prioritizes the fusion of cutting-edge technologies with ethical and environmental principles. Blockchain technology, originally used in Industry 4.0 to ensure secure transactions and supply chain transparency, is now being enhanced in Industry 5.0 to promote secure and sustainable practices in smart cities and the wider industrial ecosystem [18].

Application scope and impact
Industry 4.0 was predominantly utilized in the fields of manufacturing, logistics, and supply chain management. The primary objectives were to enhance process efficiency, minimize downtime by using predictive maintenance, and facilitate mass customization by utilizing smart manufacturing techniques. Industry 4.0 has facilitated the development of smart cities and autonomous systems by incorporating the IoT and AI into urban infrastructure and transportation networks. Industry

5.0 expands the use of these technologies to other areas, highlighting the importance of human creativity in manufacturing and innovation. Industry 5.0, which prioritizes human involvement and innovation, is applicable not only in manufacturing but also in education, healthcare, and service industries. In addition, Industry 5.0 facilitates the adoption of sustainable practices in industrial processes, leading to a reduction in environmental impact and an improvement in resource efficiency. This is achieved through the integration of technologies that support a circular economy.

Ethical and social considerations

One notable difference between Industry 4.0 and Industry 5.0 is in the incorporation of ethical and social factors in the latter. Industry 4.0 largely prioritizes economic efficiency and technical innovation, often resulting in job displacement caused by automation [19]. Although it resulted in notable improvements in productivity, it also sparked apprehensions on the future of employment, privacy, and the ethical utilization of AI and data [20].

On the other hand, Industry 5.0 tackles these challenges by giving priority to the welfare of workers and society [21]. It promotes a comprehensive and enduring industrial model that takes into account the social consequences of technological progress. Industry 5.0 advocates for the concept of "technology for good," in which innovation is in harmony with human values, ethical principles, and environmental sustainability. This entails ensuring that workers possess not only the necessary skills to collaborate with cutting-edge technology but also that they get significance and contentment from their work.

5.1.3 Evolution from Education 4.0 to Education 5.0

The shift from Industry 4.0 to Industry 5.0 carries significant practical consequences for industries, policymakers, educational institutions, and workers. Industries must reconsider their production processes to prioritize human creativity and sustainability as fundamental elements. For politicians, the task entails developing systems that promote ethical AI, safeguard data privacy, and foster job growth in response to growing automation. Industry 5.0 offers workers the opportunity to improve current skills and gain new ones, enabling them to thrive in a collaborative and innovative industrial environment. Educational institutions must modify their approach to talent development, encompassing educational orientation, values, and frameworks, to ensure that the students they cultivate are sufficiently equipped to meet the demands of future industries.

5.1.3.1 Education 4.0

Core features

Education 4.0 is defined by a set of distinct characteristics that differentiate it from prior educational models (Figure 5.5):

- Student-centered learning: Education 4.0 prioritizes the involvement of students in the learning process, transforming educators from mere conveyors of knowledge

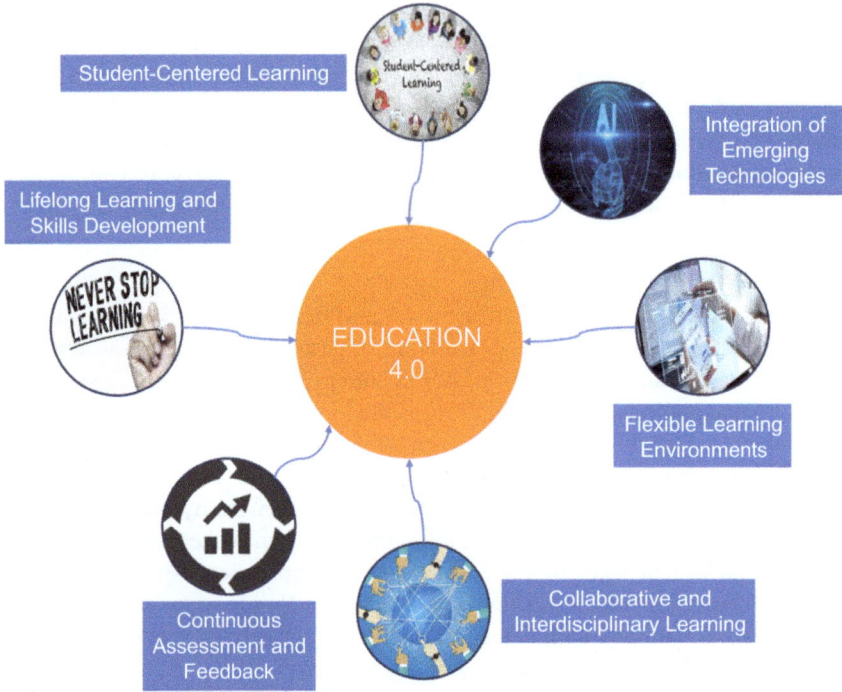

Figure 5.5   Core features of Education 4.0

to facilitators of learning. This strategy fosters student engagement in their education, fostering self-directed learning, creativity, and critical thinking. The focus is on cultivating abilities that have direct relevance to practical situations, ensuring that students are more equipped to tackle future obstacles [22].

- Integration of emerging technologies: The incorporation of cutting-edge technologies is a defining characteristic of Education 4.0. These technologies are not only tools for learning but also subjects of study, preparing students to work with and innovate in these fields. AI, IoT, VR, and AR are incorporated into the learning environment to generate immersive and interactive experiences. These technologies not only improve the learning process but also equip students with the necessary skills for the technical requirements of the contemporary workforce [23].

- Flexible learning environments: Education 4.0 advocates for the creation of versatile and adjustable learning environments that can accommodate the varied requirements of students. This encompasses the utilization of online and hybrid learning approaches, flipped classrooms, and project-based learning (PBL) [24]. The flipped classroom model, where students engage with instructional content outside the classroom and use class time for interactive, hands-on activities, is a key component of Education 4.0. This approach is often complemented by blended learning, which combines online digital media with traditional in-person teaching. One of the hallmarks of Education 4.0 is its focus on personalized learning paths. Adaptive learning systems use data analytics to tailor educational experiences to the needs and preferences of individual students, ensuring that learning is both effective and engaging [25]. These settings facilitate customized learning experiences, wherein students can advance at their own speed and utilize resources specifically designed to match their unique learning preferences.

- Collaborative and interdisciplinary learning: Collaboration and interdisciplinarity are essential elements of Education 4.0. Students are urged to collaborate on projects that encompass various fields of study, reflecting the interrelatedness of contemporary industries [26]. This method facilitates the cultivation of a diverse set of skills, encompassing collaboration, effective communication, and the capacity to apply information across different situations.
- Continuous assessment and feedback: Continuous assessment is a vital aspect of Education 4.0. Instead of exclusively depending on conventional exams, this model integrates continuous assessments that offer immediate feedback to pupils. This facilitates the recognition of areas where learning is lacking and provides the chance to swiftly rectify those gaps, thus improving the overall learning process.
- Lifelong learning and skills development: Education 4.0 emphasizes the importance of lifelong learning, recognizing that the rapid pace of technological change requires continuous skills development [27]. Educational institutions are encouraged to provide opportunities for upskilling and reskilling, ensuring that learners remain competitive in the job market throughout their careers.

Supporting technologies

Education 4.0 integrates cutting-edge technology and innovative educational methods to prepare students for the rapidly changing global landscape. The technological infrastructure supporting Education 4.0 is vast and varied, encompassing a range of tools and platforms that facilitate new forms of teaching and learning:

- AI: AI is essential in Education 4.0 as it facilitates customized learning experiences. AI algorithms have the capability to analyze data on student performance in order to offer customized feedback and recommendations, thereby assisting students in concentrating on areas where they require enhancement. AI facilitates the automation of administrative activities, enabling educators to allocate more time to teaching and mentoring.
- IoT: IoT devices are employed to establish interconnected classrooms, where diverse educational tools and resources are seamlessly integrated to facilitate a conducive learning environment. These gadgets have the capability to monitor and improve student involvement, keep track of attendance, and facilitate remote learning by granting access to instructional material at any time and from any location.
- VR/AR: VR and AR technologies provide immersive learning experiences that can vividly illustrate abstract topics. VR can be utilized to replicate historical

events or scientific investigations, enabling students to investigate these settings inside a regulated and interactive setting. AR may superimpose digital information onto the real world, offering immersive learning experiences that are extremely captivating.

- Cloud computing and big data: Cloud computing facilitates the storage and management of extensive educational data by providing the required infrastructure [28]. It facilitates the scalability of educational resources, enabling the simultaneous delivery of content to a large number of pupils. Big data analytics, however, can be utilized to examine student performance and learning patterns, assisting instructors in refining their teaching tactics and enhancing results.

- Blockchain technology: Blockchain is increasingly being explored as a tool for secure and transparent record-keeping in education [29]. It can be used to store and verify academic credentials, track the progress of learning achievements, and ensure the integrity of educational records. This technology supports the decentralization of educational data, making it accessible and verifiable by all stakeholders involved.

5.1.3.2   Education 5.0

Core features

Education 5.0 is characterized by several distinct features that set it apart from its predecessors (Figure 5.6):

- Human-centered approach: Education 5.0 shifts the focus toward a human-centered educational model, where the development of human values and

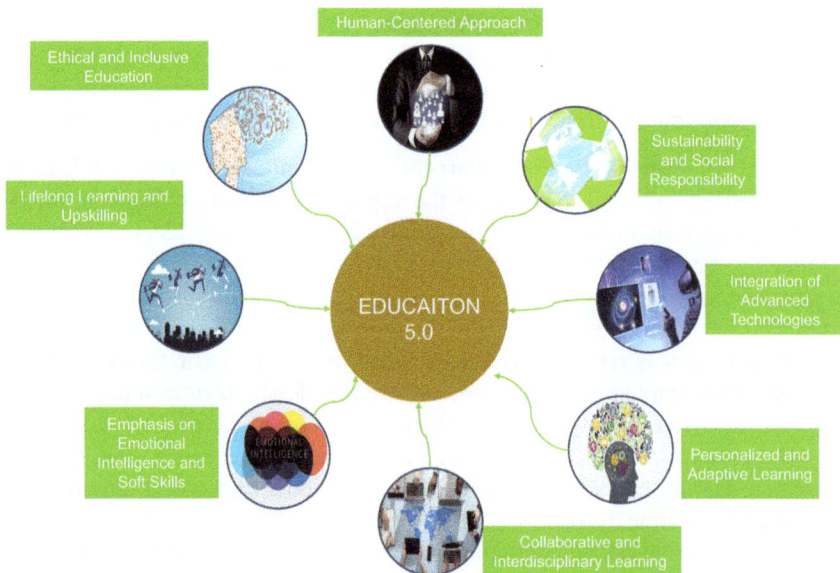

Figure 5.6   Core features of Education 5.0

emotional intelligence is prioritized alongside technical skills. This approach is rooted in the understanding that the future workforce needs to be not only technologically proficient but also capable of empathy, ethical reasoning, and social responsibility. The emphasis is on creating well-rounded individuals who can thrive in a complex, interconnected world. This human-centered approach also integrates the concept of lifelong learning, where education is seen as a continuous process that adapts to the evolving needs of individuals throughout their lives.

- Sustainability and social responsibility: A key component of Education 5.0 is its alignment with the Sustainable Development Goals (SDGs), particularly those related to education and equality. The model promotes the integration of sustainable practices and social responsibility into the curriculum, encouraging students to engage with global challenges such as climate change, inequality, and sustainable development. This focus aims to cultivate a generation of learners who are not only aware of these issues but are also equipped with the skills and knowledge to contribute meaningfully toward solving them.

- Integration of advanced technologies: Similar to its predecessor, Education 5.0 heavily relies on the integration of cutting-edge technologies. However, it goes further by embedding these technologies into the fabric of education in a way that supports personalized learning paths, adaptive learning environments, and the development of digital competencies [30]. The aim is to create a seamless blend of human and machine capabilities, where technology enhances rather than replaces human intelligence and creativity.

- Personalized and adaptive learning: Education 5.0 emphasizes the need for personalized learning experiences that cater to the individual needs, preferences, and learning paces of students [31]. Adaptive learning technologies are employed to provide customized content, real-time feedback, and tailored educational pathways. This approach ensures that students are not left behind or unchallenged, but rather, are constantly engaged at a level that suits their unique abilities and interests.

- Collaborative and interdisciplinary learning: Collaboration is a cornerstone of Education 5.0, where interdisciplinary learning is encouraged to reflect the interconnected nature of modern challenges. Students are often engaged in PBL that requires teamwork across various disciplines, fostering a deeper understanding of how different fields intersect and influence each other [32]. This approach not only builds collaborative skills but also prepares students for real-world problem-solving, where issues are rarely confined to a single discipline.

- Emphasis on emotional intelligence and soft skills: In addition to technical knowledge, Education 5.0 places a significant emphasis on developing students' emotional intelligence (EQ) and soft skills. This includes skills such as communication, leadership, teamwork, adaptability, and critical thinking. The development of these skills is seen as crucial in preparing students to navigate the complexities of modern workplaces and societal interactions.

- Lifelong learning and upskilling: Recognizing the rapid pace of technological change, Education 5.0 promotes the concept of lifelong learning, where continuous education and upskilling are essential. Educational institutions are encouraged to provide opportunities for learners to update their skills throughout their careers, ensuring they remain competitive in the job market and capable of adapting to new technologies and methodologies.
- Ethical and inclusive education: Education 5.0 is also characterized by its commitment to ethical practices and inclusivity. This involves creating learning environments that are accessible to all students, regardless of their background, abilities, or socio-economic status. It also means integrating ethical considerations into the curriculum, ensuring that students understand the moral implications of their knowledge and actions, particularly in relation to technology and innovation.

Supporting technologies

The implementation of Education 5.0 is underpinned by a range of advanced technologies that facilitate its key features:

- AI and ML: AI and ML are used to create adaptive learning systems that can tailor educational content to the needs of individual learners. These technologies also enable the automation of administrative tasks, allowing educators to focus more on teaching and mentoring.
- VR and AR: VR and AR technologies offer immersive learning experiences, allowing students to engage with content in a more interactive and impactful way. For example, they can simulate real-world environments for practical training in fields such as medicine, engineering, and architecture.
- Blockchain: Blockchain technology is being explored for its potential to securely store and share educational credentials, ensuring that students' achievements are verifiable and tamper-proof. This could revolutionize the way academic qualifications are recognized and transferred across borders.
- Big data and learning analytics: These tools are used to gather and analyze vast amounts of educational data, providing insights into student performance and helping educators develop more effective teaching strategies. They also support the personalized learning approach by identifying learning gaps and suggesting targeted interventions.

5.1.3.3   Relationship

The transition from Education 4.0 to Education 5.0 reflects a shift from a technology-centric to a human-centric approach to education. While Education 4.0 focuses on integrating advanced technologies into the educational process to prepare students for Industry 4.0, Education 5.0 seeks to harness these technologies to enhance human potential and address broader societal challenges, including sustainability and equity. The following points outline the relationship between Education 4.0 and Education 5.0.

Theoretical foundations and evolution

Education 4.0 arose as a reaction to Industry 4.0, which is characterized by the incorporation of CPS, the IoT, and digitalization in different industries. The program emphasizes the development of students' abilities in essential technology skills, such as digital literacy, complex problem-solving, and critical thinking, in order to prepare them for a world where these talents are of utmost importance.

Education 5.0 is a dynamic idea that corresponds to the projected Industry 5.0, which prioritizes technologies centered around humans and sustainability. This phase transcends mere technological competence, incorporating ethical, emotional, and social dimensions into education, with the goal of achieving a harmonious blend of human and machine cooperation.

Core objectives and philosophies

Education 4.0's main objective is to provide students with the necessary abilities for a highly technological society. The approach prioritizes student-centered learning, collaborative projects, and the incorporation of digital tools into teaching and learning methods. The fundamental ideology is to equip students with the necessary skills and knowledge to excel in a globally linked and technology-driven economy.

Education 5.0 emphasizes a comprehensive and all-encompassing approach to education. The objective is to cultivate not just technical proficiency, but also emotional intelligence, ethical comprehension, and knowledge of sustainability. The objective is to cultivate graduates who possess not only a high level of proficiency in utilizing technology but also the ability to lead with empathy and responsibility in a swiftly evolving environment.

Technological integration and its role

Technology is a key factor in the transformation of educational processes in Education 4.0. AI, VR, and learning analytics are extensively utilized, leading to the emergence of novel prospects for personalized and adaptive learning. The emphasis is placed on developing a workforce that possesses the necessary digital skills and knowledge to thrive in the context of Industry 4.0.

Education 5.0 expands upon the technological framework established by Education 4.0 but places greater emphasis on integrating these technologies in a manner that improves the overall welfare of individuals. This encompasses the ethical utilization of AI, promoting cooperation between humans and machines, and guaranteeing that technology is aligned with society's objectives such as sustainability and social fairness. In Education 5.0, technology is regarded as a means to augment human abilities rather than supplant them.

Pedagogical approaches and learning environments

Education 4.0 incorporates novel teaching methods like flipped classrooms, blended learning, and PBL. These strategies are specifically designed to enhance interactivity and ensure alignment with the requirements of the digital era. The learning environment in Education 4.0 typically combines physical and digital environments, enabling both face-to-face and online interactions.

Education 5.0 enhances these teaching methods by integrating elements of emotional intelligence and ethical deliberation into the educational program. The learning environments of Education 5.0 are characterized by increased flexibility, inclusivity, and a deliberate emphasis on achieving a harmonious integration of technical and humanistic education. There is a heightened focus on social learning and collaborative environments that equip students to effectively collaborate in varied, multidisciplinary teams.

Skills and competencies
Education 4.0 emphasizes the cultivation of skills associated with Industry 4.0, including digital literacy, critical thinking, and problem-solving. The World Economic Forum's compilation of essential talents for the Fourth Industrial Revolution, encompassing intricate problem-solving and emotional intelligence, is frequently referenced in relation to Education 4.0 [33].

Education 5.0 enhances these skills by placing greater importance on ethical decision-making, emotional intelligence, and the capacity to handle intricate social and environmental issues. The program equips students with the necessary skills and knowledge to assume leadership positions, enabling them to make informed decisions that effectively reconcile technological progress with the welfare of both humans and the environment.

Challenges and barriers
Education 4.0 encounters obstacles including the digital divide, educators' reluctance to change, and the requirement for substantial investment in technology and infrastructure. There exists a disparity in skills between what educational institutions impart and what Industry 4.0 requires.

Education 5.0 has more difficulties in incorporating humanistic principles into education systems that are heavily reliant on technology. Significant barriers include inadequate teacher preparation in emotional and ethical education, the necessity for interdisciplinary teamwork, and the difficulty of assuring equal access to sophisticated technologies. To overcome these obstacles, it is necessary to implement structural modifications and demonstrate a dedication to reevaluating conventional educational frameworks.

Future directions and implications
Education 4.0 is expected to further develop as educational institutions increasingly use digital technology and as the demands of Industry 4.0 become more prominent. The primary emphasis will continue to be on bridging the skills disparity and guaranteeing that graduates are adequately prepared for a digital and automated society.

Education 5.0 signifies the forthcoming trajectory of education, wherein the task will be to use technology in a manner that amplifies human capabilities while upholding ethical principles. Educators must develop new teaching methodologies that effectively integrate technical skills with emotional and ethical education. Additionally, they must engage in continual professional development to stay updated on these evolving developments (Figure 5.7).

Figure 5.7 Transition from Industry 4.0 to Industry 5.0 and Education 4.0 to Education 5.0

5.2 Insights for cultivation of engineering talent

The emphasis on human–machine collaboration in Industry 5.0 will require a transformation of the workforce. Workers will need to develop new skills to interact with advanced technologies and contribute to creative problem-solving. To ensure that Education 5.0 aligns with the evolving needs of Industry 5.0, several strategic adjustments must be made in the areas of educational content, teaching methodologies, and skill development. Industry 5.0 emphasizes a human-centric approach that integrates advanced technologies such as AI, robotics, and automation while prioritizing sustainability and human well-being. This shift requires a corresponding transformation in education to prepare the workforce of the future effectively.

5.2.1 Integration of human-centric skills

Industry 5.0 represents a pivotal evolution in the integration of advanced technologies with a strong emphasis on human-centric values, particularly prioritizing human well-being and ethical considerations. This shift demands a transformative approach in educational systems, especially under the Education 5.0 framework, which must balance the development of technical skills with interpersonal and ethical competencies. The development of an ethical framework within this context is essential, requiring the creation and instruction of models that guide students in making ethical decisions in their professional lives, particularly when interacting with AI and automation. Such preparation is crucial for navigating the complex moral landscapes that are increasingly prevalent in future industries.

As Industry 5.0 prioritizes sustainability and ethical practices, it becomes imperative that Education 5.0 integrates these values across all aspects of learning. This could include embedding sustainability within the curriculum, promoting ethical decision-making in technology development, and encouraging students to consider the environmental and social impacts of their work [34]. Given the increasing presence of technologies like AI, robotics, and IoT in various sectors, educational curricula must evolve to equip students not only with the technical knowledge needed to collaborate effectively with these technologies but also with the soft skills necessary for leadership, teamwork, and communication.

The dual focus of Education 5.0 on both technical expertise and soft skills is essential for fostering successful human–machine collaboration. Competencies such as leadership, teamwork, communication, emotional intelligence, and critical thinking are becoming as vital as technical proficiency in this new industrial landscape. Moreover, it is paramount that ethical considerations are woven into the curriculum. As students are trained to work alongside advanced technologies, they must also be educated on the moral implications of these tools, including issues like data privacy, algorithmic bias, and the broader societal impacts of automation. This ethical framework ensures that students are not only capable of leveraging technology but are also mindful of its potential consequences, thus preparing them to navigate the complex moral landscapes of the future. Real-world case studies and scenario analysis should be integral components of the curriculum to explore the ethical dimensions of technology in industry. These practical learning experiences help students understand the real-world implications of their decisions and the importance of ethical leadership in a technologically advanced society. For example, students in a business ethics course might analyze a scenario in which a company faces a dilemma between maximizing profits and protecting user privacy. By examining the ethical considerations involved in such a decision, students can develop the critical thinking and ethical reasoning skills necessary to navigate the complexities of Industry 5.0. Similarly, students in a computer science course might explore the ethical implications of AI and ML, considering how these technologies can be used responsibly and how they might impact society. Ethical leadership is particularly important in the context of Education 5.0, where decisions about technology use can have far-reaching consequences. As students learn to apply ethical principles to their decision-making, they will be better equipped to lead with integrity in a future shaped by continuous technological innovation.

The emphasis on human–machine collaboration is particularly critical in Industry 5.0, where the synergy between human creativity and machine efficiency is key. Educational institutions must prioritize training that develops these dual competencies, thereby preparing students to engage with AI systems in ways that enhance human potential rather than replace it. As education continues to evolve in response to the demands of Industry 5.0, integrating these elements into the curriculum will be crucial. This approach ensures that graduates are well-rounded individuals equipped to lead in a technology-driven world while upholding the values of empathy, responsibility, and ethical conduct.

In essence, Education 5.0 must embrace human-centric technologies not only as tools but as subjects of study themselves, encouraging students to innovate and drive future developments. By cultivating soft skills alongside technical knowledge, Education 5.0 can produce graduates who are not just users but creators of next-generation solutions, capable of leading in a world where human values and technological advancements go hand in hand.

5.2.2   Project-based learning and real-world application

To effectively prepare students for the demands of Industry 5.0, Education 5.0 should prioritize project-based and experiential learning methodologies. These approaches are crucial in bridging the gap between theoretical knowledge and practical application, fostering the innovative and critical thinking required in a rapidly evolving technological landscape. PBL immerses students in hands-on projects that involve AI-driven tools, allowing them to engage directly with the practical aspects of human–machine collaboration [35]. Through these projects, students not only enhance their technical skills but also develop critical interpersonal skills such as teamwork, communication, and ethical decision-making, which are essential in navigating the complexities of Industry 5.0.

In addition to fostering individual competencies, PBL encourages collaboration with industry partners, ensuring that the curriculum remains relevant and aligned with real-world needs. This collaboration can take the form of industry placements, internships, and co-designed courses, providing students with invaluable hands-on experience and insights into current industry challenges. By working on real-world problems in a technologically integrated environment, students can apply their theoretical knowledge to practical situations, thereby enhancing their creativity and problem-solving abilities.

Furthermore, to meet the interdisciplinary demands of Industry 5.0, education systems must break down traditional disciplinary boundaries. Curriculum design should integrate knowledge from various fields such as engineering, social sciences, and ethics, promoting a holistic understanding of how technology impacts society. This interdisciplinary approach fosters innovative thinking and prepares students to tackle complex, multifaceted problems that are characteristic of Industry 5.0.

Flexible learning environments also play a critical role in this educational transformation. By accommodating various learning styles and schedules, and leveraging digital platforms, education systems can provide ongoing, relevant education that adapts to the rapidly changing industry landscape. This flexibility ensures that students are not only equipped with the necessary technical skills but also possess the adaptability required to thrive in the dynamic environment of Industry 5.0.

5.2.3   Digital infrastructure and access to technology

The global education landscape is undergoing a significant transformation as we move toward Education 5.0. This evolution is not merely about integrating technology into education but about creating a holistic system where technology, pedagogy, and sustainability converge to prepare students for the complexities of Industry 5.0.

Incorporating emerging technologies into pedagogy is essential for enhancing teaching and learning within the context of Industry 5.0. The integration of AI can personalize learning experiences, catering to individual student needs and improving educational outcomes. For instance, AI can analyze students' performance data and provide insights that help educators tailor their teaching strategies to better support each student. VR/AR technologies also play a significant role in Education 5.0. These technologies can create immersive learning environments that engage students more deeply than traditional methods. For example, in a biology class, students can use VR to explore the human body in 3D, gaining a better understanding of anatomy than they would through textbooks alone. Similarly, AR can overlay digital information onto the physical world, allowing students to interact with educational content in new and innovative ways. Data analytics is another critical component of Education 5.0, enabling educators to track student progress and outcomes more effectively. By analyzing data from various sources, educators can make data-driven decisions that enhance the learning process. For instance, if data shows that a particular group of students is struggling with a specific concept, educators can intervene early, providing additional resources or adjusting their teaching methods to better support those students. These technologies not only make education more adaptive and efficient but also prepare students to work in a future where collaboration with AI and other advanced systems is commonplace. This collaboration is a key aspect of Industry 5.0, where human workers are expected to work alongside intelligent machines, combining their strengths to achieve optimal outcomes.

To effectively transition to Education 5.0, educational institutions must invest significantly in digital infrastructure. This investment is crucial for ensuring the seamless integration of technology into the learning process. The infrastructure required includes not only hardware and software but also the necessary bandwidth and support systems that allow for the uninterrupted use of these technologies. In regions where digital infrastructure is lacking, such investments can be transformative, enabling access to educational resources that were previously unavailable. Digital infrastructure encompasses providing access to advanced tools for digital literacy, fostering online collaboration, and utilizing AI-based learning platforms. These tools are pivotal in creating an environment where students can develop the skills necessary for the future workforce. For instance, AI-driven platforms can offer personalized learning experiences, adapting content and pacing to meet the individual needs of each student [36]. This not only improves learning outcomes but also ensures that no student is left behind in their educational journey.

However, the equitable distribution of these technologies is crucial in preventing disparities in educational outcomes and equipping all students with the necessary skills for Industry 5.0. One of the most significant challenges in the transition to Education 5.0 is the digital divide, which refers to the gap between those who have access to digital technologies and those who do not. This divide can be seen both within and between countries, with disadvantaged communities often having limited access to the necessary digital infrastructure and resources. Addressing the digital divide is essential for ensuring that all students have the opportunity to benefit from the advancements in Education 5.0. This requires a multifaceted approach that includes providing access to affordable internet, supplying digital devices to students who need them, and offering training and support to educators and students alike. In many developing countries, the lack of digital infrastructure is a significant barrier to the implementation of Education 5.0. For instance, in rural areas, students may have limited or no access to the internet, making it difficult for them to participate in online learning [37]. To overcome this barrier, governments and educational institutions must work together to invest in the necessary infrastructure, such as expanding broadband coverage and providing digital devices to students in need. Furthermore, it is essential to address the digital literacy gap, ensuring that all students and educators have the skills they need to use digital technologies effectively. This includes not only basic computer skills but also more advanced competencies, such as coding, data analysis, and digital content creation. By providing training and support in these areas, educational institutions can help to level the playing field, giving all students the opportunity to succeed in the digital age.

Sustainable technology empowerment is another critical aspect of Education 5.0. As educational institutions integrate more technology into their curricula, it is essential to do so in a way that is sustainable and aligned with the SDGs. This means considering the environmental and societal impacts of technology use and making decisions that balance technological advancement with ethical and environmental responsibility. For instance, institutions must consider the energy consumption of their digital infrastructure and strive to use energy-efficient technologies wherever possible. They should also be mindful of the environmental impact of the digital devices they purchase, opting for products that are designed to be durable and recyclable. Additionally, educational institutions should educate students about the environmental and societal impacts of technology, helping to cultivate a generation of professionals who are not only proficient in using advanced technologies but also conscious of their responsibilities as global citizens. Embedding sustainability into the curriculum is essential for producing graduates who are capable of making decisions that consider the long-term impacts of their actions. For example, engineering students should learn about the environmental impacts of different materials and technologies, while business students should study the social implications of corporate decisions related to technology use [38]. By integrating these considerations into their education, students will be better prepared to lead in a future where sustainability is a key concern.

Education 5.0's alignment with Industry 5.0 is crucial for producing graduates who are not only technically proficient but also ethically conscious and prepared to face the challenges of a rapidly evolving technological landscape. However, there are several barriers to technology empowerment that must be addressed to ensure the successful implementation of Education 5.0. One of the most significant barriers is inequitable access to technology, as discussed earlier. In addition to addressing the digital divide, institutions must also focus on providing adequate support for educators and students as they adapt to new technologies. This includes offering professional development opportunities for educators, ensuring that they have the skills and confidence to integrate technology into their teaching effectively. Another barrier is the lack of digital literacy among both educators and students. As mentioned, digital literacy is essential for the effective use of technology in education, and institutions must prioritize training and support in this area. This can be achieved through workshops, online courses, and other professional development opportunities that focus on digital skills. Furthermore, the integration of technology into the curriculum can be challenging, particularly for educators who are accustomed to traditional teaching methods. Institutions must provide clear guidance and support for educators as they navigate this transition, offering resources and training that help them incorporate technology in a way that enhances the learning experience rather than detracting from it. Finally, institutions must also address the issue of excessive workload, which can be a significant barrier to the successful implementation of Education 5.0. Educators are often required to take on additional responsibilities as they integrate new technologies into their teaching, which can lead to burnout and reduce the effectiveness of their teaching. To overcome this barrier, institutions must provide adequate support and resources for educators, ensuring that they have the time and energy to focus on their primary responsibility: educating students.

5.2.4   Lifelong learning and upskilling

Given the rapid pace of technological advancement, Education 5.0 emphasizes the importance of lifelong learning and continuous upskilling. This approach is crucial to ensure that the workforce remains adaptable and capable of meeting the evolving challenges posed by Industry 5.0. Educational programs should be designed to facilitate ongoing education, allowing individuals to continuously update their skills and knowledge throughout their careers.

Education 5.0 represents a significant shift from traditional educational paradigms. Unlike previous educational frameworks, which often emphasized rote learning and standardized testing, Education 5.0 is centered around the development of a dynamic, adaptable, and technologically proficient workforce. This shift is driven by the recognition that the skills required by modern industries are constantly evolving, necessitating a more flexible and responsive educational approach. As technological advancements continue to disrupt traditional industries, there is a growing need for workers who are not only skilled in their current roles but also capable of quickly adapting to new technologies and methodologies. Education 5.0 seeks to address this need by promoting lifelong learning and continuous upskilling as fundamental components of the educational process.

One of the key tenets of Education 5.0 is the promotion of lifelong learning. This concept recognizes that education does not end with formal schooling but continues throughout an individual's life. In this context, lifelong learning involves the continuous acquisition of new skills and knowledge, enabling individuals to remain relevant in an ever-changing job market. To support lifelong learning, Education 5.0 advocates for the creation of flexible educational pathways. These pathways allow individuals to pursue further education at any stage of their career, whether they are just entering the workforce, seeking to advance in their current role, or transitioning to a new field. Flexible pathways may include modular courses, micro-credentials, and online learning platforms, which provide learners with the opportunity to acquire.

Continuous education programs are a critical component of Education 5.0. These programs are designed to provide professionals with ongoing opportunities to update their skills in line with the latest technological advancements. By establishing continuous learning pathways, educational institutions can ensure that their graduates are equipped to meet the demands of Industry 5.0, where adaptability and innovation are paramount. Moreover, continuous education programs are not limited to technical skills alone. They also encompass the development of soft skills, such as critical thinking, problem-solving, and effective communication, which are increasingly recognized as essential in the modern workplace. By integrating these skills into continuous education programs, Education 5.0 prepares individuals not only for the technical challenges of the future but also for the interpersonal and organizational challenges that come with working in a rapidly changing environment.

In addition to supporting continuous education programs, Education 5.0 emphasizes the importance of adaptability. In an era where technological advancements can render entire industries obsolete in a matter of years, the ability to adapt is crucial. Education 5.0 promotes adaptability by encouraging a mindset of lifelong learning, where individuals are constantly seeking to improve and expand their skill sets. This mindset is particularly important for educators, who play a key role in preparing the next generation of workers. Continuous professional development for educators is essential to ensure that they are equipped with the latest teaching methods and technological tools. By investing in the ongoing education of educators, Education 5.0 ensures that they can effectively teach the skills needed for Industry 5.0 and beyond.

While the shift toward Education 5.0 offers significant opportunities for enhancing the adaptability and skill sets of the workforce, it also presents several challenges. One of the primary challenges is ensuring equitable access to lifelong learning opportunities. As educational institutions increasingly adopt digital technologies, there is a risk that individuals who lack access to these technologies may be left behind. To address this challenge, Education 5.0 advocates for the development of inclusive educational policies that ensure all individuals, regardless of their socioeconomic background, have access to the tools and resources needed for lifelong learning. This includes providing access to affordable internet, digital devices, and online learning platforms, as well as offering financial support for individuals who wish to pursue further education. Another challenge is the need to balance the demands of continuous learning with the pressures of work and personal life. Many individuals may find it difficult to dedicate time to further education while juggling the demands of a full-time job and family responsibilities. To overcome this challenge, Education 5.0 promotes the development of flexible learning options, such as part-time courses, evening classes, and self-paced online learning modules, which allow individuals to learn at their own pace and on their own schedule.

The transition from Industry 4.0 to Industry 5.0 requires a corresponding evolution in educational systems. Education 5.0 must integrate a human-centric approach that emphasizes both technical and interpersonal skills, supported by innovative teaching methods, continuous professional development, and robust digital infrastructure. By aligning educational content and methodologies with the needs of Industry 5.0, we can better prepare students for the future workforce, ensuring they possess the competencies necessary to thrive in an increasingly automated and interconnected world.

References

[1]   Krasnonosova, O., Mykhailenko, D., and Yaroshenko, I. (2022). Reproduction of human capital as a strategic priority for sustainable development of regions. *Problems of Sustainable Development*, 17(1), 293–300.

[2]   Azevedo, J. P., Hasan, A., Goldemberg, D., Geven, K., and Iqbal, S. A. (2021). Simulating the potential impacts of COVID-19 school closures on schooling and learning outcomes: A set of global estimates. *The World Bank Research Observer*, 36(1), 1–40.

[3]   Ahmad, T., Baig, M., and Hui, J. (2020). Coronavirus disease 2019 (COVID-19) pandemic and economic impact. *Pakistan Journal of Medical Sciences*, 36(COVID19-S4), S73.

[4]   Leong, W. Y., Chuah, J. H., and Tuan, T. B. (eds). (2020). *The Nine Pillars of Technologies for Industry 4.0*. Stevenage: Institution of Engineering and Technology.

[5]   Shi, Z., Xie, Y., Xue, W., Chen, Y., Fu, L., and Xu, X. (2020). Smart factory in Industry 4.0. *Systems Research and Behavioral Science*, 37(4), 607–617.

[6]   Leogn, W. Y., Leong, Y. Z., and Leong, W. S. (2023, October). Human-machine interaction in biomedical manufacturing. In *2023 IEEE 5th Eurasia Conference on IOT, Communication and Engineering (ECICE)*. Piscataway, NJ: IEEE; pp. 939–944.

[7]   Leong, W. Y. (2023, August). Digital technology for ASEAN energy. In *2023 International Conference on Circuit Power and Computing Technologies (ICCPCT)*. Piscataway, NJ: IEEE; pp. 1480–1486.

[8]   Möller, D., Vakilzadian, H., and Haas, R. E. (2022). From Industry 4.0 towards Industry 5.0. In *IEEE International Conference on Electro Information Technology (eIT)*.

[9]   Mubarok, K. (2020). Redefining Industry 4.0 and its enabling technologies. *Journal of Physics: Conference Series*, 1569(3), 032025.

[10] Leong, W. Y., Leong, Y. Z., and Leong, W. S. (2023). Virtual reality in education: Case studies and applications. In *IET International Conference on Engineering Technologies and Applications (ICETA)*. Yunlin, Taiwan, pp. 186–187.

[11] Shaikh, E., and Mohammad, N. (2020). Applications of blockchain technology for smart cities. In *2020 Fourth International Conference on Inventive Systems and Control (ICISC)*. Piscataway, NJ: IEEE; pp. 186–191.

[12] Groshev, M., Guimarães, C., Martín-Pérez, J., and Oliva, A. D. (2021). Toward intelligent cyber-physical systems: Digital twin meets artificial intelligence. *IEEE Communications Magazine*, 59(8), 14–20.

[13] Ali, M. I., Patel, P., Breslin, J. G., Harik, R., and Sheth, A. (2021). Cognitive digital twins for smart manufacturing. *IEEE Intelligent Systems*, 36(2), 96–100.

[14] Leong, W. Y., and Kumar, R. (2023). 5G intelligent transportation systems for smart cities. In Kumar, R., Jain, V., Wai Yie, L., and Teyarachakul, S. (eds), *Convergence of IoT, Blockchain, and Computational Intelligence in Smart Cities*. Boca Raton, FL: CRC Press; pp. 1–25.

[15] Xu, H., Wu, J., Pan, Q., Guan, X., and Guizani, M. (2023). A survey on digital twin for industrial Internet of Things: Applications, technologies and tools. *IEEE Communications Surveys & Tutorials*, 25(4), 2569–2598.

[16] Jefroy, N., Azarian, M., and Yu, H. (2022). Moving from Industry 4.0 to Industry 5.0: What are the implications for smart logistics? *Logistics*, 6(2), 26.

[17] Shadravan, A., and Parsaei, H. (2023). The paradigm shift from industry 4.0 implementation to industry 5.0 readiness. *Application of Emerging Technologies*, 115, 1–10.

[18] Suciu, M.-C., Pleşea, D., Petre, A., *et al.* (2023). Core competence—As a key factor for a sustainable, innovative and resilient development model based on Industry 5.0. *Sustainability*, 15(9), 7472.

[19] Ivanov, D. (2023). The Industry 5.0 framework: Viability-based integration of the resilience, sustainability, and human-centricity perspectives. *International Journal of Production Research*, 61, 1683–1695.

[20] Longo, F., Padovano, A., and Umbrello, S. (2020). Value-oriented and ethical technology engineering in Industry 5.0: A human-centric perspective for the design of the factory of the future. *Applied Sciences*, 10(12), 4182.

[21] Mourtzis, D., Angelopoulos, J., and Panopoulos, N. (2022). A literature review of the challenges and opportunities of the transition from Industry 4.0 to Society 5.0. *Energies*, 15(17), 6276.

[22] Elena, T., and Lilia, R. (2023). Education 4.0: The concept, skills, and research. *Journal of Language and Education*, 9(1), 5–11.

[23] Singh, K., Singh, P., Kaur, G., Khullar, V., Chhabra, R., and Tripathi, V. (2023). Education 4.0: Exploring the potential of disruptive technologies in transforming learning. In *2023 International Conference on Computational Intelligence and Sustainable Engineering Solutions (CISES)*. Piscataway, NJ: IEEE; pp. 586–591.

[24] Bostock, J. R. (2020). The application of a flexible learning model to enhance engagement with digital technologies and augmented reality in language learning. *Journal of Perspectives in Applied Academic Practice*, 8(1), 15–21.

[25] Al-Abri, A., Al-Balushi, M., Al-Shemali, F., and Tahir, N. M. (2023). A framework to support the delivery and assessment of practical-based concepts on online learning. In *2023 IEEE 13th International Conference on*

Control System, Computing and Engineering (ICCSCE), Penang, Malaysia, pp. 293–296.

[26] Molina Gutiérrez, A., Miranda, J., Chavarría, D., *et al.* (2018). Open innovation laboratory for rapid realisation of sensing, smart and sustainable products: Motives, concepts and uses in higher education. In Camarinha-Matos, L., Afsarmanesh, H., and Rezgui, Y. (eds), *Collaborative Networks of Cognitive Systems. PRO-VE 2018*. Cham: Springer; pp. 156–163.

[27] Costan, E., Gonzales, G. G., Enriquez, L., *et al.* (2021). Education 4.0 in developing economies: A systematic literature review of implementation barriers and future research agenda. *Sustainability*, 13(22), 12763.

[28] Mahariya, S. K., Kumar, A., Singh, R., *et al.* (2023). Smart Campus 4.0: Digitalization of university campus with assimilation of Industry 4.0 for innovation and sustainability. *Journal of Advanced Research in Applied Sciences and Engineering Technology*, 32(1), 120–138.

[29] Lutfiani, N., Aini, Q., Rahardja, U., Wijayanti, L., and Nabila, E. A. (2021). Transformation of blockchain and opportunities for Education 4.0. *International Journal of Education and Learning*, 3(3), 222–231.

[30] Ahmad, S., Umirzakova, S., Mujtaba, G., Amin, M. S., and Whangbo, T. (2023). Education 5.0: Requirements, enabling technologies, and future directions. arxiv preprint *arxiv:2307.15846*.

[31] Alamri, H. A., Watson, S., and Watson, W. (2021). Learning technology models that support personalization within blended learning environments in higher education. *TechTrends*, 65(1), 62–78.

[32] Vieira, R., Monteiro, P., Azevedo, G., and Oliveira, J. (2023, June). Society 5.0 and Education 5.0: A critical reflection. In *2023 18th Iberian Conference on Information Systems and Technologies (CISTI)*. Piscataway, NJ: IEEE; pp. 1–6.

[33] Konst, T., and Scheinin, M. (2020). Why Education 4.0 is not enough: Education for sustainable future. In *EDULEARN20 Proceedings: 12th International Conference on Education and New Learning Technologies, 6–7 July 2020*. Valencia: IATED, International Association of Technology, Education and Development.

[34] Alves, J., Lima, T. M., and Gaspar, P. D. (2023). Is Industry 5.0 a human-centred approach? A systematic review. *Processes*, 11(1), 193.

[35] Rožanec, J. M., Novalija, I., Zajec, P., *et al.* (2023). Human-centric artificial intelligence architecture for Industry 5.0 applications. *International Journal of Production Research*, 61(20), 6847–6872.

[36] Mourtzis, D., and Angelopoulos, J. (2023). Development of an extended reality-based collaborative platform for engineering education: Operator 5.0. *Electronics*, 12(17), 3663.

[37] Mendoza-Lozano, F. A., Quintero-Peña, J. W., and García-Rodríguez, J. F. (2021). The digital divide between high school students in Colombia. *Telecommunications Policy*, 45(10), 102226.

[38] Leong, W. Y. (eds.) (2022). *Human Machine Collaboration and Interaction for Smart Manufacturing: Automation, Robotics, Sensing, Artificial Intelligence, 5G, IoTs and Blockchain*. Stevenage: Institution of Engineering and Technology.

Chapter 6

Medical equipment engineering in Industrial 5.0

Nantha Genasan[1]

During the first industrial revolution (Industry1.0) in the 17th and 18th centuries, a dramatic change occurred when new methods and procedures were made possible by machines. Originating in 1760 in England, it made its way to America by the late 17th or early 18th century. Many different industries, including mining, textile, agricultural, glass, and others, were affected by the transition from a handicraft economy to one dominated by machinery, which was marked by Industry 1.0. Industry 2.0 refers to the subsequent change that occurred in the manufacturing sector between 1871 and 1914 and enabled the rapid exchange of both people and new ideas. An increase in corporate productivity during this revolution led to a spike in the unemployment rate as machines supplant humans in manufacturing.

The automation of memory-programmable controllers and computers sparked the digital revolution, sometimes known as Industry 3.0, in the 1970s of the 20th century [1]. Digital logic and integrated circuit chips were key to this stage's mass production, which in turn led to the development of related technologies such as computers, digital cellular phones, and the internet. Both conventional goods and corporate practices are being revolutionized by technological advancements. One aspect of the digital revolution is the digitization of formerly analogue technologies. The term "Industry 4.0" describes the coming together of traditional assets with cutting-edge technology like AI, the Internet of Things (IoT), robotics, 3D printing, cloud computing (CC), and so on. Companies that have embraced 4.0 are adaptable and ready to make decisions based on data. Industry 5.0 refers to the next generation of technology that was developed for smart and efficient machinery. The progression of industries from 1.0 to 5.0 is depicted in Figure 6.1.

Humans collaborating with robots and other intelligent machines is what is meant by the term "Industry 5.0." Through the utilization of cutting-edge technologies such as big data analytics, robots assisting humans in improving their productivity. Industry 5.0 is a revolution that means that humans and machines are working together to improve the efficiency of industrial production. People and machines are working together. The productivity of the industrial business is increasing as a result of the combination of human workers with universal robots. In order to fulfil their duties, the

[1]Perdana University Graduate School of Medicine (PuGSOM), Perdana University, Malaysia

Industry 5.0: Potential
Opportunities and Adoption
Challenges

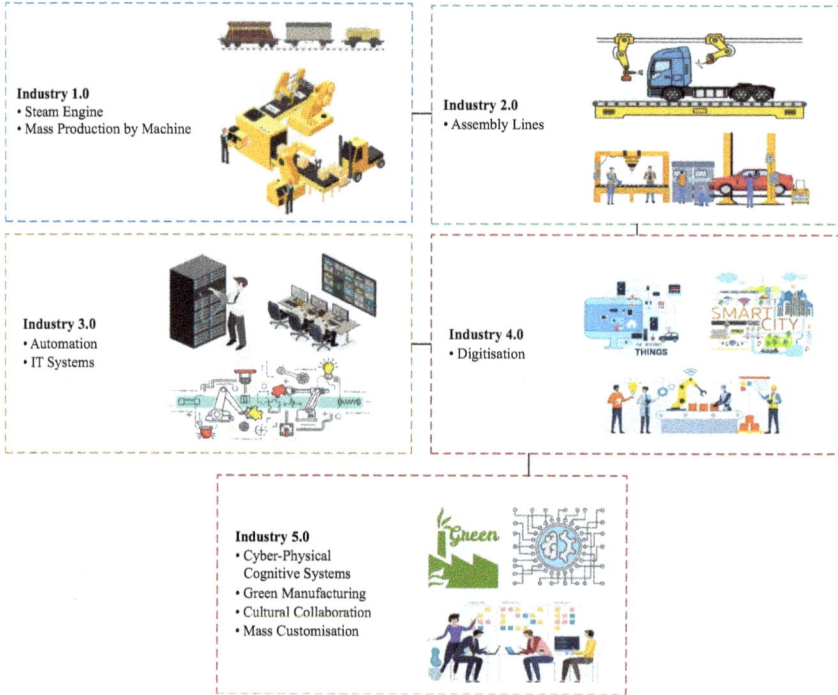

Industry 1.0
• Steam Engine
• Mass Production by Machine

Industry 2.0
• Assembly Lines

Industry 3.0
• Automation
• IT Systems

Industry 4.0
• Digitisation

Industry 5.0
• Cyber-Physical
  Cognitive Systems
• Green Manufacturing
• Cultural Collaboration
• Mass Customisation

Figure 6.1   Industrial evolution from Industry 1.0–5.0 [1]

executive teams of the manufacturing company are necessary to first define the production line, then adhere to the key performance indicators, and last make certain that the operations are operating without any difficulties. The production of robots and industrial robots is the path that manufacturing will take in the future of industry 5.0. With the development of artificial intelligence (AI) and cognitive computing technologies, the manufacturing industry is accelerating at a rapid pace, which is leading to an increase in the efficiency of corporate operations.

6.1   Industrial Revolution 5.0 and state of the art

In IR 5.0, the performance of the production process is optimized through the collaboration of both humans and machines. In a surprising turn of events, it is possible that the fifth Industrial Revolution is already in progress among businesses that are only beginning to implement IR 4.0 concepts. The reason for this is that when companies use modern technology, they do not abruptly dismiss large portions of their workforce and switch to an operation that is made entirely of

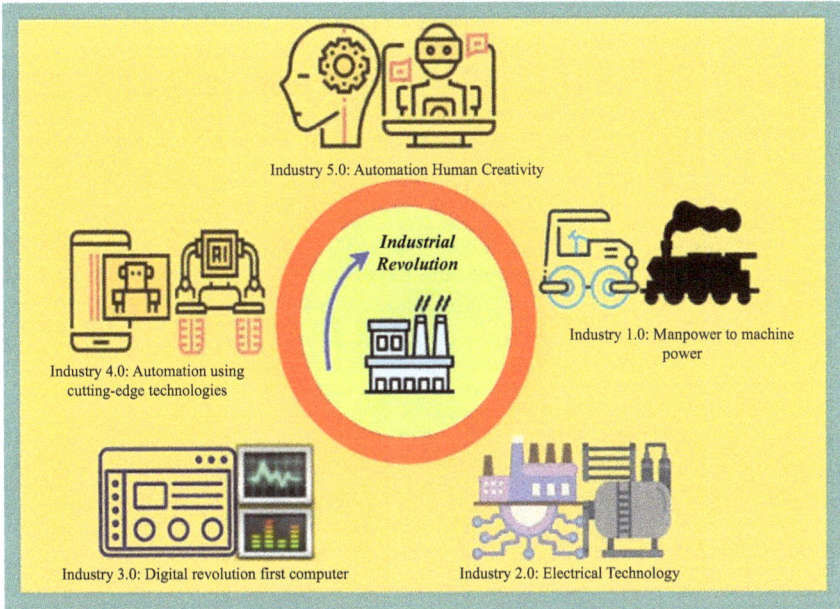

Figure 6.2   Industrial revolution [2]

automated processes. Therefore, IR 5.0 brings about an improvement in manufacturing quality by assigning jobs that require critical thinking to people and outsourcing activities that are repetitive and dull to robots or machines. Since specialists operate the machines, IR 5.0 is able to support jobs that require a higher level of expertise than IR 4.0.

In addition, it places an emphasis on enhancing the satisfaction of customers by establishing a cooperative relationship between humans and robots within the company. The fact that IR 5.0 offers more environmentally friendly solutions than conventional industrial transformations, which do not place an emphasis on the protection of the environment, is still another advantage of this technology. To sum up, whereas the present revolution is all about turning factories into smart facilities connected to the IoT, IR 5.0 is supposed to be more about bringing human intelligence and talents back into the industrial framework. The IRs that have been identified thus far are graphically shown in Figure 6.2.

6.1.1   Definitions

Different definitions have been proposed by several researchers and practitioners since IR 5.0 is still in the process of developing. The following examples of definitions that are being considered [2]:

(i)   Definition 1: By merging work processes and intelligent systems, Industrial Revolution 5.0 (IR 5.0) returns the human workforce back to the factory. In this workplace, humans and machines work together to improve the efficiency of operations by utilizing human intelligence and creativity.

(ii)   Definition 2: Synergetic factories can be built with the help of IR 5.0, which combines human intelligence with cyber-physical production systems (CPPS). To top it all off, with staffing levels hit hard by IR 4.0, authorities are on the lookout for human-centered design solutions that are innovative, ethical, and creative.

(iii) Definition 3: The return of labor to factories, distributed production, intelligent supply chain management, and hyper-customization are all components of IR 5.0, a paradigm for the next stage of industrialization. These elements combine to gradually provide a customized consumer experience.

6.1.2   Pillars of IR 5.0

Reiterating the industry's social responsibility, IR 5.0 prioritizes stakeholder value over shareholder value. Figure 6.3 shows the main components of IR 5.0, which are listed below [2].

(i)   Human-centric: Within the framework of IR 5.0, human invention and craftsmanship are merged with the speed, efficiency, and consistency that robots already possess. As a result, it encourages the empowerment of individuals, talent, and variety.

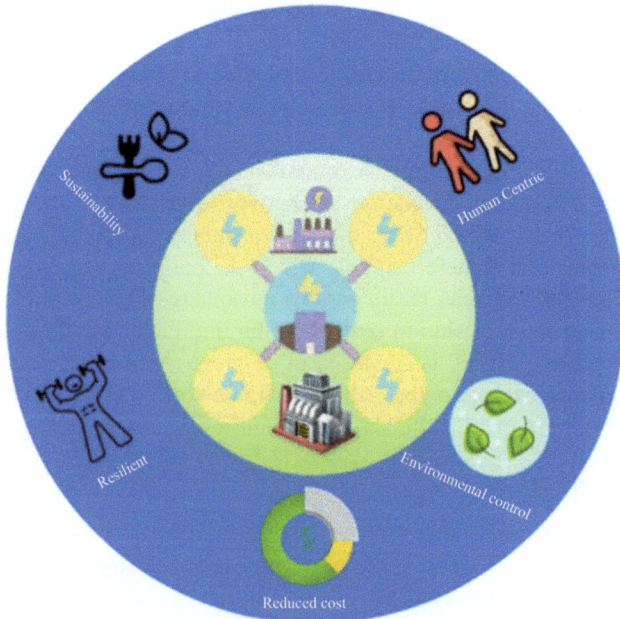

Figure 6.3   Pillars of Industrial Revolution 5.0 [2]

(ii) Sustainability: One of the most noteworthy aspects of IR 5.0 is additive manufacturing, which is also commonly referred to as 3D printing. This form of manufacturing is utilized to make the production of things more environmentally friendly. The purpose of additive manufacturing in IR 5.0 was to increase the level of satisfaction experienced by customers by infusing advantages into products and services.

(iii) Resilient: The concept of "resilience" refers to the requirement to enhance the robustness of industrial production by providing it with improved protection against disruptions and ensuring that it can provide and maintain essential infrastructure during times of crisis. It is possible to achieve a high level of resilience through the collaboration of humans and robots.

(iv) Reduced cost and environmental control: Real-time monitoring and prediction of climate, humidity, temperature, and energy consumption are achieved through the utilization of intelligent, networked sensors and specialized algorithms. This is particularly advantageous for agricultural operations that are heavily reliant on meteorological conditions. A productive output could be increased, and costly errors avoided if one were to anticipate and anticipate where to act.

6.1.3 *Opportunities of IR 5.0*

(i) Better employment: Increased automation resulting from the implementation of technologies of the next generation will positively impact employment in a variety of sectors.

(ii) Customization: Enhanced customization alternatives are provided to clients by means of highly automated manufacturing processes.

(iii) Improved human efficiency: Increased human productivity is made possible by the introduction of IR 5.0, which makes it possible for creative people to have more opportunities to come and work.

(iv) Employee safety: The ability of cooperation robots (COBOTs) to perform potentially dangerous jobs has increased workplace safety.

(v) Customer satisfaction: Businesses can increase their profitability and market share by providing more personalized products and services to their clients. This not only increases customer satisfaction and loyalty but also brings in new customers.

(vi) Better opportunities: To the extent that sufficient funds and infrastructure are accessible, it provides start-ups and entrepreneurs operating in creative and imaginative industries with significant chances to produce new products and services that are associated with IR 5.0.

(vii) Human–machine interaction: An expanded research and development platform is provided by IR 5.0, which places a larger emphasis on human-machine interaction.

(viii) Quality service: By leveraging the capabilities of IR 5.0, it is possible to deliver high-quality services to remote regions, specifically in the healthcare industry, where robotic surgical procedures can be conducted in rural locales.

(ix)   Frequent follow-up: With the help of IR 5.0, the customer will receive digital assistance in managing frequent follow-up tasks. This will be accomplished by making machines responsive to the requirements of the employees.

(x)   Higher value-job: Because individuals are granted the autonomy to take responsibility for creating once more, IR 5.0 creates occupations that are of a better value than those that were previously available.

6.1.4   Comparison of IR 4.0 and IR 5.0

Now, we are residing in the world of IR 4.0, which is swiftly progressing towards IR 5.0. The fourth infrared, which is also commonly referred to as IR 4.0, was built on top of the third infrared to make room for superior technologies. The entire world became "smarter" throughout this time. IoT, CC, CPPS, and cognitive computing were examples of some of the most important technologies that were discussed. According to the information presented in, IR 4.0 can connect systems, components, and individuals over a network, which results in the production process becoming more automated and efficient.

The objective of the fifth revolution, sometimes referred to as IR 5.0, is for humans to work together with emerging technologies to achieve greater levels of efficiency. It is essential to devise strategies for human–robot integration to achieve optimal benefits. This is because the integration of humans and robots will attempt to meet the growing need for persons who have exceptional customization and modifications. The effectiveness of human interaction with intelligent machines will be improved, which will lead to enhanced productivity. In addition to this, because of this change, there will be an increase in the number of positions that offer high salaries. For the purpose of enabling organizations to take advantage of the benefits that IR 5.0 has to offer, Table 6.1 provides a summary of the comparison between IR 4.0 and 5.0.

6.2   Artificial intelligence in healthcare

In recent years, the healthcare industry has undergone a transition from a hospital-centric perspective to a patient-centric perspective, which enables patients to exercise control over their own health care operations. Emerging disruptions in AI, IoT, big data, and aided fog and edge networks come together to make this transformation a reality and provide support for it. To summarize, digital health is characterized by the presence of intelligent sensors that can generate real-time prediction models and commercial analytics.

The term "Healthcare 4.0" refers to this patient-centered and sensor-driven analytical perspective, which enables patients to get care that is both intelligent and connected. The healthcare business has adapted its operations to line with the vision of healthcare 4.0; nevertheless, soon, the healthcare industry will be at the beginning of another paradigm shift. A smart control system, interpretable healthcare analytics, three-dimensional view models, augmented and virtual reality, and other technologies would be incorporated into the transition, which would be

Table 6.1   Comparison between IR 4.0 and IR 5.0 [2]

Industry 4.0	Industry 5.0
Process automation is the end aim.	The objective is to achieve a balance between the way humans and machines interact with one another.
The most important factor was technological advancement	The partnership between humans and machines is the most important one.
Every aspect of the environment is a virtual one.	The transition back to reality.
The number of employees was decreased due to the use of new smart technology.	The amount of human interaction with machines is on the rise.
Intelligent machines that are better integrated into business processes.	A merging of cognitive computing with human intellect is taking place.
Personalization and customization of the product are not possible in any form.	Each product can be enhanced and adapted to the individual's demands through the availability of personalization and customization options.
It's still tossing back and forth between renewable and non-renewable energy sources.	It is more environmentally friendly since renewable energy sources will be used more often.

referred to as healthcare 5.0. In this way, healthcare would be pervasive, highly individualized, dynamic, and based on reason-based analytics, which would promote innovative business solutions within the health industry.

In the future of healthcare, medical science technology envisions the connectivity of millions of IoT-based sensors. These sensors would communicate data using fifth generation (5G) network infrastructure to provide digital wellness, smart healthcare, and improved healthcare metrics. IoT and 5G, when combined with AI, create a scenario in which intelligent mobile wearables are merged with mobile communication and medical technologies to facilitate the delivery of healthcare in a remote and convenient manner. The IoT devices that are attached to patients can collect medical vitals, monitor progress, and diagnose health concerns without requiring a large amount of human engagement. When it comes to the IoT, 5G promises to deliver a throughput of 10 gigabits per second (Gbps), a latency of less than 10 milliseconds, additional cellular coverage, improved network performance, and an over 90% increase in battery lifetime. AI algorithms, such as convolutional neural networks or deep neural networks (DNN), can execute sophisticated operations on enormous data sets that have been generated. These operations include picture and word recognition, imaging, and the ability to accurately anticipate and identify diseases as well as provide remote health treatment [3].

The widespread adoption of AI has caused a stir, and many are wondering how this technology will affect people's lives and the world at large. As a result, ethics, transparency, and responsibility must guide the proper application of AI. Responsible AI was born out of this need. For AI-enabled diagnosis and analysis, a method that guarantees the properties is explainable artificial intelligence (EXAI). Therefore, healthcare outcome tracing and model improvement will be enhanced. The foundation of EXAI is feature extraction, which allows the model to be explainable and interpretable. To understand and predict the behavioral features of machine learning (ML)/deep learning (DL) models, EXAI designed a self-explanatory framework with design principles. The advent of medical AI has prompted many to reevaluate their roles and duties in the healthcare system, as well as in social policymaking. Clinical support systems, medication delivery, medical image segmentation, healthcare technology maintenance and evolution, robotic-assisted surgery, transportation, sales, finance, human resources, and a host of other healthcare-related applications are all areas where EXAI has found use. Ensuring end-to-end patient-doctor transparency in a clinical support system requires defining explainability to address specific viewpoints such as technological, legal, medical, and patient.

When it comes to healthcare operations, EXAI is utilized across a variety of clinical decision models to handle trustworthy analytics. Managing medical data, clinical diagnosis, reducing bias in healthcare sensor data, disease classification, and segmentation are all tasks that can be accomplished with its assistance. It makes it possible to easily debug models that have been trained and improves the performance of trained models by adding a module that supports the decision that the model makes regarding its output. As a result, EXAI provides transparency to AI algorithms to validate the predictions made by the model. EXAI ensures that machine learning and deep learning algorithms comply with the criteria that have been defined, and it gives end users perfect explanations of the reasoning behind its decisions and forecasts. Additionally, it enables the system to optimize the algorithm to eliminate bias by overturning the judgement that the system previously made. As a result, it makes healthcare models more secure and equitable, and it gives individuals the ability to have faith in the decision-making process. The EXAI framework can be applied to a wide range of machine learning approaches, including decision trees, artificial neural networks, random forests, and many others. A compound annual growth rate of 18.4% is indicated by the worldwide EXAI prediction that is presented in Figure 6.4. This forecast covers the period from the year 2020 to 2030. Table 6.2 shows the real-world industrial projects of EXAI in healthcare domain.

The field of AI in medicine is still in its infancy all around the world. In the international market, some of the most prominent providers include DynaMed, UpToDate, and VisualDx. These companies all offer evidence-based decision support solutions that can be deployed and utilized on smartphones and tablets. With a low clinician to population ratio (2.4 clinicians per 1000 people), Japan's healthcare ecosystem is one of a kind, making it an ideal location for research and development as well as the testing of innovative AI technology. To date, it appears that the majority of Japan's technological development efforts have been concentrated on the automation of pre-existing diagnostic modalities, with the potential for widespread deployment in hospitals and clinics. Researchers used resting-state MEG signals to categorize a variety of neurological disorders using a unique DNN

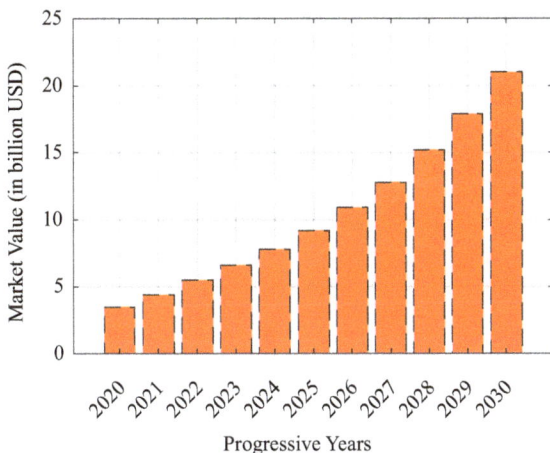

Figure 6.4 Global EXAI forecast [3]

dubbed "MNet," producing very high specificity, in a study done by Osaka University, JST PRESTO, the University of Tokyo, and RIKEN [4].

Using MNet as a classifier could lead to higher specificity and better neurological diagnoses. Because MEG-based diagnosis in critical care can be time-consuming and needs a great deal of experience, it is anticipated that using AI to analyze MEG signals will greatly lessen the workload for physicians. The application of AI in the classification of diseases extends to other fields, such as oncology. When compared with a single tumor marker, the prediction model that was created by the University of Tokyo, Shimadzu Corporation, and Juntendo University resulted in a reduction of around 50% in the rate of illness misclassification. In a similar manner, numerous institutions in Japan, Germany, the United States, and Chile have collaborated in order to improve the histological classification of breast tumors by utilizing small morphological distinctions of myoepithelial cell nuclei in the vicinity of the microenvironment [4]. It is possible that the use of these technologies in rural or resource-constrained regions of the country might guarantee a good level of treatment while also lowering the number of specimens that need to be transferred between different institutions. There is a possibility that the use of AI in pathology will spur the development of other telemedicine solutions.

The creation of these novel tools and algorithms is merely the start of innovation; more testing, approval, and verification are required. For this, Japan has implemented two regulations: First, the use of AI software and devices as clinical decision support tools—that is, for diagnosis and treatment—requires clearance under the "Act on Securing Quality, Efficacy and Safety of Products Including Pharmaceuticals and Medical Devices." Second, the MHLW declared in December 2018 that AI systems would only be used to help with diagnosis and treatment; all further acts would only be considered medical decisions made by doctors.

Table 6.2   Real-world industrial projects of EXAI in healthcare domain [3]

Project name	Objective	Duration	Company	Potential outcomes
IBM Cloud Pak	IBM and Geisinger Health System have collaborated to develop a prototype that incorporates an artificial intelligence model and determines the degree of high risk among patients who have been diagnosed with unique coronavirus disease-2019 (COVID-19).	2021–present	IBM	Specific EXAI modules to study the impact of sepsis and COVID-19 mortality rates.
Argumentation-driven explainable artificial intelligence for digital medicine (antidote) project	In order to comprehend low-level features of deep learning algorithms, which are paired with high-level feature sets of human augmentation in healthcare datasets, an integrated EXAI module has been developed.	2022–present	Wimmics with a consortium of six universities in Europe	A dialogue-based module would be set up to collect data from patients and provide interpretable analysis and predictions that justify the healthcare diagnostics.
Centre of Healthcare Informatics, Australian Institute of Health Innovation	In addition to designing models and frameworks to assess the EXAI outputs over DL models, the project's objective is to simplify the fundamental EXAI needs encountered in the healthcare industry.	2021–present	Macquarie University, Sydney, Australia	Validation tools over standard healthcare datasets.
Human-centered AI project	Create techniques that simulate the machine learning decision-making process and that help understand learning and knowledge extraction on efficient healthcare interfaces.	2020–present	University of Natural Resources and Life Sciences, Vienna, Austria	EXAI-based knowledge extraction and decision-learning module via human–computer interaction (HCI) interface.

The computer-aided diagnosis system, which uses an AI engine to identify typical breast cancer patterns on mammography, serves as an example of the difficulties with this technique. Within this context, a system could try to alert patients to make an appointment as soon as it detects a high probability of cancer, but the need for human oversight could impede or prevent this process entirely [4].

Table 6.3 shows the initiatives taken by Japanese government for example collaborating with industries and universities to implement as well expand the usage of AI in medical field. AI is being extended beyond the bedside and into the system level because of collaborations between hospitals and industry. The company Fujitsu Ltd., for example, has been working towards the goal of incorporating AI into both the management of medical data and administrative processes. In Japan's medical AI startup arena, Allm Inc. was an early success, and the company developed a platform for the field triage of patients who were suffering from acute ischemic stroke. The platform is a centralized database that contains all of the regional stroke centers that have been appraised based on their capacity to perform

Table 6.3   The Japanese Ministry of Health, Labor, and Welfare's (MHLW) initiatives to advance artificial intelligence in medicine [4]

Field	Initiatives spearheaded by MHLW
Genomic medicine	• Establishment of the Center for Cancer Genome Information Management within the National Cancer Center and aggregate genome information. • Creation of a central information database from which clinical and genetic information would be analyzed by the Cancer Genome Information Management Center.
Image diagnosis support	• Creation of a diagnostic image database through a collaboration with various academic societies (Japanese Society of Pathology, the Japanese Society of Gastrointestinal Endoscopy, the Japanese Society of Radiology, and the Japanese Society of Ophthalmology, etc.). • Implementation of guidelines through the Medical Practitioners Act and the Pharmaceuticals and Medical Devices Act.
Diagnostic and treatment support	• Build an information infrastructure that covers a wide range of intractable diseases with research funding from the Japan Medical Research and Development Organization (AMED). • Implementation of guidelines through the Medical Practitioners Act and the Pharmaceuticals and Medical Devices Act.
Drug development	• Creation of a knowledge database to locate drug targets with the National Institute of Biomedical Innovation and Health and Nutrition (NIBIO). • Matching pharmaceutical and IT companies with support from the National Institute of Biomedical Innovation and Health and Nutrition, RIKEN, and Kyoto University.
Surgical support	• Provision of grants for the standardization of the interface for interlinking surgical data.

endovascular or systemic thrombolytic treatment [4]. The triage of patients is accomplished by the utilization of a clinically validated questionnaire, real-time traffic information, and the capabilities of the nearest stroke center. To manage administrative chores, such as patient registration and invoicing, ALMEX Inc.'s "Sma-pa TERMINAL" offers an AI assistant. This assistant makes use of face recognition for patient registration while simultaneously synchronizing visit information to the patient's mobile device. The implementation of these tools has the potential to liberate both financial and human resources, hence reducing the administrative load that is present at all levels of healthcare [4].

The intensive care unit (ICU) is an interesting and difficult environment in which to use AI because of the need for quick decision-making, quick communication, and continuous information flow. Due to time restrictions and insufficient information, doctors are forced to make important judgements based on limited information since it is difficult to monitor the physiological and pathophysiological systems that influence patient circumstances. Thus, through a variety of initiatives to lessen the workloads of physicians, the application of AI in the modification, extension, and integration of traditional decision-making tools has been investigated for several decades. As a developed nation with a technologically sophisticated population, a well-established healthcare system based on universal coverage, and prior academic, government, and industry cooperation alignments, Japan is particularly positioned for the implementation of medical algorithmic technologies. The availability of pertinent, trustworthy data at scale is essential to the development of a strong AI ecosystem, and the ongoing establishment of shared healthcare databases across Japan highlights this enormous potential. These statistics have a great potential to spark a healthcare revolution in Japan if they are appropriately anonymized, made easily accessible to researchers, and combined with progressive teaching programs (Figure 6.5).

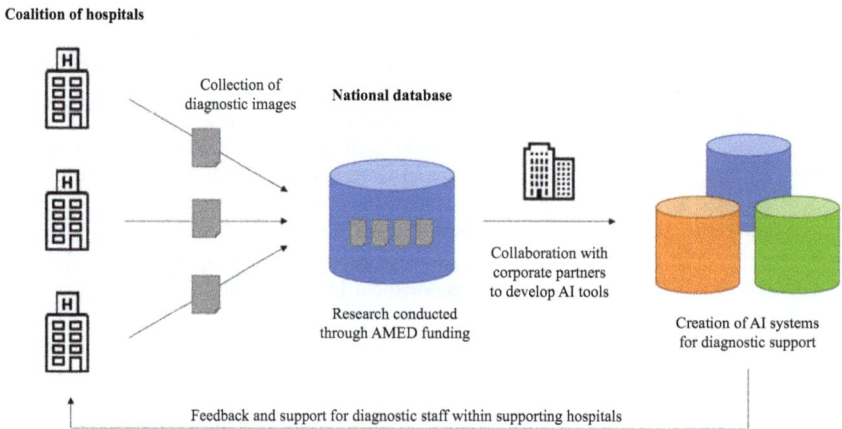

Figure 6.5   A clinical diagnostic database example to encourage the creation of additional AI solutions for the healthcare industry [4]

6.3 Sensors in healthcare

As new digital technologies emerge, they present health systems around the world with unparalleled opportunity to enhance the delivery of healthcare services. The medical field has made great strides. Nevertheless, a new technology that integrates intelligent sensors into health systems is necessary due to the absence of emotive recognition, personalized and ubiquitous health apps, and emotive smart gadgets. To enhance transportation, medical treatment, nanotechnology, mobile devices, virtual and augmented reality, and AI, sensors are utilized all over the world. Since their introduction, sensor technologies have brought about a significant improvement in medical care. To capture and convert health data into electrical impulses that may be observed, sensors have been utilized extensively. Data pertaining to health, including but not limited to heart rate, blood sugar level, stress rate, oxygen saturation rate, temperature, weight, and blood pressure, are often collected by sensory smart devices and then transferred as electrical pulses for further processing.

During the COVID-19 pandemic, sensors were continuing to revolutionize health care systems all over the world and open new doors for the provision of care by means of virtual means. To remotely collect and process health data, sensors have been effectively integrated with smartphones, smart wearable devices, and the Internet of Medical Things (IoMT). To improve clinical diagnoses and monitor biological molecules, sensors—particularly biosensors—are quickly becoming essential tools in the medical industry. One usage of wearable sensors has been in healthcare, where they can be either flexible or non-flexible and utilized for remote patient monitoring. Health care 5.0 can make use of a wide variety of sensors, as shown in Table 6.4. In healthcare 5.0, there are various functionalities of sensors that can be investigated to enhance and perhaps provide cost-effective healthcare service delivery. Wearable flexible sensors gather physiological data by establishing direct contact with the human skin. They gather and transmit data instantly utilizing wireless communication technologies, such as Bluetooth and wireless networking technology (Wi-Fi), for subsequent analysis. To track the patient's physiological indicators, smart devices can be employed to gather real-time bodily signals using sensors. These sensors may include smart bands, smart vests, smart shoes, smart socks, smartphones, smart watches, smart clothes, smart disinfection tunnels, smart face masks and shields, and smart helmets, among others [5].

Using a variety of digital health technologies and communication channels, remote patient monitoring and tracking is a way to give patients healthcare services. Smart wearables, sensor-based intelligence gadgets, and smart health apps can all help achieve this. For example, physiological factors required for COVID-19 detection have been monitored using sensor-based smart wearable devices. As a result, a variety of application domains, including blood oxygen saturation, temperature, activity levels, sleep, and respiration monitoring, have adopted remote patient monitoring. When a patient's physiological changes become concerning,

Table 6.4   Sensors in healthcare [5]

Sensor type	Vital physical signs	Observations
Skin electrodes	Electrocardiograph	Heart rate, heart rate variability
Piezoelectric sensor	Respiration	Breathing rate, physical activity, inspiration and expiration
Accelerometer	Activity, mobility and fall	Body posture, limb movement
Cuff pressure sensor	Blood pressure	Status of the cardiovascular system, Hypertension
Scalp-placed electrodes	Electroencephalogram	Electrical activity of the brain, brain potential
Glucose meter	Blood glucose	Amount of glucose in the blood for glucose monitoring
Skin electrodes	Electromyography	Muscle activity
Woven metal electrodes	Galvanic skin response	Skin electrical conductivity
Electrochemical sensors	Activity and mobility	Sweat analysis, biomolecule analysis
Temperature sensor	Temperature	Skin temperature, health state
Sphygmomanometer sensor	Blood pressure	Blood pressure
The galvanic skin response sensor	Activity and mobility	Monitor sweating
Infrared sensors	Blood flow and pressure	Measure heat radiation

they can be informed, which typically avoids hospital admissions. Virtual clinics arrange for patients and medical experts to communicate digitally over the phone, video connection, or other web-based platforms to provide remote clinical consultations. Virtual clinics can help prevent the spread of highly dangerous diseases, save patient wait times, and enhance medical care by minimizing direct patient contact.

Implantable devices that can respond to changes in their physical, chemical, or biomechanical environments are known as SMART implants (Self-Monitoring Analysis and Reporting Technology). The use of sensors allows for the measurement of these signals or changes. Thus, a combination of sensor technology and SMART implants has been developed to offer orthopedic surgery diagnostic, therapeutic, and monitoring applications. The fields of total knee and hip arthroplasty, fracture healing, and orthopedic infection evaluation have all found uses for them. During the intraoperative, postoperative, and rehabilitative phases of a total knee arthroplasty (TKA), smart sensor technologies can be utilized (Figures 6.6 and 6.7). The basic principle of Industry 5.0, which is the collaboration of surgeons and revolutionary technologies, is supported by the intra-operative reliability of establishing optimum soft tissue and ligament balancing. Potential future uses include the ability to release medicines locally, detect implant-related illnesses earlier, and self-manage their treatment. Therefore, it is anticipated that sensor-based technology would enhance clinical outcomes and patient satisfaction [6].

Figure 6.6   *An illustration of the use of sensor technology in total hip replacement (THR) surgery, showing the emission of radiofrequency (RF) waves for computational purposes. Signals are captured by mechano-acoustic sensors, which examine osseointegration, identify loosening, and assess total hip replacement implants [6]*

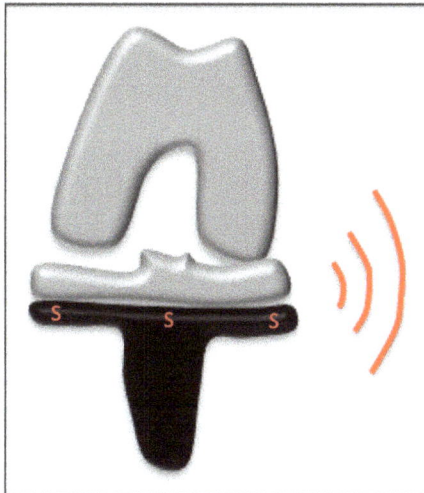

Figure 6.7   *Diagrammatic illustration of a sensor device implant in the tibial tray component of a total knee replacement (TKR). A micro-force sensor is integrated into the tibial base plate to measure the contact force and location ("S"). The load-detecting computer system located in the operating room complex receives wireless radiofrequency (RF) waves from the trial implant sensor "S" to deliver instructions for the placement of TKR components, GAP balancing, and "soft tissue" ligaments [6].*

6.4   Wireless network from 5G to 6G in healthcare

Right now, there is a limited amount of information available regarding the standards of 6G. By the year 2030, however, it is anticipated that the worldwide standardization agencies would have sorted out the specifications for sixth-generation wireless networks. The work that has been done at a few of the research centers has demonstrated that by the year 2035, 6G will be able to transmit a signal at a level that is comparable to the processing capabilities of humans. Even though the rollout of 5G is still in progress, researchers from all over the world have begun working on developing a new generation of wireless networks. Based on the vision of 6G wireless networks, Figure 6.8 presents a provisional timeline for the adoption of 5G, B5G, and 6G standards by international standardization organization. This timeline is given in relation to the vision of 6G wireless networks.

As part of the 5G network standards, the International Telecommunication Union Radiocommunication sector (ITU-R) published the criteria of International Mobile Telecommunications-2020 (IMT-2020 Standard) in the year 2015. While this was going on, the 3GPP released R13, which was a set of specifications for 5G. It is anticipated that the International Telecommunication Union (ITU) would finish the standardization of 6G (ITU-R IMT-2030) by the end of the year 2030, whilst the Third Generation Partnership Project (R23) will complete its standardization of 6G. In July of 2018, the International Telecommunication Union (ITU) launched a focus workgroup with the purpose of investigating the system technologies for B5G/6G systems. The Academy of Finland established 6Genesis in 2018, which is a flagship program that focuses on technology related to 6G networks. In a similar fashion, China, the United States, South Korea, Japan, and Russia have all initiated research for the development of B5G and 6G communication technologies [7].

Currently, smart healthcare applications are enabled by health care services worldwide using fourth generation (4G) network systems. However, in this fast-paced industrial revolution, the restricted bandwidth of these systems will not be able to satisfy the requirements of these applications in the future. The rollout of 5G networks has not yet met the evaluation standards for healthcare technologies.

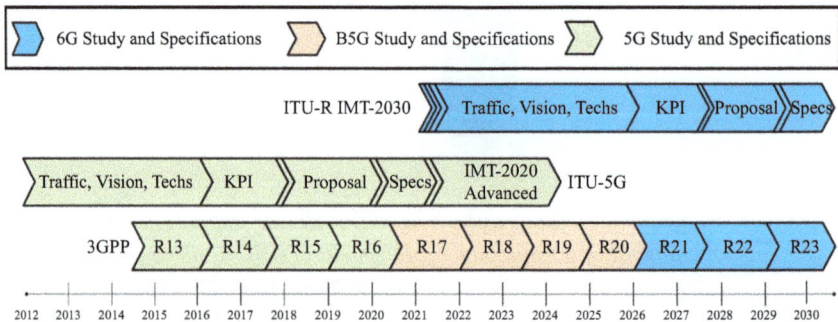

Figure 6.8   A tentative timeline of standards development for 5G, B5G, and 6G [7]

Enhanced mobile broadband (eMBB), ultrareliable low latency communications (URLLC), and massive machine type communications (mMTC) are the three main groups into which 5G services typically fall. Through mMTC, 5G services in the medical area offer dependable, high-speed internet connectivity to devices connected to the IoMT [8]. Through improved visualization made possible by eMBB, telemedicine, augmented reality (AR), and virtual reality (VR) all play a role in the diagnosis and treatment of patients. Drones and autonomous vehicles can be aided by URLLC in emergency medical surveillance—for instance, in the event of a disease epidemic. Figure 6.9 depicts the user plan-to-core-plan comparison of current and future sixth-generation networks.

The uses of sixth generation (6G) networks in healthcare services can be classified into three main areas: hospital services, remote healthcare, and disaster response units. Efficient functioning of hospitals necessitates continuous and efficient communication between hospital managers and their staff, which includes doctors, medical assistants, nurses, and patients who are getting services. Recently, there have been advancements in patient record-keeping methods, along with the implementation of AI-based services. These developments aim to improve disease diagnosis and prognostic decision-making. The utilization of robotic systems, such as the Da Vinci surgical system, has enabled the attainment of a high level of accuracy in surgical procedures [8]. Because of a lack of resources, such as medical personnel and beds, hospitals might become overwhelmed with patients. The transmission of infectious diseases among medical staff during a pandemic can

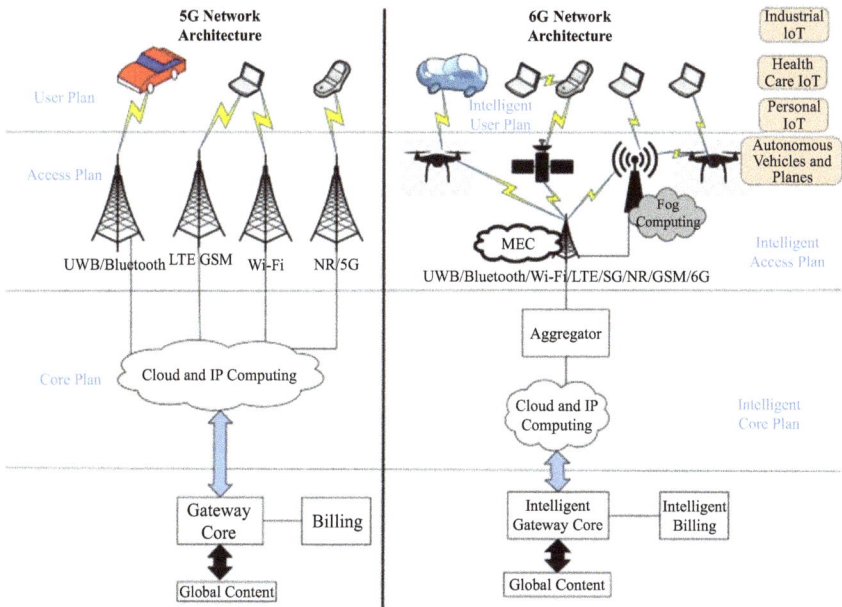

Figure 6.9   Analysis of the differences and similarities between the architectures of 5G and 6G networks [7]

amplify the breakout cycle, making already dangerous situations even worse. Consequently, the use of autonomous robotics and the IoMT in healthcare settings can lessen the need for face-to-face contact between patients and doctors. By utilizing IoMT wearable devices and sensors, patients can be closely watched, and data can be sent to clinicians online, resulting in a decrease in face-to-face encounters. As a bonus, they can assist with the delivery of medical supplies, medical examinations, and the cleaning and sanitization of patients' areas.

These technological developments will be heavily relied upon in the near future, necessitating a high data rate, dependable, and secure communication link. Wi-Fi network usage is limited because medical equipment is susceptible to electromagnetic interference (EMI) and because hospital buildings' intricate architecture prevents high frequency signals from penetrating, which can lead to network coverage problems. Sixth generation (6G) technology, which combines advanced physical security measures with ultra-reliable and low latency communication, will make it easier to meet communication requirements for both capacity and connection. Hackers and unapproved individuals can be prevented from accessing sensitive clinical data and patient privacy by using security bridges. The growth of privacy and security concerns in wireless systems is depicted in Figure 6.10.

One of the biggest problems facing the medical community across the world is meeting the health care needs of people who live in remote areas with limited access to transportation. The field of telemedicine has been assisting medical professionals in monitoring their patients remotely through videoconferencing for the purposes of diagnosis and therapy, rather than in-person consultations. The present approach of high-quality video teleconferencing may soon be replaced by holographic representation due to the rapid development of holographic technology. Wireless body area networks (WBANs), which use technology from the

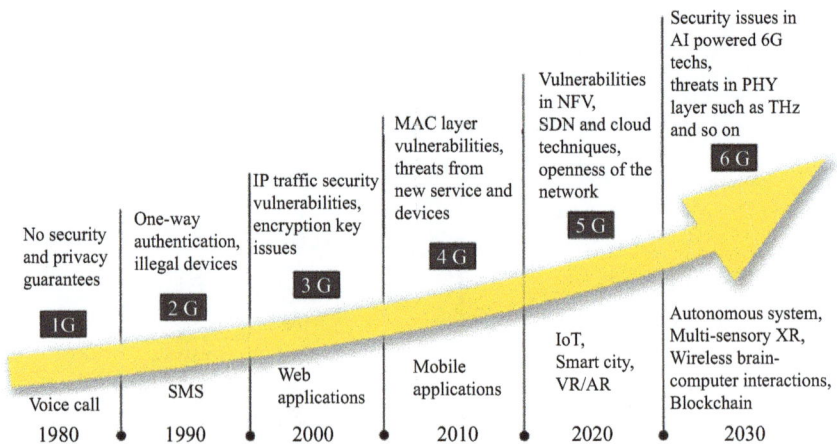

Figure 6.10   The development of new threats to wireless data privacy and security [9]

IoMT to connect wearable devices, are also seeing widespread deployment [9]. The vast, dependable, and enormous communication infrastructure is the backbone of these remote health care services. Further advancements in IoMT-based devices, biosensors with ultra-low power communication, and protected networks in remote healthcare free from jamming and spoofing will be possible with the adoption of 6G networks in the next months or years. Figure 6.11 provides examples of how 6G network system-based technologies could be used in the future of healthcare.

A medical emergency disaster response unit serves as the primary mechanism for delivering healthcare services to the public during emergencies. This is done through the deployment of field hospitals, disaster management centers, and emergency mobile services (EMS). EMS will offer essential services outside of the hospital setting, providing prehospital healthcare by ground or air ambulances to ensure the safety of emergency situations. Modern ambulances are now equipped with advanced medical technology to enable faster diagnosis and treatment by paramedics. This includes the ability to consult medical professionals through high-quality video calls. The disaster response unit must establish robust communication

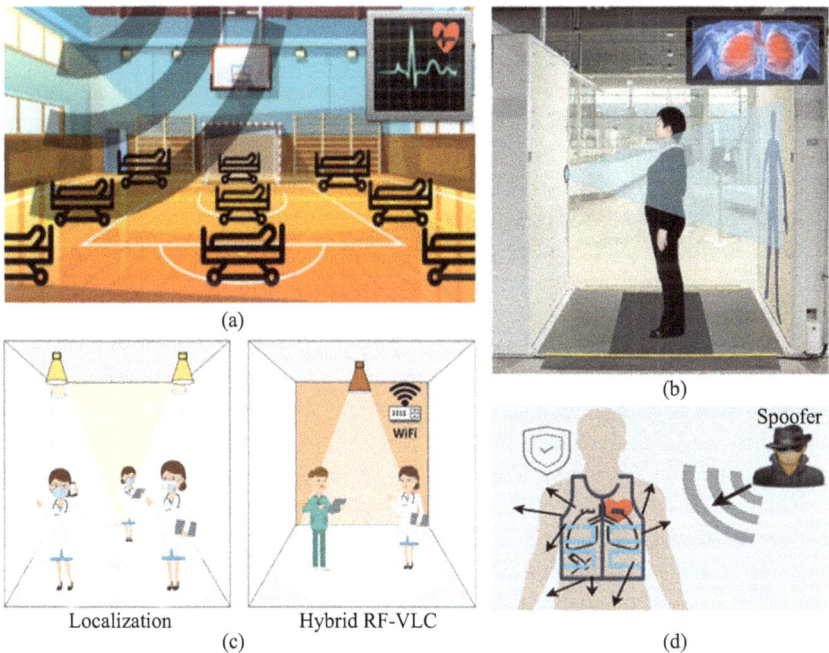

Figure 6.11   Applications of 6G network system-based technologies in the healthcare system [8]. (a) Wireless sensing in field hospitals, (b) THz for COVID-19 screening (a pandemic example), (c) VLC for localization and capacity enhancement, and (d) smart vest for WBAN security.

channels with other departments, including the fire brigade, police department, and insurance firms. Therefore, a more resilient cellular network for communication is required. In a pandemic crisis, quarantined patients require appropriate care, which is being facilitated using robots and drone technology in field hospitals. These technologies facilitate the transportation of food and medical kits, collection of test samples, and sanitization of patient spaces, among other functions. Smart ambulances necessitate high data rate communication to convey medical imaging data. The wide range of bandwidth available in the sixth generation (6G) network will contribute to addressing the existing issues in delivering comprehensive medical care. Remote health and disaster response services require high-quality coverage, connectivity with minimal latency, and high reliability. For a seamless deployment in the healthcare system, the integration of medical equipment and improvements in cellular networks must be exact and synchronized.

6.5   Blockchain technology in healthcare

A new opportunity has arisen with the rise of blockchain technology (BCT) to usher in business 5.0 in the healthcare business. BCT allows for decentralized and secure data sharing, which improves patient privacy and data security while also expediting healthcare processes. A safe and unhackable system for medical records can be constructed using BCT. Patients can choose who can view their medical records, and doctors can safely share patient information with other businesses. Through the provision of a method that is both secure and decentralized for the management of trial data, BCT has the potential to enhance both the transparency and the efficiency of clinical trials. This has the potential to improve the accuracy of data, prevent fraud, and speed up the development of drugs. By utilizing BCT, it is possible to create a drug tracking system that is both transparent and secure, which might be of assistance in preventing the introduction of counterfeit pharmaceuticals into the supply chain. This has the potential to improve patient safety while also reducing the costs of healthcare. The use of BCT allows for the tracking of medical devices and supplies as they make their way from the makers to the patient, thereby ensuring that they are of high quality and genuine. BCT has the potential to enhance the effectiveness of health insurance by providing a system that is both safe and open to scrutiny for the management of insurance claims and coverage payments. This can reduce the expenses associated with fraud and administration while also improving the outcomes for patients. An illustration of a process for healthcare applications that is based on BCT can be found in Figure 6.12.

By utilizing BCT, it is possible to establish a safe and decentralized system for the storage and exchange of private medical records. Because of this, data security and privacy would be improved, and healthcare providers would be able to share information with one another in a manner that is both more efficient and effective. A safe and tamper-proof method of storing and exchanging medical records is something that can be provided by BCT. Medical records that are saved on a blockchain are protected against hacking and other forms of cybercrime because they are

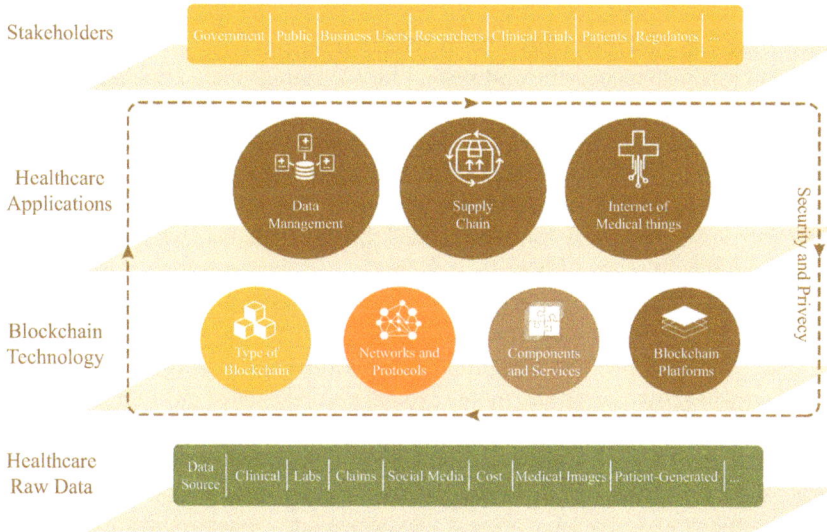

| Stakeholders | Government | Public | Business Users | Researchers | Clinical Trials | Patients | Regulators | ... |

Healthcare Applications — Data Management, Supply Chain, Internet of Medical things

Blockchain Technology — Type of Blockchain, Networks and Protocols, Components and Services, Blockchain Platforms

Security and Privacy

| Healthcare Raw Data | Data Source | Clinical | Labs | Claims | Social Media | Cost | Medical Images | Patient-Generated | ... |

Figure 6.12   A set of blockchain-based healthcare apps that work together. There are four major parts to the workflow: healthcare raw data, blockchain technology, healthcare applications, and stakeholders [10].

encrypted and dispersed across several nodes [11]. The medical records of a patient are capable of being accessed in real time by healthcare practitioners, independent of the location where the records are housed. This has the potential to increase both the rates of diagnosis and the accuracy of treatment. BCT can assist in lessening the administrative load related to keeping track of medical records. Blockchain eliminates the need for administrative workers to handle and exchange medical records by allowing people to take control of their records and authorize access to healthcare providers. Patients can heal faster and receive better care with faster and more precise diagnosis and treatment. BCT can facilitate the healthcare industry's regulatory compliance process. By utilizing BCT, healthcare providers may guarantee that their medical records adhere to privacy and security requirements, hence mitigating the likelihood of regulatory infractions and the corresponding penalties.

When a drug is traded with the intention to conceal or simulate its provenance, validity, or even effectiveness, it is counterfeit. Also, if the drug has incorrect ingredients, it is deemed to be counterfeit. Counterfeiting of products and drugs has a significant impact on supply chain management. The success of these companies in the pharmaceutical sector is a competitive aspect that has a significant impact on the efficiency, authenticity, and robust profitability of a specific healthcare industry. By allowing for enhanced accountability and reducing the risk of counterfeit or tainted drugs, BCT can be utilized to monitor the movement of pharmaceuticals from the manufacturers to the patients. Using BCT, each transaction that occurs throughout the supply chain for drugs is recorded on a ledger that cannot be altered.

This makes it much simpler to track the movement of drugs and identify the origin of any issues or errors that may occur. Apart from this, by utilizing BCT, it is possible to establish a decentralized system for the storage and exchange of genetic and other health data, which enables the development of treatment programs that are more personalized and effective. Sharing of data that is both secure and standardized and thus, BCT can support the sharing of data that is both secure and standardized between various healthcare providers. This allows for the exchange of patient data that is necessary for personalized therapy. Personalized treatments may benefit from this in terms of both their accuracy and their quality.

On the other hand, the application of BCT in the medical field presents several difficulties. As a result of the frequently massive and complicated nature of healthcare data, scalability presents a significant problem for BCT. It is possible that the blockchain will grow slower and more expensive to operate as the number of transactions that are registered on it increases. Existing electronic health record (EHR) systems and other legacy systems are utilized by a multitude of healthcare organizations, and it is possible that these systems are not compatible with BCT. Integration can be a difficult and expensive process that necessitates considerable modifications to the infrastructure that is already in place. Because BCT is still in its infancy, there are no established regulatory frameworks in place to govern its application in the medical field currently. Due to the absence of regulation, there is a potential for confusion, which in turn might restrict adoption by healthcare organizations [12].

BCT may be utilized in conjunction with one or more medical devices in certain circumstances. In situations like these, the utilization of BCT can be subject to additional regulatory requirements, such as those imposed by the Food and Drug Administration (FDA) of the United States or other agencies that meet equivalent standards in other jurisdictions [11]. It is possible that the usage of BCT will include the creation of intellectual property or the transfer of such property, such as copyrights or patents. To ensure that they respect the intellectual property rights of others and preserve their own intellectual property, healthcare organizations that use BCT are required to take the necessary precautions. Smart contracts, which are agreements that can carry out their own execution and have the conditions of the contract put into code, are often used to manage transactions that take place on a blockchain. To ensure that their smart contracts are legally binding and enforceable, healthcare organizations that use BCT must take the necessary precautions. The healthcare industry stands to benefit greatly from utilizing BCT. Healthcare 5.0 is all about putting the patient first and using technology to boost health results. BCT opens new possibilities for medical applications by facilitating safe, transparent, and efficient data administration and sharing.

References

[1]   Adel, A. (2022). Future of industry 5.0 in society: Human-centric solutions, challenges and prospective research areas. *Journal of Cloud Computing*, 11 (1), 40. https://doi.org/10.1186/s13677-022-00314-5.

[2]   Humayun, M. (2021). Industrial revolution 5.0 and the role of cutting edge technologies. *International Journal of Advanced Computer Science and Applications*, 12(12), 605–615. https://doi.org/10.14569/ijacsa.2021.0121276.

[3]   Saraswat, D., Bhattacharya, P., Verma, A., *et al.* (2022). Explainable AI for healthcare 5.0: Opportunities and challenges. *IEEE Access*, 10, 84486–84517. https://doi.org/10.1109/access.2022.3197671.

[4]   Ishii, E., Ebner, D. K., Kimura, S., Agha-Mir-Salim, L., Uchimido, R., and Celi, L. A. (2020). The advent of medical artificial intelligence: Lessons from the Japanese approach. *Journal of Intensive Care*, 8(1), 35. https://doi.org/10.1186/s40560-020-00452-5.

[5]   Mbunge, E., Muchemwa, B., Jiyane, S., and Batani, J. (2021). Sensors and healthcare 5.0: Transformative shift in virtual care through emerging digital health technologies. *Global Health Journal*, 5(4), 169–177. https://doi.org/10.1016/j.glohj.2021.11.008.

[6]   Iyengar, K. P., Zaw Pe, E., Jalli, J., *et al.* (2022). Industry 5.0 technology capabilities in trauma and orthopaedics. *Journal of Orthopaedics*, 32, 125–132. https://doi.org/10.1016/j.jor.2022.06.001.

[7]   Akhtar, M. W., Hassan, S. A., Ghaffar, R., Jung, H., Garg, S., and Hossain, M. S. (2020). The shift to 6G communications: Vision and requirements. *Human Centric Computing and Information Sciences*, 10(1), 53. https://doi.org/10.1186/s13673-020-00258-2.

[8]   Wang, M., Zhu, T., Zhang, T., Zhang, J., Yu, S., and Zhou, W. (2020). Security and privacy in 6G networks: New areas and new challenges. *Digital Communications and Networks*, 6(3), 281–291. https://doi.org/10.1016/j.dcan.2020.07.003.

[9]   Janjua, M. B., Duranay, A. E., and Arslan, H. (2020). Role of wireless communication in healthcare system to cater disaster situations under 6G vision. *Frontiers in Communications and Networks*, 1, 610879. https://doi.org/10.3389/frcmn.2020.610879.

[10] Khezr, S., Moniruzzaman, M., Yassine, A., and Benlamri, R. (2019). Blockchain technology in healthcare: A comprehensive review and directions for future research. *Applied Sciences*, 9(9), 1736. https://doi.org/10.3390/app9091736.

[11] Leong, W. Y., and Zhang, J. B. (2023). Engineer 5G to transform healthcare industry. *ASM Science Journal*, 18, 1–9. https://doi.org/10.32802/asmscj.2023.1339.

[12] Leong, W. Y. (2024). *ESG Innovation for Sustainable Manufacturing Technology: Applications, designs and standards*. Stevenage: The Institution of Engineering and Technology.

Chapter 7

Revolutionising healthcare through generative AI in Industry 5.0

Norana Abdul Rahman[1], Tan Ee Xion[2] and Vaikunthan Rajaratnam[3]

7.1 Introduction

The advent of Industry 5.0 marks a significant evolutionary step in the trajectory of industrialisation, characterised by a deliberate shift from the automation and efficiency-driven ethos of Industry 4.0 to a more nuanced, human-centric approach. At its core, Industry 5.0 embraces three fundamental principles: human-centricity, collaboration and sustainability [1]. Unlike its predecessor, Industry 5.0 places human needs, values and creativity at the forefront of technological innovation, advocating for a paradigm where technology can augment human capabilities rather than replace them. In other words, it emphasises technological transformation through digitalisation and connectivity, facilitating smart factories that leverage data and machine learning for optimised operations [2].

This transition fosters a synergy between humans and intelligent machines, emphasising the enhancement of human work rather than its diminishment. Moreover, it prioritises sustainability and eco-conscious practices, aiming for a balanced coexistence between technological advancement and environmental stewardship. This approach is likely to minimise waste and enhance the quality of life for workers, promoting a healthier work environment [3].

Economically, Industry 5.0 is poised to have a positive impact by creating more job opportunities than it displaces. This expansion of new roles, particularly in technology-driven settings, is expected to contribute to a more balanced and inclusive economy. The integration of innovative technologies with human-centric approaches in Industry 5.0 holds the promise of a progressive industrial future that harmonises technological advancement with human welfare [4].

Within this evolving landscape, the transformative potential of Generative Artificial Intelligence (Generative AI) stands out, mainly through models such as

[1]Centre for Research Excellence, Perdana University, Malaysia
[2]Department of Digital Health and Health Informatics, IMU University, Malaysia
[3]Khoo Teck Puat Hospital, Singapore, Singapore

GPT-3 and its successors. These advanced AI systems diverge significantly from traditional AI by exhibiting unprecedented creativity, adaptation and efficiency. Generative AI or Gen AI's ability to synthesise original text, images, code and other forms of data introduces a new dimension to AI's role in innovation. Its capacity to learn from extensive datasets and tailor its outputs to specific contexts enables a level of adaptation and personalisation previously unattainable. Moreover, by automating tasks and accelerating processes once bottlenecked by human effort, Gen AI offers a path to enhanced productivity and innovation [5].

The introduction of generative AI-based conversational assistants in customer support was highlighted in a working paper by Brynjolfsson *et al.* [6], examining the practical implications of AI deployment in organisations. The study emphasised that such tools significantly enhanced worker productivity, benefiting less experienced employees by enabling them to resolve more customer issues per hour [6]. The improvement also boosted morale and reduced the financial burdens associated with the frequent hiring and training of new staff, thereby enhancing the overall operational efficiency of firms. Regardless of their skill level, Gen AI tools make employees more accessible to the quality of information [6].

The healthcare sector, faced with multifaceted challenges, emerges as a prime candidate for the disruptive innovation that Gen AI promises within the framework of Industry 5.0. The unsustainable trajectory of healthcare spending and systemic inefficiencies underscores a pressing need for transformative solutions. Furthermore, disparities in access to care, influenced by geography, socioeconomic status and demographics, highlight the urgency for more equitable healthcare delivery models [7]. The healthcare workforce, burdened by high workloads and stress, faces significant burnout rates, underscoring the necessity for interventions that alleviate these pressures. Additionally, the limitations of a "one-size-fits-all" approach in treatment call for a paradigm shift towards personalisation, aiming to improve patient outcomes through tailored care strategies [8].

Kuzlu *et al.* [9] explored the exciting potential of Gen AI in healthcare, examining its current applications, potential benefits and challenges. They also highlighted the need to foster responsible Gen AI use in the future. This exploration aligns perfectly with the goals of Industry 5.0, which emphasises human-centric innovation alongside automation. When Gen AI and Industry 5.0 work hand-in-hand, they can revolutionise healthcare by transforming medical imaging, accelerating drug development and enhancing patient care and treatment. By combining the power of Gen AI with the human-centric approach of Industry 5.0, healthcare can benefit from increased productivity, efficiency and, ultimately, improved patient outcomes [9].

In summary, the convergence of Generative AI and Industry 5.0 principles offers a promising avenue for addressing these challenges. It heralds a new era in healthcare characterised by enhanced human–machine collaboration, innovation and a steadfast commitment to sustainability and personalisation. This chapter explores the potential of this convergence in revolutionising healthcare, providing a blueprint for a future where technology and human ingenuity merge to create a more efficient, accessible and personalised healthcare system.

7.2   Generative AI applications in healthcare

Generative AI or Gen AI emerges as a pivotal force set to reshape the healthcare landscape. This advanced technology harbours the capacity to refine and expedite the processes involved in drug discovery significantly, offer profound insights from medical imagery, inaugurate a personalised approach to medicine and bolster clinical decision-making with unparalleled precision. This section delves into the multifaceted applications of generative AI, highlighting its capability to fast-track the identification of viable drug candidates, enhance the accuracy of medical image analysis, devise customised treatment regimens that reflect the individual nuances of each patient and furnish healthcare professionals with instantaneous support in navigating the complexities of various patient scenarios.

The advent of generative AI in healthcare signifies a leap towards optimising existing methodologies and opens avenues for groundbreaking innovations. By leveraging its ability to generate novel data and insights, generative AI is a cornerstone technology bridging gaps in current medical practices, transforming them into more efficient, accurate and personalised healthcare solutions. Through this exploration, we will uncover the profound impact generative AI is poised to have on the healthcare industry, marking a significant stride towards the future of medical science and patient care [10,11].

7.2.1   Drug discovery and development: the impact of generative AI

Generative AI is revolutionising the field of drug discovery and development by enhancing efficiency and precision at multiple stages of the process. This innovative approach has the potential to significantly reduce the time and cost associated with bringing new drugs to market, thereby addressing one of the most critical challenges in the pharmaceutical industry [10].

7.2.2   Accelerating candidate identification

One of the most promising applications of generative AI in drug discovery is its ability to swiftly sift through vast datasets of molecular structures and biological pathways to identify potential new drug candidates. A notable instance of this application is the work by Insilico Medicine, which utilised generative AI to pinpoint preclinical candidates for fibrosis in an unprecedented timeframe, as highlighted in the study by Zhavoronkov *et al.* [12], published in Nature Biotechnology. This capability to rapidly generate and evaluate new compounds can significantly accelerate the early stages of drug development, reducing the initial discovery phase from years to months or even weeks [12].

7.2.3   Designing novel molecules

Beyond identifying existing molecules with therapeutic potential, generative AI extends its utility to designing entirely novel molecules tailored for specific diseases. This approach opens the door to more targeted therapies and treatments with potentially fewer side effects. The research demonstrates the capacity of these AI models to generate molecules with optimised properties for combating specific conditions [13], including their application in proposing potential treatments for COVID-19 through drug repurposing.

7.2.4 *Predicting drug–target interactions*

A critical step in drug development is understanding how drug molecules interact with their target proteins within the human body. Predicting these interactions accurately in the early stages can vastly improve the selection process for viable drug candidates, streamlining the path to clinical trials. Generative AI models are adept at evaluating potential drug–target interactions, significantly reducing the reliance on costly and time-consuming laboratory experiments. The study by Stokes *et al.* [14] in Cell underscores the effectiveness of deep learning in identifying new antibiotic compounds, showcasing the model's ability to predict interactions that researchers had not previously considered. This predictive capacity not only aids in the early assessment of drug efficacy and safety but also enriches our understanding of disease mechanisms and pharmacodynamics [14].

In summary, integrating generative AI into drug discovery and development heralds a new era of pharmaceutical innovation, characterised by speed, efficiency and a heightened focus on creating more effective and personalised medications. By leveraging the capabilities of AI to identify promising candidates, design novel molecules and predict drug–target interactions, researchers and developers are poised to navigate the complex landscape of drug development with greater agility and insight.

7.2.5 *Synthetic data creation: enhancing drug development with generative AI*

An increasingly pivotal application of generative AI in the pharmaceutical industry is the generation of synthetic data. This technology addresses a critical bottleneck in drug development: the scarcity of high-quality, diverse datasets often required for robust drug discovery and testing phases. Synthetic data generated by AI can simulate realistic, complex biological and chemical data, providing researchers with valuable datasets for testing hypotheses and validating the efficacy and safety of potential drugs without the need for extensive, time-consuming data collection processes [15].

7.2.6 *Facilitating comprehensive simulation and analysis*

Generative AI models, through the creation of synthetic datasets, enable the simulation of drug interactions with human biology in a controlled, virtual environment. This capability is invaluable for predicting the pharmacokinetics and pharmacodynamics of new drug candidates, offering insights into how these substances are absorbed, distributed, metabolised and excreted by the body. Synthetic data can encompass a wide range of patient demographics, genetic backgrounds and disease states, ensuring that the drugs developed are effective and safe across diverse populations.

7.2.7   Bridging data gaps and enhancing privacy

One of the significant advantages of synthetic data is its ability to bridge gaps in existing datasets, especially in areas where data is scarce or difficult to obtain due to privacy concerns or rare conditions. Moreover, synthetic data can be generated to maintain the statistical properties of actual data without compromising patient privacy. This aspect is particularly crucial in adhering to regulations such as the General Data Protection Regulation (GDPR) and the Health Insurance Portability and Accountability Act (HIPAA), as it allows for the continuation of essential research while respecting and protecting individual privacy [16].

7.2.8   Accelerating preclinical research

In preclinical research, synthetic data generated by Generative AI can drastically reduce the reliance on animal testing and laboratory experiments. By accurately simulating biological responses to new compounds, researchers can narrow the list of promising drug candidates more efficiently before proceeding to more costly and time-intensive in vivo testing. This accelerates the drug development process and contributes to more ethical research practices by reducing the need for animal subjects.

7.2.9   Enhancing drug repurposing efforts

Generative AI's ability to create synthetic data also plays a crucial role in drug repurposing efforts. By simulating how existing drugs interact with targets related to different diseases, researchers can identify new therapeutic uses for established medications. This approach can significantly shorten development timelines and reduce costs, as these drugs have already undergone extensive safety testing [17].

In conclusion, the generation of synthetic data by Generative AI represents a transformative shift in drug development, offering a powerful tool for enhancing the efficiency, scope and ethical standards of pharmaceutical research. Generative AI paves the way for faster, more effective drug discovery and development processes by providing researchers access to comprehensive, diverse and privacy-compliant datasets.

7.2.10   Advancing diagnostic precision with generative AI in medical imaging

Medical imaging, encompassing X-rays, CT scans and MRIs, constitutes a cornerstone of modern diagnostic procedures. Despite their critical role, these imaging techniques often grapple with challenges such as noise, low resolution or artefacts, which can significantly impede the accuracy and reliability of interpretations. Generative AI emerges as a transformative solution to these limitations, offering advanced image enhancement, analysis and synthetic data generation methodologies.

7.2.11   Image enhancement and super-resolution

Generative AI technologies have made substantial strides in enhancing the quality of medical images. By employing models capable of super-resolution, these AI systems can significantly improve the visibility of CT scans, MRI images and X-rays, effectively reducing noise and clarifying details crucial for accurate diagnosis. This enhancement bolsters diagnostic clarity and mitigates the necessity for repeated scans, thereby minimising patient exposure to potentially harmful radiation and streamlining the diagnostic process [18].

7.2.12   Image segmentation and analysis

Generative AI further extends its utility to medical image segmentation and precise analysis. This application aids significantly in the detection of tumours, classification of tissue types and the measurement of anatomical structures, thereby facilitating prompt and accurate diagnoses. Moreover, it assists in tailoring personalised treatment plans by providing detailed insights into the patient's condition. In the case of cardiac MRI, one may delineate the left ventricular cavity and myocardium, with the objective of measuring blood volumes and contraction rates. In recent years, machine learning and deep learning garnered a large interest from the medical imaging community due to their unprecedented achievements in a large swath of computer vision tasks [19].

7.2.13   Synthetic data generation for AI training

A pivotal contribution of Generative AI in the realm of medical imaging is its ability to generate synthetic yet realistic images. This innovation addresses a significant challenge in AI development: the scarcity of accessible real-world medical data, often compounded by privacy concerns and data sensitivity. By producing large datasets of anonymised medical images, Generative AI expands the resources available for training diagnostic models and ensures that these models are robust, versatile and capable of generalising across diverse clinical scenarios. The study by Frid-Adar *et al.* [20], highlighted in the IEEE International Symposium on Biomedical Imaging, underlines the importance of synthetic data augmentation using Generative Adversarial Networks (GANs) for enhancing liver lesion classification, emphasising the invaluable role of generative AI in developing proficient AI diagnostic tools [20].

Incorporating generative AI into medical imaging signifies a leap forward in diagnostic technology, offering enhanced image quality, precise image analysis and synthetic data generation to overcome current limitations. These advancements promise to elevate the accuracy and efficiency of diagnoses and expand the capabilities of AI-powered diagnostic systems, paving the way for more reliable, accessible and personalised medical care.

7.2.14   Personalised medicine: a generative AI-driven
approach

In the evolving healthcare landscape, the shift towards personalised medicine represents a paradigm shift, aiming to transcend the traditional "one-size-fits-all" approach to treatment. This bespoke approach to healthcare leverages the power of Generative AI to sift through and analyse vast arrays of data unique to each patient

– including genetic profiles, medical histories and a plethora of relevant personal health information. The goal is to craft individualised treatment plans that are tailored to the specific nuances of each patient's condition and carry a higher likelihood of success. Another domain where the application of AI-based models has largely been used is single-cell RNA sequencing (sc-RNAseq) analysis. A new method based on transfer learning (TL) and parameter optimisation is introduced to enable efficient, decentralised, iterative reference building and the contextualisation of new datasets with existing single-cell references without sharing raw data. In addition, few methods have emerged around genetic perturbations of outcomes at the single-cell level in cancer treatments [21].

Beyond the customisation of treatment regimens, generative AI facilitates the creation of patient-specific simulations. These virtual models offer a groundbreaking way to foresee disease progression and gauge how an individual might react to various treatment options. This predictive capability empowers healthcare professionals and patients with the information necessary to make more informed and precise medical decisions.

Moreover, the scope of personalised medicine extends into patient education and engagement. Generative AI stands at the forefront of this domain, with the ability to produce tailored educational content, lifestyle guidance and even virtual health coaching services. These AI-generated resources are designed to resonate deeply with the personal health journeys of individuals, accommodating their unique needs, preferences and learning styles. This level of customisation ensures that patients are well-informed about their health conditions and treatment options and are more engaged and proactive in managing their health.

The integration of Generative AI into personalised medicine heralds a new era in healthcare, where treatments are no longer generalised but are finely tuned to address the individual health nuances of each patient. This approach promises to enhance the efficacy of medical interventions and transform the patient experience, making healthcare more responsive, engaging and, ultimately, more effective [22].

7.2.15　Clinical decision support

Generative AI stands at the forefront of revolutionising clinical decision support, offering sophisticated tools that enhance the precision and efficacy of healthcare delivery. Its capabilities span various aspects of medical practice, from diagnosis and treatment planning to real-time patient monitoring, fundamentally transforming the landscape of clinical care.

7.2.16　Data synthesis for diagnosis

The modern medical environment is characterised by an overwhelming influx of data, encompassing patient records, laboratory results, imaging studies and a continually expanding repository of medical research. Generative AI navigates this deluge of information, employing advanced algorithms to synthesise structured and unstructured data swiftly. It identifies pivotal insights, formulates possible diagnoses and recommends subsequent investigative steps. This aspect of generative AI is invaluable, especially when tackling complex medical cases or encountering patients whose symptoms diverge from conventional disease patterns. By ensuring that no critical detail is missed, generative AI is an indispensable diagnostic tool, offering a comprehensive analysis that might elude even the most diligent human scrutiny [23].

7.2.17   Suggesting treatment pathways

Furthermore, generative AI significantly contributes to the identification of treatment options. These models can propose evidence-based therapeutic strategies by digesting various inputs, including clinical guidelines, existing research literature and specific patient data. This capability is an essential adjunct to physicians' expertise and introduces them to potentially novel or unconventional treatment avenues. Such AI-driven suggestions can mitigate inherent biases in clinical judgment, broadening the spectrum of considered treatments and elevating the likelihood of selecting the most effective intervention for the patient [24].

7.2.18   Real-time support in critical care

Perhaps one of the most critical applications of generative AI is providing real-time support during high-stakes clinical scenarios, such as surgery or intensive care. By continuously analysing patient data streams – including vital signs, laboratory results and other pertinent information – generative AI can detect subtle but significant changes in a patient's condition, offering timely alerts to healthcare professionals. This system can recommend modifications to ongoing treatment plans in response to evolving clinical situations and even anticipate potential complications before they manifest. This dynamic, context-sensitive guidance level acts as a digital safeguard, significantly enhancing the capacity for rapid, informed decision-making in environments where every second counts.

Integrating generative AI into clinical decision-support systems represents a pivotal advancement in healthcare technology. This innovation promises to enhance the diagnostic process, expand treatment options and improve patient care in real time. By synthesising vast datasets, generative AI is reshaping the foundation of clinical decision-making, steering it towards unprecedented accuracy and patient-centric care. It suggests comprehensive treatment pathways and provides unwavering support in critical care settings, ultimately refining the healthcare experience [10].

Generative AI's influence extends far beyond data analysis. It offers real-time support in critical care situations, suggesting comprehensive treatment pathways tailored to individual patient needs. More importantly, it empowers healthcare professionals to deliver highly personalised care with heightened accuracy and efficiency. This empowerment is a key aspect of its potential to transform healthcare delivery and redefine clinical decision-making.

7.2.19   The role of generative AI as an assistive tool

It is crucial to underscore that generative AI in clinical decision support is an assistive technology rather than a substitute for human expertise. While it offers profound benefits in data analysis, treatment suggestions and real-time monitoring, the ultimate responsibility for patient care remains in medical professionals' hands. Generative AI is designed to augment, not replace, the critical thinking and clinical judgment that are the hallmarks of professional healthcare delivery. By leveraging this technology, clinicians can enhance their diagnostic and therapeutic capabilities, ensuring that patient care is precise and personalised. Yet, the essence of medical practice – the human touch and expert judgment – remains paramount [23].

7.3   Challenges and considerations in implementing generative AI in healthcare

The integration of generative AI into healthcare, while promising to revolutionise diagnosis, treatment planning and patient care, brings forth a complex array of challenges and considerations that must be navigated carefully. As we delve into the transformative potential of this technology, it becomes imperative to address critical concerns surrounding data privacy and security, bias and explainability, and the regulatory landscape governing AI applications in healthcare (Figure 7.1). These areas are crucial not only for the ethical and effective implementation of generative AI but also for maintaining public trust and ensuring that advancements in healthcare technology genuinely enhance patient outcomes [25].

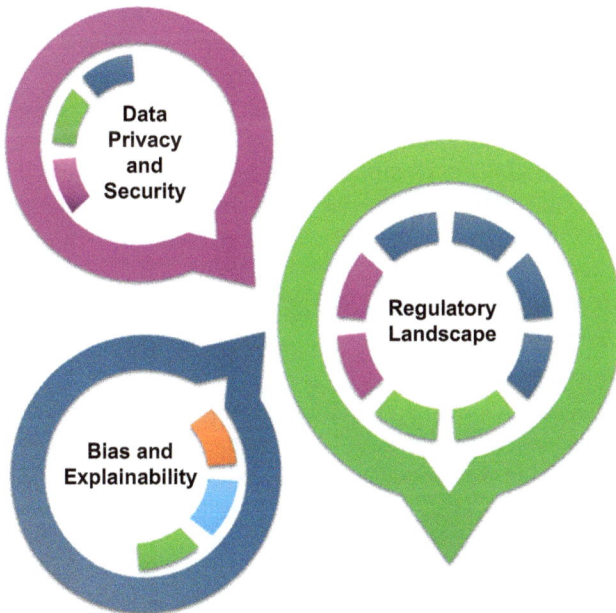

Figure 7.1   Challenges of AI in healthcare

7.3.1   Data privacy and security

The foundation of generative AI's success in healthcare hinges on the availability and quality of data. However, using sensitive patient information raises significant concerns regarding privacy and security. It is essential to establish robust data handling protocols that protect against unauthorised access and ensure the ethical use of data. This entails a commitment to safeguarding patient confidentiality while leveraging the power of generative AI to improve healthcare delivery.

7.3.2   Bias and explainability

Another critical challenge is the potential for bias within the datasets used to train generative AI models, which can lead to skewed or unfair outcomes. Addressing this issue requires a concerted effort to ensure diversity and representativeness in training data, alongside developing AI models that are transparent and explainable. Enhancing the explainability of generative AI models is crucial for fostering trust among healthcare providers and patients, ensuring that AI-supported decisions are fair and understandable.

7.3.3   Regulatory landscape

The rapid advancement of AI in healthcare also poses challenges in terms of regulatory compliance. The evolving nature of regulatory frameworks necessitates that healthcare providers and AI developers stay abreast of current standards and guidelines. Ensuring compliance with these regulations is paramount to foster innovation and guarantee that AI applications in healthcare are safe, reliable and beneficial to patient care.

As we explore the challenges and considerations associated with implementing generative AI in healthcare, it becomes clear that navigating these issues is essential for harnessing the full potential of this technology. By addressing these concerns head-on, the healthcare sector can move forward to maximise the benefits of generative AI while mitigating its risks, ultimately leading to more effective, equitable and personalised patient care.

7.3.4   Data privacy and security: safeguarding sensitive healthcare information

7.3.4.1   The challenge

The healthcare industry's digital revolution, aimed at improving care, has had an unintended consequence: patient records, a treasure trove of personal information, have become a magnet for cyberattacks [26]. While technologies like AI and the Internet of Things (IoT) offer advantages in healthcare management, they also introduce a new challenge: securing vast amounts of data during storage and transfer [27].

Healthcare data, including diagnoses, medications, treatments, medical history and more, ranks among the most confidential and sensitive personal information we possess. It reveals a complete picture of an individual's health. Protecting this data from unauthorised access and maintaining patient trust is crucial. The consequences of a data breach can be severe, causing both personal and financial harm. This could involve stolen identities, discrimination based on medical history or manipulated insurance claims.

7.3.4.2 A stark reminder

The landscape of healthcare data breaches has become increasingly concerning in recent years, impacting millions of individuals and underlining the urgent need for more robust data security measures. Studies reveal a worrisome trend: a significant rise in breaches over time, such as the infamous Anthem data breach of 2015. This data breach exposed inherent weaknesses in how healthcare data is managed in electronic health records (EHR). This incident compromised the personal health records of nearly 80 million individuals [28].

Such breaches can severely jeopardise patient confidentiality, eroding trust in healthcare systems and highlighting the paramount importance of stringent cybersecurity safeguards [26].

7.3.4.3 Considerations for enhanced security

To navigate these challenges, healthcare organisations and developers of generative AI models are urged to implement privacy-by-design principles [29].
Key strategies illustrated in Figure 7.2 include:

- **Anonymisation and de-identification:** Whenever feasible, personal identifiers should be removed from datasets to mitigate risks to patient privacy, which will still allow for necessary data analysis and research without revealing personal information.
- **Robust encryption:** Data at rest and in transit should be encrypted using state-of-the-art methods to protect against unauthorised access. For example, Privacy Enhancing Technologies (PETs) and differential privacy, which obfuscate data to protect patient privacy [30].

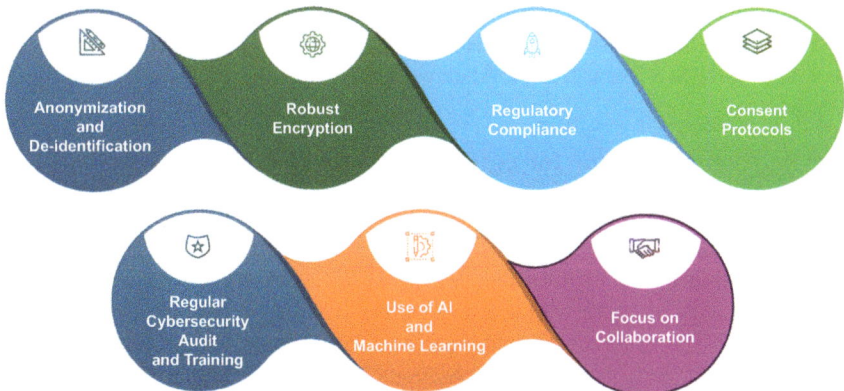

Figure 7.2 Strategies for enhanced security of data in healthcare

- **Regulatory compliance:** Adherence to legal frameworks such as HIPAA in the United States or GDPR in the European Union is non-negotiable, ensuring standardised protections across jurisdictions.
- **Consent protocols:** Explicit patient consent for using their data is essential, reinforcing the ethos of transparency and respect for patient autonomy.
- **Regular cybersecurity audit and training:** Regular audits, staff training and best practices by establishing clear policies and procedures can lessen cyber risks and keep patient data secure to prevent unauthorised access and sharing.
- **Use of AI and machine learning (ML):** Leveraged the advancement of AI and ML to improve the security and privacy of healthcare systems quickly. This will lead to a more secure environment for patient information [29].
- **Focus on collaboration:** the importance of the role of a multidisciplinary team, including experts in telecommunication, instrumentation and computer science, to manage EHRs effectively and enable the secure exchange of medical data across different geographic regions [31].

7.3.5   Bias and explainability: ensuring fairness and transparency

7.3.5.1   The challenge

The integrity of datasets used to train generative AI models is critical. Bias within these datasets can manifest in the models' outputs, potentially resulting in skewed diagnoses or treatment recommendations that exacerbate healthcare disparities. Moreover, the imperative for explainability – understanding how AI models derive their conclusions – is central to identifying and rectifying biases and fostering user trust.

7.3.5.2   Highlighting existing biases

Instances of gender and racial biases in commercial healthcare algorithms have been documented, such as a study revealing an algorithm's underestimation of the health needs of Black patients [32]. The intersectional field of fairness in ML is crucial in addressing these biases, emphasising the need for algorithmic fairness and strategies like decentralised learning and model explainability to mitigate biases in healthcare AI systems. This evidence points to the urgent need for vigilance and corrective measures in AI development to prevent the perpetuation of inequities.

7.3.5.3   Considerations for equity and clarity

The success of AI in healthcare hinges on two key factors: fairness and transparency. The following shows the strategies to confront these challenges:

- **Minimising bias and data privacy:** AI models are only as good as the data they are trained on. If the data lacks diversity, the AI can inherit biases that lead to unfair or inaccurate outcomes for certain demographics. Using data that represents a wide range of backgrounds helps the AI learn from a broader spectrum, resulting in fairer and more accurate results for all patients. While diverse real-world data is ideal, patient privacy is paramount. Techniques like anonymisation and de-identification involve carefully removing personal identifiers from patient data. This allows researchers to use the data while minimising the risk of someone being re-identified [33].

- **Leveraging synthetic data for diverse training**: While real-world data with diverse demographics is ideal for training AI models, patient privacy is paramount. This is where synthetic data generation techniques come in. Synthetic data is artificially created data that statistically resembles real data but does not contain any actual patient information. It offers a valuable alternative for training AI models while ensuring privacy protection [34].

    Two main techniques are used for generating synthetic data in healthcare:

    o **Generative adversarial networks (GANs)**: Imagine two AI doctors locked in a competition. In a healthcare context, one GAN (the generator) tries to create realistic anonymised chest X-rays, while the other GAN (the discriminator) tries to distinguish the synthetic X-rays from real ones. Through this ongoing competition, the generator progressively improves its ability to create realistic X-rays that can be used to train AI models for tasks like pneumonia detection.

    o **Variational autoencoders (VAEs)**: VAEs work differently. Think of a VAE as an AI that compresses medical records. First, it takes a real patient record containing information like blood pressure, medications and diagnoses. The VAE then condenses this data into a smaller, coded format that captures the key characteristics of the patient's health. Next, the VAE learns to decode this code and generate new, similar patient records. While not exact replicas, these synthetic records maintain the underlying statistical properties of real data. For example, a VAE might use anonymised data from diabetic patients to create new synthetic records with similar blood sugar patterns and risk factors.

    Both GANs and VAEs offer valuable tools for generating synthetic healthcare data while protecting patient privacy. The specific technique chosen depends on the type of data and the desired outcome for the AI model and when real training examples are limited [15].

- **Data augmentation: creating a diverse training landscape for healthcare AI**

    Data augmentation is a powerful technique used to artificially increase the amount and diversity of data available for training AI models in healthcare [35]. Imagine you are training an AI to analyse chest X-rays for pneumonia. With a limited dataset, the AI might struggle to recognise pneumonia in images with slightly different lung positioning or varying levels of image noise. This is where data augmentation comes in.

    Data augmentation involves manipulating existing medical data to create variations from which the AI model can learn. Here are some common techniques applied to healthcare data:

    o **Geometric transformations**: This involves techniques like rotating, flipping or scaling X-ray images. This helps the AI learn to recognise abnormalities regardless of the patient's position during the X-ray or slight variations in camera angle.

o **Adding noise**: A small amount of random noise can be strategically added to medical images to simulate real-world variations. For example, adding noise to a mammogram can help the AI model become more robust to slight variations in image quality caused by different mammography machines.

o **Intensity adjustments**: Techniques like adjusting brightness or contrast in medical images can help the AI model learn to identify relevant features even under different lighting conditions during image capture.

o **Bias mitigation techniques**: Employing advanced methodologies during model development to detect and neutralise bias. Building fairness into AI requires vigilance during development. For instance, [36] highlights the importance of integrating diversity and transparency into the data sourcing and integration processes to reduce bias, ensure fairness and protect privacy. This emphasises the role of involving end-users directly in the process as a strategy for mitigating bias. This highlights the critical aspects of mitigation strategy for AI by targeting different stages of interaction with the data and AI systems [37].

Several advanced methods have been developed to detect and mitigate bias at the dataset collection stage. It's essential to employ a framework that can identify and rectify biases in datasets before they are utilised for training AI systems. For example, the Functional Dependencies to Discover Data Bias (FAIR-DB) framework plays a pivotal role in identifying and correcting biases within datasets prior to their use in AI model training [38]. This approach helps diminish biases, improve data integrity and prevent discriminatory practices. Hence, by proactively removing these biases before training and deploying the AI models, we can ensure AI systems operate fairly across all situations in healthcare.

o **Advancements in explainable AI**: Developing AI systems that offer clear insights into their reasoning processes, enhancing user understanding and trust. For AI to be truly reliable, we need to understand how it makes decisions. Explainable AI (XAI) focuses on creating transparent systems that clearly explain their reasoning and conclusion. Besides, transparency is also important for trust-building in AI judgments and for developers to continuously refine the systems on insightful feedback.

Textual XAI, particularly within healthcare, current research involves creating models that can generate counterfactual explanations tailored to patient interactions. For example, an AI system used in predicting patient risk for diseases such as diabetes could provide explanations that specify how slight changes in lifestyle or treatment parameters might alter a patient's risk profile [39]. A counterfactual explanation in this context might illustrate what would happen if a patient reduced their body mass index (BMI) or maintained a more consistent medication regimen, showing how these changes could potentially shift the prediction from high risk to lower risk.

For image-based XAI, research has expanded to include counterfactual visual explanations [40]. These methods involve modifying parts of an input image to show how these changes would affect the AI's decision. This could be particularly useful in medical imaging, where counterfactual visual explanations could help clinicians understand why a model identifies a specific area as indicative of a disease and how changes in the image (like the appearance or size of a tumour) might alter the diagnosis.

Such explanations not only help in making the AI system's predictions more transparent but also empower patients and healthcare providers with actionable insights. These insights can guide patient behaviour toward better health outcomes and assist clinicians in understanding and trusting AI-driven advice, thus enhancing the overall effectiveness of healthcare delivery.

7.3.6 Regulatory landscape: navigating the evolving frameworks

7.3.6.1 The challenge

The pace at which AI is being integrated into healthcare outstrips the evolution of regulatory frameworks, posing challenges for ensuring the safety, efficacy and ethical deployment of generative AI technologies.

7.3.6.2 Regulatory responses

Efforts by bodies such as the FDA to guide the regulation of AI-based medical software exemplify the ongoing work to create coherent regulatory landscapes. These guidelines ensure that AI technologies enhance patient care without compromising safety or ethical standards [41].

7.3.6.3 Considerations for regulation

Effective regulation will require (as shown in Figure 7.3):

Collaborative Efforts

Innovation-Safety Balance

Ethical Oversight

Educational and Awareness Programs

Figure 7.3   Regulatory requirements

- **Collaborative efforts:** Engaging regulators, healthcare entities and AI developers in dialogue to forge comprehensive guidelines. A coherent regulatory landscape requires active engagement with various stakeholders, including AI developers, users, impacted communities, ethicists and academia. This engagement ensures that the regulations remain practical and that they effectively address the concerns of all parties involved. Stakeholder inputs can lead to the iterative refinement of regulations.
- **Innovation-safety balance:** Crafting policies that encourage innovation while prioritising patient safety. Given the rapid development of AI technologies, regulatory frameworks must be adaptable and scalable. This means that they should be designed to accommodate new advancements without necessitating complete overhauls. Regular reviews and updates to the regulations should be institutionalised as part of the regulatory process.
- **Ethical oversight:** The establishment of review boards to oversee the ethical aspects of AI development and application in healthcare settings.
- **Educational and awareness programs:** To effectively navigate and evolve the regulatory landscape, ongoing education and awareness programs are essential for both regulators and the regulated. These programs help stakeholders understand the implications of AI regulations and ensure they can adapt to legal requirements as they evolve.

By addressing these multifaceted challenges, the healthcare sector can fully leverage the potential of generative AI to transform patient care, ensuring advancements are groundbreaking and grounded in ethical, secure and equitable practices.

7.4   The future of generative AI in healthcare

7.4.1   Expanding the horizons of human–AI collaboration in healthcare

7.4.1.1   The vision

The essence of generative AI in healthcare is to serve as an augmentative force, not a replacement for the human touch and expertise of medical professionals. This vision of collaboration between humans and AI is predicated on the premise that AI, with its superior capacity for data analysis and synthesis, can unlock insights and patterns within vast datasets – insights invaluable to physicians. Doing so aims to enrich the physician's expertise, intuition and decision-making process, offering a composite view that blends the best of artificial intelligence with human judgment. By leveraging health informatics and analytics, this synergy not only facilitates more accurate diagnoses and personalised treatment plans but also holds the commitment to improved patient outcomes through a more nuanced understanding of complex medical data [42].

7.4.1.2 Benefits

The benefits of such a symbiotic relationship are manifold. Physicians supported by AI can achieve faster, more accurate diagnoses, formulate personalised treatment plans that reflect the patient's unique health profile, significantly reduce the margin for medical errors and address the growing concern of physician burnout by easing the workload. Projects like IBM Watson for Oncology have already begun to illustrate the practicality of this vision, showing how AI can act as an advisor, offering evidence-based recommendations while the physician retains final decision-making authority [43].

7.4.2 Virtual health assistants: bridging gaps in patient care

7.4.2.1 Potential

Generative AI brings to the fore the potential for creating virtual health assistants that excel in natural language understanding and generation. These AI-powered assistants can conduct initial patient consultations, effectively gather patient history and symptomatology, and triage cases based on urgency. Furthermore, they provide invaluable continuous support for patients with chronic conditions, facilitating better disease management [44].

7.4.2.2 Benefits

The deployment of virtual health assistants promises substantial benefits, including relieving the primary care system's burdens, enhancing healthcare accessibility – especially in remote or underserved regions – and offering consistent, reliable follow-up for chronic disease management. This technological advancement represents a significant leap towards more inclusive, efficient and patient-centred healthcare delivery.

7.4.2.3 Considerations

However, implementing virtual health assistants necessitates rigorous validation to prevent misdiagnosis and ensure that these systems complement rather than replace the irreplaceable value of direct physician-patient interactions. Ensuring the accuracy and reliability of these assistants is paramount to their success and integration into healthcare.

7.4.3 Revolutionising medical research through generative AI

7.4.3.1 Accelerating discovery

In medical research, Generative AI stands as a catalyst for innovation, potentially sifting through existing research findings to unveil novel research avenues, predict drug interactions, optimise clinical trial design and facilitate the identification of suitable trial participants. This capability could drastically streamline research methodologies, reduce operational costs and hasten the journey of groundbreaking treatments from conception to clinical application [45].

7.4.3.2   Benefits

The application of Generative AI in research not only accelerates the discovery process but also opens new frontiers in the development of treatments, promising a future where medical breakthroughs occur at an unprecedented pace. Initiatives like Benevolent AI exemplify how AI can mine scientific literature to identify new therapeutic targets or repurpose existing medications, demonstrating the transformative impact of AI on medical research.

7.4.3.3   A balanced approach

As we venture into the future of healthcare shaped by generative AI, it is crucial to navigate this journey with a balanced mix of optimism and caution. The path forward should be marked by responsible AI development, meticulous evaluation and an unwavering commitment to ethical considerations and patient-centric care. By adhering to these principles, the healthcare community can unlock the full potential of generative AI, ensuring that its integration translates into tangible benefits for patient care and medical research.

7.4.3.4   Evidence synthesis

Generative AI, a pivotal innovation in the evolution towards Industry 5.0, is set to redefine healthcare through its applications in medical imaging, drug development, patient care and treatment planning. This assertion is supported by Kuzlu and colleagues [9], who highlight the transformative potential of Generative AI within these domains [9]. However, the adoption of this technology is not without challenges. Issues such as data privacy, security and adherence to regulatory frameworks have been identified as significant hurdles, as noted by Saraswat and colleagues [46].

The advent of advanced generative AI models, primarily transformers and diffusion models, has markedly improved clinical diagnostics, data reconstruction and the synthesis of drugs, according to Shokrollahi *et al.* [47]. These advancements are part of a broader integration of AI into healthcare, which has been observed to enhance the efficiency and accuracy of drug discovery, clinical trials and patient care processes [48].

Moreover, the synergy between generative AI and large language models, like ChatGPT, offers promising prospects for managing data and information in healthcare settings, a development elucidated by Wang and team [49], including Yu [50]. This integration is crucial for the transition towards Healthcare 5.0, which seeks to leverage AI, the IoT and 5G communications to significantly improve healthcare quality, as envisioned by [51]. Looking forward, the role of AI in medicine is anticipated to be transformative, enabling precision in diagnosis, proactive disease prevention, customisation of treatment plans, real-time patient monitoring and enhancements in medical imaging techniques. Harry [52] underlines these potential advancements, signalling a future where healthcare is more personalised and responsive. Nonetheless, the integration of AI into healthcare practices brings to the fore ethical and legal considerations that must be meticulously addressed [53].

This introduction sets the stage for a detailed exploration of generative AI's role in advancing healthcare, acknowledging its transformative potential and the imperative to navigate the accompanying ethical, legal and practical challenges.

7.4.3.5   Main findings

The synthesis of the main findings from the provided document underscores the transformative impact of generative AI on healthcare, highlighting advancements in clinical diagnostics, drug development and patient care. These findings, derived from recent research, suggest a promising horizon for healthcare innovation, albeit accompanied by challenges that necessitate careful consideration and strategic policy interventions.

Generative AI's role in revolutionising healthcare is primarily noted through its potential to enhance medical imaging, accelerate drug development and improve treatment planning and patient care. Kuzlu and co-workers [9] emphasise generative AI's transformative capabilities, suggesting that such technologies reshape the healthcare landscape by enabling more precise diagnostics and tailored treatments. Similarly, Shokrollahi *et al.* [47] highlight the significant advancements in clinical diagnostics and drug synthesis by generative AI models, particularly noting the efficiency and accuracy these technologies introduce to the medical field [9,47].

The research further indicates that integrating generative AI with other technological innovations, such as large language models (LLM) and deep learning algorithms, can enhance data management and information processing within healthcare settings. For instance, Yu *et al.* [50] discuss how deep learning-powered generative AI and language models offer new avenues for managing healthcare data, facilitating more efficient patient care and information dissemination [50].

However, the findings also illuminate challenges associated with implementing Generative AI in healthcare, including data privacy, security and regulatory compliance concerns. Saraswat and team point out the critical importance of addressing these issues to leverage the full potential of generative AI in improving clinical health outcomes [46].

Moreover, the documents suggest a future where healthcare is further personalised and efficient, driven by AI's capabilities for precision diagnosis, proactive disease prevention and real-time monitoring. Harry [52] envisions a healthcare system where AI-driven technologies play a central role in advancing medical imaging and treatment planning, aligning with the broader goals of Healthcare 5.0.

In summary, the collected research underscores the significant promise of Generative AI in enhancing healthcare delivery and outcomes. The main findings from the reviewed studies advocate for adopting Generative AI technologies in healthcare, emphasising the need for strategic policies and frameworks to address accompanying challenges effectively. This synthesis, rooted in the evidence provided by Kuzlu *et al.* [9], Shokrollahi *et al.* [47], Yu *et al.* [50], Saraswat *et al.* [46] and Harry [52], highlights the pivotal role of generative AI in shaping the future of healthcare towards more personalised, efficient and innovative practices.

7.4.3.6   Policy recommendations

The synthesis of the policy recommendations from the documents underscores the need for strategic, informed approaches to harness the potential of generative AI in healthcare while addressing ethical, legal and practical challenges. These recommendations, informed by recent research, emphasise the importance of regulatory frameworks, stakeholder engagement and the integration of ethical considerations in deploying generative AI technologies.

One of the pivotal recommendations highlights the necessity of developing robust regulatory frameworks that ensure data privacy, security and compliance with legal standards. Saraswat *et al.* [46] advocate for the importance of XAI in healthcare, suggesting that transparent, understandable AI processes are crucial for gaining trust and ensuring ethical use [46]. Supported by Ali and team, a critical review of clinical data emphasises the importance of evaluating the effectiveness of the XAI model [39]. This is echoed by Yu and colleagues [50], who emphasise organisational and regulatory support for leveraging deep learning and large language models in healthcare, indicating that clear guidelines and ethical standards are essential for effective implementation [50].

Furthermore, the recommendations stress the importance of stakeholder engagement in developing and deploying generative AI technologies. The need for collaboration between healthcare professionals, patients, policymakers and technology developers is highlighted as a critical factor in realising the benefits of generative AI in healthcare settings. This involves proactive dialogue and partnership to align technological advancements with healthcare needs and ethical standards.

Shaheen [48] suggests specific policy recommendations related to AI-driven drug discovery and patient care, highlighting the role of technology in automating clinical trials and enhancing patient engagement. These recommendations point to the potential of generative AI to streamline healthcare processes and improve outcomes but also note the necessity of policies that support innovation while safeguarding ethical principles and patient privacy.

In addressing the challenges associated with generative AI implementation in healthcare, the document also points to the need for continuous research, development and assessment of AI technologies. This includes monitoring the impact of Generative AI on healthcare delivery, patient outcomes and the healthcare workforce, ensuring that the benefits of technology are realised without compromising ethical standards or exacerbating inequalities.

The synthesised policy recommendations call for a balanced, thoughtful approach to integrating generative AI into healthcare. They advocate for establishing comprehensive regulatory frameworks, stakeholder engagement and prioritising ethical considerations in technology deployment. These recommendations, informed by the insights of various authors, like Shaheen [48], Saraswat *et al.* [46], Yu *et al.* [50] and other contributors to the documents, reflect a consensus on the need for strategic policies to navigate the complexities of Generative AI in healthcare, ensuring its potential is harnessed responsibly and effectively.

7.4.4   Conclusion: embracing the Industry 5.0 paradigm through generative AI in healthcare

As we stand at the cusp of integrating generative AI within the healthcare sector, it is pivotal to recognise this journey as symbolic of the broader transition towards Industry 5.0. This transition is not merely technological; it signifies a holistic shift towards systems that emphasise human-centricity, sustainability and enhanced collaboration between humans and machines. Generative AI, in this context, emerges as a crucial enabler, possessing the transformative potential to address the intricate challenges plaguing modern healthcare systems. Its role in personalising patient care, streamlining diagnostic and treatment processes, and pioneering unprecedented advances in medical research underscores the essence of Industry 5.0 – melding technological innovation with human insight to foster well-being and efficiency.

The realisation of generative AI's full potential in healthcare is contingent upon fostering synergistic collaborations among medical professionals, AI technologists and policymakers. Such multidisciplinary partnerships are vital for navigating the ethical, technical and regulatory landscapes that define the healthcare industry today. They ensure that the deployment of AI technologies is both responsible and aligned with the overarching objectives of human-centric healthcare.

Moreover, the integration of generative AI into healthcare highlights a critical imperative for engineers and AI developers: to grasp the profound implications of this technology on patient care and healthcare delivery. Understanding these implications is not just about harnessing the power of AI but about ensuring that a deep sensitivity to healthcare's needs, challenges and ethics guides its application. This understanding necessitates including healthcare professionals in the teams responsible for designing and developing AI-driven systems and devices. Their insights, drawn from the front lines of patient care, are invaluable in shaping solutions that are not only technologically advanced but also attuned to the practicalities and nuances of clinical practice.

Incorporating the principles of Industry 5.0 and fostering interdisciplinary collaboration highlight the path forward for the successful integration of generative AI in healthcare. They underscore the need for a balanced approach that combines innovation with empathy, technical excellence with ethical stewardship and automation with human insight. As we venture into this future, we must continue championing research and innovation aimed at the responsible development and application of generative AI. This endeavour is not just about advancing healthcare but about reimagining it in a way that is more responsive, equitable and tailored to the diverse needs of the global population.

In conclusion, the journey of embedding generative AI into the healthcare landscape, guided by the principles of Industry 5.0, offers a promising vista of enhanced healthcare delivery and medical research. For engineers and developers, understanding the impact of generative AI and embracing the expertise of healthcare professionals are essential steps in creating systems and devices that genuinely meet the needs of patients and clinicians alike. Through collective effort, innovation and a commitment to ethical principles, we can harness the full potential of generative AI, ushering in a new era of healthcare that epitomises the ideals of human-centric technology and collaborative innovation.

References

[1] Zizic, M. C., Mladineo, M., Gjeldum, N., and Celent, L. (2022). From Industry 4.0 towards Industry 5.0: A review and analysis of paradigm shift for the people, organization and technology. *Energies*, 15(14), 5221. https://doi.org/10.3390/en15145221.

[2] Zalozhnev, A. Y., and Ginz, V. N. (2023). Industry 4.0: Underlying technologies. Industry 5.0: Human–computer interaction as a tech bridge from Industry 4.0 to Industry 5.0. *2023 9th International Conference on Web Research (ICWR)*, pp. 232–236. https://doi.org/10.1109/ICWR57742.2023.10139166.

[3] Nahavandi, S. (2019). Industry 5.0—A human-centric solution. *Sustainability*, 11(16), 4371. https://doi.org/10.3390/su11164371.

[4] Hyunjin, C. (2020). A study on application of generative design system in manufacturing process. *IOP Conference Series: Materials Science and Engineering*, 727(1), 012011. https://doi.org/10.1088/1757-899x/727/1/012011.

[5] Epstein, Z., Hertzmann, A., Akten, M., *et al.* (2023). Art and the science of generative AI. *American Association for the Advancement of Science*, 380 (6650), 1110–1111. https://doi.org/10.1126/science.adh4451.

[6] Brynjolfsson, E., Li, D., and Raymond, L. (2023). *Generative AI at Work*. Cambridge, MA: National Bureau of Economic Research. https://doi.org/10.3386/w31161.

[7] Speer, M., McCullough, J. M., Fielding, J. E., Faustino, E., and Teutsch, S. M. (2020). Excess medical care spending: The categories, magnitude, and opportunity costs of wasteful spending in the United States. *American Journal of Public Health*, 110(12), 1743–1748. https://doi.org/10.2105/AJPH.2020.305865.

[8] Wiederhold, B. K., Cipresso, P., Pizzioli, D., Wiederhold, M., and Riva, G. (2018). Intervention for physician burnout: A systematic review. *Open Medicine*, 13(1), 253–263. https://doi.org/10.1515/med-2018-0039.

[9] Kuzlu, M., Xiao, Z., Sarp, S., Catak, F. O., Gurler, N., and Guler, O. (2023). The rise of generative artificial intelligence in healthcare. *2023 12th Mediterranean Conference on Embedded Computing (MECO)*.

[10] Soferman, R. (2019). The transformative impact of artificial intelligence on healthcare outcomes. *Journal of Clinical Engineering*, 44(3), E1. https://doi.org/10.1097/JCE.0000000000000345.

[11] Bajwa, J., Munir, U., Nori, A., and Williams, B. (2021). Artificial intelligence in healthcare: Transforming the practice of medicine. *Future Healthcare Journal*, 8(2), e188–e194. https://doi.org/10.7861/fhj.2021-0095.

[12] Zhavoronkov, A., Ivanenkov, Y. A., Aliper, A., *et al.* (2019). Deep learning enables rapid identification of potent DDR1 kinase inhibitors. *Nature Biotechnology*, 37(9), 1038–1040. https://doi.org/10.1038/s41587-019-0224-x.

[13] Insilico Medicine. (2021). Insilico: Linking target discovery and generative chemistry AI platforms for a drug discovery breakthrough. Retrieved from https://www.nature.com/articles/d43747-021-00039-5 (accessed Dec 4, 2024).

[14] Stokes, J. M., Yang, K., Swanson, K., *et al.* (2020). A deep learning approach to antibiotic discovery. *Cell*, 180(4), 688–702.e13. https://doi.org/10.1016/j.cell.2020.01.021.

[15] Chen, R. J., Lu, M. Y., Chen, T. Y., Williamson, D. F. K., and Mahmood, F. (2021). Synthetic data in machine learning for medicine and healthcare. *Nature Biomedical Engineering*, 5(6), 493–497. https://doi.org/10.1038/s41551-021-00751-8.

[16] Mbonihankuye, S., Nkunzimana, A., and Ndagijimana, A. (2019). Healthcare data security technology: HIPAA compliance. *Wireless Communications and Mobile Computing*, 2019(1), 1927495. https://doi.org/10.1155/2019/1927495.

[17] Ye, C., Swiers, R., Bonner, S., and Barrett, I. (2022). A knowledge graph-enhanced tensor factorisation model for discovering drug targets. *IEEE/ACM Transactions on Computational Biology and Bioinformatics*, 19(6), 3070–3080. https://doi.org/10.1109/TCBB.2022.3197320.

[18] Young, T., Dowling, J., Rai, R., *et al.* (2023). Clinical validation of MR imaging time reduction for substitute/synthetic CT generation for prostate MRI-only treatment planning. *Physical and Engineering Sciences in Medicine*, 46(3), 1015–1021. https://doi.org/10.1007/s13246-023-01268-x.

[19] Skandarani, Y., Jodoin, P.-M., and Lalande, A. (2023). GANs for medical image synthesis: An empirical study. *Journal of Imaging*, 9(3), 69. https://doi.org/10.3390/jimaging9030069.

[20] Frid-Adar, M., Diamant, I., Klang, E., Amitai, M., Goldberger, J., and Greenspan, H. (2018). GAN-based synthetic medical image augmentation for increased CNN performance in liver lesion classification. *Neurocomputing*, 321, 321–331. https://doi.org/10.1016/j.neucom.2018.09.013.

[21] Morilla, I. (2021). Repairing the human with artificial intelligence in oncology. *Artificial Intelligence in Cancer*, 2(5), 60–68. https://doi.org/10.35713/aic.v2.i5.60.

[22] Abernethy, A., Adams, L., Barrett, M., *et al.* (2022). The promise of digital health: Then, now, and the future. *NAM Perspectives*. Discussion Paper, National Academy of Medicine, Washington, DC. https://doi.org/10.31478/202206e.

[23] Alowais, S. A., Alghamdi, S. S., Alsuhebany, N., *et al.* (2023). Revolutionising healthcare: The role of artificial intelligence in clinical practice. *BMC Medical Education*, 23(1), 689. https://doi.org/10.1186/s12909-023-04698-z.

[24] Sivaraman, V., Bukowski, L. A., Levin, J., Kahn, J. M., and Perer, A. (2023). Ignore, trust, or negotiate: Understanding clinician acceptance of AI-based treatment recommendations in health care. *Proceedings of the 2023 CHI Conference on Human Factors in Computing Systems*, pp. 1–18. https://doi.org/10.1145/3544548.3581075.

[25] Pasricha, S. (2023). AI ethics in smart healthcare. *IEEE Consumer Electronics Magazine*, 12(4), 12–20.

[26] Seh, A. H., Zarour, M., Alenezi, M., *et al.* (2020). Healthcare data breaches: Insights and implications. *Healthcare*, 8(2), 133. https://doi.org/10.3390/healthcare8020133.

[27] Williams, C. M., Chaturvedi, R., and Chakravarthy, K. (2020). Cybersecurity risks in a pandemic. *Journal of Medical Internet Research*, 22(9), e23692. https://doi.org/10.2196/23692.

[28] McCoy Jr., T. H., and Perlis, R. H. (2018). Temporal trends and characteristics of reportable health data breaches, 2010–2017. *JAMA*, 320(12), 1282–1284. https://doi.org/10.1001/jama.2018.9222.

[29] Obaid, O. I., and Salman, S. A.-B. (2022). Security and privacy in IoT-based healthcare systems: A review. *Mesopotamian Journal of Computer Science*, 29–40. https://doi.org/10.58496/MJCSC/2022/007.

[30] Chong, K. M. (2021). Privacy-preserving healthcare informatics: A review. *ITM Web of Conferences*, 36, 04005. https://doi.org/10.1051/itmconf/20213604005.

[31] Keshta, I., and Odeh, A. (2021). Security and privacy of electronic health records: Concerns and challenges. *Egyptian Informatics Journal*, 22(2), 177–183.

[32] Obermeyer, Z., Powers, B., Vogeli, C., and Mullainathan, S. (2019). Dissecting racial bias in an algorithm used to manage the health of populations. *Science*, 366(6464), 447–453. https://doi.org/10.1126/science.aax2342.

[33] Rocher, L., Hendrickx, J. M., and de Montjoye, Y.-A. (2019). Estimating the success of re-identifications in incomplete datasets using generative models. *Nature Communications*, 10(1), 3069. https://doi.org/10.1038/s41467-019-10933-3.

[34] Giuffrè, M., and Shung, D. L. (2023). Harnessing the power of synthetic data in healthcare: Innovation, application, and privacy. *Npj Digital Medicine*, 6 (1), 1–8. https://doi.org/10.1038/s41746-023-00927-3.

[35] Suganyadevi, S., Seethalakshmi, V., and Balasamy, K. (2022). A review on deep learning in medical image analysis. *International Journal of Multimedia Information Retrieval*, 11(1), 19–38. https://doi.org/10.1007/s13735-021-00218-1.

[36] Firmani, D., Tanca, L., and Torlone, R. (2019). Ethical dimensions for data quality. *Journal of Data and Information Quality*, 12(1), 2:1–2:5. https://doi.org/10.1145/3362121.

[37] Mei, A., Saxon, M., Chang, S., Lipton, Z. C., and Wang, W. Y. (2023). Users are the North Star for AI transparency. *arXiv*. arXiv:2303.05500. https://doi.org/10.48550/arXiv.2303.05500.

[38] Azzalini, F., Criscuolo, C., and Tanca, L. (2021). FAIR-DB: FunctionAl dependencIes to discoveR Data Bias. *Proceedings of the EDBT/ICDT 2021 Joint Conference*. https://ceur-ws.org/Vol-2841/PIE+Q_4.pdf.

[39] Ali, S., Abuhmed, T., El-Sappagh, S., *et al.* (2023). Explainable artificial intelligence (XAI): What we know and what is left to attain trustworthy artificial intelligence. *Information Fusion*, 99, 101805. https://doi.org/10.1016/j.inffus.2023.101805.

[40] Mothilal, R. K., Sharma, A., and Tan, C. (2020). Explaining machine learning classifiers through diverse counterfactual explanations. *Proceedings*

of the 2020 Conference on Fairness, Accountability, and Transparency, pp. 607–617. https://doi.org/10.1145/3351095.3372850.

[41] Yaeger, K. A., Martini, M., Yaniv, G., Oermann, E. K., and Costa, A. B. (2019). United States regulatory approval of medical devices and software applications enhanced by artificial intelligence. *Health Policy and Technology*, 8(2), 192–197. https://doi.org/10.1016/j.hlpt.2019.05.006.

[42] Gennatas, E. D., and Chen, J. H. (2021). Chapter 1 – Artificial intelligence in medicine: Past, present, and future. In L. Xing, M. L. Giger, and J. K. Min (eds.), *Artificial Intelligence in Medicine*. New York: Academic Press; pp. 3–18. https://doi.org/10.1016/B978-0-12-821259-2.00001-6.

[43] Wani, S. U. D., Khan, N. A., Thakur, G., *et al.* (2022). Utilisation of artificial intelligence in disease prevention: Diagnosis, treatment, and implications for the healthcare workforce. *Healthcare*, 10(4), 608. https://doi.org/10.3390/healthcare10040608.

[44] Preum, S. M., Munir, S., Ma, M., *et al.* (2021). A review of cognitive assistants for healthcare: Trends, prospects, and future directions. *ACM Computing Surveys*, 53(6), 130:1–130:37. https://doi.org/10.1145/3419368.

[45] Xiao, Y., Bi, M., Guo, H., and Li, M. (2022). Multi-omics approaches for biomarker discovery in early ovarian cancer diagnosis. *eBioMedicine*, 79, 104001. https://doi.org/10.1016/j.ebiom.2022.104001.

[46] Saraswat, D., Bhattacharya, P., Verma, A., *et al.* (2022). Explainable AI for healthcare 5.0: Opportunities and challenges. *IEEE Access*, 10, 84486–84517.

[47] Shokrollahi, Y., Yarmohammadtoosky, S., Nikahd, M. M., Dong, P., Li, X., and Gu, L. (2023). A comprehensive review of generative AI in healthcare. *arXiv*. arXiv:2310.00795. https://doi.org/10.48550/arXiv.2310.00795.

[48] Shaheen, M. Y. (2021). AI in healthcare: Medical and socioeconomic benefits and challenges. *ScienceOpen Preprints*. https://doi.org/10.14293/S2199-1006.1.SOR-.PPRQNI1.v1.

[49] Wang, G., Yang, G., Du, Z., Fan, L., and Li, X. (2023). ClinicalGPT: Large language models fine-tuned with diverse medical data and comprehensive evaluation. *arXiv*. arXiv:2306.09968. https://doi.org/10.48550/arXiv.2306.09968.

[50] Yu, P., Xu, H., Hu, X., and Deng, C. (2023). Leveraging generative AI and large language models: A comprehensive roadmap for healthcare integration. *Healthcare*, 11(20), 2776.

[51] Mohanta, B., Das, P., and Patnaik, S. (2019, May). Healthcare 5.0: A paradigm shift in digital healthcare system using artificial intelligence, IOT and 5G communication. *2019 International Conference on Applied Machine Learning (ICAML)*.

[52] Harry, A. (2023). The future of medicine: Harnessing the power of AI for revolutionizing healthcare. *International Journal of Multidisciplinary Sciences and Arts*, 2(1), 36–47.

[53] Dave, T., Athaluri, S. A., and Singh, S. (2023). ChatGPT in medicine: An overview of its applications, advantages, limitations, future prospects, and ethical considerations. *Frontiers in Artificial Intelligence*, 6, 1169595. https://doi.org/10.3389/frai.2023.1169595.

Chapter 8

Transforming healthcare: the synergy of medical equipment engineering and Industry 5.0 technologies

Wendy Wai Yeng Yeo[1] and Chia Chao Kang[2]

8.1 Introduction

The emergence of Industry 5.0 which builds upon the paradigm shift introduced by Industry 4.0 has brought a synergy between healthcare professional, patients and machines in healthcare system. Together with the advancements in medical equipment engineering, this integration of cutting-edge technologies alongside with human skills and creativity has revolutionised the healthcare delivery, leading towards improved patient outcomes and enhances quality of care. This enables seamless communication and collaboration between healthcare professionals and machines, resulting in more efficient diagnosis and treatment processes. This includes the remarkable advancements in adoption of revolutionary technologies such as artificial intelligence (AI), robotic, internet of things (IoT) as well as data analytics and machine learning under Industry 5.0 framework.

8.2 Remote monitoring and diagnosis

The human-centric approach of Industry 5.0 encourages the development of medical equipment which engages patients in their own healthcare. This includes the wearable health devices that continuously monitor a patient's vital signs and health metrics using smart watches, fitness trackers and others [1] (Figure 8.1). Wearable devices continuously collect health data and transmit it to connected platforms typically via Bluetooth or Wi-Fi. This data is sent to smartphones, tablets or cloud-based servers to be accessed and analysed.

Remote monitoring and diagnosis via wearable devices utilise technology equipped with sensors and communication capabilities to continuously track and evaluate an individual's health status in real time, as shown in Figure 8.2. Examples

[1]School of Pharmacy, Monash University Malaysia, Malaysia
[2]School of Electrical Engineering and Artificial Intelligence, Xiamen University Malaysia, Malaysia

Figure 8.1   Wearable technology. Source: https://qubika.com/blog/wearable-technology-in-healthcare/.

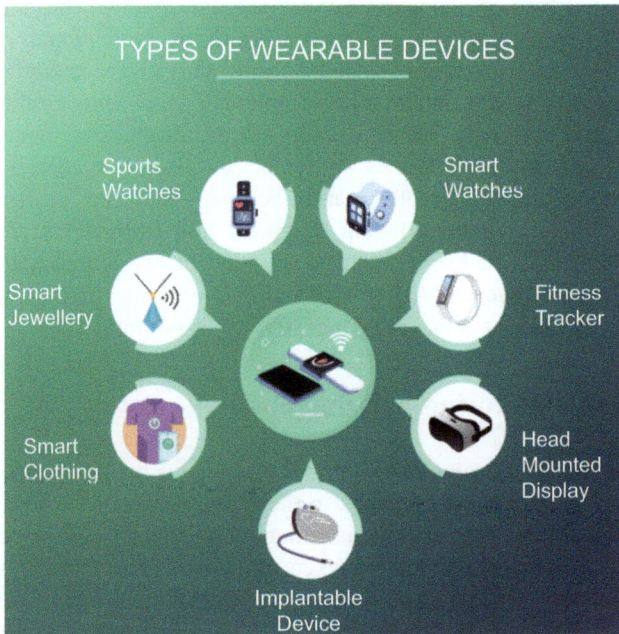

Figure 8.2   Different wearable technologies currently available in the market. Source: https://www.appventurez.com/blog/wearable-designing.

of such devices include smartwatches, fitness trackers, chest straps and specialised medical wearables. These devices are designed to be worn on the body and monitor various physiological parameters, including:

i.   Heart rate: Measures the number of heart beats per minute.
ii.   Blood pressure: Tracks systolic and diastolic blood pressure.
iii. Body temperature: Monitors core body temperature.
iv.   Respiratory rate: Records the number of breaths per minute.
v.   Blood oxygen levels (SpO2): Measures the oxygen saturation in the blood.
vi.   Electrocardiogram (ECG): Captures the electrical activity of the heart.
vii. Blood glucose levels: Tracks glucose levels in the blood for diabetic patients.

One of the primary benefits of wearable devices is real-time monitoring. The continuous data flow allows for immediate detection of abnormalities or critical health events. The system will alert the patient or healthcare providers when certain thresholds are crossed, such as extremely high heart rate or low blood oxygen levels. The data collected is analysed using advanced algorithms and AI. This analysis can identify and recognise trends and patterns that might indicate the onset of a medical condition and predict potential health issues before they become critical. The integration of AI and advanced algorithms enhances this process by deeply analysing vast amounts of data to identify patterns or risks that may not be immediately apparent, improving diagnostic accuracy and fostering personalised healthcare.

Lastly, wearable devices can be integrated with telehealth services to conduct real-time and historical health data during virtual visits. It enhances the interaction between patients and providers by making care more informed and efficient. While during the appointments, health centre providers have the access to live readings of the patients from wearable devices or other means of diagnosis. As a result, they can make timely and factual decision in relation to data. This will reduce self-reported symptoms that might capture some relevant information and thus providing only a small perspective of the patient's health.

These wearables allow the continuous monitoring of human physical activities, physiological as well as biochemical parameters during daily life. Hence, it reduces the frequency of in-person visits required for patient with chronic conditions, allowing for more efficient continuous monitoring of their health status remotely. This not only eases patient experience but also optimises healthcare resources and potentially reduces overall cost [2]. Interestingly, biometric data from wearable devices can also be used to predict infection risks or identify early potential outbreaks for public health and thus, provide timely intervention [3,4].

8.3   Precision surgery

With the implementation of Industry 5.0, healthcare professionals can now benefit from increased automation and connectivity in medical equipment engineering. The collaborative robot (cobot) is now gaining popularity as a form of robotic automation designed to work with human workers in a shared and collaborative workspace. In the healthcare industry, a cobot has a crucial role in surgical assistance, allowing for precision and minimally invasive surgery. Moreover, these robots which equipped with sensors and vision systems that enable them to detect and respond to human presence in real-time, ensuring a safe collaborative environment [5].

On the other hand, Chourasia *et al.* (2023) highlighted the transformative potential of Industry 5.0 in precision surgery that emphasising the integration of human-robot collaboration [6]. While robotic assistance can significantly enhance surgical precision and reduce human error, it also raises concerns about the dependency on technology and the potential for technical failures [6]. The success of such systems relies heavily on the seamless integration of AI and robotics with human expertise, ensuring that surgeons remain in control and can intervene when necessary. This balance is crucial to maintaining patient safety and achieving optimal surgical outcomes.

The recently launched next generation robotic exoscope, Modus X, developed by Synaptive Medical in Canada, with 4K 3D optics, 27-times magnification, automated image focus and voice guidance enables fluorescence-guided surgery for neurosurgery, spine, ear-nose-throat (ENT) and reconstructive microsurgery (Figure 8.3). Various studies highlighted the benefits of using exoscope, including reduced operative time, shorter hospitalisation stay, reduced blood loss with favourable clinical outcomes in addition to ergonomically visual technology particularly advantageous for spinal surgeries [7,8].

The adoption of Mako's system in the robotic-assisted surgery total hip replacement, total knee replacement and partial knee replacement surgery elevates the ability to address surgeons' technical imperfections and ensures consistent surgical results [9] (Figure 8.4). It has been reported that Asian patients have a higher incidence of constitutional varus deformity compared to patients from Western countries, due to their distinct body features and anatomical characteristics [10]. Thus, this technology allows surgeon to provide personalised surgical

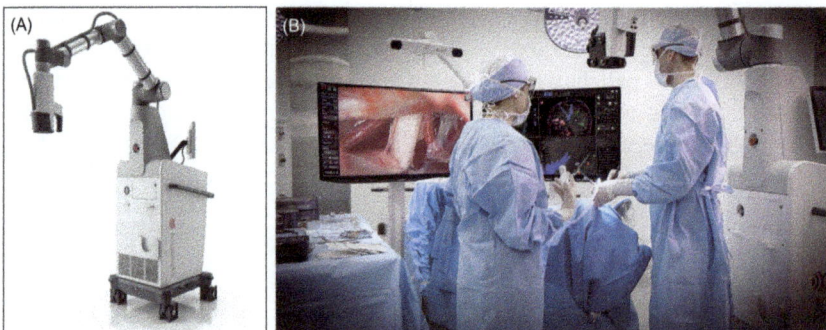

Figure 8.3   (A) Fourth generation of robotic exoscope, Modus X. (B) Modus V robotic system enables surgeons to visualise structures in the brain with high resolution. Source: https://www.synaptivemedical.com/ products/modus-x/.

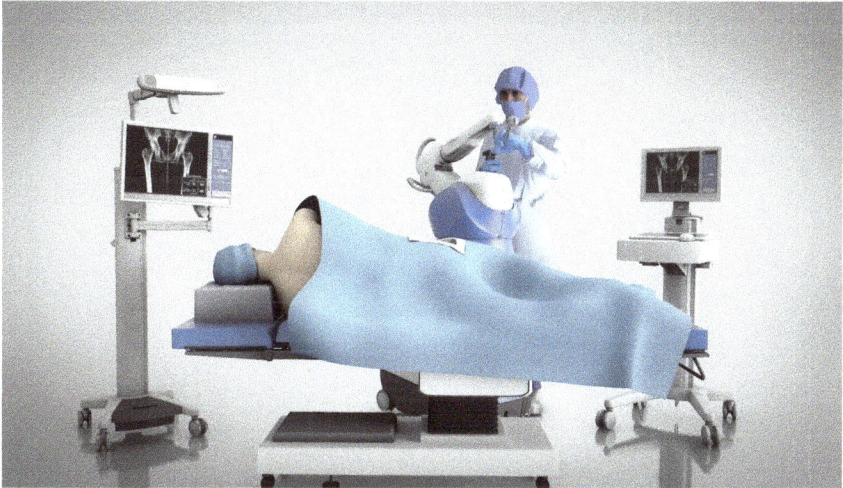

Figure 8.4   Total hip replacement surgery with Mako. Source: https://patients. stryker.com/hip-replacement/options/mako-robotic-arm-assisted.

experience based on the diagnosis and anatomy of individual patient which helps to provide more accurate placement of the implant.

With the ongoing advancements in robotic capabilities, coupled with the integration of AI and machine learning, minimally invasive techniques in precision surgery have been further enhanced. The key components including robotic arms, surgical console and vision system in robotic surgery increase the precision and safety in various surgical tasks such as dissecting, cauterising and suturing [11]. For example, the robotically assisted minimally invasive coronary artery bypass grafting (CABG) has gained wide popularity as it is less invasive which is associated with low mortality and morbidity as compared to open cardiac surgery [12,13] (Figure 8.5).

Overall, the numerous advantages of a minimally invasive approach and precision surgery will continue to drive adoption of robotic surgery in the future with the advent of Industry 5.0. This progress enables healthcare professionals to improve patients' mobility and functionality, leading to faster recovery and better quality of life. Although robots excel in consistency and precision, they lack of flexibility, adaptability and critical thinking skills. This necessitates a strong emphasis on collaboration between humans and machines to enhance human abilities, experiences and creativity.

Notably, one of the key research gaps that need to be addressed before fully harnessing Industry 5.0 in the realm of precision surgery is the long-term efficacy and safety of integrating AI and robotic systems in surgical procedures. For example, it is crucial to investigate how AI and robotic-assisted surgeries impact patient health in the long-term benefits or complications which may necessitate monitoring patients' recovery over time. Additionally, analysis of the need for

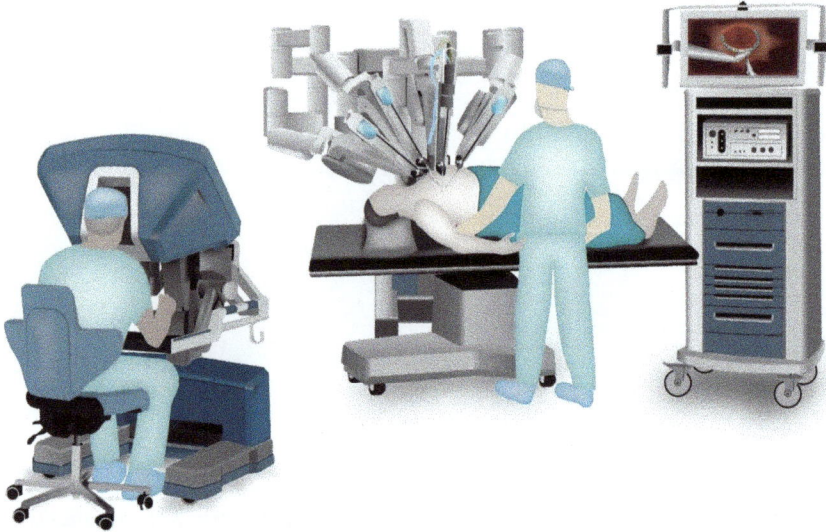

Figure 8.5   Minimally invasive coronary artery bypass grafting with the use of robotic systems. Source: https://www.acc.org/Latest-in-Cardiology/ Articles/2020/05/08/08/11/Robotic-Cardiac-Surgery-Review, Image courtesy of author, T. Sloane Guy, MD, FACC.

follow-up surgeries or exposure of new risks due to the use of AI and robotics warrant further investigation.

8.4   Precision healthcare

In the era of Industry 5.0, the new industrial paradigm shifts from automation to integration of technology into our lives, emphasises sustainable, human-centric. This shift plays a pivotal in personalised medicine, particularly in the field of medical equipment engineering that focuses on the design of an innovative development model and production of highly customisable medical devices. These devices can be tailored for individual patients as the disease, with the ultimate goal to increasing quality of life and improving health outcomes.

Personalised medical devices such as prostheses which enable the healthcare providers to provide treatments tailored to each patient's unique physiology and needs are becoming common. One of the development models is focuses on the prototyping and testing with the growing trend in 3D printing technology. This allows for the creation of personalised prosthetics of the custom-fit prosthetic limbs and orthopaedic implants [14,15] that conducted through clinical trials to ensure efficacy, safety and regulatory compliance. The 3D printing technology is one of the key components that offer new strategies or models for innovation development. For

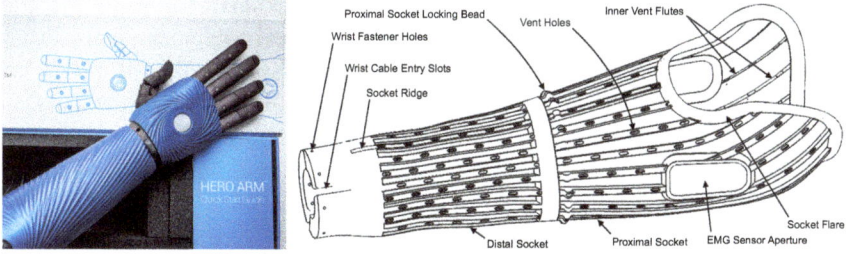

Figure 8.6 Hero Arm^TM, a 3D printed-upper limb prosthesis. Source: https://openbionics.com/en/3d-printing-in-upper-limb-prosthesis/.

example, the world's first medically certified 3D-printed multi-grip bionic arm, Hero Arm^{TM} from Open Bionics, UK which is a lightweight and affordable myoelectric prosthesis (Figure 8.6). The myoelectric prosthesis uses sensors that detect muscular contractions from specific muscle groups, amplifies these signals and converts them into intuitive and proportional movements for the bionic hand [16,17].

Next, leveraging the vast amounts of patient data using generative AI and state-of-art medical equipment enables healthcare professionals to customise treatments based on a patient's unique genetic [4,18]. A notable example is the prediction risk of diabetes using patient network model combined with machine learning techniques [19,20]. This chronic disease risk prediction involves the analysis of patient's genetic profile, lifestyle factors, socio-demographic and behaviour characteristics using AI models which can be trained on large datasets. Additionally, this predictive capability also leverages continuous real-time data from wearable devices by incorporating diverse health data such as physical activity, sleep patterns and daily diets.

8.5 Predictive maintenance

With the revolutionary technologies, it is imperative to implement predictive maintenance (PdM), also known as condition-based maintenance in Industry 5.0 framework to enhance medical equipment reliability as well as ensure patient and users safety [21]. PdM models forecast equipment issues before they occur and monitor parameters such as temperature, vibration and others, allowing maintenance to be scheduled proactively rather than reactively.

PdM systems use state-of-the-art technology such as sensors, IoT devices, data analytics, machine learning algorithms and AI to track the condition of medical equipment in real time. A study claims that PdM management solutions based on machine learning and the IoT can reduce diagnostic and maintenance costs on medical equipment in a hospital setting by as much as 25% [22]. This has reduced unnecessary labour and costs. Besides that, it is also enabled better planning and resource allocation for maintenance. Moreover, it also has significantly enhanced the efficiency of medical equipment and the efficacy of maintenance [22].

In order to minimise disruptions to clinical operations and ensure the continuous availability and extended lifespan of vital medical equipment, maintenance activities need to be strategically scheduled. In Industry 5.0, PdM will play a critical role in the engineering of medical equipment. It guarantees that medical facilities have dependable and well-maintained equipment to provide high-quality care to the patients.

8.6   Privacy, security and ethical concerns

Industry 5.0 may cause a significant problem in the medical equipment engineering of the digital healthcare system. There are moral issues that need to be resolved such as possible biases in AI algorithms which may cause clinical outcomes to be interpreted incorrectly besides privacy and security issues about patient data [23,24]. It is important to emphasise that on personal health information even with a simple device like a fitness tracker. Due to that personal health information is so sensitive and private, there are major ethical and privacy concerns when it is potential to be misused or unauthorised access.

Therefore, it is important that ensuring the proper use and protection of personal health data which requires the implementation of legal frameworks for privacy and data protection, such as the General Data Protection Regulation (GDPR) and the Health Insurance Portability and Accountability Act (HIPAA) [25–27]. Moreover, techniques like encryption and anonymisation can be used to shield patient identification from the information that medical devices have collected [28]. Furthermore, all data that is transferred from medical devices to central systems and stored in databases should use robust encryption techniques to prevent unauthorised access. Regular security audits and vulnerability assessments can also be performed to identify and address any security vulnerabilities in medical systems and equipment [29,30].

8.7   Conclusion

The combination of advances in medical equipment engineering and Industry 5.0 concepts presents a transformative opportunity for the delivery of healthcare services. The widespread application of robotic automation in parallel with human expertise promotes a patient-centric approach by streamlining healthcare procedures and improving patient experience. Several significant advantages include remote monitoring and diagnosis, precision surgery and precision healthcare. Nevertheless, this brings up issues related to patient's data security and privacy. These concerns can be effectively mitigated by implementing robust legal frameworks and stringent data protection measures that safeguard sensitive patient information.

The integration of AI, robotics, IoT and machine learning with medical equipment in Industry 5.0 will revolutionise patient care, treatment and diagnosis in the future. Thus, to ensure that medical equipment meets the specific requirements of medical professionals and patients, stakeholders in the healthcare, technology and regulatory sectors should work together proactively.

References

[1] S. Canali, V. Schiaffonati, and A. Aliverti, "Challenges and recommendations for wearable devices in digital health: Data quality, interoperability, health equity, fairness," *PLoS Digit Health*, vol. 1, no. 10, p. e0000104, 2022, doi:10.1371/journal.pdig.0000104.

[2] D. M. Hilty, C. M. Armstrong, A. Edwards-Stewart, M. T. Gentry, D. D. Luxton, and E. A. Krupinski, "Sensor, wearable, and remote patient monitoring competencies for clinical care and training: Scoping review," *J Technol Behav Sci*, vol. 6, no. 2, pp. 252–277, 2021, doi:10.1007/s41347-020-00190-3.

[3] M. M. Raza, K. P. Venkatesh, and J. C. Kvedar, "Intelligent risk prediction in public health using wearable device data," *NPJ Digit Med*, vol. 5, no. 1, p. 153, 2022, doi:10.1038/s41746-022-00701-x.

[4] C. C. Kang, T. Y. Lee, W. F. Lim, and W. W. Y. Yeo, "Opportunities and challenges of 5G network technology toward precision medicine," *Clin Transl Sci*, vol. 16, no. 11, pp. 2078–2094, 2023, doi:10.1111/cts.13640.

[5] P. Biswas, S. Sikander, and P. Kulkarni, "Recent advances in robot-assisted surgical systems," *Biomed Eng Adv*, vol. 6, p. 100109, 2023, doi:10.1016/j.bea.2023.100109.

[6] S. Chourasia, A. Tyagi, Q. Murtaza, R. S. Walia, and P. Sharma, "A critical review on Industry 5.0 and its medical applications," in R. P. Singh, M. Tyagi, R. S. Walia, and J. P. Davim (eds.), *Advances in Modelling and Optimization of Manufacturing and Industrial Systems*, Singapore: Springer Nature, 2023//2023, pp. 251–261.

[7] D. M. Kusyk, S. Jeong, E. Fitzgerald, *et al.*, "Surgical posture with microscopic versus exoscopic visualization in anterior cervical procedures," *World Neurosurg*, vol. 181, pp. e562–e566, 2024, doi:10.1016/j.wneu.2023.10.094.

[8] N. L. A. Nawabi, B.F. Saway, C. Cunningham, *et al.*, "Intraoperative performance with the exoscope in spine surgery: An institutional experience," *World Neurosurg*, vol. 182, pp. 208–213, 2024, doi:10.1016/j.wneu.2023.12.004.

[9] M. Roche, "The MAKO robotic-arm knee arthroplasty system," *Arch Orthop Trauma Surg*, vol. 141, no. 12, pp. 2043–2047, 2021, doi:10.1007/s00402-021-04208-0.

[10] H. J. Jung, M. W. Kang, J. H. Lee, and J. I. Kim, "Learning curve of robot-assisted total knee arthroplasty and its effects on implant position in asian patients: A prospective study," *BMC Musculoskelet Disord*, vol. 24, no. 1, p. 332, 2023, doi:10.1186/s12891-023-06422-w.

[11] K. Reddy, P. Gharde, H. Tayade, M. Patil, L. S. Reddy, and D. Surya, "Advancements in robotic surgery: A comprehensive overview of current

utilizations and upcoming frontiers," *Cureus*, vol. 15, no. 12, p. e50415, 2023, doi:10.7759/cureus.50415.

[12] J. Bonatti, S. Wallner, I. Crailsheim, M. Grabenwöger, and B. Winkler, "Minimally invasive and robotic coronary artery bypass grafting—A 25-year review," *J Thorac Dis*, vol. 13, no. 3, pp. 1922–1944, 2021, doi:10.21037/jtd-20-1535.

[13] C. Cao, C. Harris, B. Croce, and C. Cao, "Robotic coronary artery bypass graft surgery," *Ann Cardiothorac Surg*, vol. 5, no. 6, p. 594, 2016, doi:10.21037/acs.2016.11.01.

[14] S. Safali, T. Berk, B. Makelov, M. A. Acar, B. Gueorguiev, and H. C. Pape, "The possibilities of personalized 3D printed implants: A case series study," *Medicina (Kaunas)*, vol. 59, no. 2, p. 249, 2023, doi:10.3390/medicina59020249.

[15] M. Javaid, A. Haleem, R. P. Singh, and R. Suman, "3D printing applications for healthcare research and development," *Glob Health J*, vol. 6, no. 4, pp. 217–226, 2022, doi:10.1016/j.glohj.2022.11.001.

[16] N. Smita and D. R. Kumar, "Application of artificial intelligence (AI) in prosthetic and orthotic rehabilitation," in S. Volkan, Ö. Sinan, and B. Pınar Boyraz (eds.), *Service Robotics*, Rijeka: IntechOpen, 2020, Ch. 2, pp. 1–21.

[17] Z. Chen, H. Min, D. Wang, Z. Xia, F. Sun, and B. Fang, "A review of myoelectric control for prosthetic hand manipulation," *Biomimetics (Basel)*, vol. 8, no. 3, p. 328, 2023, doi:10.3390/biomimetics8030328.

[18] I. Ghebrehiwet, N. Zaki, R. Damseh, and M. S. Mohamad, "Revolutionizing personalized medicine with generative AI: A systematic review," *Artifl Intell Rev*, vol. 57, no. 5, pp. 1–41, 2024, doi:10.1007/s10462-024-10768-5.

[19] N. P. Tigga and S. Garg, "Prediction of type 2 diabetes using machine learning classification methods," *Procedia Comput Sci*, vol. 167, pp. 706–716, 2020, doi:10.1016/j.procs.2020.03.336.

[20] H. Lu, S. Uddin, F. Hajati, M. A. Moni, and M. Khushi, "A patient network-based machine learning model for disease prediction: The case of type 2 diabetes mellitus," *Appl Intell*, vol. 52, no. 3, pp. 2411–2422, 2021, doi:10.1007/s10489-021-02533-w.

[21] O. Manchadi, F.-E. Ben-Bouazza, and B. Jioudi, "Predictive maintenance in healthcare system: A survey," *IEEE Access*, vol. 11, pp. 61313–61330, 2023, doi:10.1109/access.2023.3287490.

[22] A. Shamayleh, M. Awad, and J. Farhat, "IoT based predictive maintenance management of medical equipment," *J Med Syst*, vol. 44, no. 4, p. 72, 2020, doi:10.1007/s10916-020-1534-8.

[23] R. Franco D'Souza, M. Mathew, V. Mishra, and K. M. Surapaneni, "Twelve tips for addressing ethical concerns in the implementation of artificial intelligence in medical education," *Med Educ Online*, vol. 29, no. 1, p. 2330250, 2024, doi:10.1080/10872981.2024.2330250.

[24] L. A. Jawad, "Security and privacy in digital healthcare systems: Challenges and mitigation strategies," *Abhigyan*, vol. 42, no. 1, pp. 23–31, 2024, doi:10.1177/09702385241233073.

[25]   Consolidated text: Regulation (EU) *2016/679 of the European Parliament and of the Council of 27 April 2016 on the protection of natural persons with regard to the processing of personal data and on the free movement of such data, and repealing Directive 95/46/EC (General Data Protection Regulation) (Text with EEA relevance)*, 2016.

[26]   Health Insurance Portability and Accountability Act of 1996, *Public Law*, 104–191 (HIPAA), 1996.

[27]   D. Xiang and W. Cai, "Privacy protection and secondary use of health data: Strategies and methods," *Biomed Res Int*, vol. 2021, p. 6967166, 2021, doi:10.1155/2021/6967166.

[28]   K. Abouelmehdi, A. Beni-Hessane, and H. Khaloufi, "Big healthcare data: Preserving security and privacy," *J Big Data*, vol. 5, no. 1, pp. 1–18, 2018, doi:10.1186/s40537-017-0110-7.

[29]   C. M. Mejia-Granda, J. L. Fernandez-Aleman, J. M. Carrillo-de-Gea, and J. A. Garcia-Berna, "Security vulnerabilities in healthcare: An analysis of medical devices and software," *Med Biol Eng Comput*, vol. 62, no. 1, pp. 257–273, 2024, doi:10.1007/s11517-023-02912-0.

[30]   R. Hasan, S. Zawoad, S. Noor, M. M. Haque, and D. Burke, "How secure is the healthcare network from insider attacks? An audit guideline for vulnerability analysis," *Presented at the 2016 IEEE 40th Annual Computer Software and Applications Conference (COMPSAC)*, 2016.

Chapter 9

Trends for additive manufacturing in the Industry 5.0 era

Palaniswamy Mohankumar[1] and Nida Jafri[2]

Industrialisation brought about numerous innovations and altered the logistics, wellness and medical sectors. Since the Neolithic era, the industrial revolution has been widely regarded as the greatest significant shift in human history. Since the First Industrial Revolution affected practically every element of everyday life worldwide, it is seen as a significant turning point in world history. The revolutions were termed 'Mechanical Revolution' for the first, 'Electrical Revolution' for the second, 'Automated Revolution' for the third and 'Digital Revolution' for the fourth.

9.1 Industrial revolutions

Historians continue to disagree over the exact beginning and conclusion of the First Industrial Revolution. The adoption of steam power and industrial automation in the 18th century marked the start of the First Industrial Revolution. The industrial revolution started in Britain in the 1780s, but it wasn't truly felt till the 1830s or 1840s, according to academic Eric Hobsbawm [1]. While some researchers place it in the era from around 1740 to anytime between 1820 and 1840, others, like T. S. Ashton, suggest that it happened approximately between 1760 and 1830 [2]. New manufacturing techniques were adopted throughout Europe, the USA and the rest of the world during the First Industrial Revolution. But what was this First Industrial Revolution actually about? The shift from manual to mechanical production methods, the invention of new ways for producing chemicals and iron, the growing use of water and steam power, the creation of machine tools and the emergence of a mechanised factory structure [3].

9.1.1 First Industrial Revolution

A significant historical turning point, the First Industrial Revolution had an impact on nearly every part of daily life. Specifically, average earnings and population

[1]SRM College of Physiotherapy, Faculty of Medicine and Health Sciences, SRM Institute of Science and Technology, Kattankulathur, India
[2]New Energy Research Centre, Turku University of Applied Sciences, Finland

started to show previously unheard-of persistent rise. Most historians of economics concur that the First Industrial Revolution's beginnings marked the most significant development in the history of mankind since the gradual domestication of both animals and plants. Massive changes were brought about by innovations like the steamship and the steam-powered locomotive, which made it possible for people and products to travel long distances in a shorter amount of time. The First Industrial Revolution was characterised by the shift from labour to steam- or water-powered machinery for manufacturing.

9.1.2   Second Industrial Revolution

The Second Industrial Revolution, often referred to as the Technological Revolution, took place between 1870 and 1914 and was a time of intense modernisation and standardisation. The introduction of new technical systems included the introduction of telephones and electricity, in particular. It turned factories into cutting-edge production lines that produced enormous amounts of productivity and substantial economic development. With some significant inventions, things began to move more quickly. Consider automobiles, aircraft and chemical fertilisers. All inventions allowed us to do more and move more quickly. Inventions like electric lights, radios and telephones changed how people lived and interacted with one another as urbanisation increased. When you give it some thought, the Second Industrial Revolution was the one that made the contemporary world possible.

9.1.3   Third Industrial Revolution

Through little automation utilising processors and memory-programmable controllers, the Third Industrial Revolution got underway in the 1950s, just a few years after World War II concluded. Since the development of these technologies, it is possible to automate a whole production process without the need for human intervention. Robots that follow preprogrammed sequences without human assistance are known instances of this. The Automation Revolution is another name for the Third Industrial Revolution. Semiconductors, mainframe computers, laptop computers and the Internet were all brought about by the Third Industrial Revolution. Analog devices have been replaced by digital ones. Production became automated during the Third Industrial Revolution as a result of the use of communication technologies and field-level devices like programmable logic controllers in the manufacturing process [4].

9.1.4   Fourth Industrial Revolution

The Fourth Industrial Revolution, also known as Industry 4.0 (translated from Industrie 4.0 in German), started in 2011 as a project inside the German government's high-tech policy. The year 2011 saw official introduction of the phrase 'Industrial 4.0' at the Hannover Fair. Industry 4.0 is based on the advancements made during the Third Industrial Revolution, particularly in the area of digitalisation. Another name for it is the Digitation Revolution [5]. What does the phrase

'Digitation Revolution' actually mean? A network link expands production systems with digital computer technology, giving them a virtual digital twin on the Internet. These enable information about themselves to be produced as well as communication with additional services. The Fourth Industrial Revolution has the potential to alter how individuals work. It has the ability to draw people into more intelligent networks, which leads to more productive work. More adaptable ways of conveying the correct information to the right person at the right time are made possible by the digitisation of the production environment. Production systems, in the form of Cyber Physical Production Systems, were able to make intelligent choices in the Industry 4.0 era through real-time communication and collaboration amongst 'manufacturing things', allowing for the flexible mass production of individualised high-quality goods. The collaborative Industry 4.0 Platform was formed by the professional associations BITKOM, VDMA and ZVEI in order to advance the project and provide a synchronised, cross-sectoral technique [6]. Similar strategic efforts have been implemented globally by several nations, including the Made in China 2025 program, the Industrial Internet Consortium (USA), Industria 4.0 (Italy), Produktion 2030 (Sweden) and Society 5.0 (Japan), to mention a few [7].

9.1.5   Fifth Industrial Revolution

The Fourth Industrial Revolution is currently leading us to the Fifth Industrial Revolution. Industry 5.0 will enable us to better automate the production process, resulting in the ingestion of real-time field data. Industry 5.0's primary feature is that every step of the process will be greatly dependent on the 'person' doing it. That's why it's also known as the 'Personalised Revolution'. Industry 5.0 is a revolution that seeks to leverage human specialists' creativity along with specific, smart and efficient robots. Consequently, Industry 5.0 'brings workers back to the workplace' by combining operations with intelligent technologies to harness the creative and mental capacities of humans and machines to improve process efficiency. The available literature lacks a unified definition of Industry 5.0, even in the face of converging suggestions. It is believed that independent manufacturing and human intelligence are working together in a loop to bring about the Fifth Industrial Revolution or Industry 5.0. The premise behind Industry 5.0's arrival is that Industry 4.0 prioritises digitalisation and AI-driven technologies over the original values of sustainability and social justice in order to increase industrial productivity and versatility. Dispersed scholarly initiatives have been advocating for the Fifth Industrial Revolution since 2017. The Fifth Industrial Revolution, or Industry 5.0, was explicitly called for by the European Commission in 2021 following meetings with representatives from financing agencies, research and technology groups and other European institutions. On 4 January 2021, the paper 'Industry 5.0: Towards a Sustainable, Human-centric, and Resilient European Industry' was formally issued [8].

Three interrelated basic values – human-centricity, sustainability and resilience – are at the heart of Industry 5.0. The human-centric method replaces technology-driven advancement with a purely human-centric and society-centric approach, placing fundamental human necessities and interests at the centre of the manufacturing process. The industry must be sustainable in order to respect the limits of the planet. In order to create an economy with improved resource efficiency and effectiveness, it must create circular processes that reuse, re-purpose and recycle resources from nature in addition to lowering waste and negative environmental effects. The term 'resilience' describes the necessity for industrial production to become more resilient in order to better withstand interruptions and guarantee that vital infrastructure can be provided for and supported during times of crisis. Future industries must be able to quickly adapt to changes in the (geo-) political landscape as well as natural disasters.

'Human–robot co-working' – Humans and robots will collaborate whenever and wherever it is feasible under this scenario. Robots will handle the repetitive duties while humans concentrate on those needing ingenuity. Collaborative robots, or cobots as some businesses are calling them, are one of the technologies that will help Industry 5.0 become more human-centric by enabling real-time machine–human cooperation. The bioeconomy is an additional vision of Industry 5.0. An equilibrium between the economy, industry and environment can be reached via the wise utilisation of natural resources for industrial uses. '*The production of renewable biological resources and the conversion of these resources and waste streams into value-added products, such as food, feed, bio-based products, and bioenergy*', is what the European Commission defines as the bioeconomy [7]. It encompasses portions of the chemical, biotechnology and energy sectors as well as agriculture, forestry, fisheries, food and pulp and paper manufacturing.

Keep in mind that Industry 5.0 could focus on the bioeconomy as well as human–robot collaboration. In addition, other challenges including space industries, space mining and space life may be the next wave of change or a component of it. Scientists have previously advised us to use space resources with caution. Perhaps the next 'gold rush' will be in space mining. Utilising Industry up-cycling for waste management is one of Industry 5.0's primary goals. Interoperability of Network sensor data is part of Industry 5.0, which is an enhanced version of Industry 4.0 with new capabilities. Among the aspects are the introduction of collaborative robots, cyber-physical cognitive systems, hyper customisation in the industry, predictive maintenance and smart additive manufacturing. The known technologies of Industry 4.0 and 5.0 are summarised in Table 9.1.

Table 9.1   Known technologies of Industry 4.0 and 5.0 [9]

Industry 4.0	Meaning
Additive manufacturing (AM)	AM is able to create things with more advanced features and sophistication. Additionally, it can aid in the creation of product prototypes.
Artificial intelligence (AI)	The use of computers to simulate intelligent behaviour with little to no human interaction is known as artificial intelligence.
Augmented and virtual reality	Technology that aims to make a display more engaging and connected in reality, therefore enhancing the real-life experiences of viewers.

(Continues)

Table 9.1 (Continued)

Industry 4.0	Meaning
Cloud services	Cloud computing has the potential to remove the industry's requirement for computer hardware maintenance.
Computer aided designs (CAD)	CAD is an innovation that allows for the digital design and production of goods.
Cyber physical systems (CPS)	An emerging category of physically and computationally connected technologies that may interact with people in a variety of creative ways.
Global positioning systems (GPS)	GPS devices calculate locations with accuracy. It accomplishes this by employing satellites that carry out the signal transmission.
Internet of Things (IoT)	IoT is the term used to describe how sensors and things are connected to computers and networks.
Nanotechnology	Nanotechnology has greatly enhanced the control over the single atoms or molecules that go into creating a macroscale item.
Sensors and actuators	Devices that can react to a variety of environmental stimuli, such as temperatures sound, light, magnetic fields and many more, and then create a response.
Big data	Collections of organised, unorganised and semi-organised data that are incredibly vast and varied, and they are growing rapidly over time. Conventional data management systems are unable to store, handle and evaluate these datasets because to their enormous size and complexity in volume, velocity and diversity.
Autonomous robots	Intelligent machines that require little to no human involvement to operate over extended periods of time with superhuman efficiency and the ability to learn about their surroundings.
Simulation	Through the use of models, a simulation replicates how systems or processes in the real-world work. The simulation shows how the model changes over time under various situations, while the model itself depicts the essential behaviours and features of the chosen process or system.
Industry 5.0	Meaning
Cobots	These robots take up tedious jobs from humans rather than replacing them, which helps manufacturers improve output. Cobots thereby contribute to worker safety and increased productivity.
Cyber physical cognitive systems	It belongs to the trusted autonomy subgroup. It helps humans make smarter decisions, classify items and use it for a variety of defence purposes, among other things.
Hyper customisation	The idea behind hyper customisation is to employ AI and big data to prepare more specialised products based on the demands of the consumer in real-time.
Predictive maintenance	By using predictive maintenance, one may schedule tasks based on how well the machinery is doing. As a result, it can lower maintenance expenses and raise output and quality.
Smart additive manufacturing	Additive manufacturing with customisation and data processing at system level and technology level.

9.2   Additive manufacturing

One new manufacturing technique that enables unparalleled and versatile designs to accomplish lean and sustainable aims is additive manufacturing. Rapid prototyping, freeform fabrication or 3D printing are other names for Additive Manufacturing (AM). In contrast to subtractive manufacturing techniques, it is described as 'the process of joining materials to make objects from 3D model data, usually layer upon layer'. AM is very different from most traditional production techniques, which get more challenging as geometric complexity increases. For other manufacturing processes, a complete and in-depth analysis of the component geometry is required to determine the optimal order for fabricating certain features, tools and methods that are required and whether more fixtures are required to complete the component. Comparatively speaking, AM just needs a few basic dimensions data and a basic understanding of how the AM machine and the materials used to create the component work. In addition to the source material, operational factors affect the strength, porosity, ductility and surface quality of the final component. AM lowers waste and energy consumption, enabling mass customisation and sustainability [10]. Because of its decentralised structure, it relies less on centralised factories and encourages local production. With less time frames and process procedures, AM provides a number of options for producing better, customised, efficient and effective parts. These advantages are expected to lessen adverse environmental effects, raise societal needs like consumer-centric components and boost financial advantages like return on investment.

AM has been in use since the 1980s. Chuck Hull filed the first patent application for 3D printing invention in the 1980s. Hull created the stereolithography (SLA) technology, which uses a laser to harden a liquid photopolymer. Fused deposition modelling (FDM) is an invention of Scott Crump in the 1990s. It uses a heated extruder to deposit layers of molten plastic to create three-dimensional (3D) objects. The early 2000s saw the development of selective laser sintering (SLS), a process that uses a laser to melt powdered material to create 3D objects. Metal AM was created in the middle of the 2000s using a process called direct metal laser sintering (DMLS), which forms 3D objects by fusing metal powder with a laser. There were several developments in the AM throughout the history. Significant achievements are mentioned in Table 9.2. The principles of AM are mentioned in Figure 9.1.

There are several stages involved in printing a model/structure through AM. They are:

- **Design** – A CAD software is required to design the structure which is to be printed or reverse engineering is done on a 3D scan. No holes or pores are left open.
- **Preparation** – Prepare or convert the CAD design into the language that a AM machine can understand. Generally, it is Standard Surface Tessellation Language (STL).

Table 9.2   Milestones achieved in AM [11]

Year	Milestones
1986	Chuck Hull invented SLA
1992	Carl Deckard developed SLS
1993	Scott Crump invented FDM
1995	First commercial 3D printer – ZPrinter was launched by Z Corporation
2002	DMLS technology was developed
2005	First 3D-printed prosthetic leg was developed in New Zealand
2007	First commercial 3D metal printer was launched by EOS GmbH
2008	First self-replicating 3D printer was launched by RepRap Project
2011	First 3D-printed organ was implanted into a patient
2014	First 3D-printed car – Strati was produced by Local Motors
2016	HP entered into 3D printing with MJF technology
2020	Several medical equipment was printed and used during COVID-19 pandemic
2024	First 3D-printed engine – Agnikul was used to launch a rocket in India

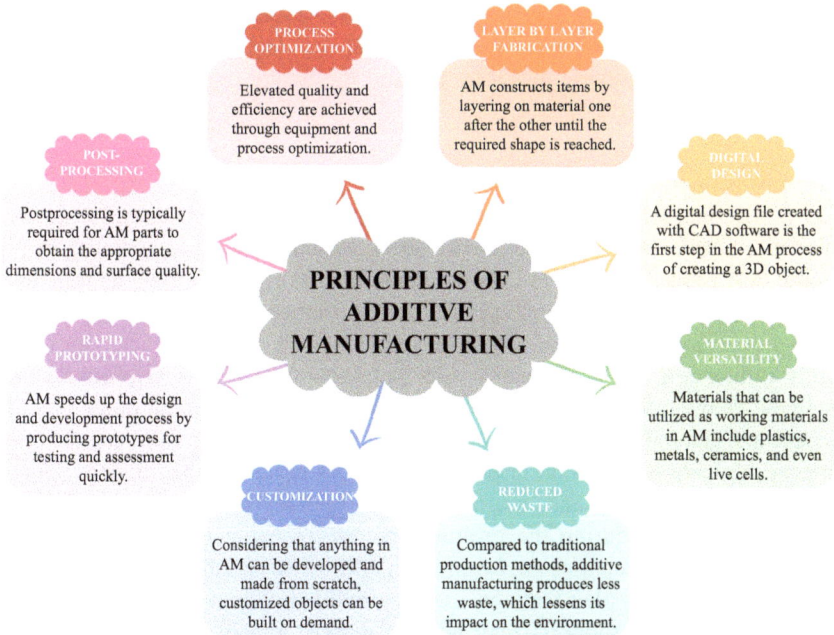

Figure 9.1   Principles of AM

Table 9.3   Post-processing techniques used in AM [12]

Mechanical	Physical	Chemical	Biological
Blasting	Chemical polishing	Acid	Biomimetic
Grinding	Ion implantation	Alkaline	Cell seeding
Machining	Microwave	Chemical vapour deposition	Chemical conjunction
Polishing	Physical vapour deposition	Hydrogen peroxide	Cross-linking
Shot peening	Thermal spray	Sol gel	
		UV irradiation	

- **Print** – AM machine prints according to the CAD design. Material used is determined by the user.
- **Post-process** – Once the structure is printed, remove the support structure or powder. The outer surface may require sanding, polishing, heat treatment. The post-processing techniques are mentioned in Table 9.3.
- **Inspection** – Inspect the printed structure whether it meets the requirements. If necessary, CT or X-ray inspection may be done.
- **Iteration** – If the printed structure did not meet the requirements, the CAD model has to be redesigned and the material used for printing shall be changed.

There are several printing methods used today. They are power bed fusion, directed energy deposition, sheet lamination, material extrusion, material jetting, binder jetting and Vat photopolymerisation. Cold spray additive manufacturing and Friction based printing are recently invented methods. The methods, techniques and other details are mentioned in Table 9.4. and the image representations are provided in Figure 9.2.

9.2.1   Materials used in AM

In AM, materials serve as the fundamental input for creating objects and structures, with their form – solid, powder, wire, sheet, liquid or slurry – dictated by the specific AM process employed. The selection of materials is crucial as they must meet the desired properties and application requirements, which can include variations in mechanical, thermal, chemical, optical or electrical characteristics, ultimately influencing the performance and functionality of the final product. Core materials are the main components used in AM, providing essential properties like strength and durability, with examples including thermoplastic filaments in fused filament fabrication and metal or polymer powders in powder bed fusion. Support materials, on the other hand, are temporary structures that ensure stability and prevent deformation during the printing process, essential for successful object formation. Support structures are essential in AM for successful printing of objects with overhangs or complex features, as they provide necessary stability during the build process. These supports may possess distinct properties compared to the primary material, with options like water-soluble or breakable supports tailored to specific AM techniques and core materials. For instance, material jetting often

Table 9.4 Review of AM methods, technologies, materials and their advantages

Method	Mechanism	Technology	Material type and mechanism	Materials used	Printing resolution	Advantages
Power bed fusion	Uses electron or laser beam to build 3D objects	Direct metal laser sintering (DMLS) Electron beam melting (EBM) Selective laser melting (SLM) Selective laser sintering (SLS)	Powder and laser with optical	Metallics, ceramics, polymers, composites, thermoplastic polyurethane, carbon nanotube, argon, aluminium, titanium, stainless steel, Inconel, glass	80–250 μm	High-dimensional accuracy, large material database, high value products, no solvent required, wide range of materials used
Directed energy deposition	Using laser beam to build 3D objects	Direct metal deposition (DMD) Electron beam free form fabrication (EBF3) Laser deposition welding (LDW) Laser engineered net shaping (LENS) Laser metal deposition (LMD) Wire arc additive manufacturing (WAAM) Wire laser additive manufacturing (WLAM)	Powder & electron beam, powder & laser, arc wire filament & extrusion, powder & injection	Metallics, ceramics, polymers, Inconel, stainless steel, steel, aluminium, alloy, glass	250 μm	Large material database, strong interface strength, high value products, fast, joining dissimilar metals
Sheet lamination	Sheets of materials are joined one after the other either	Friction stir additive manufacturing (FSAM) Laminated object manufacturing (LOM)	Sheet & chemical/ mechanical	Metals, plastics, aluminum, zirconium, copper, stainless steel, nickel, paper	Varies	Low cost, easy to use, joining dissimilar metals

(Continues)

Table 9.4 (Continued)

Method	Mechanism	Technology	Material type and mechanism	Materials used	Printing resolution	Advantages
	by adhesive, wielding or thermal bonding	Selective deposition lamination (SDL) Ultrasonic consolidation (UC) Ultrasound additive manufacturing (UAM)				
Material extrusion	Material is drawn through a nozzle, heated and deposited layer by layer	Fused deposition modelling (FDM) Fused filament fabrication (FFF)	Filament & thermal extrusion	Polymers, food, living cells, glass, hydrogels, bioinks, ceramics, thermoplastics, acrylo-nitrile butadiene styrene (ABS) filament, polypropylene pellets, stainless steel powder, low density polyethylene, poly-ether ether ketone (PEEK)	50–200 μm	Large material database, low cost, bio-composite materials
Material jetting	Printhead dispenses droplets of photosensitive material that solidifies layer-by-layer, under ultraviolet (UV) light	Drop-on-demand (DOD) Material jetting (MJ) Multijet modelling Nanoparticle jetting (NPJ)	Liquid & optical, liquid & laser	Polymers, ceramics, composite biologicals, copper, bronze, zirco-nia, graphene, bioinks	5–200 μm	Printing of multiple materials simultaneously, good printing resolution, high cell proliferation, smooth surface, low wastage

(Continues)

	Description	Technology	Basis	Materials	Resolution	Advantages
Binder jetting	A liquid binding agent is selectively deposited to join powder particles	Binder jetting (BJ)	Powder & chemical/mechanical	Metals, sands, polymers, ceramics, composites, titanium, aluminium, stainless steel, tungsten	13–16 μm	Large material database, affordable, Fast, multicolour, create complex parts quickly
Vat photopolymerisation	Selectively curing of photoreactive polymers by using a laser, light or ultraviolet (UV)	Continuous direct light processing (CDLP) Digital light processing (DLP) Digital light synthesis (DLS) Direct UV printing (DUP) Stereolithography (SLA)	Liquid & laser with optical	Graphene oxide, polymer resin, biomaterials paste, ceramic resin, bio-resin, Inconel paste, photosensitive resin, polymer, iron oxide	10 μm	High-dimensional accuracy, high resolution, fast, no nozzle, smooth surfaces
Cold spray additive Manufacturing	After being driven in a compressed gas stream with high velocity, fine powder particles form a layer by bonding and deforming upon impact with a substrate.	Cold spray additive manufacturing	Powder & compressed gas	Titanium, invar, aluminium, copper	Data not available	Researches are still in early stages
Friction based	Fabricate parts in solid state below the melting temperature.	Additive friction stir deposition (AFSD) Friction stir additive manufacturing (FSAM)	Friction stir with material feeding	Aluminium, metals	From ~ μm to ~ nm	Researches are still in early stages

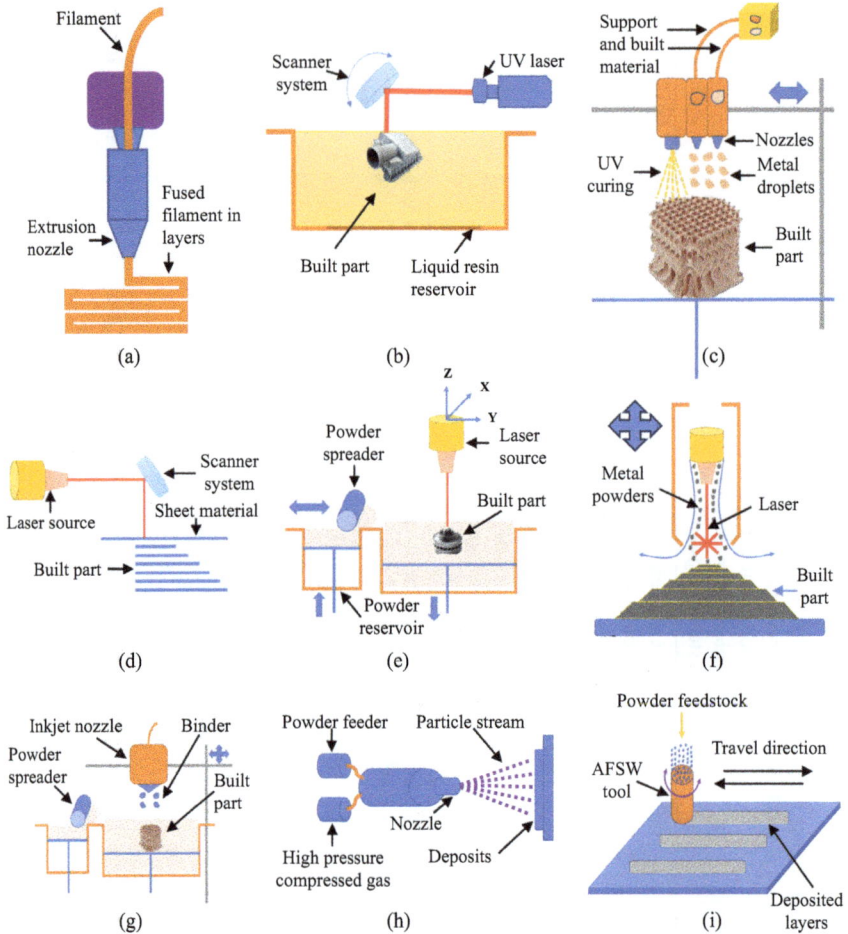

Figure 9.2   AM methods: (a) material extrusion; (b) vat photopolymerisation; (c) material jetting; (d) sheet lamination; (e) powder bed fusion; (f) direct energy deposition; (g) binder jetting; (h) cold spraying; and (i) additive friction stir deposition [11]

employs removable support materials for easy extraction, while powder bed fusion methods leverage the surrounding unbound powder as a natural support medium. AM is continually evolving, supported by ongoing research that broadens the range of materials utilised in the process. Currently, five primary categories dominate the landscape: polymers, which offer versatility and lightweight options; metals and alloys, known for their strength and durability; ceramics and glass, valued for their heat and chemical resistance; cement and concrete, which are crucial for construction applications; and biomaterials, which are increasingly significant in medical and environmental fields, each bringing distinct properties and potentials to various applications [13]. A list of common materials used in AM is provided in Table 9.5.

Table 9.5 Common materials used in AM [11]

Metals	Polymers	Composites	Ceramics	Biochemicals	Sand
Aluminum	Acrylonitrile butadiene	Bio-composites	Alumina	Alginate	Ceramic sand
Bronze	High- and Low-density	Carbon fibre reinforced polymers	Glass ceramics	Chitosan	Chromite sand
	Polyethylene				
Carbon steel	Polyacrylamide	Ceramic matrix composites	Hydroxyapatite	Collagen	Olivine sand
Cobalt chrome	Polyamide	Conductive composites	Porcelain	Fibrin	Silica sand
Copper	Polycarbonate	Glass fibre reinforced polymers	Powdered glass	Gelatin	Zircon sand
Ferrous	Polyetherimide	Hybrid composites	Silicon carbide	Hyaluronic acid	
Gold	Polylactic acid	Metal matrix composites	Titanium dioxide	Lignin	
Magnesium	Polypropylene	Nanocomposites	Tricalcium phosphate	Polyethylene glycol	
Molybdenum	Polyvinyl alcohol		Zirconia		
Nickel alloy	Thermoplastic				
	polyurethane				
Platinum					
Silver					
Stainless steel					
Titanium					
Tungsten					
Zinc					
Zirconium					

Note: A combination of metal and ceramic are used in AM in the form of 'Cermet'.

9.2.2   Multi-material AM

Considered a keystone for industry 4.0, AM has transformed the industrial sector. AM has been widely embraced by a number of companies in the electronics, automotive, aerospace and medical sectors. It became well known when it was possible to create intricate 3D graphics layer by layer using just one material. Producing 3D structures with multi-material qualities is currently in demand due to the manufacturing industry's ongoing evolution. This means creating a single 3D object with several material properties in certain regions. With the use of multi-material additive manufacturing (MMAM) technology, objects with the required qualities may now be produced in key places according to the functional needs of the finished items [14,15].

International Organization for Standardization (ISO) with the American Society for Testing and Materials (ASTM) defined MMAM as 'a layer-by-layer fabrication technique that intentionally modifies process parameters and gradationally varies the spatial of material(s) organization within one component to meet intended function'. When two distinct materials are utilised, a hybrid AM 3D component is created; when more than two materials are used, an MMAM 3D component is created. The ability to manage the variation of material components was made possible by MMAM technology, which represented a paradigm shift away from conventional manufacturing techniques [16,17].

9.2.3   Combining dissimilar materials in AM

One frequent and traditional technique for combining different materials is welding. It has been suggested that several welding processes, including friction stir welding, arc welding, high energy beam welding, resistance spot welding and ultrasonic welding, be used to fuse not compatible metals. Although welding processes are economical, there are several significant problems with them when combining incompatible materials. Among these problems include the development of detrimental intermetallic compounds (IMC) and the existence of a sizable heat-affected zone (HAZ) at the interface, which can cause cracking and poor interface bonding. In an effort to address these problems, researchers are now focusing on AM. In contrast to conventional techniques, AM builds structures layer by layer, exhibiting exceptional potential for unparalleled design freedom and rapid production timelines. This technology makes it possible to fabricate complex structures with near-net form accuracy, which is frequently not possible with other techniques. Regulating the heat input and powder composition during dissimilar metal joining is one of AM's key benefits, as it helps to optimise the mechanical characteristics of the interface by limiting the creation of detrimental IMCs. Distinct metals have been joined using various AM processes. These include friction stir additive manufacturing, ultrasonic additive manufacturing, laser metal deposition, electron beam powder bed fusion and WAAM [18]. The list of similar and dissimilar metals used in AM are mentioned in Table 9.6.

Table 9.6 Dissimilar and similar metals used in AM [19]

Dissimilar metals	Similar metals
Inconel alloy 625 (IN625) and cobalt chrome molybdenum (CoCrMo)	Aluminium and aluminium
Inconel and cast iron	Nickel and nickel
Inconel and cobalt nickel chromium aluminium yttrium (CoNiCrAlY)	Steel and steel
Inconel and copper	Titanium and titanium
Nickel and titanium	
Nickel titanium (NiTi) and copper	
Nickel titanium (NiTi) and titanium aluminium vanadium alloy (Ti6AI4V)	
Stainless steel and titanium aluminium vanadium alloy (Ti6AI4V)	
Steel and aluminium	
Steel and copper	
Steel and Inconel	
Titanium aluminium vanadium alloy (Ti6AI4V) and nickel chromium alloy (IN718)	
Titanium aluminium vanadium alloy (Ti6AI4V) and tungsten (W)	
Titanium and aluminium	
Titanium and cobalt chrome molybdenum (CoCrMo)	
Titanium and copper	
Titanium and tantalum	
Tungsten and copper	
Tungsten and steel	
Tungsten carbide cobalt (WC-Co) and thermocouples	

9.2.4 Developments in AM

- **Multi-material AM** – MMAM allows for the creation of items with a range of properties and materials, leading to the production of increasingly intricate and practical items of value [20].
- **Large-scale printing** – Thanks to advancements in technology and software, larger items, such as construction materials, parts for airplanes and even entire vehicles, may now be printed. New developments in AM have led to the release of printers with larger build volumes [21].
- **Improved speed and efficiency** – A new age of quicker print speeds has been brought about by the development of numerous distinctive AM methods. Modern printing techniques like SLA, DLP and CLIP (continuous liquid interface production) have notably increased the rate at which items are created [22].

- **Improved precision and accuracy** – Significant improvements in technology, software and materials allowed AM to operate more precisely and accurately, allowing for the production of components with complex geometries and tight tolerances. With remarkable detail, items with complex forms and precise dimensions may now be created. One of the most important factors in achieving perfect outcomes throughout the printing process is precise motion control [23].
- **Material choices** – AM can now work with a variety of materials when compared with the early stages thanks to the introduction of new ones and the improvement of existing ones. As AM materials metals, polymers, ceramics and composites are all utilised [11,24].
- **Withstand high temperature** – The ability of AM technology to operate with high-temperature materials such as metals and ceramics opens up new possibilities for the automobile and aerospace sectors. The development of new high temperature alloys for titanium, nickel and steel has led to a significant advancement in metal additive manufacturing in recent years. These alloys are used in energy and aerospace applications for turbine blades and exhaust systems because they can withstand temperatures of up to 1500 °C. High temperature alloys, such as nickel-based alloys that can withstand temperatures of up to 1200 °C, have been made possible via the powder bed fusion process [25].
- **Bioprinting** – Developments in bioprinting make it possible to create 3D-printed organs and tissues from living cells, which might revolutionise the medical sector. Researchers are developing new printing materials that are more suitable for usage with living tissues and cells. For example, hydrogels made of organic materials such as collagen and alginate are used as printing substrates because they closely mimic the properties of biological tissues [26].
- **Automation** – AM is becoming more automated with the use of robots and artificial intelligence, which lowers human error and boosts production efficiency and speed. The AM process is being automated with the development of robotic technologies. These devices are capable of loading and unloading parts, orienting and positioning them for printing and carrying out additional duties including quality assurance and inspection. The AM process is being optimised with AI [27].
- **Sustainability** – Compared to traditional production methods, AM uses less material and produces less waste, thus it's becoming a more ecologically responsible choice. Scholars are exploring the use of recycled materials, including metals, polymers and other materials, in AM. Additionally, they are constantly working on creating biodegradable materials, especially for AM [28].

9.3   Recent trends and future scope of AM

9.3.1   Hybrid AM

Despite its benefits, AM is not always the best option for certain industrial applications because to its part size restrictions and more expensive and time-consuming construction procedure. Consequently, AM and conventional manufacturing can be integrated using hybrid manufacturing (HM), a technique that combines two manufacturing processes, to maximise the benefits of each strategy while solving its drawbacks. The synergistic fusion of AM method with other procedures or sources of power in a parallel or successive arrangement is known as hybrid additive manufacturing (HAM). HAM is essentially a sequential process that involves either the additional application of an AM process on a prefabricated substrate or subtractive manufacturing to an AM product, particularly when mixing additive and subtractive processes [19]. HAM is used to apply subtractive operations to the AM part in order to either accomplish geometric perfection or enhanced surface finish or to create a complicated geometry at a lower cost or manufacturing time.

9.3.2   Support structures

Solid, tree-like and other types of structures are examples of general support structures. Distinct printing and sintering procedures in ceramic additive manufacturing present difficulties for the removal of solid supports; also, depending only on a single lattice or tree support might not provide adequate backing. We categorised support solutions into contact supports, contactless supports, support-free techniques and combined supports based on the type of support, material morphology, manufacturing process and the interaction between components and support by compiling the literature on supports used in ceramic additive manufacturing [29].

Contact support – A straight connection between the component's surface and the support structure is made possible by the contact support. Solid blocks, grids, shells, point pillars and tree shapes are examples of common contact supports. This kind of support is commonly used in the majority of component printing methods.

Contact less support – The precision of the surface may be damaged by removing contact supports. The component can be partially supported in the SLA or SLS process when a support bath, such as a ceramic slurry or powder bed, is utilised. Researchers have discovered that it is possible to guarantee component printing while enabling support removal by keeping a relatively narrow support space between the component and support. We refer to this kind of support as a contactless support.

Support free techniques – The design and installation of support structures are required regardless of the kind of support – contact or contactless – which results in the consumption of raw materials. To do 3D printing without extra supports, two methods can be used. In the first, the component's structure is altered to create what are referred to as self-supporting structures, which enable the component to support itself. Using the qualities of the raw material, such as the powder bed and slurry/paste, or other media, including magnetic fields, as so-called non-structural supports, the second method involves printing without the need for structural supports.

9.3.3   4D printing

This process is a 3D printing variation that uses smart materials (SM) that may change on their own in response to mechanical, electrical, magnetic, thermal or other environmental stimuli (see Figure 9.3). The addition of time is the fourth dimension (4D) that causes a 3D-printed item to alter in response to environmental changes. A clever design is made to program SM's behaviour in response to a specific stimulus. This design facilitates comprehension of the product's form change, allowing consideration of the printing process. The object that is printed has multiple shapes, each having distinct qualities that can be changed in response to a stimulus, hence altering its physical form or attributes. '4D' stands for temporal dimension, indicating that the stimulus causes the object to change shape after printing. The AM printing technique, material, stimuli, modelling and interaction mechanism are the five primary components of the 4DP process [30].

9.3.4   5D printing

This technology made it possible to precisely print 3D items with curved or concave shapes. The printer technique was achieved by incorporating two more axes: the extruder head and the printing platform rotation (see Figure 9.4). This novel approach was named 5DP. Similar technologies are used by 3D and 5DP processes, such as the use of 3D CAD files and similar printing supplies. Pre-printing, printing and post-printing are the three stages of standard 5DP operations, each of which needs to be verified and examined for model conformity in order to guarantee the intended results. Analysing the printing process, creating trajectories, needs, model orientation and digital model conformance are important factors [30]. The materials used and the printing technologies used in 4D and 5D printing are listed in Table 9.7.

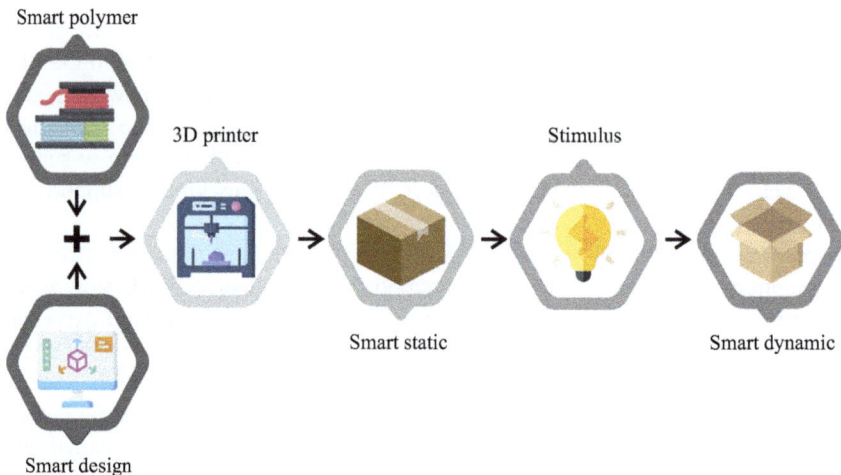

Figure 9.3   4D printing mechanism [31]

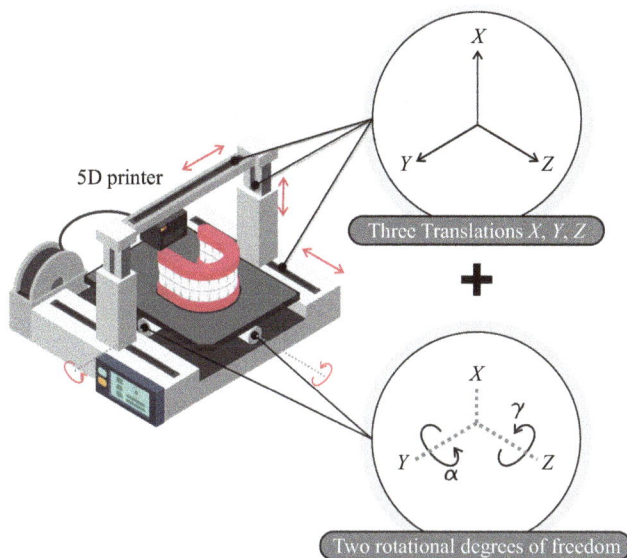

Figure 9.4   5D printing mechanism [30]

Table 9.7   General materials used in 4D and 5D printing [30]

Trend	Technology	Material used
4D printing	Material extrusion	Polylactic acid, iron oxide, decylamine (DA), bisphenol A diglycidyl ether (BDE), graphene, nanoplatelets, n-butylamine, poly (propylene glycol) bis (2-aminopropyl ether) (PEA), diacrylate 1,4-bis-[4-(6-acryloyloxyhexyloxy) benzoyloxy]-2-methylbenzene (DBBM), 2-benzyl-2-(dimethylamino)-4-morpholinobutyrophenone (BDM)
	Vat photo-polymerisation	Polycaprolactone (PCL), soybean oil epoxidised acrylate (SOEA), acetone, bis(2,4,6-trimethylbenzoyl)-phenylphosphineoxide (BTMP), carbon porous, nanocookies, Irgacure 819
5D printing	Material extrusion	Alginate
	Vat photo-polymerisation	Researches are ongoing
6D printing	Researches are ongoing	Researches are ongoing

9.3.5   6D printing

The consequence of combining 4DP and 5DP is this novel printing technique (see Figure 9.5). The final product is produced using five degrees of freedom; thus, the

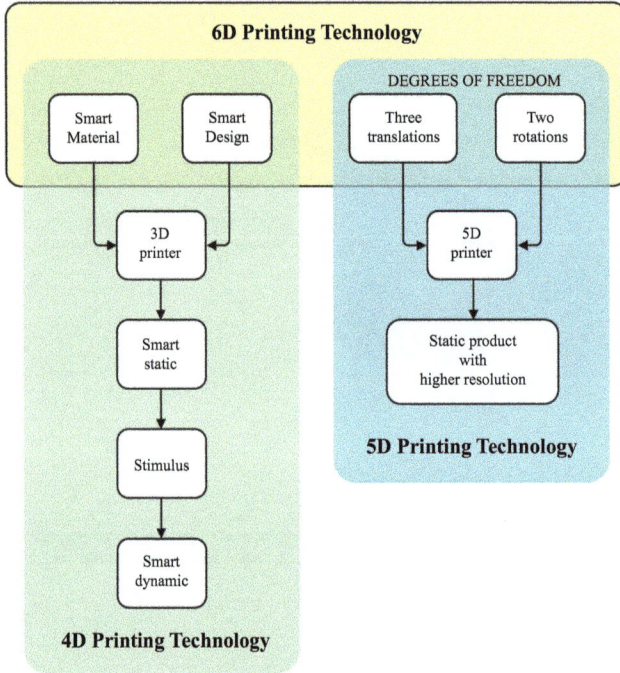

Figure 9.5   6D printing mechanism [30]

designer must come up with a clever design that takes shape memory (SM) features and modifications into account. This material is able to change its properties or shape in reaction to outside stimuli. In addition to conserving time and materials during printing, the things created using this technique should be stronger and of a higher calibre than those made with the previously described procedures. Additionally, it is anticipated that 6DP-printed buildings and objects would behave intelligently and have distinctive stimuli response characteristics. By integrating computational optimisation techniques, the materials' intelligence can be enhanced.

9.4   AM in healthcare

Printed tissues and organs – The bioprinting of tissues and organoids is one of the main applications of 3D printing in medicine. Bioprinters employ a computer-guided pipette to layer living cells, or bio-ink, on top of one another to construct fabricated living tissue, as opposed to printing with plastic or metal. The subject of 3D bioprinting tissues and organoids is one that is developing quickly and has tremendous possibilities for both drug research and regenerative medicine [32].

   AM in surgery preparation – In order to prepare doctors for complex surgical procedures, 3D printing is also used to create duplicates of organs unique to each patient. The original digital model used in the 3D printer is created using medical imaging techniques. This method works well to expedite surgical procedures while reducing patient trauma. It's already been effectively used in a number of surgical applications and its popularity is growing [33].

AM in surgical instruments – Tools that are sanitary, precise and incredibly compact can be produced using 3D printing and used in tiny spaces without harming patients needlessly. If 3D printing is used instead of traditional manufacturing, production costs might also be lower. An emerging field that has the potential to revolutionise surgical tool manufacturing and use is 3D printing of surgical instruments [34].

AM in oral healthcare – Researchers looked at creating biomaterials specifically designed for UV-cured 3D printing in the dental field. The antibacterial and anti-pollution qualities of 3D-printed oral healthcare products have historically shown flaws, which has limited their usefulness. In order to overcome this difficulty, they created antibacterial compounds by combining hydroxyethyl methacrylate, choline chloride and isophorone diisocyanate to create quaternary ammonium groups. Subsequently, these agents were combined with diluents, resins and photoinitiators to provide raw materials appropriate for UV-curing photosensitive resins used in 3D printing. The resins were then 3D printed using digital light processing (DLP) in order to cure. Testing for both mechanical and antibacterial properties was done on the final 3D-printed material. Our results show that the 3D-printed compounds showed remarkable hardness, tensile qualities, antimicrobial properties and long-lasting antimicrobial durability after being cured using this method. These advancements hold out the prospect of broad acceptance and use in the realm of oral healthcare [35].

AM in splints – Finger splinting is the standard conservative treatment for tendinous mallet finger injuries; it immobilises the affected finger in an extended position to promote healing of the ruptured extensor tendon. Present-day splints, such as the Zimmer and Stack, have a high failure rate of about 50%. The reasons for non-compliance with splint regimens are linked to discomfort, skin problems and poor splint fit. Researchers created and constructed a customisable, adjustable finger splint that is 3D printed in order to address the aforementioned difficulties. The assessment of the finger extension angle, splint fit, splint comfort and skin maceration was conducted using angle measurement and questionnaire response from the individuals. The distal interphalangeal joint may be kept at an extended angle of 8.1 degrees with the help of a 3D-printed finger splint. Their design makes it easier to use and accomplishes the purpose of holding the fingertip extended [36].

AM in orthosis – Although cast immobilisation is beneficial for fractures, there are a number of drawbacks. Consequently, 3D-printed orthoses have been presented as a possible good substitute. Researchers investigated the idea that, in short-arm casts for the immobilisation of distal radial fractures, 3D-printed polylactic acid (PLA) orthoses might offer better biomechanical qualities than plaster of Paris. Human models were used to create the circular casts and the 3D braces using modified mannequin forearms. Five 3D-printed braces and five traditional plaster casts were developed which are used to treat distal radius fractures. After that, each sample was put through a three-point bend load test on an Instron 68SC2 testing apparatus. When compared to plaster of Paris short-arm casts intended for the immobilisation of distal radial fractures, 3D-printed polylactic acid short-arm orthoses showed better biomechanical qualities. Preclinical evidence, when combined with data from earlier research, indicates that PLA 3D-Braces may preserve distal radius fracture alignment and stability with potential benefits over traditional casts in terms of biomechanical characteristics, post-fabrication modification, patient sanitation, convenience and everyday activities [37].

AM in prosthetics – Prosthetic limbs made specifically for the wearer are another product of 3D printing. Amputation victims must typically wait a long period to receive prosthetics through the conventional method. With less time and money, 3D printing produces far more affordable goods that give patients the identical or enhanced features. Particularly for kids, whose prosthetic limbs can become antiquated quickly, the price tag is appealing. With the use of 3D printing, prosthetics can be made into incredibly individualised and customised items that are suited to each person's specific requirements and anatomy. This enhances the prosthetic device's fit and functionality, improving the patient's mobility and quality of life [38].

AM in simulation – It is a medical emergency to treat compartment syndrome. Prompt diagnosis is imperative and therapeutic measures must be implemented to prevent limb ischemia. It is crucial to measure the pressure inside each compartment. Accurate diagnosis of acute compartment syndrome requires expertise with the pressure monitoring device and an understanding of compartments. Through simulations, one can become familiar with the tool and put the process into practice. 3D-printed material simulations are becoming more and more common due to their cheaper cost and potential for frequent replication. Using syringes, 3D-printed tibia and fibula, silicone-based lower leg soft tissue, a Foley catheter and syringes, a team of researchers presented a basic low-fidelity model. With frequent practice, their low-fidelity simulator helps to enhance procedural skills and recall of information [39].

AM in drugs – Printing pharmaceuticals are arguably one of the most fruitful applications of 3D printing. With the use of this technology, customised medications with exact dosages catered to the need of each patient are possible. The flexibility to alter dosage and formulation is a significant benefit as it can enhance therapeutic efficacy and lower the chance of adverse responses. The pharmaceutical sector could undergo a transformation thanks to the speedy creation of individualised medications made possible by 3D printing. It can be utilised to produce exact dosages and specially formulated products that are catered to each person's particular need. By guaranteeing that treatments are more efficient and have fewer adverse effects, this can enhance patient results [40].

AM in drug delivery – Complicated drug delivery systems that release medications at precise intervals or to certain parts of the body are also made using 3D printing technologies. The delivery of pharmaceuticals to patients could be revolutionised by the 3D printing of intricate drug delivery systems. It can be utilised to produce extremely accurate and precisely designed devices that are suited to each person's particular medical requirements. By making sure that drugs are given at the appropriate time and amount, this can increase their efficacy and safety. It can cut lead times and expenses related to conventional manufacturing processes, which sometimes entail costly and time-consuming specialised fabrication [41].

AM in medical models – Medical teaching models can also be created via 3D printing. Preparing patient-specific models for surgical preparation and training is made possible with 3D printing, as was previously indicated. All of the medical models that are now utilised in education might be improved upon with the use of 3D printing. Using 3D printing, complex and detailed models that are useful for training and education can be produced. Anatomical models that can be used to instruct students on medical procedures and human anatomy can be produced using this technology [42].

Intelligent materials and biological structure in AM – Intelligent materials and biological structure are combined by AM to create intelligent bionic structures that can self-assemble, self-drive or self-mending. We refer to this cutting-edge manufacturing technique as 4D printing. Smart bionic structures respond to external stimuli like temperature, humidity, light, electricity, pH, sound, mechanical force and magnetic fields by simulating biological responses that result in material transformations. Through interdisciplinary efforts, smart bionic structures have been created, offering insightful knowledge and experience for innovation across a range of fields [43].

9.4.1   Organ-on-a-chip

In the last few years, there has been a significant increase in the use of 2D cell cultures and animal models for disease modelling, drug discovery and in vivo investigations of the physiological processes and functions of organs. However, due to substantial discrepancies in their physiological functions, 2D cell cultures are unable to replicate 3D human physiological models. Animal experimentation is also expensive, time-consuming, unethical and the data it produces is not always accurate enough to anticipate outcomes from human clinical trials. Preclinical models tailored to humans have been developed in an attempt to address these issues. Organ-on-a-chip (OOC) technology is a viable approach to the in vitro development of an organ's physiological microsystems. Using materials science, microfluidics and tissue engineering, the method replicates the exchange process of substances between the organ and external environment. Nutrients and medications are precisely infused using microfluidic chips. Furthermore, it is possible to cultivate cells and replicate the microenvironment of human organs using biomimetic tissues that have an extracellular matrix (ECM) and other biochemical gradients in OOC [44]. Organ on a chip technique is used in:

Heart-on-a-chip – Heart-on-a-chip has been widely investigated in recent decades as a potent tool to mimic the microenvironment of cardiac tissues in vitro and evaluate both contractility and electrophysiology. These implanted sensor devices can be employed for harmless electronic readouts of tissue mechanical stresses within cell incubator conditions, drug response studies and guiding the self-assembly of physiomimetic laminar cardiac tissues [45].

Vessel-on-a-chip – The utilisation of vessel-on-a-chip in 3D ECM micro-environments has been extensively studied due to its potential for studying inter-cellular communication, micro-scale fluid dynamics and biological chemical delivery at the tissue level. With an outside layer made of smooth muscle and an inner layer made of endothelial cells, the bio-printed vascular construct demonstrated superior vasoactivity, appropriate biomechanics, favourable perfusability, permeability and the capacity to produce vascularised tissue [46].

Lungs-on-a-chip – Lungs-on-a-chip have been widely used to assess respiratory disorders linked to pollution, tobacco use, infections, nanoparticles and medications. They have also been used as a potentially ground-breaking and precise respiratory model paradigm to close the knowledge gap between cell culture and preclinical trials. The lung tissue was exposed to outside air in the bio-printed models, which successfully assured culture medium perfusion to the tissues and supplied air-liquid interface culture conditions [47].

Intestine-on-a-chip – Intestine-on-a-chip can be utilised to reproduce natural 3D intestinal cell and tissue morphology or reinstate other important intestine differentiated functions, in addition to studying barrier function and modelling drug absorption in a human intestinal epithelial cell line. It is possible to 3D print sensors in addition to tissues [48].

Liver-on-a-chip – The liver, which is the body's largest solid organ, keeps blood sugar levels in check, filters pollutants out of the blood and controls blood coagulation. The liver-on-a-chip has proven to be a useful tool in recent decades for simulating the operation of the human liver, which has aided in the creation of medications that treat liver ailments. A liver-on-a-chip model demonstrated improved liver-specific gene expression and an effective medication response when compared to a chip without the biliary system [49].

Pancreas-on-a-chip – With the use of the pancreas-on-a-chip, researchers can examine the endocrine portion of the organ on a microfluidic chip, which is intended to serve as a uniform platform for assessing the potency and quality of islets [50].

Kidney-on-a-chip – The kidneys, being the main components of the urinary system, play a vital role in regulating the volume of different bodily fluids, acid-base equilibrium, elimination of toxins, fluid osmolality and concentrations of different electrolytes. A kidney-on-a-chip using in vitro renal unit micro-tissue is getting a lot of attention across a variety of sectors since it can mimic and replicate the fundamental physiological traits and activities of the kidney [51].

Brain-on-a-chip – The blood–brain barrier (BBB) serves as a physiological barrier that separates the peripheral blood circulation from the central nervous system. As a result, a lot of work has gone into creating brain-on-a-chip systems that precisely replicate the BBB microenvironment and build medication delivery systems with precisely defined physicochemical properties of the brain [52].

Together with the structures already described, additional OOC models, such as bone-on-a-chip, breast-on-a-chip and skin-on-a-chip models, have also been created using 3D printing [53]. The skeletal system is a persistent venue for the spread of several malignancies, including lung, breast, prostate and melanoma

Table 9.8   Commonly used AM technologies and materials in healthcare [73]

Technologies	Printing resolution	Materials used for bone printing
Electron beam melting (EBM)	50–100 µm	Titanium aluminium vanadium alloy
Extrusion based printing (EBP)	20–100 µm	Tri calcium phosphate and Sodium alginate composite with silver nanoparticles, β-Tri calcium phosphate and polycaprolactone blend, Polylactic acid and hydroxyapatite blend
Inkjet printing (IJP)	20–100 µm	Hydroxyapatite, β-Tri calcium phosphate and Calcium sulphate blend, Vascular endothelial growth factor-loaded gelatine, alginate and β-tricalcium phosphate composite
Multi-head deposition system (MHDS)	100–200 µm	Poly-ε-caprolactone, growth factors encapsulated in hydrogel, human dental pulp stem cells, bone morphogenic protein-2 and vascular endothelial growth factor, Poly-ε-caprolactone, poly (lactic-co-glycolic acid), β-tricalcium phosphate and human nasal inferior turbinate tissue-derived mesenchymal stromal cell-laid extracellular matrix
Poly-jet printing (PJP)	30–40 µm	VeroBlackPlus and TangoPlus blend
Selective laser melting (SLM)	20–50 µm	Titanium, Stainless steel, Polyamide, Titanium aluminium vanadium alloy
Selective laser sintering (SLS)	50–100 µm	Poly-ε-caprolactone, Poly-ε-caprolactone and hydroxyapatite blend, Titanium aluminium vanadium alloy
Stereolithography (SLA)	70–300 µm	45S5 Bioglass powder, acrylate-based monomer, Nanoscale hydroxyapatite and polyamide blend, Hydroxyapatite and polyvinyl alcohol blend, Calcium carbonate and tricalcium phosphate with phosphoric acid polypropylene glycol and a light absorber and photoinitiator blend
Three-dimensional powder printing (3DPP)	40–50 µm	Hydroxyapatite and Tri calcium phosphate blend, Titanium aluminium vanadium alloy, Corn starch and Gelatine blend

carcinomas. For this reason, bone-on-a-chip can help research both bone defects/fractures and these tumours. The technologies and materials used in the medical AM are provided in Table 9.8.

9.5   Applications of AM in energy

Advanced manufacturing plays a crucial role in addressing climate change by enabling the rapid and scalable deployment of clean energy technologies like nuclear, solar, wind and energy storage [54,55]. Unlike traditional manufacturing methods that are labour and capital-intensive and lack flexibility, advanced manufacturing techniques are designed to be labour-saving, adaptable and quick to implement. This adaptability is vital to transforming the global energy sector, which requires significant shifts in production capabilities within the next decade or two to meet the urgent demands of climate mitigation and transition towards sustainable energy sources. AM technologies are gaining traction for their ability to significantly lower energy consumption associated with traditional manufacturing processes by minimising material waste and reducing the need for extensive machining. Literature suggests that their broader adoption could decrease global energy demand by up to 27% [56]. Recently, AM has been increasingly utilised across different energy sectors, enhancing material performance and energy efficiency, and positioning it as a key solution for the future of energy generation, conversion and storage. Indeed, AM technologies are revolutionising the energy sector by enabling the production of optimised components that enhance efficiency and performance. Their ability to create complex geometries allows for lighter and more durable parts, which is particularly beneficial in applications such as renewable energy generation and storage systems. As industries continue to embrace these innovations, AM is poised to play a crucial role in advancing sustainable energy solutions and reducing overall environmental impact. This chapter overviews AM technologies and their specific applications within the major energy domains – nuclear energy, batteries and fuel cells. While acknowledging AM's presence in other renewable energy sectors like wind, solar and hydroelectric power, the focus remains on critical energy consumption and storage systems. It emphasises the necessity for rigorous validation and qualification of AM processes to ensure the reliability of energy systems. Additionally, the chapter outlines future research directions to enhance the integration and efficiency of AM technologies across these essential energy sectors.

9.5.1   AM in nuclear energy

Nuclear energy is increasingly exploring AM to enhance design flexibility, reduce costs and improve production timelines for power plant components, despite AM being a newer approach compared to established methods like powder metallurgy. AM offers significant advantages, including real-time process monitoring, which enhances manufacturing control and provides valuable data on component properties. Additionally, AM allows for smoother transitions between materials and the creation of functionally graded compositions, optimising the microstructure-properties relationship more effectively than traditional joining techniques such as welding or brazing [57]. Material transitions in nuclear plants necessitate the use of diverse materials due to fluctuations in pressure, temperature and radiation exposure. AM has proven effective in producing graded compositions across a variety of materials, including Inconel, copper and titanium alloys, tailored for specific applications [58]. Advances in modelling and simulation have further facilitated the development of these functionally graded materials, enhancing the potential for 3D-printed micro composites to meet the demanding conditions in nuclear environments.

9.5.2 AM in batteries

AM presents a transformative approach to manufacturing lithium-ion batteries (LIBs) by allowing for the integrated production of complete batteries with packaging in one step, circumventing the need for assembly or secondary processing. This streamlined process not only reduces production time and minimises waste but also maximises material usage. Furthermore, AM enables the creation of intricate 3D structures that facilitate improved lithium-ion diffusion, potentially enhancing the overall electrochemical performance of LIBs, and thus driving advancements in energy storage technology.

Reyes *et al.* [59] in 2018 successfully fabricated a complete LIB) using fused filament fabrication (FFF), incorporating distinct filaments for the electrodes, a hybrid electrolyte and current collectors. They developed a hybrid electrolyte by infusing polylactic acid (PLA) with $LiClO_4$ to enhance ionic conductivity, achieving an increase from 8.2×10^{-11} mS cm^{-1} to 0.085 mS cm^{-1}, although this remained below the values of liquid electrolytes. Additionally, they employed lithium titanate (LTO) as the negative electrode with graphene to improve conductivity and lithium manganese oxide (LMO) for the positive electrode with multi-walled carbon nanotubes (CNTs). However, despite their innovations, the electrodes exhibited low capacities, yielding only 3.34 mAh cm^{-3} for the negative electrode and 6.99 mAh cm^{-3} for the positive electrode, significantly lower than the theoretical capacities of their active materials (600 mAh cm^{-3} for LTO and 596 mAh cm^{-3} for LMO). The comparison of single-print and assembled LIBs revealed that the single-print LIB, infused post-printing with $LiClO_4$-salt solution, exhibited significantly lower capacity (1.16 mAh cm^{-3}) than the assembled LIB (3.91 mAh cm^{-3}), likely due to insufficient electrolyte infusion into the electrodes. The use of high PLA content in the electrode filaments, while necessary for mechanical properties and printability, inhibited electrical contact between conductive particles and active materials, contributing to low capacities and notable irreversible capacity loss from solid electrolyte interphase (SEI) formation during cycling. This indicates that while Fused Filament Fabrication (FFF) can produce functional LIBs in a single step, the electrochemical performance is compromised by high polymer content and elevated electrical resistivity compared to conventional LIBs.

Wei *et al.* [60] in 2018 developed a custom direct ink writing (DIW) printer to create semisolid electrodes and separators for LIBs, utilising glassy carbon substrates as current collectors. They produced thick biphasic electrodes with lithium iron phosphate (LFP) as the positive material and LTO as the negative, incorporating Ketjen black as a conductive additive. The electrode inks, which combined active materials and conductive particles with LiTFSI and PC solution, were optimised for printability through careful adjustments to their rheological properties and the inclusion of a non-ionic dispersant to prevent flocculation. The study by

Wei *et al.* demonstrates a successful method for creating a fully 3D-printed LIB using components like UV-curable epoxy and ethoxylated trimethylolpropane triacrylate layers, which were printed in sequence on a glassy carbon substrate and cured with UV light. Despite achieving a respectable areal capacity of 4.45 mAh cm^{-2} and a minimal capacity loss after two cycles, the LIB exhibited lower capacity than a non-packaged version at 14.5 mAh cm^{-2}, indicating that while DIW is viable for LIB fabrication, the necessity for UV-curing complicates the efficiency of the process due to the specific ink compositions used.

Maurel *et al.* [61] in 2019 demonstrated an innovative approach to fabricating a complete LIB using a FDM printer, where they achieved layer-wise production by varying the filament materials for the negative electrode, separator and positive electrode. Following the AM process, the printed structure was infused with a liquid electrolyte, showcasing a potential advancement in the efficiency of battery production. In the study by Maurel *et al.*, graphite-PLA and LFP-PLA filaments were developed for use as negative and positive electrodes, respectively, in energy storage applications. The incorporation of carbon black as a conductive additive enhanced the conductivity of both electrode materials, while adding poly (ethylene glycol) dimethyl ether as a plasticiser allowed for high loading of active materials, achieving 49.2 wt.% graphite in the negative electrode and 49 wt.% LFP in the positive electrode, thereby optimising the overall performance of the electrodes. The use of plasticisers in PLA matrix filaments enhances ductility and printability, allowing for a higher active material content in the SiO$_2$-PLA composite separator, which improves liquid electrolyte uptake. Despite achieving a capacity of 30 mAh g^{-1} for the single-print battery, the electrochemical performance was hampered by the limitations of the FDM printer resolution, necessitating a layer thickness greater than 200 μm. This increased thickness adversely affected Li-ion and electron transport, resulting in a lower capacity of 15 mAh g^{-1} when the cell was assembled. Consequently, the study emphasises the critical role of printing resolution in AM and its impact on battery efficiency.

9.5.3   AM in fuel cells and electrolysers

Fuel cells serve as efficient energy converters by transforming chemical energy from fuels into electricity through electrochemical reactions, with AM technologies enhancing their component fabrication. Specifically, biological fuel cells, including microbial and enzymatic varieties, have benefitted from AM techniques that produce high-performance components such as chambers, membranes and electrodes designed for optimal biocompatibility and conductivity. Studies have shown that AM can significantly improve the efficiency and design of microbial fuel cells (MFCs), leading to innovations like adaptable designs and superior electrode materials, while polymer electrolyte membrane fuel cells are positioned for various applications ranging from automotive to stationary energy systems due to their high efficiency and clean byproduct profile. Overall, AM methods streamline the manufacturing process of fuel cells, making them more viable for both practical use and

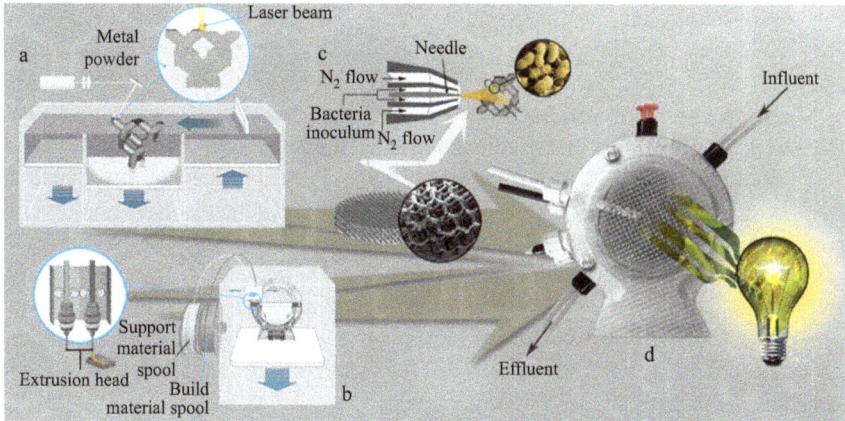

Figure 9.6 *AM techniques for microbial fuel cells (MFC). (a) Selective laser melting (SLM). (b) Fused deposition modelling (FDM). (c) Anest-Iwata spray gun for bacteria spray-coating. (d) MFC assembly [65].*

advanced applications. AM has emerged as a transformative technology for fabricating components in fuel cells and electrolysers, allowing for enhanced design flexibility and efficiency. A schematic representation of AM in MFC cell is presented in Figure 9.6.

Research by Bourell *et al.* [62] in 2011 highlighted the influence of carbon fibre content on the strength and electrical conductivity of printed graphite bipolar plates (BPs), while porous metal/carbon foams demonstrate desirable properties like high porosity and effective thermal/electric conductivity for fuel cell cathodes. Various AM techniques, including electron beam melting (EBM) and DMLS, have been employed to create intricate structures such as titanium-alloy BPs with optimised flow channels and micro-fuel cells designed for portable applications. Additionally, advancements in fabricating polymer composite plates that can be converted to carbon plates open new avenues for microfluidics and fuel cell technology, showcasing AM's potential in the energy sector. Mo *et al.* [63] in 2016 utilised EBM to create perforated titanium plates as porous transport layers, highlighting the significance of precise digital design in achieving optimal pore structures (20–100 µm) for efficient liquid supply and gas removal. Simultaneously, Ambrosi *et al.* [64] explored selective laser melting (SLM) and FDM to produce metallic electrodes and components for liquid/gas management in electrolysers.

9.6 Current challenges in AM

Material limitations – There is currently a restricted range of materials that can be used in AM. There are now significantly fewer materials accessible for AM than there are for traditional manufacturing techniques, despite the fact that the diversity of materials is constantly growing. Not every material can be used with every AM method [66].

Strength and compatibility – Two major issues arise when comparing objects made using AM to those made with traditional manufacturing methods: compatibility and strength. The discontinuities and variable layer adhesion introduced by AM's layer-by-layer process affect the mechanical characteristics and overall strength of the final product. Material, dimensional and surface compatibility are among the challenges of compatibility in AM. It's frequently required to post-process, test thoroughly and validate in order to integrate AM components into current systems. This is particularly important in sectors where tight regulations guarantee functionality, safety and seamless integration, including aircraft and healthcare [67].

Cost – Even with larger-scale production runs, many businesses may still find AM to be too expensive. Not only does AM require expensive tools and materials, but it also requires a significant investment of time and knowledge to maintain the technology and software utilised by the machines. The cost of materials is one of AM's main financial constraints. When employed in AM, specialised polymers, metals, composites and other materials can be much more expensive than when used in conventional production methods [68].

Quality control – It can be difficult to maintain consistent quality in AM, especially when working with large or intricate components. differences in temperature, humidity and machine calibration can affect the printing process and result in differences in the final product. The properties of the materials used in AM are influenced by a number of factors, including as the printing technique, the supply of materials and the post-processing phases [69].

Post-processing – For many 3D-printed items to have the necessary functionality and polish, post-processing is necessary. This may entail extra stages like coating, painting or sanding, which would lengthen the production process and increase costs. In order to prevent deformation or collapse, the majority of 3D-printed objects require support structures throughout the printing process, which might be difficult to remove [70].

Intellectual property – AM has the enormous potential to upend established manufacturing procedures and supply networks, posing issues with intellectual property rights and counterfeiting. The protection of intellectual property rights will become a more urgent issue when AM becomes more widely available [71].

Sustainability – The effects of AM on the environment are now being studied and evaluated by researchers. The creation and disposal of 3D-printed items may have an adverse effect on the environment even though AM has the ability to use less energy and produce less trash than existing manufacturing processes [72].

The challenges of AM in relation to IR 5.0 are provided in Table 9.9.

Table 9.9 Challenges in AM and IR 5.0 [74]

	Challenges	Meaning
Social	Change in working environment Lifelong learning	An adaptable workspace surrounded by smart machinery will unavoidably alter the dynamics of the industrial team. Training individuals to use the latest innovations must take place at the same time that technology is adapted to them.
	Managing human resources	The human resources department will encounter new difficulties as robots become more integrated into the business. They will have to determine what jobs will be taken over by robots along with their existing duties.
	Skilled workers	The primary talent shortage and cost limitations in offering sufficient training to those using cobots are the key challenges.
	Trust and accepting technology	The task of constructing sympathetic computers that can operate consistently necessitates substantial progress in cognitive science.
Technical	Access to large amount of data	An obstacle in assessing these devices lies in the vast volume of raw data they generate, which hampers efficient assessment.
	Appearance of cobots	Cobots will exhibit significant variations. Due to the incorporation of human-like capabilities such as grasping, squeezing and interaction influenced by desire and contextual circumstances.
	Cobot maintenance	It is important to ensure the safe maintenance and upgrading of cobots. Safeguarding cobotic information will provide a formidable challenge, surpassing the complexity of safeguarding an organisation's existing IT systems.
	Data analysis	Industry 5.0 is expected to generate a significant amount of data in the areas of Human Machine Interaction and computational human factors, which will require analysis.
	Data quality	High-quality data is essential for the successful implementation of Industry 5.0. By integrating artificial intelligence (AI) and ontologies, engineers may establish complete connectivity between machine tools and enable the creation of objects on the Internet.
	Human centred and personalised AI	AI systems should be customised to accommodate variations in individuals' talents, job preferences, mental and physical health and professional goals.
	Logistics support	Rethinking and reevaluating technical changes is necessary for resilience and sustainability in smart logistics. In this context, the role of man in technological transformation – which is necessary to assure sustainability on the social, environmental and economic fronts – should be the primary focus.
	Performance	It is necessary to create new performance measures for human–machine systems that account for the effort put in and the collaboration shown in embracing the change.

(Continues)

Table 9.9   (*Continued*)

	Challenges	Meaning
	Production	The change from system-oriented manufacturing to human-oriented manufacturing is a challenge for existing production systems.
	Scalability	Scalability, as it relates to Industry 5.0, is the capacity of a system to function under varying operating conditions, regardless of the quantity of hyper-connected systems in the network growing or shrinking.
	Transparency	People must be aware of the reasoning and decision-making processes used by the AI system in order to trust its judgment. It is important for manufacturers of machines to maintain transparency with users on the data they gather in order to make judgments. Users ought to be able to choose which data gathering sources to exclude, knowing the consequences of doing so.
Safety	Access	In order to guarantee that only authorised stakeholders have access to sensitive resources like intellectual property, access control methods must be established in the Industry 5.0 ecosystems.
	Audit	An essential need for evaluating how well service performance aligns with regulatory compliance requirements is the capacity to audit. Investigation of dispute resolution situations is also necessary for audit records.
	Authentication	To build reciprocal confidence in the ecosystem, authentication of a vast array of diverse stakeholders – including IoT nodes, machines, fog nodes, communication nodes and collaborating partner nodes – is essential.
	Collaboration	Humans are returning to the manufacturing plants in Industry 5.0, where they will coexist with cobots. Although this appears to be a successful method of producing customised goods, there are some aspects of human–robot collaboration to take into account.
	Integrity	Integrity is a key problem from an Industry 5.0 data security standpoint since third-party networks will supply control instructions and monitoring data.
	Privacy	In Industry 5.0, machines and humans are connected through data transferred over the Internet. To maintain the confidence of the cloud production ecosystem, such data cannot be viewed by malevolent users on the Internet.
Legal	Regulatory compliance	The key prerequisites for each industrial revolution to really take hold are laws and regulations. Although there are generally available standards for automation, innovation policy and industrial policy, additional particular requirements for the Industry 5.0 age must be upheld.
	Sustainability	Industry needs to take into account social, environmental and technological factors in order to become a provider of true experts. The core of Industry 5.0 is the synergy of three segments: ecological, social and technical.

9.7 AM in ESG

9.7.1 AM and environment

Due to industrialisation, dependency between urban and rural areas, and environmental degradation, rapid population increase in metropolitan areas poses a hazard to the environment. As a result, both the production and consumption of energy have increased quickly. Due to rising levels of consumption, concerns about the effects on the environment, such as air pollution, ground and surface water contamination, and global warming, are mounting. Unusual rural-to-urban migration has led to serious problems for the economy, society and environment. AM and the recycling sector are potential solutions to end the imbalanced reliance between urban and rural areas. The ease of use of home-based businesses in 3D printing may be the primary factor promoting suburban and rural growth. Furthermore, as AM is less reliant on transportation and supply networks, its localisation may be appropriate for rural growth. Particularly in suburban or rural settings, AM-based projects can minimise CO_2 emissions, lead times, logistics and storage costs while improving process handling efficiency [75]. AM tilts the scales in favour of outsourcing over insourcing and in favour of localisation over globalisation.

An increasing number of businesses are striving to increase their ecological footprint as a result of decades-long emphasis on industrial sustainability. In this situation, AM is becoming more popular. Actually, AM is a technology that allows for the very precise fabrication of complex-shaped products with minimal material waste – or at least substantially less than with traditional production methods. The literature highlights the many environmental benefits of common AM technologies, including increased ecological efficiency. Without a doubt, the ability to create with the least amount of energy, raw material and emission waste is a significant benefit. As previously indicated, these advantages are made feasible by the circular economy approach that informs a product's whole life cycle [76].

9.7.2 AM and society

In recent decades, relocation from village to towns and cities has caused issues that have affected economies, societies and ecosystems. Above all, the biggest threat to sustainability may be relocation brought on by poverty. The UN Human Settlements Program has therefore recommended that migrant flow be critically examined. Temporary labour sourcing may meet demands for labour quickly, but often leaves difficulties behind. For example, there has been no discernible improvement in the worldwide rural population ratio over the past 20 years, despite a 65% decline. Instead, it has exacerbated crime, disrupted the ecological balance, destroyed rural development, harmed businesses and spawned slums and homeless. Segregation and an imbalance in the population between urban and rural areas have resulted in extreme poverty, which has a negative influence on society, the environment and global economy. Since health, education and standard of living are all equally weighted, the multidimensional poverty index (MPI) is frequently used to gauge how well people are living in rural areas. The MPI reports that 85% of those who are poor reside in rural areas. Rural locations with low MPI populations and congestion can combine to create a strong labour force that can support sustainable manufacturing. The relocation of job prospects to suburban or rural areas may serve as a catalyst for jobless people to enter the labour market. AM manufacturing and material recycling, in contrast to the service sector, shall not be appropriate for cities. This speaks to the idea that moving from urban to suburban or rural areas might help with concerns of overpopulation, urban-rural interdependence and sustainable manufacturing in these locations.

There is currently little research on the social impact of AM, despite the literature review highlighting the critical need of taking social resources into account in production and company management processes.

In particular, two concerns about the societal effects of applying AM technology are highlighted. First, attention is drawn to the potential impact on working conditions. In that it minimises long-term exposure to challenging work environments, AM production methods undoubtedly offer health advantages over traditional ones. However, there are also concerns regarding the handling of the specific moulding materials, as some of them prove to be extremely dangerous for employees' health. The second is the possible advantages of moving production elsewhere. In terms of social sustainability, this opportunity may present significant challenges, even though it may be advantageous from an economic one [77].

9.7.3   AM and governance

Every industrial business has used lean management (LM) within the past three decades. Industries may use lean thinking to eliminate waste and promote corporate development by optimising manufacturing, encouraging customer interaction and making the most of service resources. The goal of LM is to eliminate wasteful processes in order to save expenses, enhance productivity, boost flexibility and optimise value creation for clients. Toyota Motor Corporation invented the term 'lean' in the 1940s in Japan in an effort to transform production methods; this approach is known as lean manufacturing. From an asset-level viewpoint, there are eight main categories of operational waste that can arise from non-value-adding operations. The waste that has been detected is known as TIMWOODS and covers transportation, inventory, movement, waiting, overproduction, overprocessing, defects and skills. Taiichi Ohno created the first seven categories of trash as part of TPS. Following their acceptance of the Toyota Production Systems, the Western world established the eighth waste type. A popular technique for adopting lean manufacturing is the 5S method, which stands for 'sort', 'set', 'shine', 'standardise' and 'sustain'. The technique aids in standardising everyday operations to achieve and preserve ongoing cleanliness and orderliness at work. The term '5S' refers to a five-step visual management process that helps to encourage cleanliness, orderly marks, labelling of work areas and proper maintenance in addition to helping to eliminate unnecessary procedures.

9.7.4   AM and economics

AM is a technology that has a major strategic impact on the manufacturing industry. It influences not only operational strategy but also entire corporate strategy and sustainability perspectives. As recent study [78] made obvious, AM is in fact altering corporate operations, business models and strategies while also offering a number of advantages for sustainability in all of its forms. In terms of economic sustainability specifically, AM is different from traditional production in that it avoids the impact of economies of scale because the cost per unit of the products produced is independent of production volume. AM manufacturing offers true flexibility in terms of quantity and the capacity to effectively fulfil each customer's unique needs, especially with regard to product modification and time-to-market. Additionally, storing raw materials can be done for a lot less money when AM methods are used. In particular, design for AM makes it possible to create digital warehouses, where components are manufactured rather than kept in response to client requests. Compared to conventional decentralised production systems, which are conveniently positioned close to clients, smaller storage rooms are therefore needed.

Businesses benefit further from this in terms of environmental and economic sustainability. If AM service bureau chooses to place the manufacturing location close to the customer, it can significantly lower transportation expenses. Additionally, by lowering emissions from transportation, it can benefit the environment. The ability to design a product using a continuous and sustainable logic – that is, from a process perspective where rework and recycling processes are anticipated – allows companies to be guided toward a circular business model, which is another intriguing feature of these technologies. In conclusion, if AM technology were also widely used, it would enable a drastic change in the value chain's sustainability from an economic standpoint [79].

9.8 AM in SDG

The three main disciplines that make up sustainability are economy, environment and society [75]. These disciplines are interdependent. Reusing materials shields the environment from pollution, contamination, material recovery and saving, and is inextricably linked to all of these important elements. For this reason, material recycling and sustainable manufacturing have become essential topics in the discourse on sustainability. Reusing materials also boosts economies and promotes peaceful civilisations.

SDG 7 emphasises that improving people's communities and quality of life depends on increasing the availability of electricity and other energy sources. AM works to manage resources for development in a sustainable manner by advancing renewable energy sources and energy efficiency [80].

SDG 8 places a strong emphasis on the creation of well-paying, high-quality jobs through long-term economic growth and increased economic productivity, as well as resource efficiency in both production and consumption. AM is in favour of full employment, decent jobs for all, the creation of labour rights, safe and secure workplaces and the prohibition of forced labour, human trafficking and child labour [80].

SDG 9 calls for efforts to develop resilient infrastructure, support fair and sustainable industry and foster innovation. Businesses want to use 3D printing to create more inventive product production processes, boost flexibility and enhance their supply networks. more investment in 3D printing and the realisation that it can be a competitive alternative to traditional manufacturing have led to a strong need for more network collaboration to support the technology's wider adoption [80].

The establishment of sustainable production and consumption practices, which are essential to maintaining the level of survival for both the current and future generations, is the emphasis of SDG 12. The fundamental causes of the three issues faced by our planet today – pollution, biodiversity loss and climate change – are unchecked. Technical advancements have made it possible to recycle and reintegrate polymers, such as PC and acrylonitrile ABS, that are widely used in the automobile industry. Thus, it was suggested that recyclable PC and ABS might be extruded and used to produce parts using 3D printing. After doing a lifecycle evaluation, Kreiger *et al.* [81] found that a distributed model for producing filament from trash uses up to 80% less energy than traditional manufacture, mostly because of transportation.

The goal of SDG 13 is to keep the rise in global temperatures to less than 2 °C. This is due to the fact that a temperature rise of more than 2 °C would have detrimental effects on the community at large. Because of rising greenhouse gas emissions, the average global temperature has already risen by 1.1 °C over pre-industrial levels. Sea levels are rising as a result of this temperature increase, as are more frequent extreme weather events. The use of 3D printing has resulted in significant drops in CO_2 emissions and energy consumption, from 38% to 75%. Many disadvantages of conventional manufacturing techniques include waste of materials, residual tensions, complexity, long supply chains, rigid design and higher CO_2 emissions. Twenty percent or so of the CO_2 emissions in the world are caused by these processes. On the other hand, these problems are mostly resolved by 3D printing-based manufacturing, which also makes it possible to employ recyclable and biodegradable materials in production [82].

Despite AM being used in the reality for a long time in different industries and sectors like glass manufacturing, aerospace industry, automotive industry, medical industry, food industry, recycling industry and much more, its role in ESG and SDG is not exactly measured. To understand more about the role of AM in ESG and SDG, more research is required. Currently, there is a lack of relevant literatures in the primary fields like sustainability and global value chains.

References

[1]   E. J. Hobsbawn, *Industry and Empire: From 1750 to the Present Day*, The New Press, New York, 1999.

[2]   A. Ashton, *The Industrial Revolution: 1760–1830*, Oxford University Press, Oxford, 1997.

[3]   P. P. Groumpos, "A critical historical and scientific overview of all industrial revolutions," *IFAC PapersOnLine*, vol. 54, no. 13, pp. 464–471, 2021.

[4]   Y. Lu, X. Xu and L. Wang, "Smart manufacturing process and system automation – A critical review of the standards and envisioned scenarios," *Journal of Manufacturing Systems*, vol. 56, pp. 312–325, 2020.

[5] K. Schwab, *The Fourth Industrial Revolution*, Portfolio Penguin, New York, 2017.

[6] "Plattform Industrie 4.0," [Online]. Available: https://www.plattform-i40.de/IP/Navigation/DE/Home/home.html [accessed 12 June 2024].

[7] X. Xu, Y. Lu, B. Vogel-Heuser and L. Wang, "Industry 4.0 and Industry 5.0 – Inception, conception and perception," *Journal of Manufacturing Systems*, vol. 61, pp. 530–535, 2021.

[8] M. Breque, L. D. Nul and A. Petridis, "Industry 5.0 – Towards a sustainable, human-centric and resilient European industry," European Commission, Brussels, 2021.

[9] M. Khan, A. Haleem and M. Javaid, "Changes and improvements in Industry 5.0: A strategic approach to overcome the challenges of Industry 4.0," *Green Technologies and Sustainability*, vol. 1, p. 100020, 2023.

[10] P. Mohankumar, "3D processing for human–machine interaction and additive manufacturing," in W.Y. Leong (ed), *Human Machine Collaboration and Interaction for Smart Manufacturing*. Stevenage: The Institution of Engineering and Technology, 2022, pp. 193–222.

[11] K. Kanishka and B. Acherjee, "Revolutionizing manufacturing: A comprehensive overview of additive manufacturing processes, materials, developments, and challenges," *Journal of Manufacturing Process*, vol. 107, pp. 574–619, 2023.

[12] A. Aufa, M. Z. Hassan, Z. Ismail, *et al.*, "Current trends in additive manufacturing of selective laser melting for biomedical implant applications," *Journal of Materials Research and Technology*, vol. 31, pp. 213–243, 2024.

[13] F. Cavallo, J. Ceulemans, B. Gasner, *et al.*, *Innovation Trends in Additive Manufacturing*, European Patent Office, Munich, 2023.

[14] B. Lu, H. Lan and H. Liu, "Additive manufacturing frontier: 3D printing electronics," *Opto-Electronic Advances*, vol. 1, no. 1, p. 170004, 2018.

[15] E. Edri, N. Armon, E. Greenberg, *et al.*, "Laser printing of multilayered alternately conducting and insulating microstructures," *ACS Applied Materials & Interfaces*, vol. 13, no. 30, pp. 36416–36425, 2021.

[16] T. C. Dzogbewu and D. d. Beer, "Powder bed fusion of multimaterials," *Journal of Manufacturing and Materials Processing*, vol. 7, no. 1, p. 15, 2023.

[17] T. C. Dzogbewu, N. Amoah, S. Afrifa Jnr, S. K. Fianko and D. J. d. Beer, "Multi-material additive manufacturing of electronics components: A bibliometric analysis," *Results in Engineering*, vol. 19, p. 101318, 2023.

[18] S. Razzaq, Z. Pan, H. Li, S. Ringer and X. Liao, "Joining dissimilar metals by additive manufacturing: A review," *Journal of Materials Research and Technology*, vol. 31, pp. 2820–2845, 2024.

[19] I. H. Zainelabdeen, L. Ismail, O. F. Mohamed, K. A. Khan and A. Schiffer, "Recent advancements in hybrid additive manufacturing of similar and dissimilar metals via laser powder bed fusion," *Materials Science & Engineering A*, vol. 909, p. 146833, 2024.

[20] R. Nandhakumar and K. Venkatesan, "A process parameters review on selective laser melting-based additive manufacturing of single and multi-material: Microstructure, physical properties, tribological, and surface roughness," *Materials Communication Today*, vol. 35, p. 105538, 2023.

[21] R. Robayo-Salazar, R. M. d. Gutierrez, M. A. Villaquitan-Caicedo and S. D. Arjona, "3D printing with cementitious materials: Challenges and opportunities for the construction sector," *Automation in Construction*, vol. 146, p. 104693, 2023.

[22] R. Parhi, "Recent advances in 3D printed microneedles and their skin delivery application in the treatment of various diseases," *Journal of Drug Delivery Science and Technology*, vol. 84, p. 104395, 2023.

[23] Y. Li, C. Su and J. Zhu, "Comprehensive review of wire arc additive manufacturing: Hardware system, physical process, monitoring, property characterization, application and future prospects," *Results in Engineering*, vol. 13, p. 100330, 2022.

[24] K. Copenhaver, T. Smith, K. Armstrong, *et al.*, "Recyclability of additively manufactured bio-based composites," *Composites Part B: Engineering*, vol. 255, p. 110617, 2023.

[25] A. Sinha, B. Swain, A. Behera, *et al.*, "A review on the processing of aero-turbine blade using 3D print techniques," *Journal of Manufacturing and Materials Processing*, vol. 6, no. 1, p. 16, 2022.

[26] A. Harding, A. Pramanik, A. Basak, C. Prakash and S. Shankar, "Application of additive manufacturing in the biomedical field-A review," *Annals of 3D Printed Medicine*, vol. 10, p. 100110, 2023.

[27] A. Haleem, M. Javaid, R. Pratap Singh, R. Suman and S. Khan, "Management 4.0: Concept, applications and advancements," *Sustainable Operations and Computers*, vol. 4, pp. 10–21, 2023.

[28] N. R. Madhu, H. Erfani, S. Jadoun, M. Amir, Y. Thiyagarajan and N. P. S. Chauhan, "Fused deposition modelling approach using 3D printing and recycled industrial materials for a sustainable environment: A review," *The International Journal of Advanced Manufacturing Technology*, vol. 122, pp. 2125–2138, 2022.

[29] M. Srivastava, S. Rathee, V. Patel, A. Kumar and P. G. Koppad, "A review of various materials for additive manufacturing: Recent trends and processing issues," *Journal of Materials Research and Technology*, vol. 21, pp. 2612–2641, 2022.

[30] J. L. Amaya-Rivas, B. S. Perero, C. G. Helguero, *et al.*, "Future trends of additive manufacturing in medical applications: An overview," *Heliyon*, vol. 10, p. e26641, 2024.

[31] S. J. Trenfield, A. Awad, C. M. Madla, *et al.*, "Shaping the future: Recent advances of 3D printing in drug delivery and healthcare," *Expert Opinion on Drug Delivery*, vol. 16, no. 10, pp. 1081–1094, 2019.

[32] P. Rawal, D. M. Tripathi, S. Ramakrishna and S. Kaur, "Prospects for 3D bioprinting of organoids," *Bio-Design and Manufacturing*, vol. 4, pp. 627–640, 2021.

[33]   B. Chang, A. Powell, S. Ellsperman, *et al.*, "Multicenter advanced pediatric otolaryngology fellowship prep surgical simulation course with 3D printed high-fidelity models," *Otolaryngology–Head and Neck Surgery*, vol. 162, no. 5, pp. 658–665, 2020.

[34]   D. Shilo, O. Emodi, O. Blanc, D. Noy and A. Rachmmiel, "Printing the future – Updates in 3D printing for surgical applications," *Rambam Maimonides Medical Journal*, vol. 9, no. 3, p. e0020, 2018.

[35]   T. Wu, X. Ding, T. Liu, G. Lai, X. Zhang and Q. Chen, "Advancements in developing biomaterials for 3D printing photosensitive resins containing quaternary ammonium molecules for enhanced oral healthcare," *Next Materials*, vol. 4, p. 100211, 2024.

[36]   S. L. Teng, Y. R. Wong, P. P. Hoon Lim and D. A. McGrouther, "An adjustable and customised finger splint to improve mallet finger treatment compliance and outcomes," *Annals of 3D Printed Medicine*, vol. 13, p. 100142, 2024.

[37]   M. P. D'Amado, J. B. d. Albuquerque II, W. Bezold, B. D. Crist and J. L. Cook, "Biomechanical comparison of traditional plaster cast and 3D-printed orthosis for external coaptation of distal radius fractures," *Annals of 3D Printed Medicine*, vol. 14, p. 100146, 2024.

[38]   A. Manero, P. Smith, J. Sparkman, *et al.*, "Implementation of 3D printing technology in the field of prosthetics: Past, present, and future," *International Journal of Environmental Research and Public Health*, vol. 16, no. 9, p. 1641, 2019.

[39]   A. Duvvi, E. Yates, S. Seligson, *et al.*, "A hybrid 3D-printed model for lower extremity compartment syndrome simulation," *Annals of 3D Printed Medicine*, vol. 14, p. 100153, 2024.

[40]   X. Zhu, H. Li, L. Huang, M. Zhang, W. Fan and L. Cui, "3D printing promotes the development of drugs," *Biomedicine & Pharmacotherapy*, vol. 131, p. 110644, 2020.

[41]   S. Beg, W. H. Almalki, A. Malik, *et al.*, "3D printing for drug delivery and biomedical applications," *Drug Discovery Today*, vol. 25, no. 9, pp. 1668–1681, 2020.

[42]   J. Garcia, Z. Yang, R. Mongrain, R. L. Leask and K. Lachapelle, "3D printing materials and their use in medical education: A review of current technology and trends for the future," *BMJ Simulation & Technology Enhanced Learning*, vol. 4, no. 1, pp. 27–40, 2018.

[43]   X. Li, S. Zhang, P. Jiang, *et al.*, "Smart bionic structures: Connecting nature and technology through additive manufacturing," *Additive Manufacturing Frontiers*, vol. 3, p. 200137, 2024.

[44]   J. Yan, Z. Li, J. Guo, S. Liu and J. Guo, "Organ-on-a-chip: A new tool for in vitro research," *Biosensors and Bioelectronics*, vol. 216, p. 114626, 2022.

[45]   T. Agarwal, G. M. Fortunato, S. Y. Hann, *et al.*, "Recent advances in bioprinting technologies for engineering cardiac tissue," *Materials Science and Engineering: C*, vol. 124, p. 112057, 2021.

[46] H. Cui, W. Zhu, Y. Huang, *et al.*, "In vitro and in vivo evaluation of 3D bioprinted small-diameter vasculature with smooth muscle and endothelium," *Biofabrication*, vol. 12, no. 1, p. 015004, 2020.

[47] I. Francis, J. Shrestha, K. R. Paudel, P. M. Hansbro, M. E. Warkiani and S. C. Saha, "Recent advances in lung-on-a-chip models," *Drug Discovery Today*, vol. 27, no. 9, pp. 2593–2602, 2022.

[48] W. J. Kim and G. H. Kim, "Intestinal villi model with blood capillaries fabricated using collagen-based bioink and dual-cell-printing process," *ACS Applied Materials & Interfaces*, vol. 10, no. 48, pp. 41185–41196, 2018.

[49] H. Lee, S. Chae, J. Y. Kim, *et al.*, "Cell-printed 3D liver-on-a-chip possessing a liver microenvironment and biliary system," *Biofabrication*, vol. 11, no. 2, p. 025001, 2019.

[50] M. Jang and H. N. Kim, "From single- to multi-organ-on-a-chip system for studying metabolic diseases," *BioChip Journal*, vol. 17, pp. 133–146, 2023.

[51] T. T. Nieskens and M. J. Wilmer, "Kidney-on-a-chip technology for renal proximal tubule tissue reconstruction," *European Journal of Pharmacology*, vol. 790, pp. 46–56, 2016.

[52] I. Raimondi, L. Izzo, M. Tunesi, M. Comar, D. Albani and C. Giordano, "Organ-on-a-chip in vitro models of the brain and the blood-brain barrier and their value to study the microbiota-gut-brain axis in neurodegeneration," *Frontiers in Bioengineering and Biotechnology*, vol. 7, p. 435, 2020.

[53] X. Wu, W. Shi, X. Liu and Z. Gu, "Recent advances in 3D-printing-based organ-on-a-chip," *EngMedicine*, vol. 1, p. 100003, 2024.

[54] A. T. Nelson, "Prospects for additive manufacturing of nuclear fuel forms," *Progress in Nuclear Energy*, vol. 155, no. 3, p. 104493, 2023.

[55] A. H. Alami, M. Mahmoud, H. Aljaghoub, A. Mdallal, M. A. Abdelkareem and S. K. Kamarudin, "Progress in 3D printing in wind energy and its role in achieving sustainability," *International Journal of Thermofluids*, vol. 20, p. 100496, 2023.

[56] L. A. Verhoef, B. W. Budde, C. Chockalingam, B. G. Nodar and A. J. v. Wijk, "The effect of additive manufacturing on global energy demand: An assessment using a bottom-up approach," *Energy Policy*, vol. 112, pp. 349–360, 2018.

[57] K. A. Terrani, "Advanced manufacturing and materials to enable advanced nuclear energy," *American Nuclear Society*, vol. 118, no. 1, pp. 1575–1575, 2018.

[58] L. Bobbio, B. Bocklund, R. Otis and J. P. Borgonia, "Characterization of a functionally graded material of Ti–6Al–4V to 304L stainless steel with an intermediate V section," *Journal of Alloys and Compounds*, vol. 742, pp. 1031–1036, 2018.

[59] C. Reyes, R. Somogyi, S. Niu, *et al.*, "Three-dimensional printing of a complete lithium ion battery with fused filament fabrication," *ACS Applied Energy Materials*, vol. 1, no. 10, pp. 5268–5279, 2018.

[60] T.-S. Wei, B. Y. Ahn, J. Grotto and J. A. Lewis, "3D printing of customized Li-ion batteries with thick electrodes," *Advanced Materials*, vol. 30, no. 16, p. e1703027, 2018.

[61] A. Maurel, S. Grugeon, B. Fleutot, *et al.*, "Three-dimensional printing of a LiFePO4/Graphite battery cell via fused deposition modeling," *Scientific Reports*, vol. 9, no. 1, p. 18031, 2019.

[62] D. Bourell, M. Leu, K. Chakravarthy, N. Guo and K. Alayavalli, "Graphite-based indirect laser sintered fuel cell bipolar plates containing carbon fiber additions," *CIRP Annals*, vol. 60, no. 1, pp. 275–278, 2011.

[63] J. Mo, R. R. Dehoff, W. H. Peter, T. J. Toops, J. B. Green Jr and F.-Y. Zhang, "Additive manufacturing of liquid/gas diffusion layers for low-cost and high-efficiency hydrogen production," *International Journal of Hydrogen Energy*, vol. 41, no. 4, pp. 3128–3135, 2016.

[64] A. Ambrosi and M. Pumera, "Multimaterial 3D-printed water electrolyzer with earth-abundant electrodeposited catalysts," *ACS Sustainable Chemistry & Engineering*, vol. 6, no. 12, pp. 16968–16975, 2018.

[65] F. Calignano, T. Tommasi, D. Manfredi and A. Chiolerio, "Additive manufacturing of a microbial fuel cell – A detailed study," *Scientific Reports*, vol. 5, p. 17373, 2015.

[66] P. Awasthi and S. S. Banerjee, "Design of ultrastretchable and super-elastic tailorable hydrophilic thermoplastic elastomeric materials," *Polymer*, vol. 252, p. 124914, 2022.

[67] N. Zohdi and R. Yang, "Material anisotropy in additively manufactured polymers and polymer composites: A review," *Polymers*, vol. 13, no. 19, p. 3368, 2021.

[68] T. D. Ngo, A. Kashani, G. Imbalzano, K. T. Nguyen and D. Hui, "Additive manufacturing (3D printing): A review of materials, methods, applications and challenges," *Composites Part B: Engineering*, vol. 143, pp. 172–196, 2018.

[69] A. M. Mirzendehdel, M. Behandish and S. Nelaturi, "Topology optimization for manufacturing with accessible support structures," *Computer-Aided Design*, vol. 142, p. 103117, 2022.

[70] M. Ali, A. Almotari, A. Algamal and A. Qattawi, "Recent advancements in post processing of additively manufactured metals using laser polishing," *Journal of Manufacturing and Materials Processing*, vol. 7, no. 3, p. 115, 2023.

[71] P. Patpatiya, K. Chaudhary and V. Kapoor, "Reverse manufacturing and 3D inspection of mechanical fasteners fabricated using photopolymer jetting technology," *MAPAN*, vol. 37, pp. 753–763, 2022.

[72] H. A. Colorada, E. I. Gutierrez Velasquez and S. N. Monteiro, "Sustainability of additive manufacturing: The circular economy of materials and environmental perspectives," *Journal of Materials Research and Technology*, vol. 9, no. 4, pp. 8221–8234, 2020.

[73] C. Dong, M. Petrovic and I. J. Davies, "Applications of 3D printing in medicine: A review," *Annals of 3D Printed Medicine*, vol. 14, p. 100149, 2024.

[74] J. Pizon, M. Witczak, A. Gola and A. Swic, "Challenges of human-centered manufacturing in the aspect of Industry 5.0 assumptions," *IFAC PapersOnLine*, vol. 56, no. 2, pp. 156–161, 2023.

[75]   H. Wu and H. Yabar, "Impacts of additive manufacturing to sustainable urban–rural interdependence through strategic control," *Results in Control and Optimization*, vol. 5, p. 100066, 2021.

[76]   D. Son, S. Kim and B. Jeong, "Sustainable part consolidation model for customized products in closed-loop supply chain with additive manufacturing hub," *Additive Manufacturing*, vol. 37, p. 101643, 2021.

[77]   E. J. Parry, J. M. Best and C. E. Banks, "Three-dimensional (3D) scanning and additive manufacturing (AM) allows the fabrication of customised crutch grips," *Materials Today Communications*, vol. 25, p. 101225, 2020.

[78]   M. K. Niaki and F. Nonino, "The value for business and operations strategy," in *The Management of Additive Manufacturing: Enhancing Business Value*. Cham: Springer, 2018, pp. 91–129.

[79]   B. Bigliardi, E. Bottani, E. Gianatti, L. Monferdini, B. Pini and A. Petroni, "Sustainable additive manufacturing in the context of Industry 4.0: A literature review," *Procedia Computer Science*, vol. 232, pp. 766–774, 2024.

[80]   A. H. Alami, A. G. Olabi, A. Alashkar, *et al.*, "Additive manufacturing in the aerospace and automotive industries: Recent trends and role in achieving sustainable development goals," *Ain Shams Engineering Journal*, vol. 14, p. 102516, 2023.

[81]   M. Kreiger, M. Mulder, A. Glover and J. Pearce, "Life cycle analysis of distributed recycling of post-consumer high density polyethylene for 3-D printing filament," *Journal of Cleaner Production*, vol. 70, pp. 90–96, 2014.

[82]   B. Louman, R. J. Keenan, D. Kleinschmit, *et al.*, "SDG 13: Climate action – Impacts on forests and people," in P. Katila, C. J. Pierce Colfer, W. de Jong, G. Galloway, P. Pacheco, and G. Winkel (eds), *Sustainable Development Goals: Their Impacts on Forests and People*. Cambridge: Cambridge University Press, 2019, pp. 419–444.

Chapter 10

Immersive technologies: a new dimension for Industry 5.0

Nassirah Laloo[1] and Mohammad Sameer Sunhaloo[1]

Industry 5.0 aims to bring a transformative shift in the industrial sector, marked by the integration of human ingenuity with intelligent machinery to foster production processes that are more personalised, efficient and sustainable [1]. In contrast to Industry 4.0, which primarily concentrated on automating manufacturing and production through cyber-physical systems, the Internet of Things (IoT) and big data analytics, Industry 5.0 reinstates the human component as a pivotal figure within the production ecosystem (Figure 10.1).

Whereas Industry 4.0 sought to optimise productivity and efficiency through machine-to-machine interactions and data-driven strategies, Industry 5.0 prioritises the synergy between humans and machines [2]. It harnesses human creativity, critical thinking and problem-solving skills in conjunction with advanced technologies such as robotics, artificial intelligence (AI) and immersive technologies. This collaborative framework aspires to cultivate a more dynamic and adaptable industrial landscape, wherein technology serves to enhance human capabilities rather than replace them (Figure 10.2).

Industry 5.0 is built around three core principles namely human-centricity, sustainability and resilience. Human-centricity recognises the intrinsic value of human creativity, innovation and emotional intelligence in the industrial process (Figure 10.3). Rather than relying solely on automation, Industry 5.0 focuses on developing technologies that enhance the human experience, ensuring that jobs become more meaningful, safe and fulfilling. It promotes a personalised approach to production and services, encouraging co-creation and customisation with consumers. Sustainability ensures compliance with the growing global demand for environmentally responsible practices, Industry 5.0 integrates sustainable strategies into its core. It adopts circular economy principles, optimising resource utilisation, minimising waste and reducing carbon footprints. Industry 5.0 aims to balance economic growth with ecological responsibility, designing products, processes and business models that ensure long-term environmental sustainability. With the increasing prevalence of global

[1]School of Innovative Technologies and Engineering, University of Technology, Mauritius

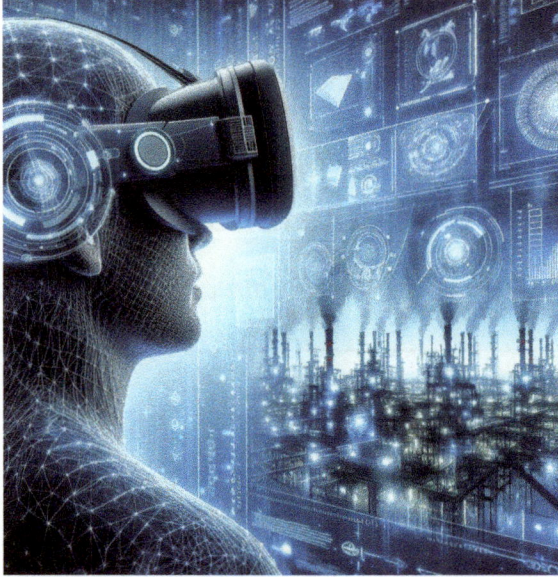

Figure 10.1   XR technologies and Industry 5.0

Figure 10.2   Industry 4.0 to Industry 5.0

disruptions such as pandemics, geopolitical conflicts and supply chain issues, resilience has become a critical priority. Industry 5.0 focuses on creating robust systems that can quickly adapt to and recover from unforeseen challenges. It

Figure 10.3   Transformative role of XR technologies

Figure 10.4   Industry future with immersive technologies

considers the adoption of advanced technologies like AI, machine learning and immersive experiences to forecast risks, simulate responses and enhance decision-making in uncertain conditions (Figure 10.4).

Figure 10.5   AR VR XR MR for Industry 5.0

Immersive technologies including Virtual Reality (VR), Augmented Reality (AR), Mixed Reality (MR) and Extended Reality (XR) support the intrinsic requirement of Industry 5.0 and are pivotal in enabling the transition to Industry 5.0 [3]. These technologies offer innovative ways to merge digital and physical environments, enhancing human–machine interaction and collaboration (Figure 10.5).

This chapter explores the transformative role of immersive technologies, particularly Virtual Reality (VR), in shaping the future of Industry 5.0. It examines how immersive tools enhance human–machine collaboration, improve training and skill development and support sustainable practices, while fostering innovation across various sectors such as manufacturing, healthcare, education and retail. By integrating these advanced technologies, Industry 5.0 aims to create a more adaptive and inclusive industrial environment where human creativity is augmented by intelligent machines, driving both economic growth and environmental responsibility.

10.1   Immersive technologies and human-centricity

AR, VR, MR and XR provide a platform to reinforce the human-centric focus of Industry 5.0 by enhancing human experiences, empowering workers and encouraging meaningful engagement between people and technology. These technologies contribute to human-centricity in different ways (Figure 10.6).

VR, AR and MR provide immersive and interactive training environments that enable workers to develop new skills and practice complex tasks without the risk of real-world dangers and consequences [4]. For instance, VR can create realistic simulations of hazardous situations for safety training, while AR can overlay step-by-step

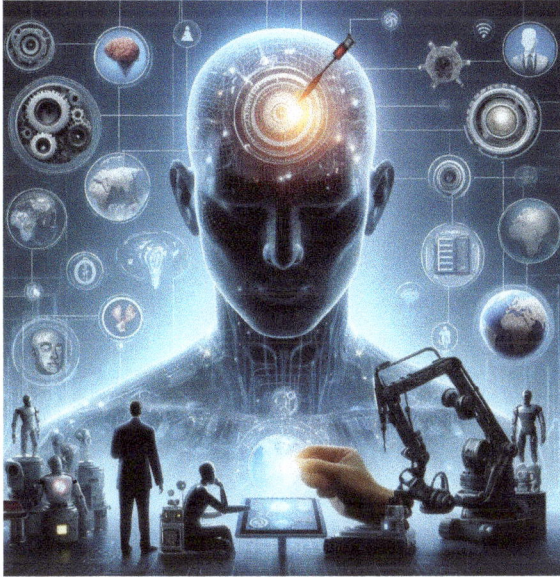

Figure 10.6   Human centricity in Industry 5.0

Figure 10.7   XR for enhanced learning and training

instructions directly onto a worker's field of view, facilitating on-the-job learning. Immersive technologies allow flexible online or remote training [5]. This personalised, hands-on approach accelerates skill acquisition and increases retention, making training more engaging and effective (Figure 10.7).

Figure 10.8   XR for improved human–machine collaboration

AR and MR enhance the way humans interact with machines by providing intuitive, real-time information and feedback [6]. For example, AR can project digital information over physical equipment, allowing workers to perform maintenance and repair tasks more accurately and efficiently. MR environments allow workers to manipulate both digital and physical objects simultaneously, blending the strengths of human intuition and machine precision to improve decision-making and problem-solving (Figure 10.8).

XR technologies enable a more personalised approach to manufacturing and product development, allowing for customisation according to individual preferences and needs [7]. VR and AR tools can simulate different design options, enabling customers to co-create products or experience services tailored to their unique requirements. This not only enhances customer satisfaction but also fosters a sense of ownership and engagement in the production process (Figure 10.9).

XR technologies make complex information more accessible by presenting it in a visual and interactive format. They can help bridge communication gaps, cater to diverse learning styles and assist people with disabilities [8]. For example, VR can provide auditory cues or tactile feedback for visually impaired users, while AR can offer real-time translations for non-native speakers. This inclusivity helps create a more equitable workplace where technology is designed to enhance the capabilities of all individuals.

Immersive technologies contribute to worker well-being by reducing the physical and mental strain associated with complex or repetitive tasks [9]. AR can provide ergonomic guidance to minimise strain, while VR can simulate stressful scenarios to train employees in managing anxiety and decision-making under

Figure 10.9   XR for enhanced accessibility and inclusivity

Figure 10.10   XR for promoting worker well-being and safety

pressure. Additionally, MR and XR applications can monitor worker health in real-time, identifying potential hazards or fatigue, thereby promoting a safer and healthier work environment (Figure 10.10).

XR technologies provide virtual spaces where teams can meet, collaborate and share information as if they were physically present, overcoming geographical and physical barriers [10]. This fosters a more flexible work environment, reduces travel-related stress and supports a better work-life balance, all while maintaining productivity and engagement. Virtual meetings, brainstorming sessions and design reviews become more interactive and dynamic, improving collaboration across global teams.

10.2   Immersive technologies and sustainability

VR and XR technologies enable the creation of virtual prototypes and digital twins, allowing designers and engineers to test and iterate on products in a fully digital environment. This reduces the need for multiple physical prototypes, significantly cutting down on material waste, energy consumption and production costs [11]. For example, automotive and aerospace companies use VR simulations to test new vehicle designs, minimising the resources required for physical models (Figure 10.11).

AR and MR provide real-time insights and data visualisation that help optimise resource use in industrial processes [12]. AR overlays can display information on energy consumption, water usage and waste generation, enabling workers and managers to identify inefficiencies and take immediate corrective actions. By using MR to visualise and analyse resource flows, companies can implement more efficient production methods and reduce their carbon footprint (Figure 10.12).

Figure 10.11   XR for reducing physical prototyping and material waste

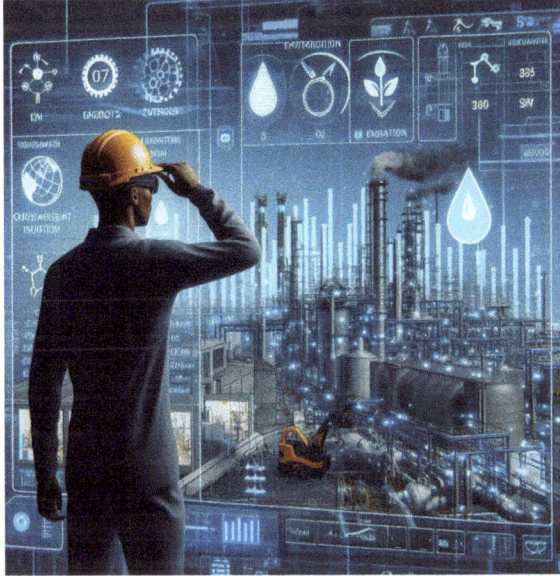

Figure 10.12   XR for optimising resource management and energy efficiency

XR technologies facilitate remote inspections, maintenance and monitoring of industrial equipment and facilities, reducing the need for travel and lowering greenhouse gas emissions [13]. For example, an engineer can use AR glasses to guide remote teams through complex repairs or maintenance tasks, minimising the need for on-site visits. This approach not only reduces the environmental impact associated with travel but also improves operational efficiency and safety (Figure 10.13).

Immersive technologies enable more sustainable product design by allowing for virtual testing and evaluation of different materials, production methods and supply chain options [14]. VR simulations can model the environmental impact of various design choices, enabling designers to prioritise eco-friendly materials and processes. This reduces the environmental impact of product development while promoting the creation of sustainable, long-lasting products (Figure 10.14).

AR, VR and XR play a crucial role in supporting circular economy models by simulating product lifecycles from design and production to recycling and reuse [15]. These technologies allow companies to explore ways to extend product life-spans, reduce waste and design for disassembly and recycling. For example, VR can simulate end-of-life scenarios for products, helping companies develop strategies for recycling or repurposing materials (Figure 10.15).

XR technologies provide immersive training environments that raise awareness about environmental issues and best practices [16]. VR experiences can simulate the impact of pollution, deforestation or climate change, helping employees and stakeholders understand the importance of sustainable practices. AR can provide interactive, real-time instructions for sustainable operations, such as reducing energy consumption, minimising waste or using sustainable materials (Figure 10.16).

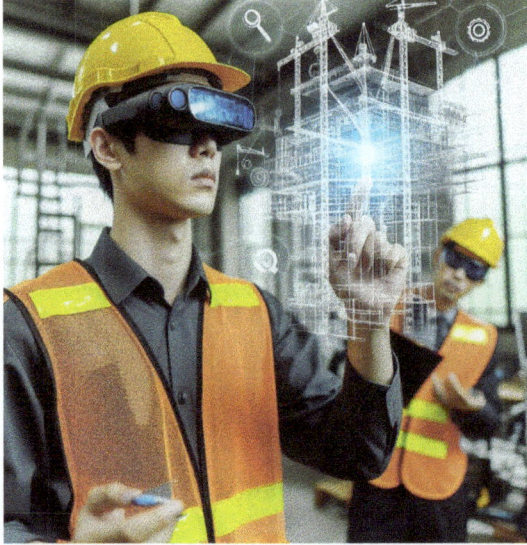

Figure 10.13   XR for enabling remote operations and maintenance

Figure 10.14   XR for supporting sustainable product design and development

XR technologies facilitate virtual meetings, collaboration and remote work, reducing the need for large physical office spaces and associated energy consumption [17]. By creating virtual workspaces, companies can lower their real estate needs, minimise commuting and reduce the overall environmental impact of

Figure 10.15   XR for facilitating circular economy practices

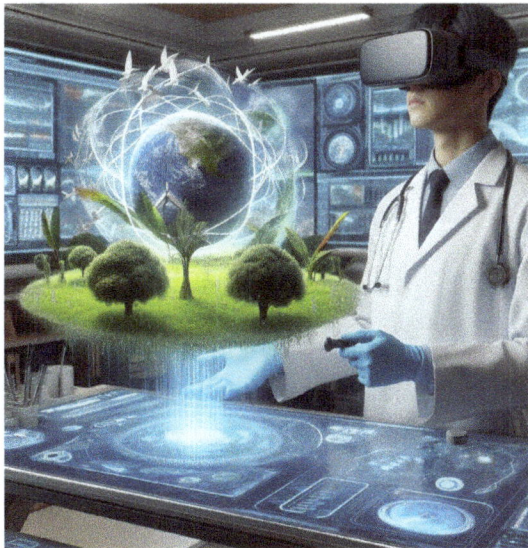

Figure 10.16   XR for enhancing environmental awareness and training

their operations. This supports a more sustainable, flexible and resilient work environment (Figure 10.17).

AR and VR technologies enable consumers to make more informed and sustainable choices by providing detailed product information, visualising the environmental impact of their purchases and offering virtual try-on experiences that

Figure 10.17   XR for enabling virtual workspace and reducing office footprints

Figure 10.18   XR for sustainable consumer behaviour

reduce returns and waste [18]. For example, AR can allow customers to visualise furniture in their homes before purchasing, reducing the likelihood of returns and minimising waste (Figure 10.18).

10.3   Immersive technologies and resilience

VR and XR can simulate a wide range of emergency scenarios, such as natural disasters, equipment failures or cyberattacks, allowing organisations to develop and practice response strategies in a controlled environment [19]. By creating realistic training simulations, these technologies help employees understand and react effectively to crises, thereby improving overall preparedness and reducing downtime in the event of an actual disaster (Figure 10.19).

AR and MR provide real-time data visualisation, enabling quick access to critical information during emergencies [20]. For instance, AR can overlay vital data like equipment status, energy consumption or safety alerts onto a worker's field of view, allowing them to make informed decisions faster. MR combines physical and digital information, facilitating collaborative decision-making across teams, even when they are geographically dispersed (Figure 10.20).

XR technologies facilitate remote monitoring, diagnostics and maintenance of equipment and facilities, ensuring operational continuity even when on-site access is limited [21]. AR can guide remote teams through complex repairs or provide real-time instructions to on-site personnel, minimising the need for expert travel and reducing response times. This remote capability enhances resilience by maintaining operations under adverse conditions, such as pandemics, travel restrictions or severe weather (Figure 10.21).

XR technologies support flexible work arrangements, allowing employees to collaborate virtually from different locations. Virtual workspaces created by VR and MR can replicate the in-office experience, fostering effective communication

Figure 10.19   XR for enhancing disaster preparedness and response

Figure 10.20   XR for supporting real-time monitoring and decision-making

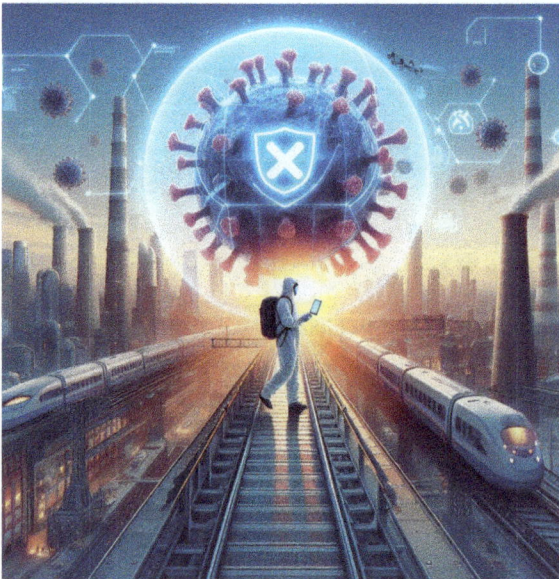

Figure 10.21   XR for enabling remote operations and maintenance

Figure 10.22 XR for building adaptive and flexible work environments

and collaboration despite physical distances [22]. This adaptability is crucial during disruptions that limit access to traditional workplaces, helping organisations maintain productivity and employee engagement (Figure 10.22).

Immersive technologies like AR and XR help companies visualise and analyse their entire supply chains in real-time, identifying potential vulnerabilities and optimising logistics [23]. AR can be used for real-time tracking of goods and inventory, while VR simulations can model different supply chain scenarios to predict disruptions and plan contingencies. These capabilities enhance supply chain resilience by enabling quicker adaptation to changes in demand, supply shortages or logistical challenges (Figure 10.23).

MR and XR technologies enable predictive maintenance by allowing real-time monitoring of equipment conditions and performance [24]. Sensors connected to AR and MR systems can detect early signs of wear and tear, triggering pre-emptive maintenance actions that prevent breakdowns and extend equipment lifespan. This proactive approach reduces the likelihood of unexpected failures and minimises downtime, ensuring continuous and reliable operations (Figure 10.24).

XR technologies also contribute to psychological resilience by providing tools for stress management and mental health support [25]. VR can offer immersive relaxation and mindfulness sessions, helping employees cope with stress and anxiety, especially during crises. AR applications can provide real-time ergonomic guidance to reduce physical strain and improve workplace safety, contributing to a healthier and more resilient workforce (Figure 10.25).

XR technologies create virtual environments for real-time collaboration, brainstorming and problem-solving, fostering a culture of continuous innovation [26]. This

Figure 10.23   XR for improving supply chain resilience

Figure 10.24   Facilitating predictive maintenance and risk management

alternative is of immense importance in times of rapid change, as it allows organisations to pivot quickly, develop new strategies and respond to emerging challenges more effectively (Figure 10.26).

Figure 10.25 XR for supporting mental resilience and employee well-being

Figure 10.26 XR for enabling real-time innovation

10.4   Conclusion

Industry 5.0 is based on human-centricity, sustainability, and resilience principles. The Industry 5.0 Market is estimated to be worth around USD 658.4 billion by 2032. As depicted in Figure 10.27 below, there is an exponential forecast in the period 2022 to 2032 (Figure 10.27) [27]. Immersive Technologies such as AR, VR, MR and XR appear promising as an enabler in supporting Industry 5.0. By enabling a seamless human-machine collaboration, optimising resource use, and promoting adaptive systems, these technologies create more engaging, efficient, and flexible industrial environments. These immersive technologies empower workers through personalised training and real-time insight, support sustainable practices through virtual prototyping and remote operations, and provide resilience by enabling rapid response to disruptions and continuous innovation. Amid the constant challenges experienced by industries, immersive technologies can play a vital role in supporting the industry's economic viability and transforming how people work, learn, and interact in this new era of industrial evolution.

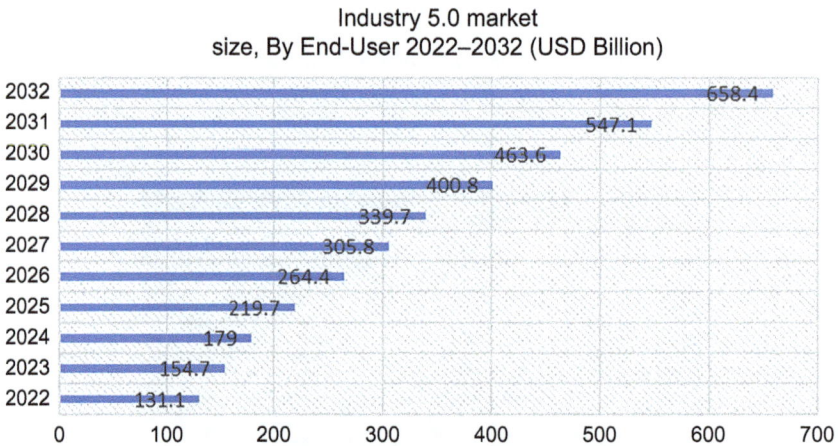

**Industry 5.0 market
size, By End-User 2022–2032 (USD Billion)**

Year	Value
2032	658.4
2031	547.1
2030	463.6
2029	400.8
2028	339.7
2027	305.8
2026	264.4
2025	219.7
2024	179
2023	154.7
2022	131.1

Figure 10.27   Industry 5.0 market size forecast (2022–2032)

References

[1]   Leng, J., Sha, W., Wang, B., *et al.*, 2022. Industry 5.0: Prospect and retrospect. *Journal of Manufacturing Systems*, 65, pp. 279–295.

[2]   Xu, X., Lu, Y., Vogel-Heuser, B. and Wang, L., 2021. Industry 4.0 and Industry 5.0. Inception, conception and perception. *Journal of Manufacturing Systems*, 61, pp. 530–535.

[3]   Martínez-Gutiérrez, A., Díez-González, J., Perez, H. and Araújo, M., 2024. Towards Industry 5.0 through metaverse. *Robotics and Computer-Integrated Manufacturing*, 89, p. 102764.

[4]   Chemerys, H., Vynogradova, A., Briantseva, H. and Sharov, S., 2021. Strategy for implementing immersive technologies in the professional training process of future designers. *Journal of Physics: Conference Series*, 1933 (1), p. 012046.

[5]   Fracaro, S.G., Glassey, J., Bernaerts, K. and Wilk, M., 2022. Immersive technologies for the training of operators in the process industry: A systematic literature review. *Computers & Chemical Engineering*, 160, p. 107691.

[6]   Kaplan, A.D., Cruit, J., Endsley, M., Beers, S.M., Sawyer, B.D. and Hancock, P.A., 2021. The effects of virtual reality, augmented reality, and mixed reality as training enhancement methods: A meta-analysis. *Human Factors*, 63(4), pp. 706–726.

[7]   Gong, L., Fast-Berglund, Å. and Johansson, B., 2021. A framework for extended reality system development in manufacturing. *IEEE Access*, 9, pp. 24796–24813.

[8]   Valakou, A., Margetis, G., Ntoa, S. and Stephanidis, C., 2023. A framework for accessibility in XR environments. In *International Conference on Human–Computer Interaction* (pp. 252–263). Cham: Springer Nature.

[9]   Khan, S., 2023. The future of XR-empowered healthcare: Roadmap for 2050. In Alam, M., Banday, S.A. and Usta, M.S. (eds), *Extended Reality for Healthcare Systems* (pp. 265–275). New York: Academic Press.

[10] Johnstone, R., McDonnell, N. and Williamson, J.R., 2022. When virtuality surpasses reality: Possible futures of ubiquitous XR. In *CHI Conference on Human Factors in Computing Systems Extended Abstracts* (pp. 1–8).

[11] Li, H., Fang, S., Chen, L., Menadue, V. and Martin, S., 2024. Extended reality (XR)—A magic box of digitalization in driving sustainable development of the construction industry: A critical review. *Sustainable Development*, 32(3), pp. 2830–2845.

[12] Donalek, C., Djorgovski, S.G., Cioc, A., *et al.*, 2014. Immersive and collaborative data visualization using virtual reality platforms. In *2014 IEEE International Conference on Big Data (Big Data)* (pp. 609–614). Piscataway, NJ: IEEE.

[13] Chu, C.H. and Pan, J.K., 2024. A systematic review on extended reality applications for sustainable manufacturing across the product lifecycle. *International Journal of Precision Engineering and Manufacturing-Green Technology*, 11(3), pp. 1017–1028.

[14] Bangcuyo, R.G., Smith, K.J., Zumach, J.L., Pierce, A.M., Guttman, G.A. and Simons, C.T., 2015. The use of immersive technologies to improve consumer testing: The role of ecological validity, context and engagement in evaluating coffee. *Food Quality and Preference*, 41, pp. 84–95.

[15] Soman, R.K., Nikolić, D. and Sanchez, B., 2024. Extended reality as a catalyst for circular economy transition in the built environment. In De Wolf, C., Çetin, S. and Bocken, N.M.P. (eds), *A Circular Built Environment in the Digital Age* (pp. 171–193). Cham: Springer.

[16] Aguayo, C. and Eames, C., 2023. Using mixed reality (XR) immersive learning to enhance environmental education. *The Journal of Environmental Education*, 54(1), pp. 58–71.

[17] Lo, T.T.S., Chen, Y., Lai, T.Y. and Goodman, A., 2024. Phygital workspace: A systematic review in developing a new typological work environment using XR technology to reduce the carbon footprint. *Frontiers in Built Environment*, 10, p. 1370423.

[18] Scurati, G.W., Bertoni, M., Graziosi, S. and Ferrise, F., 2021. Exploring the use of virtual reality to support environmentally sustainable behavior: A framework to design experiences. *Sustainability*, 13(2), p. 943.

[19] Lee, Y., Kim, S.K., Yoon, H., Choi, J., Kim, H. and Go, Y., 2021. Integration of extended reality and a high-fidelity simulator in team-based simulations for emergency scenarios. *Electronics*, 10(17), p. 2170.

[20] Zheng, M., Pan, X., Bermeo, N.V., *et al.*, 2022. Stare: Augmented reality data visualization for explainable decision support in smart environments. *IEEE Access*, 10, pp. 29543–29557.

[21] Casini, M., 2022. Extended reality for smart building operation and maintenance: A review. *Energies*, 15(10), p. 3785.

[22] Orel, M., 2022. *Collaboration Potential in Virtual Reality (VR) Office Space: Transforming the Workplace of Tomorrow*. Cham: Springer Nature.

[23] Tsang, Y.P., Yang, T., Chen, Z.S., Wu, C.H. and Tan, K.H., 2022. How is extended reality bridging human and cyber-physical systems in the IoT-empowered logistics and supply chain management? *Internet of Things*, 20, p. 100623.

[24] Mourtzis, D., Angelopoulos, J. and Panopoulos, N., 2020. Intelligent predictive maintenance and remote monitoring framework for industrial equipment based on mixed reality. *Frontiers in Mechanical Engineering*, 6, p. 578379.

[25] Gaggioli, A., Pallavicini, F., Morganti, L., *et al.*, 2014. Experiential virtual scenarios with real-time monitoring (interreality) for the management of psychological stress: A block randomized controlled trial. *Journal of Medical Internet Research*, 16(7), p. e3235.

[26] Bruzzone, A.G., Sinelshchikov, K., Massei, M., Fabbrini, G. and Gotelli, M., 2020. Extended reality technologies for industrial innovation. In *European Modeling and Simulation Symposium, EMSS* (vol. 202, pp. 1–9).

[27] Industry 5.0 Market, Report Code 40674 (2023). Available at: https://marketresearch.biz/report/industry-5-0-market/ (accessed 8 September 2024).

Chapter 11

Engineering education: does Industry 5.0 need Education 5.0?

*Shuh Huey Ho[1], Muhamad Nur Arsh Mohamad Basir[2],
Nur Adibah Raihan Affendy[3] and
Nurul Izzatul Akma Katim[1]*

Industry 5.0 (IR 5.0) is transforming the engineering profession and creating a need for new skills and knowledge. Here are some key areas where engineering education can be adapted to prepare students for the IR 5.0 era: interdisciplinary skills, digital skills, design thinking, entrepreneurship, and sustainability. By adapting engineering education to meet the needs of the IR 5.0 era, educators can prepare students for careers that leverage the latest technologies and enable them to make a positive impact on society. Additionally, industry-academic partnerships can help bridge the gap between academia and the workforce by providing students with real-world experiences and opportunities to work on cutting-edge projects.

11.1 Industrial revolution

The industrial revolution is a phenomenon that occurs due to the advancement of industrial technology. The industrial revolution started in the late 18th century with the invention of water and steam production machinery. In the early 20th century, Industrial Revolution 2.0 began. Mass production in industry and at home occurred due to the arrival of electric power. Industrial Revolution 3.0 began with the introduction of electronic systems, computer technology, and automation. Industry Revolution 4.0 referred to the intelligent networking of machines and processes in the industry with the aid of information and

[1]Department of Mechanical Engineering, Faculty of Engineering and Technology, Tunku Abdul Rahman University of Management and Technology (TAR UMT), Federal Territory of Kuala Lumpur, Malaysia
[2]Deputy Director (Academic) Office, Sungai Petani Community College, Malaysia
[3]Department of Commerce, Mukah Polytechnic, Malaysia
[4]Department of Engineering and Technology, Faculty of Information Sciences and Engineering, Management and Science University (MSU), Malaysia

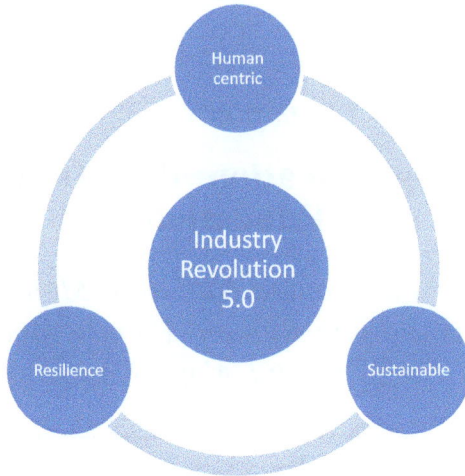

Figure 11.1   Pillars of Industry Revolution 5.0 [2]

communication technology. Industry Revolution 4.0 focused on the digital transformation of industries through automation, data exchange, and the integration of technologies such as the Internet of Things (IoT), artificial intelligence (AI), and big data. Industry Revolution 5.0 is the most recent phase which shifts the focus toward collaboration between humans and machines to create innovative solutions and achieve complex tasks. Industry revolution 5.0 involves the use of IoT-connected devices and sensors to collect data from industrial systems and the use of advanced automation technologies to handle dangerous tasks. Industry revolution 5.0 emphasized on human creativity, problem solving, emotional intelligence, and critical thinking [1]. With the influence of advanced technologies such as big data and IoT, robots help humans work faster and better. Three pillars of industry revolution 5.0 are shown in Figure 11.1.

11.1.1   Human centric

Human-centric approach focused on human needs and interest. The development of new roles for employees resulted from the shift from technology driven to a human-centric and society-centric approach. Technology is to adapt the needs and diversity of the employees. The involvement of employees in many tasks such as design, monitoring, engineering, maintenance of processes and programming is still needed as these tasks are still hard to automate. Employees can make decisions efficiently and accurately with the help of automation technologies.

11.1.2 Sustainability

Industry 5.0 focuses on minimizing waste by designing products that can be recycled, reduce waste and environmental impact, reused, and repurposed. Advanced technologies can be used to achieve these aims.

11.1.3 Resilience

Resilience refers to the need to develop a higher degree of robustness in industrial production, arming it better against disruptions and ensuring it can provide and support critical infrastructure in times of crisis [3].

11.2 Evolution in education system

For a developing country like Malaysia, Ministry of Higher Education (MOHE) has innovated the education by implementing digital technologies for a flexible learning pathway to face the future challenge [1]. Following the trend of Education 1.0 to following Education 5.0 as in Figure 11.2, the teacher-centered learning approaches in Education 1.0 causing students to be more dependent and lack of critical thinking and problem-solving skills. This method is seemly disadvantages, by means students tend to memorize the information without understand it [2]. In

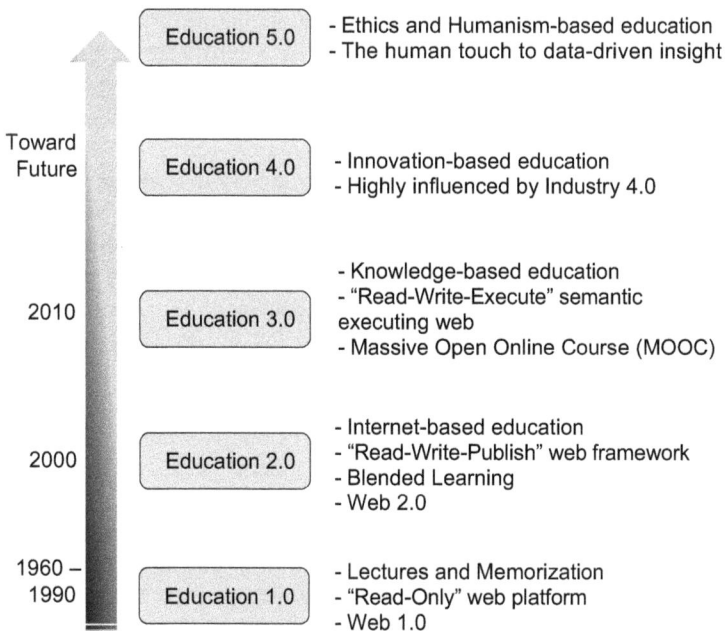

Education 5.0	- Ethics and Humanism-based education - The human touch to data-driven insight
Education 4.0	- Innovation-based education - Highly influenced by Industry 4.0
Education 3.0	- Knowledge-based education - "Read-Write-Execute" semantic executing web - Massive Open Online Course (MOOC)
Education 2.0	- Internet-based education - "Read-Write-Publish" web framework - Blended Learning - Web 2.0
Education 1.0	- Lectures and Memorization - "Read-Only" web platform - Web 1.0

Figure 11.2   Education trend

those days, the "Read-Only" web platform as tool to find a reference is seemly limited and the resource is more favor on technical use [3].

To overcome the issue with largely passive students in Education 1.0, Education 2.0 is more on transforming the passive to active students (student-centered learning). The teaching and learning process now requires the use of printed book and presentation slides with regard to increasing technology in IR 2.0. In addition, many institutions in Malaysia especially TVET has introduce learning management system (LMS) with integration of Web 2.0 applications for e-learning. However, authors in [4] highlight the challenges namely of lack lecture skills in encouraging active communication in LMS, lack of digital literacy skills, as well as issue with hardware and internet coverage. On positive aspect, the readiness of Web 2.0 allows student to find the information through peer and helps them to communicate worldwide through platform such as Gmail, Facebook, or WhatsApp [5].

Massive open online course (MOOC) is a result of merging the LMS, virtual learning environment, and Personalized Learning Support for the purpose of Education 3.0 [6]. The MOOC provide the flexible learning environment where students can customize the program taken. For example, the previous program for engineering is more on the core subjects learning but for MOOC, students can blend the engineering, art and business as one resulting another program such as technopreneurship. Starting 2015, MOHE has establish MOOC in their education and institution like Universiti Teknologi Malaysia under UTM MOOC and Micro-Credential, Universiti Sains Malaysia under USM MOOC and recently TVET institution like Polytechnic under Micro Credential platform has already implement it [7]. Upon completion, students will be given a certificate or badged indicate achievement, competency, and quality of the course taken. For the time being, the Malaysian Qualifications Agency (MQA) has acknowledged the use of MOOC for a credit transfer and accreditation of prior experiential learning (APEL) [8].

Education 4.0 is directly related to IR 4.0, namely IoT, cloud computing, big data analytics, augmented reality, and advanced robots and co-robots. The idea of hybrid learning using flipped classroom concept in Education 4.0 is the theoretical part of studies using video or e-note is conducted remotely while the practical activities like project and problem-based learning are implemented in classroom [9]. The IR 4.0, taking an example of IoT in education capable to assist teachers in monitoring students' performance such as attendance through the surveillance cameras [10]. In addition, videoconferencing using an IoT platform such as Zoom, Google Meet, and Microsoft Teams is widely adopted is facilitating educational institutions for enhanced academic experience [11].

By adopting the student-centered and sustainability-focused, Education 5.0 is a mixed of the well-established model from Education 1.0 to Education 4.0 that relying on advanced technologies and incorporating radically innovative aspects from IR 5.0 (Figure 11.3) [12]. To ensure learning process become more interesting and fun, game-based learning in Education 5.0 will be adopted. The elements such as leaderboards, rewards, and points will be given to enhance student motivation and engagement [13]. In addition, student-focused AI in Education

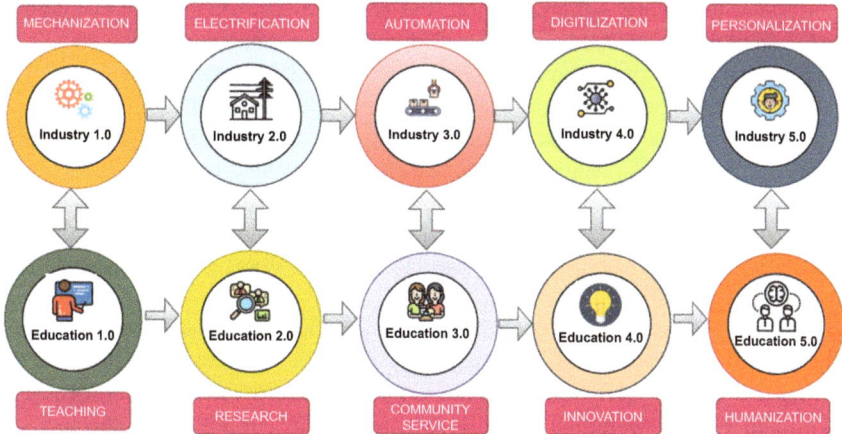

Figure 11.3   Transition from Industry 1.0 to Industry 5.0 and Education 1.0 to Education 4.0

5.0 has become a norm. AI tools such as Chat GPT [14] have been continuously used by students to help them in finding the suitable coding for projects and writing essays. The benefit of using these AI tools is the task can be completed in short duration as well as expose student with others possible solution that had not be taught in class.

11.3   Industry 5.0 in Engineering Education 5.0

Industry 5.0 focusing on the collaboration between humans and advanced technologies such as IoT, AI, and robotics. This shift aims to create more resilient, sustainable, and human-centric systems. Engineering education has been shifted from hands-on and practical emphasis to engineering science and analytical emphasis; to outcomes-based education and accreditation; to emphasizing engineering design; to applying education, learning, and social-behavioral sciences research; to integrating information, computational, and communications technology in education during the past 100 years [15]. Engineering education has been impacted as the result of advancement in technologies.

Practical hands-on experiences with IoT, AI, and robotics should be included in engineering programs as the usage of AI, machine learning, and advanced robotics is the central of Industry 5.0. Students will need to be proficient in coding, machine learning algorithms, and systems integration.

Other than that, Industry 5.0 emphasizing the need for systems that enhance human capabilities and well-being (human-centric). Students should learn to design products that not only apply advanced technology but also address human needs and improve user experience; thus, ergonomics and human factors should be included in the curriculum in engineering subjects.

Besides, teamwork, communication, and project management should be emphasized in engineering education. Students should be given group projects based on real-world scenarios to strengthen their teamwork and communication skills and with advanced technologies. Collaborative projects with industry partners are essential in the curriculum to students to ensure the students gain hands-on experience or practical experience. Additionally, students should be exposed to multidiscipline projects. This is due to Industry 5.0 requiring interdisciplinary solutions [13]. Engineering education will prioritize real-world problem-solving, encouraging students to tackle complex, open-ended problems. Courses and projects can be designed to enhance analytical skills and creative thinking.

The collaboration between industry and educator in developing curriculum will facilitate the formation of knowledge with the support of experts in the industry that can give a right direction to make enthusiastic, intelligent students and make them experts in future. Besides that, regular discussions with the industry partners can lead to improvement such as upgrading the curriculum's content which matches with the latest technology used in industry. Thus, the chances of graduates getting a job will be increased [16].

The demand for skilled workers with proficiency in new technologies and interdisciplinary expertise is increasing as Industry 5.0 continues to reshape the global economy. Traditional jobs are evolving, making it essential for the workforce to adapt to the changes. Consequently, the integration of technologies in engineering education is necessary to well prepare the students in a real working environment.

11.4   Challenges in Engineering Education 5.0 in Malaysia

The concept of Education 5.0 is characterized as student-centered that utilize advanced technology of IR 5.0. The idea of learning should be adaptable and flexible [17] aims to enhance the critical thinking and problem-solving skills of learners. Some of the major challenges in Table 11.1 faced by MOHE are explained below.

The program offered may be deferred since students prefer the studies that guarantee future jobs. Following the study in [18,19] from 2007 to 2021 for the number of unemployed first-degree graduates in Malaysia, the main reasons for employer did hire the local graduates were due to lack of technical skills required as well as the number of student enrolment in institutions kept increasing over the year. With customized study highlighted in Education 5.0, students may have a difficulty in finding a job since some of the engineering tasks required proficiency in certain skills that may not be selected by them. One of the factors for students choosing a major in studies is due to high income rate and future job guarantee [20,21], however with program offered that the future job yet exists may become a challenge in the number of enrollments [22].

The digital skills are one of the big challenges in Education 5.0 [23]. The authors in [24,25] reveal that issues such as internet speed capacity and coverage, infrastructure facilities and curriculum design, and pedagogy are still outdated for

Table 11.1   Summary of challenges in the realization of Education 5.0 in Malaysia

Reference	Challenges	Possible solution
[22]	Difficult to predict the future jobs that do not yet exist.	Collaboration between institution, industry and stakeholders should be established.
[23]	Readiness of students with the necessary digital skills in IR 5.0 world.	Education and training providers should provide an access to technology and resources as early as primary school.
[26]	Research on the effectiveness and impacts of Education 5.0 is limited.	Conducting more research by collecting the data based on selected well-established institutions.
[29]	Plagiarism issue when using AI tools.	Integrity policies should be endorsed, and action should be taken against the students that practice unethical work.
[36]	Hybrid offline and online training.	The high-speed internet coverage in rural areas like Sabah and Sarawak should be extended.
[37]	Most teachers are still lacking on the know-how of technology in teaching–learning situations.	More training should be provided, and ICT infrastructures should be upgraded in institutions so that teachers familiar with new technology.
[41]	MOHE need to bear high implementation cost.	Find alternative options such as open-source software to reduce the cost.

future IR 5.0. The absence of standardization in the technologies and platforms utilized by educational institutions can make resource sharing and exchange difficult. This could further widen the gap between students from different socio-economic backgrounds [13]. The digital divide between students' due to different facilities taking example a primary school in rural area still practices teacher-centered learning while higher education institutions are already toward MOOC.

The concept of Education 5.0 is relatively new, and the effectiveness is yet to be studied. For policymakers, segmenting the level of education, i.e. primary, secondary, and tertiary schools became difficult. Besides, the successfulness of this model cannot be guaranteed, and MOHE will not put Malaysia education at risk unless convincing research data is shown.

Ethical issues related to plagiarism right now become a major concern. A widespread use of Chat GPT has become troublesome for a cognitive assessment related to essay writing, problem solving as well as originality of technical manuscript in academic [27]. Some universities, namely Universiti Teknologi MARA, International Islamic University Malaysia, and Universiti Sains Malaysia have taken a step ahead by drafting the legal framework in maintaining the integrity of their institution [28,29]. In dealing with academic plagiarism, MOHE must intervene with the policies to produce a fair and standard provision in overcoming the loopholes in the educational institution policies.

Study conducted in [30] has proven the flexibility of the online learning with helps of variety of add-on functions in online platforms. However, the effectiveness of offline learning cultivating the experimental spirit throughout engineering students was less effective [31]. With hybrid offline and online learning could bring out whole set of comprehensive competencies for both teachers and students. For an engineering students, the limitation of practical assessment in online learning could be overcome by offline learning [32,33]. Still, MOHE need to overcome the issue of internet coverage and speed in rural areas of Malaysia [34] for this hybrid learning method to be successfully implemented [35].

Next is the challenge in teacher's competency in technology-based teaching [36]. Ishak *et al.* [37] show among 57% of teachers in Malaysia that used technology for education, the study reported only 39% of them admitted to not receiving any training [38]. Due to weaknesses of this, the online teaching tends to be short as compared to offline teaching, not to mention the difficulty of them using variety of device such as personal computers, laptops, tablets, and smartphones due to lack of competency in handling technology-based teaching. Besides, the use of CAD, CAM, and CAE tools for engineering studies requires significant computing resources and has simultaneously increased and required teachers to be experts in handling the tools. The study shows a good outcome using these tools works well and is expected to support the Engineering Education 5.0 later [39].

Right now, MOHE is the track of implementing MOOC in learning environment, and TVET institutions like polytechnic have put an effort into developing the micro credentials platform. The issue of high implementation cost can be overcome by encouraging teacher to develop their own micro credentials platform based on their expertise [40]. In addition, the work-based learning (WBL) culture with involvement of industry for engineering studies able to produce a positive outcome since their early implementation in 2015 [41]. This new learning approach was introduced to reduce the low employment rate due to lack of graduate skills. However, industry and stakeholders with low-income rates tend to isolate themselves from this program and MOHE need to put additional cost to fund this program [42].

Malaysian education system has made significant progress in recent years, with improvements in enrolment rates, literacy rates, and access to education. However, there are still challenges that need to be addressed, such as the quality of education, the relevance of the curriculum to the needs of the economy. Collaboration between industry partners and educators should be encouraged to bridging the gap between the academia and innovation.

11.5 Conclusion

Malaysian education system has made significant progress in recent years, with improvements in enrolment rates, literacy rates, and access to education. However, there are still challenges that need to be addressed, such as the quality of education, the relevance of the curriculum to the needs of the economy. Collaboration between industry partners and educators should be encouraged to bridging the gap between the academia and innovation.

References

[1] Sirat, M., and Wan, C. D. (2022). Higher education in Malaysia. In Symaco, L. P., and Hayden, M. (eds), *International Handbook on Education in South East Asia* (pp. 1–23). Singapore: Springer Nature.

[2] Altun, M. (2023). The ongoing debate over teacher centered education and student centered education. *International Journal of Social Sciences & Educational Studies*, 10(1), 106.

[3] Alharbi, A. M. (2023). Implementation of Education 5.0 in developed and developing countries: A comparative study. *Creative Education*, 14(5), 914–942.

[4] Hasnan, M. B., and Mohin, M. B. (2021). Implementation of LMS-CIDOS in polytechnic english language classroom: Issues and challenges. *Asian Journal of University Education*, 17(4), 527–537.

[5] Hursen, C. (2021). The effect of problem-based learning method supported by web 2.0 tools on academic achievement and critical thinking skills in teacher education. *Technology, Knowledge and Learning*, 26(3), 515 533.

[6] Songkram, N., Chootongchai, S., Khlaisang, J., and Koraneekij, P. (2021). Education 3.0 system to enhance twenty-first century skills for higher education learners in Thailand. *Interactive Learning Environments*, 29(4), 566–582.

[7] Ahmat, N. H. C., Bashir, M. A. A., Razali, A. R., and Kasolang, S. (2021). Micro-credentials in higher education institutions: Challenges and opportunities. *Asian Journal of University Education*, 17(3), 281–290.

[8] Malaysian Qualification Agency (MQA). (2019, May 10). *Guideline on Micro-Credential.* https://www2.mqa.gov.my/QAD/garispanduan/GuidelineonMicro-credential10Mei.pdf. Accessed 20 Jan 2024.

[9] Moraes, E. B., Kipper, L. M., Hackenhaar Kellermann, A. C., *et al.* (2023). Integration of Industry 4.0 technologies with Education 4.0: Advantages for improvements in learning. *Interactive Technology and Smart Education*, 20(2), 271–287.

[10] Huk, T. (2021). From Education 1.0 to Education 4.0-Challenges for the contemporary school. *New Educational Review*, 36–46.

[11] Correia, A. P., Liu, C., and Xu, F. (2020). Evaluating videoconferencing systems for the quality of the educational experience. *Distance Education*, 41(4), 429–452.

[12] Lantada, A. D. (2020). Engineering education 5.0: Continuously evolving engineering education. *International Journal of Engineering Education*, 36(6), 1814–1832.

[13] Ahmad, S., Umirzakova, S., Mujtaba, G., Amin, M. S., and Whangbo, T. (2023). Education 5.0: Requirements, enabling technologies, and future directions. *Education and Information Technologies*, 28(3), 1–20.

[14] Firat, M. (2023). How chat GPT can transform autodidactic experiences and open education? *International Journal of Educational Technology in Higher Education*, 20(1), 1–15.

[15] Gürdür Broo, D., Kaynak, O., and Sait, S. M. (2022). Rethinking engineering education at the age of industry 5.0. *Journal of Industrial Information Integration*, 25, 100311.

[16] Subri, U. S., Sohimi N. E., Affandi, H. M., Noor, S. M., and Yunus, F. A. N. (2022). 'Let's collaborate': Malaysian TVET-engineering institution and industry partnership. *Journal of Technical Education and Training*, 14(2), 165–176.

[17] Barak, M., and Levenberg, A. (2016). A model of flexible thinking in contemporary education. *Thinking Skills and Creativity*, 22, 74–85.

[18] Moo, K. H., and Da Wan, C. (2023). Graduate employability in Malaysia: Unpacking the concept, policy and practices. *IIUM Journal of Educational Studies*, 11(2), 3–25.

[19] Seng, L. C. (2018). Malaysia public universities' graduate employability policies: An analysis of first degree graduates unemployment and underemployment issues. *International Journal of Social Science and Humanities Research*, 6(4), 480–489.

[20] Rosalina, D., Yuliari, K., Purnamasari, W., and Zati, M. R. (2020). Factors affecting intention in accounting study program students choosing the public accountant profession. *Jurnal Akuntansi dan Bisnis: Jurnal Program Studi Akuntansi*, 6(1), 86–95.

[21] Dhingra, S., and Machin, S. J. (2020). The crisis and job guarantees in urban India. *Journal of Economic Perspectives*, 34(4), 3–24.

[22] Al-Emran, M., and Al-Sharafi, M. A. (2022). Revolutionizing education with Industry 5.0: Challenges and future research agendas. *International Journal of Information Technology*, 6(3), 1–5.

[23] Khan, N., Khan, S., Tan, B. C., and Loon, C. H. (2021). Driving digital competency model towards IR 4.0 in Malaysia. In *Journal of Physics: Conference Series*, 1793(1), 012049.

[24] Aki, W. F. H. W. M., and Adnan, A. H. M. (2022). Promoting digital literacy in higher education: Case study of a medical laboratory program in Malaysia. *International Journal of Academic Research in Progressive Education and Development*, 11(4), 1–12.

[25] Norman, H., Adnan, N. H., Nordin, N., Ally, M., and Tsinakos, A. (2022). The educational digital divide for vulnerable students in the pandemic: Towards the new agenda 2030. *Sustainability*, 14(16), 10332.

[26] Gupta, K. P., and Bhaskar, P. (2023). Teachers' intention to adopt virtual reality technology in management education. *International Journal of Learning and Change*, 15(1), 28–50.

[27] Cheah, J. K. S. (2024). Academic plagiarism in Malaysia: Perspectives. In Eaton, S. E. (ed), *Second Handbook of Academic Integrity* (pp. 31–39). Cham: Springer Nature.

[28] Zain, M. I. M., Rahmat, N. E., Zulkarnain, M. N., and Awasthi, S. (2021). Plagiarism of academic writing in Malaysian universities: A legal analysis. *Environment-Behaviour Proceedings Journal*, 6(16), 197–202.

[29] Zain, M. I. M., Rahmat, N. E., and Zulkarnain, M. N. (2021). A legal perspective on academic plagiarism of research writing in Malaysian universities. *Malaysian Journal of Social Sciences and Humanities (MJSSH)*, 6(7), 282–292.

[30] Huang, R. H., Liu, D. J., Tlili, A., Yang, J., Wang, H., and Zhang, M. (2020). *Handbook on Facilitating Flexible Learning During Educational Disruption: The Chinese Experience in Maintaining Undisrupted Learning in COVID-19 Outbreak*. Beijing: Smart Learning Institute of Beijing Normal University, 54 p.

[31] Kim, J., and Ryu, S. J. (2023). Enhancing sustainable design thinking education efficiency: A comparative study of synchronous online and offline classes. *Sustainability*, 15(18), 13293.

[32] Huang, R., Tlili, A., Wang, H., *et al.* (2021). Emergence of the online-merge-offline (OMO) learning wave in the post-COVID-19 era: A pilot study. *Sustainability*, 13(6), 3512.

[33] Dang, F. (2023). Exploring a hybrid online and offline English teaching model based on model hierarchy analysis. *Applied Mathematics and Nonlinear Sciences*, vol. 9, no. 1, pp. 1–15.

[34] Affendy, N. A. R., Ayi, T. P., and Basir, M. S. S. M. (2022, September). A study on students' online learning performance and satisfaction during the pandemic: The case for TVET Education Institution in Borneo Malaysia. In *2022 XII International Conference on Virtual Campus (JICV)* (pp. 1–4). Piscataway, NJ: IEEE.

[35] Ni, L. B. (2023). Hybrid and virtual learning: Bridging the educational and digital device. *Malaysian Journal of Social Sciences and Humanities (MJSSH)*, 8(1), e002035.

[36] Ong, Q. K. L., and Annamalai, N. (2024). Technological pedagogical content knowledge for twenty-first century learning skills: The game changer for teachers of industrial revolution 5.0. *Education and Information Technologies*, 29(2), 1939–1980.

[37] Ishak, N., Din, R., and Othman, N. (2022). Teachers' perceptions and challenges to the use of technology in teaching and learning during COVID-19 in Malaysia. *International Journal of Learning, Teaching and Educational Research*, 21(5), 281–308.

[38] Ebrahimi, S. S., and Jiar, Y. K. (2018). The use of technology at Malaysian public high schools. *Merit Research Journal of Education and Review*, 6(3), 54–60.

[39] Sola-Guirado, R. R., Guerrero-Vacas, G., and Rodríguez-Alabanda, Ó. (2022). Teaching CAD/CAM/CAE tools with project-based learning in virtual distance education. *Education and Information Technologies*, 27(4), 5051–5073.

[40] Othman, I. W., Mokhtar, S., and Esa, M. S. (2022). The stages of national education system operation: Issues, rationale, and challenges for the Ministry of Education Malaysia (MOE) in facing post pandemic norms of Covid-19. *International Journal of Education, Psychology and Counselling*, 7(47), 616–638.

[41] Moo, K. H. (2024). The implementation structure of work-based learning (WBL) in Malaysia: The perspective of industry. *IIUM Journal of Educational Studies*, 12(2), 3–20.

[42] Wan Mokhtar, W. N. N. N., Mohamad, N. H., Wan Nawawi, W. N., and Anuar, J. (2024). Stakeholders perspectives on workbased learning (WBL) implementation in Malaysia: A review. *Journal of Tourism, Hospitality and Culinary Arts*, 16(1), 597–606.

Chapter 12

An overview of ESG and Industry 5.0 integration in project sustainability

Lee Sun Heng[1]

Integrating environmental, social, and governance (ESG) criteria into project management, along with principles of Industry 5.0 (IR 5.0), is becoming a powerful approach to drive sustainability. IR 5.0, well defined by its human-centric approach, resilience, and sustainability, builds upon the technological advancements of Industry 4.0 (IR 4.0) by emphasizing collaborations between humans and machines. It provides a robust framework for integrating ESG principles into project management practices. This collaboration leads to better project management practices, where we are not only focusing on technological efficiency but also taking care of environmental, social responsibility, and uphold strong governance.

This chapter explores an overview of ESG integration in project sustainability with principles of IR 5.0 and creates lasting value for organizations and stakeholders. On the environmental criteria, it focuses on reducing the ecological impact of projects by using energy-efficient technologies, managing waste effectively, and making sustainable use of resources. Socially, it concentrates on the well-being of individuals and communities, promoting fair labor practices, health and safety, and community engagement. From the governance perspective, it highlights the importance of ethical management, transparency, and strictly adhere to the laws and regulations, in which building trust and accountability.

Furthermore, this chapter also focuses on the human-centric nature of IR 5.0, where cutting-edge technologies including artificial intelligence (AI), co-robotics (cobots), and the Internet of Things (IoT) work hand-in-hand with people to enhance their capabilities and lead to better project results. It also discusses the necessity of resilience in project management, stressing the need for adaptive solutions and risk management to deal with the gaps of interruptions with the aids of new technologies and predictive analysis. Additionally, this chapter highlights how ESG elements and IR5.0 can be integrated throughout the project lifecycle, from planning, execution to monitoring and closure. By identifying the barriers and suggesting approaches, this chapter lays out a conceptual guide for integrating ESG and Industry 5.0 in project management.

[1]HLS Pro Construction Sdn. Bhd., Penang, Malaysia

12.1   Evolution from IR 4.0 to IR 5.0

Industry 4.0 (IR4.0) enhances production efficiency by integrating machines and systems to streamline and improve the production process [1]. IR4.0 is a trend of technological revolutions and advancement, and is about big data analytics, innovation of technology, machine learning, IoT, and digitalization. IR4.0 has undoubtedly driven remarkable progress by integrating technology, new ideas, innovation, and digital transformation. This has especially boosted competitiveness in manufacturing and led to the creation of cyber-physical systems (CPS) that connect the world [1]. However, IR 4.0 has been criticized for being overly focused on technical and economic gains, and often neglecting the social and human dimensions that are equally important [2]. As a result, there is a growing recognition of the downsides of IR4.0, despite its economic success, its limited contribution to the improvement of human society, and its unsustainable social consequences [3–5].

In 2022, the European Commission (EC) introduced the term "Industry 5.0" [6], aims to rectify the shortcoming of IR4.0, whereby to enhance the human aspect, sustainability and resilience. This concept, although is relatively new to many emerging countries, it marks a very important shift in industry strategies, focusing not only from an economic point of view but also enhancing the societal well-being and environmental sustainability, which aligns with global sustainability goals. European industry plays an important role in the current economic and societal transformations. According to the EC, IR5.0 is a vision of industry that goes beyond effectiveness and productivity as the main objectives, and instead emphasizes the role of industry in societal contribution [7]. IR5.0 focuses more on the interaction between humans and machines, emphasizing how this process improves interactions, streamlines technology efficiency, and reflects a trend in human-machine integration.

As the new era continues to evolve, many stakeholders are encountered with challenges in progressing their technology implementation in various industries including manufacturing, construction, automotive, engineering, or education. According to a study by Deloitte, US manufacturing will be facing a shortage of skilled workers by 2030, with a big number of 2.1 million job vacancies [8]. In manufacturing, there are still many tasks that require the capabilities of the human brain in problem solving regarding safety, which is irreplaceable by robots. This is due to the need for human intuition and judgement to safely and efficiently perform key tasks associated with human life [1]. In the engineering field, there is increasing demand for technology adoption, especially in sectors such as electrical vehicles and strategic investment in IoT-connected vehicles [9]. This trend has huge implications for the manufacturing industry, where manufacturers are looking forward to more innovative design solutions, smart factories, and advanced data analytics solutions as part of the roadmap to prepare for a better sustainability society and engineering landscape [10]. Figure 12.1 shows the key differences from IR4.0 to IR5.0.

Industry 4.0
Digitalisation

Industry 5.0
Humanisation

Big Data Analytics,
Artificial Intelligent,
Machine Learning
Internet of Things

Efficiency,
Productivity
Sustainability

Human-centricity,
Sustainability,
Resilience

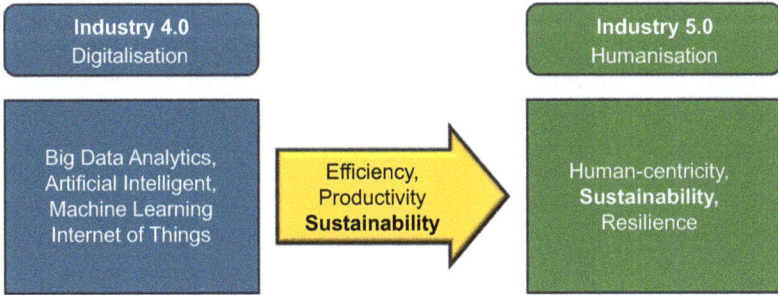

Figure 12.1   Key differences from IR4.0 to IR5.0

12.1.1   IR5.0 and ESG integration in project sustainability

IR5.0 literature is gently gaining attention from the researchers and considered as very early development and young stage [53,54]. While there are limited high-quality journal studies available, key contributions are emerging. Focusing on ESG integration and IR 5.0, Ghobakhloo *et al*. make significant contributions to the IR5.0 transition and its capability to facilitate the potential change of sustainable industry. This chapter has provided a robust theoretical write-up, identified 11 key enablers, and developed a strategy roadmap to the transformation of IR5.0 and promotes sustainable development. This is particularly relevant as it highlights the importance of human-centric elements and environmental sustainability. With the limitation callout, the future study shall extend to other industries to explore the integration of IR5.0 in delivering environmental and societal values.

In paper "Industry 5.0: Prospect and Retrospect," Leng *et al*. [54] provide a comprehensive study on the review of IR5.0. They thoroughly map out the transition of IR4.0 to IR5.0, with the integration of technologies like AI, robotics, digital twins, and human-machine collaborations to emphasize human involvement in the manufacturing process. This article mentions sustainability as its main pillar, but regrettably, it doesn't delve into the detailed integration of ESG into the principles of IR 5.0, particularly in project aspects. Apart from the theoretical discussion, the author suggests future study areas to cover the application and implementation path of IR5.0 in other industries, thereby enhancing its practical implications.

Rehman and Umar [15] investigate how technology in the IR5.0 can improve ESG elements and achieve ESG goals. They present a well-rounded study to mitigate risk associated with greenwashing, environmental impacts, social inequality, and governance issues. It also emphasizes the IR5.0 technology's advancement to prevent fraud, improve supply chain transparency, and improve governance's fraud detection. Furthermore, it also focuses on human-centric and sustainable technology, which fosters worker safety and social inclusion. However, it lacks practical examples or case studies on demonstrated ESG and IR5.0 integration.

In the paper by Pietro De Giovanni [20], he offers a strong foundation in ESG analysis and links it with a transformation of the digital environment and the metaverse in IR5.0. The paper discusses and highlights the potential revolutions and benefits gained from responsible digital technologies and human-centric goals, which have significant implications for sustainability, business strategy, and societal development, particularly as they transition into IR 5.0. However, empirical validations of project-level sustainability and governance risk management are underexplored. Subsequent studies utilizing case studies would enhance the creditability of the literature and reliability of findings.

The body of literature provides a theoretical foundation for the role that IR5.0 plays in promoting sustainability and ESG integration. However, there is an obvious need for additional practical studies and research in order to improve its credibility and its capacity to be applied in the real world.

12.1.2   Research gaps

While existing literature on IR5.0 focuses more on the technological advancement and its benefits, yet there remains a gap in exploration how ESG can integrate systematically into project sustainability, especially the empirical studies in real-world projects in an emerging market. Thus, there is limited study can be as reference. Furthermore, while some of the papers discuss extensively on the technological transformation, the governance aspect and social impact including the ethical AI and society equity are underexplored. Therefore, the existing research gaps create a room for opportunities to develop a need for the topic as proposed.

12.2   The synergy of ESG and Industry 5.0

Integrating ESG principles into IR5.0 marks a transformative shift on how we approach project sustainability. This collaborative effort not only strengthens corporate commitment but also aligns with industry practices in broader sustainability goals. By leveraging advanced technologies and human-centric strategies, the synergy between ESG and IR5.0 can create sustainable value and a robust framework that addresses critical social and environmental challenges.

The paradigm shifts from IR4.0 to IR5.0 focus on three main pillars: human-centricity, sustainability, and resilience. The well-being of humans is becoming important and set at the forefront of any process, and it entails the working environment, human rights, mental and physical health, growth, and development in order to create a more resilient environment. The United Nations (UN) describes sustainability as "meeting the needs of the present without compromising the ability of future generations to meet their own needs," back in 1987 [11]. The introduction of 17 sustainability development goals (SGDs) in 2015 further enhanced this concept. With the enhancement, the ESG factors were placed in the picture, and sustainability became more significant, getting numerous attentions from all over the world [12,13]. Resilience in the IR5.0 context is multifaceted and defined as the capacity of industries, systems, humans, and organizations to adapt and build responses to threats while keeping operations running continuously [14].

IR5.0, although still in its early phase of development, holds the potential to make a positive impact on all aspects of ESG [15]. The emergence of the IR5.0 has provided a pivotal platform and prospects for developing countries such as India, according to the study by Sharma and Gupta [16]. By prioritizing people in the industrial process and technological innovation, the next industrial revolution will create a more sustainable and equitable society. The availability of technologies is the key to transformations, providing opportunities for the creation of potential innovations. Embracing this breakthrough accelerates progress in a variety of sectors and improves the ability of individuals and organizations to adapt and respond to market disruption and volatility. By utilizing the entire spectrum of opportunities, companies can become resilient and responsive to market demands.

12.2.1   Enhancing environmental

Integrating ESG with IR5.0 activities offers significant benefits by improving the preservation of the environment. This approach encourages businesses to minimize their ecological footprint [17]. Focusing on sustainability efforts enables companies to utilize green technologies for waste reduction and resource conservation, thereby enhancing operational efficiency and long-term profitability. For example, AI-driven analytics can optimize energy usage and reduce carbon footprints, while IoT devices can monitor real-time reading and improve traceability and transparency.

Additionally, this commitment to sustainability can enhance brand reputation and attract environmentally conscious consumers, which significantly boosts the customer's satisfaction and loyalty. These technologies enable companies to not only meet ESG standards but also exceed them by continuously improving their sustainability performance. Ultimately, enhancing environmental stewardship paves the roadmap for stakeholders who value sustainability, leading to new collaborations and partnerships [18].

12.2.2   Human-centric automation and social equity

Alongside the technology's advancement, the core aspects of IR5.0 are emphasized in the human-centric automation to enhance productivity and well-being. This approach can produce a more inclusive and equitable workplace by leveraging machines and robots. By delegating hazardous and repetitive tasks to the robots, humans can focus on fulfilling roles that require more complex judgment. This approach aligns with the social component in ESG not only provides a safer workplace and reduces injury incidents, but at the same time improves job satisfaction and promotes innovation cultures. According to a 2023 Deloitte global human capital trend survey, 87% of business leaders believe that developing the right workplace model is crucial for business success [19]. Furthermore, IR5.0 could foster equality through diversity and inclusion efforts, mitigating biases in hiring and promotion in the new employment process [20].

12.2.3   Governance and ethical AI

Effective governance is essential for successful integration of ESG principles and IR5.0 technologies. In the view of IR5.0, it has the potential to significantly enhance governance practices by supporting transparency, enabling data-driven

Table 12.1   The synergy of ESG and IR5.0 implementation

Company	Synergy of ESG -1R5.0	Impact to the Society and Environment	Source
Toyota Motor Corporation	human-centric approach to manufacturing, incorporating collaborative robots (cobots)	improve efficiency while prioritizing employee well- being	[24–26]
Unilever	utilized advanced data analytics and AI to optimize supply chains	reducing waste and energy consumption across its production facilities significant reductions in carbon emissions and resource usage	[27,28]
IKEA	adopted a circular economy approach, focusing on sustainable product design and supply chain management.	significantly reduced waste, and promoted sustainable consumption	[29,30]
Siemens	integrating AI-driven predictive maintenance systems across its manufacturing plants.	resulted in lower greenhouse gas emissions and reduced waste,	[30,31]

Source: Created by author.

decision-making, thereby also aligning with the ESG objectives. For instance, predictive data analytics can significantly contribute to risk management by offering early warnings to the organization, enabling the implementation of proactive measures [21]. This approach not only strengthens better governance in all aspects of management but also helps reduce governance-related risks such as fraud, unethical behavior, and regulatory noncompliance by identifying risk [22].

The basic principles of IR5.0 are concerned with driving long-term stakeholder interest in aligning profitability and sustaining the future, as well as incorporating their concerns into strategies and implementation plans through risk management. With the integration of ESG and IR5.0, there is growth of opportunities for the businesses in creating responsible practices while maintaining profitable activities. Table 12.1 illustrates the key industry players who are collaborating and implementing IR5.0 to benefit society, the environment, and their own organizations.

12.3   How ESG impact project sustainability

Project sustainability has emerged as a vital focus in project management, driven by the increasing importance of ESG criteria. As awareness of sustainability grows among the stakeholders, integrating ESG principles into project planning and execution is no longer just a trend but a necessity. These criteria are important in assessing the long-term impact of environmental, societal, and governance practices. In the current trend, business owners may enhance corporate branding by implementing ESG principles to attract more investors and increase brand loyalty.

In today's business landscape, companies can significantly enhance their corporate branding by embedding ESG principles into their projects. Implementing this will not only attract more investors, but it will also foster greater brand loyalty. With the project goals aligning with ESG standards and addressing potential environmental and social risks, organizations can achieve long-term success while ensuring compliance with regulations and industry standards. Sustainable project management not only benefits the environment and society but also strengthens an organization's resilience and market competitiveness. It represents a holistic approach that balances the present's needs without compromising the ability of future generations to meet their own needs.

12.3.1 Integrating ESG into project life cycle

The project life cycle, as outlined in the 7th edition of the Project Management Body of Knowledge (PMBOK) Guide, focuses on sequential phases from initiation to closure, i.e., project initiation, planning, execution, monitoring, and completion [32]. Each phase is unique and necessitates specific processes and practices to ensure the project's success. Integrating ESG into the project life cycle is a game-changing strategy that restructures how projects are thought, planned, and executed. With this approach, ESG integration ensures projects are financially feasible, ecologically sustainable, socially equitable, and at the same time governed with transparency and accountability. This holistic approach to project management aligns with the growing demands for sustainability across all facets of business operations.

12.3.1.1 Initiation: clearly define ESG scope

Establishing clear ESG goals alongside the traditional project scope is essential during project initiation to ensure integration of sustainability and ethical considerations from the outset. This includes clearly identifying how the project can actively contribute to exemplary ESG practices. On the environmental front, it is crucial to define comprehensive environmental assessments and risk management strategies that address climate mitigation and biodiversity conservation [33]. This involves identifying potential environmental impacts, setting clear targets for reducing emissions, and preserving natural habitats, thereby aligning the project with broader climate goals.

Secondly, prioritize societal equity by engaging stakeholders and the local community to ensure the project meets societal needs and expectations [34]. This includes involving community members in decision-making processes and addressing social inequalities, thus fostering a sense of ownership and support for the project. Finally, thoroughly examine the governance impacts by identifying and adhering to relevant national, local, and council regulations and codes. This ensures that the project not only complies with legal requirements but also upholds the highest standards of governance, contributing to the project's long-term success and acceptance. By embedding these ESG considerations into the project's foundation, organizations can create a more sustainable, equitable, and ethically sound framework that benefits all stakeholders

12.3.1.2   Planning: creating a sustainable framework

The planning stage is a crucial and often time-intensive phase of the project life cycle. At this stage, all the concepts and ideas generated during the initiation phase are meticulously documented and transformed into a detailed, actionable plan. This includes clearly defining the assessment plan as a comprehensive framework that can be executed with minimum errors and maximum efficiency. Creating a sustainable roadmap for the next step requires the integration of ESG assessment solutions, procurement strategy, supply chain management selections, design development planning, and overall project approach into this framework.

Technology plays a key role during this stage, as it helps optimize both costs and timelines. Allocating the right resources for the ESG initiatives is crucial, as they contribute to the project's long-term impact. This includes identifying the most cost-effective options, such as design option selections, supply chain management selections, green technologies, and social programs, which can enhance project sustainability and societal benefits. Additionally, it is critical to proactively incorporate the costs associated with ESG efforts into the overall project budget during the planning phase. This ensures the inclusion of the ESG initiative, paving the path for a successful and sustainable project implementation.

12.3.1.3   Execution: implement ESG practices

A make-it-happen stage, implementing ESG practices during the execution phase is essential for translating sustainability goals into tangible actions. At this phase, the project team can actively incorporate sustainable practices, such as minimizing waste, utilizing renewable energy sources, and upholding fair labor practices. This is a repetitive task and is to be executed throughout the process until the end of the project. These efforts not only contribute to achieving the project's ESG objectives but also enhance overall project efficiency by reducing resource consumption and fostering a positive work environment.

Leveraging technology is important for enhancing efficiency, ensuring accuracy, and meeting ESG objectives. Real-time data enables the project team to make informed decisions and timely adjustments, optimize sustainability outcomes, and keep the project aligned with its ESG goals [30]. For example, consider the use of prefabrication systems, modular building design, building information modeling (BIM) [35], and the deployment of IoT sensors for real-time monitoring of energy consumption and emissions [36] on a construction site. This streamlined process not only satisfies regulatory requirements but also positively contributes to the ESG commitments.

12.3.1.4   Monitoring and controlling: ESG performance

To ensure that ESG efforts are effectively executed, real-time monitoring and controlling of performance metrics become indispensable. Leveraging technology use of sensors, by providing instant feedback on the environmental key indicators such as air quality, carbon emissions, resource usage, and social impacts as they occur. However, supply chain management can enhance transparency and accountability by using blockchain technology to monitor and track the origin and sustainability of materials. By continuously monitoring these metrics, projects can adapt to changing circumstances and optimize their sustainability outcomes. Through prioritizing ESG considerations, projects can boost stakeholder satisfaction and build trust with stakeholders who value transparency and responsible sourcing, as they are increasingly valuing ethical and environmentally responsible practices.

12.3.1.5 Closure: continuous improvement and reporting

During the closing phase of a project, the focus shifts to a thorough review of the project's performance against its ESG goals, comparing the achieved outcomes against the planned ones. At this stage, the outcomes are documented, and lessons learned are recorded, identifying both successes and paving the way for continuous improvement in ESG practices. Feedback and postmortems from stakeholders are equally important in developing an insight plan for a future project.

Transparent reporting on ESG performance to stakeholders is also crucial at this stage; it involves clearly communicating the project's impact on environmental sustainability, social responsibility, and governance standards. This not only enhances accountability and trust but also reinforces the organization's commitment to sustainable practices. The closing phase becomes a springboard for ongoing refinement and advancement in ESG practices, ensuring that each project contributes to a legacy of continuous improvement and responsible growth. This can enhance the corporate branding and strengthen brand loyalty with stakeholders, ensuring long-term support for future initiatives. Figure 12.2 shows an overview of the integration of ESG and IR 5.0 in project sustainability.

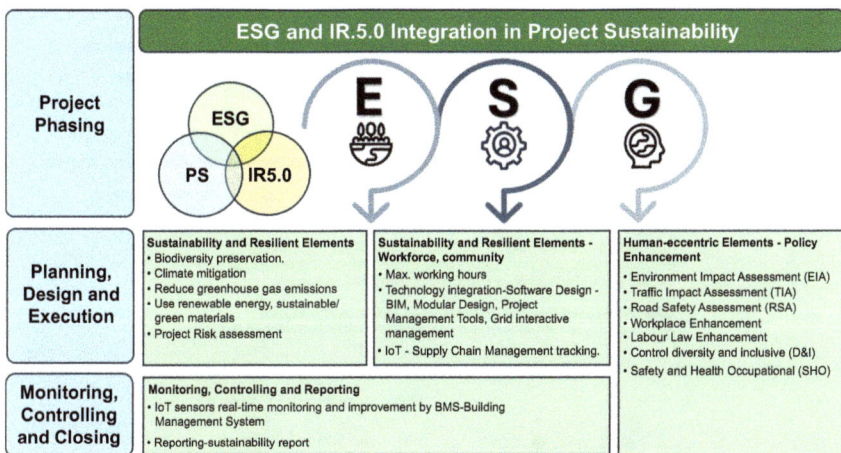

Figure 12.2   Integration of ESG and 1R 5.0 in project sustainability (PS)

12.4   Key contributions of ESG to Industry 5.0

The integration of ESG factors into the industry IR5.0 framework is crucial for advancing project sustainability and ensuring that technological advancements contribute to the broader goals of environmental protection, social equity, and responsible governance. Key contributions from the ESG are multifaceted, and in the project sustainability field, it has been widely recognized that the power of the ESG is beyond what we have seen.

12.4.1   Enhanced ESG reporting disclosure

According to recent research by Statista [37] for worldwide sustainability reporting, the Asia Pacific region has had the largest growth rate in sustainability reporting, showing a significant shift from about 50 percent to almost 90 percent of companies engaging in sustainability reporting, as depicted in Figure 12.3. These key contributions revealed that investors in many regions across the world, especially Asia Pacific, have grown in awareness and experienced the benefits of the long-term success of the ESG investment journey. Global technology leaders such as NVIDIA, Intel, Cisco, Apple, and PayPal have taken the lead in reducing their carbon footprint to net zero greenhouse gases (GHGs) or negativity, conserving resources like water and electricity, and contributing to environmental protection through transparency in sustainability reporting [38].

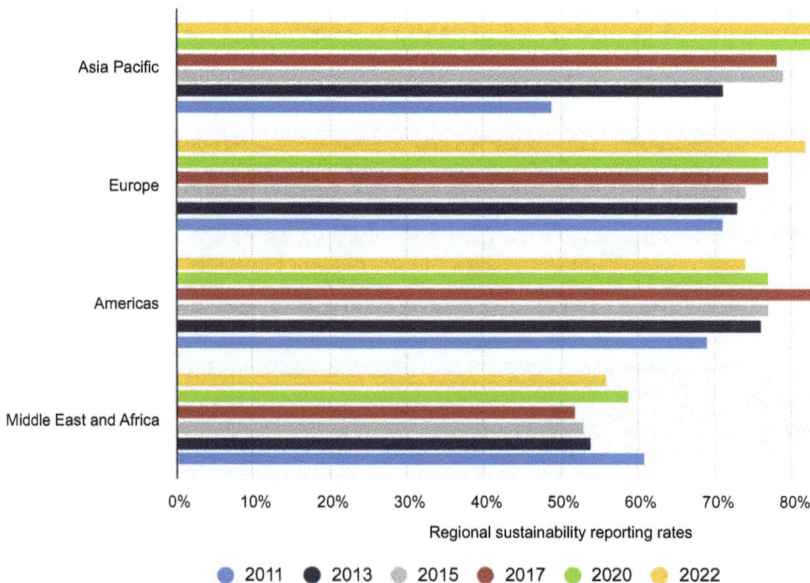

Figure 12.3   Sustainability reporting rates of firms worldwide

12.4.2   Improvement in human-centric approach

In IR5.0, ESG promotes inclusive workplaces that prioritize employee well-being, diversity, and fair labor practices, valuing and integrating the human workforce alongside advanced technologies. Cisco, in the 2023 reporting, disclosed that they have conducted an ESG materiality assessment every two years to understand the most significant ESG topics that stakeholders focus on and continuously enhance and evaluate strategies and goals through validation priorities in relation to business risks and opportunities. Figure 12.4 displays the assessment results. The assessment has identified five out of the 18 most important human-centric approaches, which include talent, human rights, diversity and inclusion, employee well-being, and employee safety and health [39]. These approaches are highly prioritized by stakeholders and significantly impact the business. This further suggests that the current business approach incorporates "human-centric" trends.

NVIDIA, in their latest 2024 report, emphasized people, diversity, and inclusion, with the company treating employees as the greatest asset in building long-term value for stakeholders. Fifty percent of their board of directors are ethically or

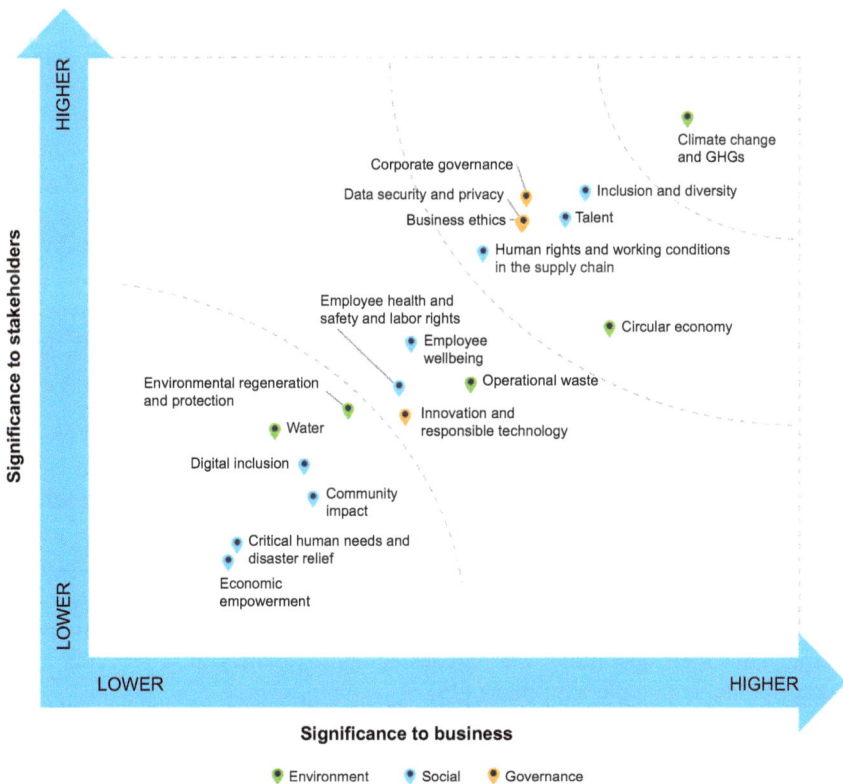

Figure 12.4   ESG materiality matrix

gender diverse, and they have invested in a diverse, talented pipeline globally, particularly among women who are underrepresented in the technology profession [40]. More and more corporate investors, including technology leaders, are concerned about people's investment alongside technological growth.

12.4.3   Reduced environmental impact

12.4.3.1   Biodiversity preservation

One of the critical components of the environmental impact in the project life cycle is the efforts to protect our ecosystem, biodiversity conservation. Any projects that involve green filed or brown field development will directly or indirectly disrupt natural habitats [41]. It can result in far-reaching consequences for ecosystems as well as the local communities. With the integration of ESG and IR5.0, potential impacts on biodiversity early in the project life cycle have been properly assessed, and the mitigation plan has been taken care of. Even in an assessing large-scale project, Environmental Impact Assessment (EIA) is a mandatory long-term monitoring measures throughout the project cycle and is include the preservation of natural habitats, restoration of degraded ecosystems, waste management, and the use of biodiversity offsets [42]. By prioritizing biodiversity conservation, projects can contribute to the protection of vital ecosystems and ensure that they do not harm the environmental balance on which both human and natural systems depend.

12.4.3.2   Climate mitigation

Reducing environmental impact through climate mitigation in the IR5.0 is a transformative shift. Industries have the potential to make significant contributions to global climate mitigation efforts by utilizing the ESG approach, implementing energy-efficient designs, renewable energy, and sustainable resource optimization. Energy efficiency design widely optimizes energy consumption across various sectors, particularly in the construction, oil and gas, and technology fields, reducing the environmental impact through net zero carbon footprint or decarbonization [43]. Many studies show that industry operations can be in line with global climate mitigation goals if they incorporate energy-efficient design to reduce a project's carbon footprint [44].

Driven by ESG with initiative IR5.0, renewable energy resources and circular economy not only reduce energy used but also at the same time reduce carbon footprints and promote climate mitigation [45]. As a market leader, Microsoft has committed to becoming "carbon negative" by 2030. They are leading the way in implementing carbon-free electricity and clean energy solutions in their business operations, as a long-term investment in protecting ecosystems [46]. Aside from that, Apple aims to achieve carbon neutrality in 2030 by sourcing 100% renewable electricity for their operations and purchasing approximately 20% recycled materials for producing Apple's products [38].

12.4.4   Technology advancement

The technological advancements in Industry 5.0, combined with ESG criteria, enable real-time monitoring and adaptation of industrial processes, which is a

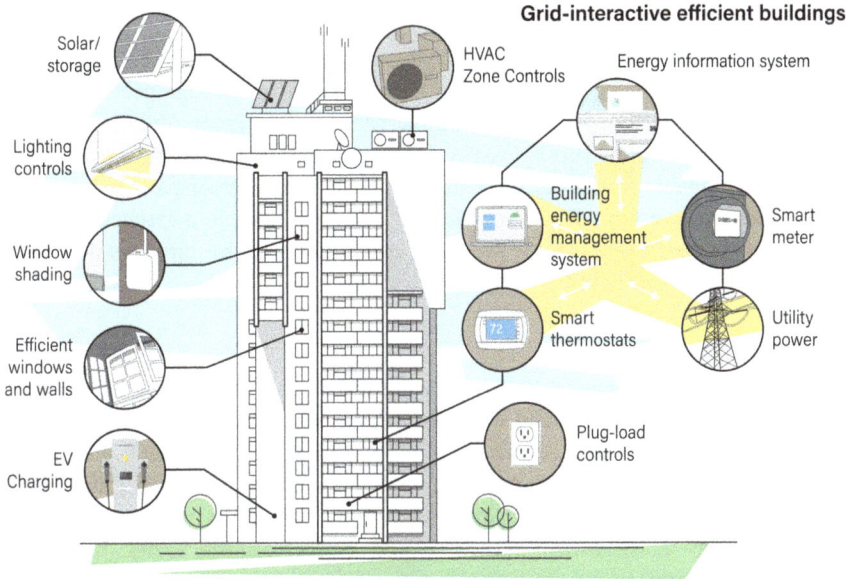

Figure 12.5   Grid-interactive efficient buildings

critical contribution to the real world. Technologies such as IoT remote sensors and AI-driven data analytics enable continuous monitoring of energy use, enabling prompt responses to potential environmental issues and necessary adjustments [5,47].

Advanced technologies, such as building energy simulations and BIM, allow for the optimization of energy consumption in buildings and infrastructure projects, which directly contributes to reducing GHG emissions [33]. These tools enable the simulation of various scenarios, particularly those involving different climate and environmental risks, thereby facilitating better risk preparation. Furthermore, the American Council for an Energy-Efficient Economy (ACEEE) has explored and engaged in a project to study grid-interactive efficient buildings to understand the opportunities and barriers and develop a program to support these efforts [48]. Grid-interactive, energy-efficient buildings consume less energy and produce less carbon. Thus, building design optimization for energy efficiency and grid integration benefits customers and system operators by providing flexible building loads to the grid. Figure 12.5 illustrates the overall concept of a grid-interactive, efficient building.

12.5   Challenges

Even though the IR5.0 is progressing toward transformation, the path will continue to develop alongside the challenges. The IR5.0 faces several key challenges that are listed below.

12.5.1   Sustainable supply chain management

Ensuring the ESG compliance in the entire supply chain management (SCM) can be difficult, especially for global networks that span different regions. Compliance with the local authority specifications, governance requirement, and global policy may be time-consuming. These possess significant challenges for the organizations that seek sustainable branding. Lack of external stakeholder or partnership in the supply chain concerning ESG knowledge, such as lack of awareness of ESG practices, experiences, and know-how, is also hindering the progress of project sustainability [49,50].

12.5.2   Fraud issues

Technological advancements in Industry 5.0, such as IoT and AI, have raised concerns about data privacy and cybersecurity, particularly when handling ESG-related information. Therefore, data reliability and integrity become an issue in data collection and quality validation moving forward. Major fraud risks related to ESG, such as regulatory non-compliances, greenwashing, money laundering, governance-related fraudulent activities, and nonperforming loans, have been observed worldwide. In the greenwashing scandal activities, DWS, the asset management arm of Deutsche Bank, was accused of excessively overstating its funding in ESG and misleading investors in sustainable investment [51]. Therefore, it is crucial to focus on generating transparency in ESG reporting.

12.5.3   Skill gaps

Successful integration of ESG principles into Industry 5.0 projects requires expertise in specific areas such as data analytics, emerging technologies, and knowledge of sustainability. The shortage of talent with these competencies can slow down ESG adoption. However, the workers' talent is still crucial to bridge the gap between the projected outcome and what actually happens in the real world. Tech leaders such as Apple, Nvidia, and Microsoft are heavily investing in upskills training and employee engagement programs to effectively meet market demands [38,40,46]. Industry 5.0 sets a new agenda for human-centric and assistive technology development and shifts the focus to a resilient society. Therefore, at the same time, awareness, education, training, and development are essential to equipping the workforce with the necessary skills and progress for sustainability.

12.5.4   Regulatory complexity

The lack of standardization and clear guidelines in regulatory compliance related to ESG practices has left progress in the midst. These challenges will necessitate collaboration between government, industry, education society, and civil society to develop standardized practices, supportive policies, and a culture of sustainability that can guide Industry 5.0 toward a more sustainable and equitable future. In Asia Pacific, approximately 60% of small and medium-sized enterprises (SMEs) view local government agencies as their primary source of guidance on ESG practices, as they are the foremost adopters of UN approaches. In this context, a recent world-wide statistic [52] suggests that a top-down ESG implementation process and compliance are among the most effective ways to drive the implementation of the ESG initiative. However, we need to establish a clear process standard and a set of ESG goals to guide us in the right direction.

12.6 Limitation and conclusion

During the writing of this chapter, a limited number of studies had been published, particularly pertaining to the sustainability of Industry 5.0 projects and their impact on ESG. Many researchers discussed principles of IR4.0 that carry forward to IR5.0. However, the integration of IR5.0 with ESG principles represents a crucial step toward building resilient, responsible, and sustainable organizations. Collaborative partnerships, continuous learning, and government support are identified as key enablers for successful adoption. By embracing Industry 5.0 technologies and methodologies, companies can not only mitigate ESG-related risks but also create value for stakeholders and contribute positively to global sustainability goals. Ultimately, this integration not only addresses current challenges but also paves the way for a more sustainable and prosperous future in project management. Future research should also focus on developing frameworks and methodologies for measuring the impact of ESG integration within IR 5.0 projects. As Rehman and Umar [15] argue, the success of IR 5.0 will depend on the ability of industries to adapt to new governance models that prioritize transparency, ethical decision-making, and social equity.

This chapter establishes a foundation for studying the implementation of ESG in alignment with IR5.0 within the project, highlighting its challenges and key contributions. In the real world, the return on investment always hinders the ESG initiative unless a set of ESG practices and compliance is firmed and regulated; otherwise, the organization's financial goal remains opaque or unreliable. Therefore, the support and commitment of stakeholders and corporate partnerships on the ESG path are relatively crucial for driving the success of a project toward long-term sustainability. By embedding ESG principles into Industry 5.0, industries can achieve balanced and resilient growth that benefits not just the economy but also society and the environment.

References

[1]   S. Grabowska, S. Saniuk, and B. Gajdzik, "Industry 5.0: Improving humanization and sustainability of Industry 4.0," *Scientometrics*, vol. 127, no. 6, pp. 3117–3144, 2022, doi:10.1007/s11192-022-04370-1.

[2]   X. Xu, Y. Lu, B. Vogel-Heuser, and L. Wang, "Industry 4.0 and Industry 5.0—Inception, conception and perception," *Journal of Manufacturing Systems*, vol. 61, pp. 530–535, 2021, doi:10.1016/j.jmsy.2021.10.006.

[3]   S. Nousala, G. Metcalf, and D. Ing, (eds.), *Industry 4.0 to Industry 5.0: Explorations in the Transition from a Techno-Economic to a Socio-Technical*

Future. Berlin: Springer, 2023. [Online]. Available at: https://doi.org/10.1007/978-981-99-9730-5.

[4]   A. R. Santhi and P. Muthuswamy, "Industry 5.0 or Industry 4.0S? Introduction to Industry 4.0 and a peek into the prospective Industry 5.0 technologies," *International Journal on Interactive Design and Manufacturing (IJIDeM)*, vol. 17, no. 2, pp. 947–979, 2023, doi:10.1007/s12008-023-01217-8.

[5]   M. Asif, C. Searcy, and P. Castka, "ESG and Industry 5.0: The role of technologies in enhancing ESG disclosure," *Technological Forecasting and Social Change*, vol. 195, p. 122806, 2023, doi:10.1016/j.techfore.2023.122806.

[6]   Industry 5.0, a transformative vision for Europe, Research and Innovation. Available at: https://research-and-innovation.ec.europa.eu/knowledge-publications-tools-and-data/publications/all-publications/industry-50-transformative-vision-europe_en (accessed June 10, 2024).

[7]   Industry 5.0, Research and Innovation, 2024. Available at: https://research-and-innovation.ec.europa.eu/research-area/industrial-research-and-innovation/industry-50_en (accessed June 22, 2024).

[8]   S. P. Wellener, V. Reyes, H. Ashton, and C. Moutray, "Creating pathways for tomorrow's workforce today," *Deloitte Insights*, 2023. Available at: https://www2.deloitte.com/us/en/insights/industry/manufacturing/manufacturing-industry-diversity.html.

[9]   M. Dennis, "Engineering industry trends: A 2024 outlook," *Planet Forward*, 2024. Available at: https://www.theplanetforward.com/engineering-industry-trends-a-2024-outlook/ (accessed June 26, 2024).

[10]   T. Lin, "Manufacturing in 2024: Key data and analytics trends shaping the industry's future – Wavicle Data Solutions," *Wavicle Data Solutions*, 2023. Available at: https://wavicledata.com/blog/manufacturing-2024-key-data-analytics-trends/ (accessed June 27, 2024).

[11]   United Nations, "Sustainability | United Nations," *United Nations*. Available at: https://www.un.org/en/academic-impact/sustainability (accessed June 26, 2024).

[12]   M. Taliento, C. Favino, and A. Netti, "Impact of environmental, social, and governance information on economic performance: Evidence of a corporate 'sustainability advantage' from Europe," *Sustainability*, vol. 11, no. 6, p. 1738, 2019, doi:10.3390/su11061738.

[13]   C. Consolandi, H. Phadke, J. Hawley, and R. G. Eccles, "Material ESG outcomes and SDG externalities: Evaluating the health care sector's contribution to the SDGs," *Organization & Environment*, vol. 33, no. 4, pp. 511–533, 2020, doi:10.1177/1086026619899795.

[14]   D. Romero and J. Stahre, "Towards the resilient Operator 5.0: The future of work in smart resilient manufacturing systems," *Procedia CIRP*, vol. 104, pp. 1089–1094, 2021, doi:10.1016/j.procir.2021.11.183.

[15]   A. Rehman and T. Umar, "Literature review: Industry 5.0. Leveraging technologies for environmental, social and governance advancement in

corporate settings," *Corporate Governance*, 2024, doi:10.1108/cg-11-2023-0502.

[16] R. Sharma and H. Gupta, "Harmonizing sustainability in Industry 5.0 era: Transformative strategies for cleaner production and sustainable competitive advantage," *Journal of Cleaner Production*, vol. 445, p. 141118, 2024, doi:10.1016/j.jclepro.2024.141118.

[17] M. Appiah, M. Li, M. A. Naeem, and S. Karim, "Greening the globe: Uncovering the impact of environmental policy, renewable energy, and innovation on ecological footprint," *Technological Forecasting and Social Change*, vol. 192, p. 122561, 2023, doi:10.1016/j.techfore.2023.122561.

[18] B. Kapeller, R. Plummer, J. Baird, and M. Jollineau, "Assessing factors of environmental stewardship success: Organizational perceptions from the Niagara region of Ontario, Canada," *Environmental Management*, vol. 70, no. 2, pp. 273–287, 2022, doi:10.1007/s00267-022-01669-w.

[19] T. Mahoutchian, M. Kamen, N. Paynter, *et al.*, "Activating the future of workplace," *Deloitte Insights*, 2023. Available at: https://www2.deloitte.com/us/en/insights/focus/human-capital-trends/2023/future-workplace-trends.html.

[20] P. De Giovanni, "Sustainability of the metaverse: A transition to Industry 5. 0," *Sustainability*, vol. 15, no. 7, p. 6079, 2023, doi:10.3390/su15076079.

[21] "How is Corporate Governance related to sustainability?," 2024. Available at: https://greenly.earth/en-us/blog/company-guide/how-is-corporate-governance-related-to-sustainability (accessed June 28, 2024).

[22] N. Nirino, E. Battisti, A. Ferraris, S. Dell'Atti, and M. F. Briamonte, "How and when corporate social performance reduces firm risk? The moderating role of corporate governance," *Corporate Social Responsibility and Environmental Management*, vol. 29, no. 6, pp. 1995–2005, 2022, doi:10.1002/csr.2296.

[23] W. Y. Leong, *ESG Innovation for Sustainable Manufacturing Technology: Applications, designs and standards*. Stevenage: The Institution of Engineering and Technology, 2024.

[24] W. Y. Leong, Y. Z. Leong, and W. S. Leong, "ESG imperatives for healthcare sustainability," *2024 IEEE 6th Eurasia Conference on Biomedical Engineering, Healthcare and Sustainability*, IEEE ECBIOS, 2024.

[25] W. Y. Leong, Y. Z. Leong, and W. S. Leong, "ESG-driven engineering in smart energy networks," *IEEE 6th Eurasia Conference on IoT, Communication and Engineering*, IEEE ECICE, 2024.

[26] M. Bechet, T. Lütke Siestrup, and A. Uhl, "Unilever case study: Implementing the real-time, digital enterprise to unlock value and enable business growth," *360° The Business Transformation Journal*, vol. 2014, Business Transformation Academy, pp. 66–79, 2014. Available at: http://hdl.handle.net/11654/9349.

[27] Manjurul, R. H., Marketing strategy and sustainable plan of Unilever. *International Journal of Scientific Research and Engineering Development*, vol. 3, no. 4, pp. 680–692, 2020. Available at: https://www.researchgate.net/publication/343539636.

[28] F. Laurin and K. Fantazy, "Sustainable supply chain management: A case study at IKEA," *Transnational Corporation Review*, vol. 9, no. 4, pp. 309–318, 2017, doi:10.1080/19186444.2017.1401208.

[29] H. Tanjil and S. Razia, "Sustainable supply chain management practices: A case study of IKEA, " in *Fundamentals of Management: Selected Readings*, M. Ali Akkas and F. Ahmed (eds.), London: ResearchGate, 2024, pp. 361–368. Available at: https://www.researchgate.net/publication/380098981.

[30] W. Y. Leong, Y. Z. Leong and W. S. Leong, "Smart manufacturing technology for environmental, social, and governance (ESG) sustainability," *2023 IEEE 5th Eurasia Conference on IoT, Communication and Engineering (ECICE)*, Yunlin, Taiwan, 2023, pp. 1–6, doi:10.1109/ECICE59523.2023.10383150.

[31] V. K. Annanth, M. Abinash, and L. B. Rao, "Intelligent manufacturing in the context of industry 4.0: A case study of siemens industry," *Journal of Physics Conference Series*, vol. 1969, no. 1, p. 012019, 2021, doi:10.1088/1742-6596/1969/1/012019.

[32] Project Management Institute, *A Guide to the Project Management Body of Knowledge (PMBOK® Guide) – 7th edn and the Standard for Project Management*. Newtown Square, PA: Project Management Institute; 2017.

[33] G. Balde, J. Patil, and S. Badgujar, "Critical discussion on the impact of Industry 5.0 on humans and ecology in stabilizing biodiversity," *International Research Journal of Modernization in Engineering Technology and Science*, vol. 5, pp. 53–65, 2023, doi:10.56726/irjmets33327.

[34] F. Di Maddaloni and L. Sabini, "Very important, yet very neglected: Where do local communities stand when examining social sustainability in major construction projects?," *International Journal of Project Management*, vol. 40, no. 7, pp. 778–797, 2022, doi:10.1016/j.ijproman.2022.08.007.

[35] J. Chungath, "Modeling Architecture to manage Energy, Social, and Governance (ESG) data of commercial buildings," M.S. thesis, Department of Computer and Systems Sciences, Stockholm University, Stockholm, Sweden, 2023. Available at: http://www.diva-portal.org/smash/get/diva2:1800599/FULLTEXT01.pdf (accessed August 05, 2024).

[36] N. H. Motlagh, M. Mohammadrezaei, J. Hunt, and B. Zakeri, "Internet of things (IoT) and the energy sector," *Energies*, vol. 13, no. 2, p. 494, 2020, doi:10.3390/en13020494.

[37] "Global sustainability reporting rates 2022, Statista," *Statista*, 2024. Available at: https://www.statista.com/statistics/1338724/sustainability-reporting-rates-worldwide/ (accessed July 01, 2024).

[38] "Top 10: ESG strategies from the world's largest companies," *Sustainability Magazine*, 2023. Available at: https://sustainabilitymag.com/top10/top-10/.

[39] "Stakeholder engagement and ESG materiality, Cisco ESG Reporting Hub." Available at: https://www.cisco.com/c/m/en_us/about/csr/esg-hub/governance/materiality.html (accessed July 02, 2024).

[40] *Nvidia Sustainability Report Fiscal Year 2024*. Available at: https://images. nvidia.com/aem-dam/Solutions/documents/FY2024-NVIDIA-Corporate-Sustainability-Report.pdf. (Accessed: 18 August 2024).

[41] S. Aich, A. Thakur, D. Nanda, S. Tripathy, and H.-C. Kim, "Factors affecting ESG towards impact on investment: A structural approach," *Sustainability*, vol. 13, no. 19, p. 10868, 2021, doi:10.3390/su131910868.

[42] E. Bohnett, A. Coulibaly, D. Hulse, T. Hoctor, and B. Ahmad, "Corporate responsibility and biodiversity conservation: Challenges and opportunities for companies participating in China's belt and road initiative," *Environmental Conservation*, vol. 49, no. 1, pp. 42–52, 2022, doi:10.1017/s0376892921000436.

[43] F. S. Hafez, B. Sa'di, M. Safa-Gamal, *et al.*, "Energy efficiency in sustainable buildings: A systematic review with taxonomy, challenges, motivations, methodological aspects, recommendations, and pathways for future research," *Energy Strategy Reviews*, vol. 45, p. 101013, 2023, doi:10.1016/j.esr.2022.101013.

[44] K. S. Woon, Z. X. Phuang, J. Taler, *et al.*, "Recent advances in urban green energy development towards carbon emissions neutrality," *Energy*, vol. 267, p. 126502, 2023, doi:10.1016/j.energy.2022.126502.

[45] C. Turner, J. Oyekan, W. Garn, C. Duggan, and K. Abdou, "Industry 5.0 and the circular economy: Utilizing LCA with intelligent products," *Sustainability*, vol. 14, no. 22, p. 14847, 2022, doi:10.3390/su142214847.

[46] Microsoft, "2024 environmental sustainability report, Microsoft CSR," Microsoft Sustainability. Available at: https://www.microsoft.com/en-us/corporate-responsibility/sustainability/report?ICID=SustainabilityReport22_SustWeb (Accessed: 20 August 2024).

[47] B. Caiazzo, T. Murino, A. Petrillo, G. Piccirillo, and S. Santini, "An IoT-based and cloud-assisted AI-driven monitoring platform for smart manufacturing: Design architecture and experimental validation," *Journal of Manufacturing Technology Management*, vol. 34, no. 4, pp. 507–534, 2022, doi:10.1108/jmtm-02-2022-0092.

[48] "Grid-interactive efficient buildings (GEBs)," ACEEE. Available at: https://www.aceee.org/grid-interactive-efficient-buildings-gebs (accessed July 28, 2024).

[49] R. R. R. Bezerra, V. W. B. Martins, and A. N. Macedo, "Validation of challenges for implementing ESG in the construction industry considering the context of an emerging economy country," *Applied Sciences*, vol. 14, no. 14, p. 6024, 2024, doi:10.3390/app14146024.

[50] B. Jonsdottir, T. O. Sigurjonsson, L. Johannsdottir, and S. Wendt, "Barriers to using ESG data for investment decisions," *Sustainability*, vol. 14, no. 9, p. 5157, 2022, doi:10.3390/su14095157.

[51] A. Sipiczki, *A Critical Look at the ESG Market*, Brussels: CEPS, 2022. Available at: chrome-extension://efaidnbmnnnibpcajpcglclefindmkaj/https://cdn.ceps.eu/wp-content/uploads/2022/04/PI2022-15_A-critical-look-at-the-ESG-market.pdf (Accessed: 23 August 2024).

[52] "Top-down ESG implementation worldwide," Statista. 2023. Available at: https://www.statista.com/topics/10953/top-down-environmental-social-and-corporate-governance-implementation-worldwide/#topicOverview (accessed July 28, 2024).

[53] M. Ghobakhloo, M. Iranmanesh, M. E. Morales, M. Nilashi, and A. Amran, "Actions and approaches for enabling Industry 5.0-driven sustainable industrial transformation: A strategy roadmap," *Corporate Social Responsibility and Environmental Management*, vol. 30, no. 3, pp. 1473–1494, 2022, doi:10.1002/csr.2431.

[54] J. Leng, W. Sha, B. Wang, *et al.*, "Industry 5.0: Prospect and retrospect," *Journal of Manufacturing Systems*, vol. 65, pp. 279–295, 2022, doi:10.1016/j.jmsy.2022.09.017.

Chapter 13

Carbon emission under Industry 5.0

Zeng Feng Tang[1] and Wai Yie Leong[2]

Industry 5.0 integrates people and machines into production to achieve more personalization, possibilities, and sustainable development. Industry 5.0 has three core elements: Humancentric, sustainability, and resilience. Humancentric emphasizes human–machine collaboration. Talent is the core of industrial operation. Technology is to serve humans better, not to replace humans. The sustainability of Industry 5.0 includes economic, environmental, social, and technological sustainability. In these aspects, Industry 5.0 has new requirements and development directions for sustainability. The carbon footprint runs through every link of the industry. Reducing carbon emissions is the requirement of Industry 5.0 for environmental sustainability. The technological innovation and progress of Industry 5.0, as well as the concept of sustainability, all affect and reduce carbon emissions to varying degrees.

13.1 What is Industry 5.0

As the world enters a more intelligent era, Industry 5.0 has become a new chapter in the development of manufacturing and technology. This concept was first proposed by the European Commission in 2021, marking the arrival of the "Fifth Industrial Revolution".

Industry 5.0 is not just a technological upgrade, it represents a new industrial concept that emphasizes human–machine collaboration, sustainability, and personalized production methods. Unlike the previous stage of Industry 4.0, which focused on automation and efficiency, Industry 5.0 pays more attention to achieving a higher value of human work and environmental sustainability of production activities while improving efficiency. This revolutionary leap will not only redefine the future of manufacturing but also have a profound impact on the world economy [1].

In 2011, Germany proposed the concept of Industry 4.0, which aims to realize the digitization and networking of production processes through Internet technology and intelligent manufacturing. Industry 4.0 has promoted the widespread

[1]Faculty of Management, Qiannan University of Economics, China
[2]Faculty of Engineering and Quantity Surveying, INTI International University, Malaysia

application of technologies such as the Internet of Things, big data, and artificial intelligence in the manufacturing industry, opening a new chapter in intelligent manufacturing and smart factories.

Industry 5.0 is not only the successor of Industry 4.0 but also its supplement and development, introducing key dimensions that were not deeply explored in the Industry 4.0 era – humanization, environmental sustainability, and flexible adaptability. This marks that society has entered a new industrial era, in which technological innovation is no longer just focused on maximizing economic benefits, but pays more attention to its contribution to social value [2].

13.1.1   Core elements of Industry 5.0

The European Commission proposed three core elements of Industry 5.0: Humancentric, Sustainability, and Resilience. Industry 5.0 has different degrees of improvement in concept compared with the existing four industrial revolutions, but it pays more attention to human–machine collaboration, personalization, and sustainability (Figure 13.1).

13.1.1.1   Humancentric: a new era of human–machine collaboration

The "Humancentric" principle emphasized by Industry 5.0 aims to reshape the relationship between people and machines and achieve harmonious collaboration between the two. The most important difference between Industry 4.0 and Industry 5.0 is the relationship between people and machines in the production process. In the Industry 5.0 stage, the most important core is "people", and more attention is paid to the combination of manpower and technology, and the sustainability of skills and training required by manpower.

Industry 5.0 emphasizes people-centeredness, that is, the core position of people in the production process. People are still the decision-makers, and machines/artificial intelligence systems are auxiliary. People's creativity and

Figure 13.1   Core values of Industry 5.0

ability to respond quickly to uncertain situations are a crucial part of the industrial production process and should be fully explored so that product innovation and quality and efficiency improvement can leap to a new level.

Industry 5.0 focuses on other social goals that the industry can achieve besides economic growth. The first is to serve people. People have always been the object and center of service in the process of industrial development and reform. Without the goal of people-centeredness, the meaning of industrial development and reform will be lost. Unlike Industry 4.0, which emphasizes automation and the Internet of Things, Industry 5.0 focuses more on the collaboration between humans and robots. Human–machine collaboration technology mainly includes human–machine collaboration at the cognitive level, human–machine collaboration at the decision-making level, human–machine collaboration at the control level, and reliable and safe interaction between humans and machines. The purpose of industrial development is not to completely replace humans with machines, but to release humans from heavy physical labor, dangerous labor, and repetitive mental labor, to fully protect industrial workers and give full play to human experience and wisdom; and to assist people in completing more challenging tasks, or to assist in improving the skills of industrial workers, provide help to everyone willing to work, and create more new jobs [3].

Industry 5.0 can personalize and customize goods at the industrial level by integrating robots and humans. Since collaborative robots perform repetitive tasks with strict and predictable efficiency, humans can supervise the process to ensure that real-time customization requests are understood and realized. Therefore, Industry 5.0 encourages customized and personalized product production, which can respond to customer needs more flexibly. This means that the production system can be quickly adjusted to produce small batches or even customized products for a single customer.

Industry 5.0 uses machines and technology to liberate humans from repetitive and complicated work and develop human potential. The application scenarios of Industry 5.0 are becoming more and more extensive. Industry 5.0 can not only further improve production efficiency in the manufacturing industry but also exert its potential in education, medical care, etc., and apply artificial intelligence, big data analysis, etc. to other fields, helping humans to further liberate and develop productivity from different fields (Figure 13.2).

13.1.1.2 Sustainability: towards green and sustainable production

Industry 5.0 focuses on environmental protection and resource conservation and strives to find the best balance between economic growth and ecological balance. Implementation strategies include the use of renewable energy, the promotion of recycling technologies, and the reduction of carbon dioxide emissions and industrial waste through technological innovation. Industry must be sustainable if it is to respect the production limits of the earth. It requires the development of a circular economy, the reuse, and recycling of natural resources, and the reduction of waste and environmental impact. Sustainability means reducing energy consumption and greenhouse gas emissions, avoiding the depletion and degradation of natural

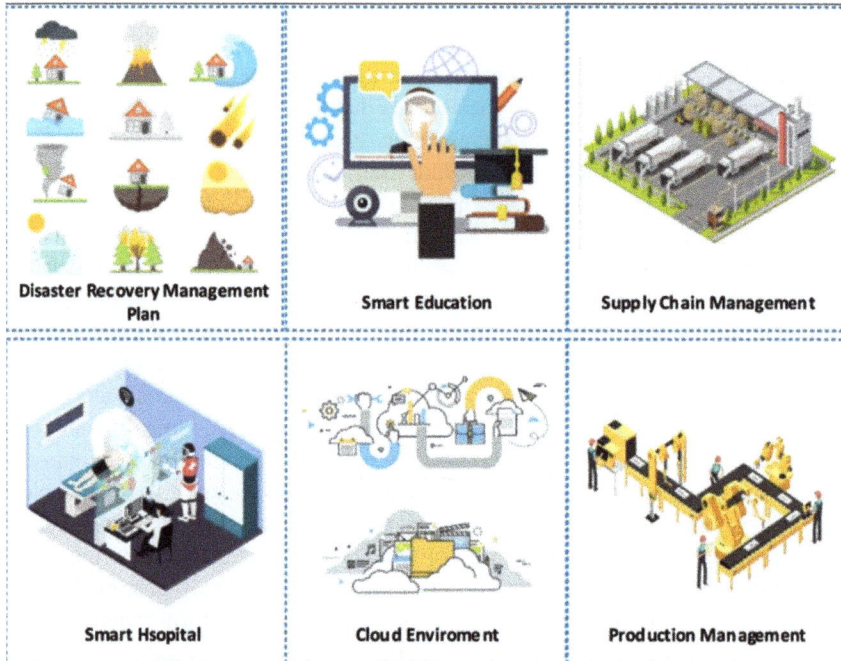

Figure 13.2   Potential application scenarios of Industry 5.0

resources, and ensuring the needs of the present generation without compromising the needs of future generations [4].

13.1.1.3   Resilience: enhance the ability to respond to challenges

Resilience refers to the ability of industries and societies to quickly adapt and recover in the face of unforeseen challenges, such as epidemics and natural disasters. The current Industrial 4.0 era is an era of the Internet of Everything. The interconnection of man, machine, object, and environment brings various uncertainties to enterprises, thus providing the possibility of improving their production efficiency and market competitiveness. However, before the outbreak of the COVID-19 pandemic, countries around the world paid little attention to the robustness of the industrial system. Therefore, Industrial 5.0 not only focuses on the ability of enterprises to cope with external uncertainties, such as market, supply chain, and customer uncertainties, but also begins to pay attention to the robustness of a wider range of industrial systems, such as the ability of the industry and even the entire industrial chain to resist unknown risks, to cope with large-scale or even global emergencies, or technological blockades by other countries. The resilience of Industrial 5.0 refers to the use of advanced digital technology and data analysis to predict potential risks and formulate effective response strategies. In the manufacturing industry, the transparency and data sharing of the entire supply chain are emphasized to improve the stability of the system and the ability to cope with emergencies, thereby enhancing the competitiveness and sustainable development capabilities of the entire industry [5].

13.1.2 Sustainability of Industry 5.0

Sustainability of Industry 5.0 refers to the use of smart technologies and innovative means to improve production efficiency and product quality while minimizing negative impacts on the environment, promoting efficient use of resources, and achieving coordinated development in the three aspects of economy, society, and environment. This concept not only involves the production process itself but also covers the entire supply chain, product design, energy use, waste management, and other links, aiming to create a more environmentally friendly and sustainable industrial ecosystem [6].

13.1.2.1 Economic sustainability

Economic sustainability refers to maintaining efficient use of resources, reducing waste, and creating long-term value for the future while ensuring economic growth and development, rather than pursuing short-term benefits. Economic sustainability focuses on the sustainable development of enterprises (i.e. profitability). Through the intelligent manufacturing of Industry 5.0, enterprises can better meet the personalized needs of consumers and provide customized products and services. This not only improves the consumer experience but also reduces resource waste and promotes sustainable consumption in society.

Enterprises need to find new solutions that can achieve sustainable consumption and production patterns. At the same time, the innovation and design of solutions should both enable and encourage individuals to adopt more sustainable lifestyles, reduce impacts, and improve well-being. Companies need to better understand the environment and study the social impact of their products and services, including the product life cycle and the impact of using these products and services on lifestyles in society [7].

For example, IKEA has been building a sustainable supply chain management system. They give priority to the use of renewable or recyclable materials in the selection of product materials and ensure that suppliers meet strict environmental and social responsibility standards. IKEA promises that all products will be made of renewable or recyclable materials by 2030. This not only reduces environmental impact but also enhances the resilience of the supply chain and the sustainable image of the company's brand (Figure 13.3) [8].

Enterprises can not only improve production efficiency and reduce costs through intelligent automation and human–machine collaboration, thereby maintaining their competitiveness and economic sustainability. Enterprises can also take advantage of technological innovation and the development of new business models to enable

Figure 13.3   IKEA's sustainability-driven self-learning supply chain

them to respond to market changes and continue to grow. Industry 5.0 can also enable enterprises to support local economic development, promote local employment, and reduce regional economic imbalances caused by globalization through localized production and supply chain management. This strategy helps stabilize the local economy and enhance local economic resilience. For example, Nestlé promotes sustainable agricultural practices in the agricultural supply chain to ensure long-term supply stability and quality. Through its "Nescafé Plan", Nestlé helps coffee farmers adopt sustainable planting methods such as smart irrigation technology, digital farm management, and sustainable soil management technology. Nestlé applies technological innovation to the production and operation of the supply chain, which not only improves farmers' income and quality of life but also ensures that Nestlé has a stable supply of high-quality coffee beans in the future, thereby achieving economic sustainability and reducing environmental impact (Figure 13.4) [9].

13.1.2.2   Social sustainability

In Industry 5.0, social sustainability means that in the process of promoting industrial and technological development, enterprises not only pay attention to economic benefits and environmental impacts but also attach great importance to contributions to social welfare. Industry 5.0 promotes the development of social sustainability through human-oriented technological innovation, technological empowerment, and an emphasis on human–machine collaboration.

Employee well-being
Industry 5.0 emphasizes human–machine collaboration, and technology is designed to enhance rather than replace human capabilities. Industry 5.0 uses advanced

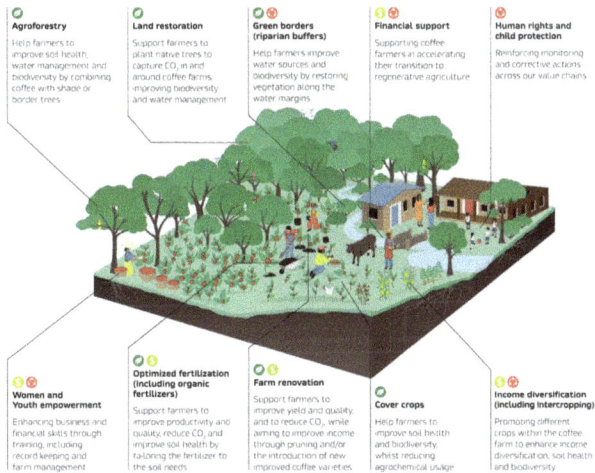

Figure 13.4   Help make coffee farming more sustainable: the Nescafé plan 2030

sensors, smart robots, the Internet of Things (IoT), and data analysis technologies to monitor workplace safety in real-time and prevent work-related injuries and accidents. In the production process, not only does it reduce workers' physical labor, but by reducing repetitive and dangerous tasks, workers' health and safety are better protected. With the improvement of the working environment, Industry 5.0 can also focus on the mental health of employees. Smart technology can identify employees' stress and fatigue levels, and companies can provide corresponding support and resources to help employees maintain mental health. Industry 5.0 optimizes the working environment, improves work efficiency, and improves employees' job satisfaction and quality of life.

Industry 5.0 brings about changes in new technologies and new work models. Enterprises and governments need to provide employees with opportunities for retraining and skill upgrading. Through continuous education and training, workers can adapt to job changes brought about by new technologies and obtain new career opportunities, thereby reducing the risk of unemployment caused by automation and technological upgrades. Although automation may reduce some traditional jobs, Industry 5.0 also creates a large number of new jobs, especially in the fields of technology development, data analysis, robot maintenance, etc. These new jobs not only require highly skilled labor but also provide more employment options and opportunities for all kinds of people in society [10].

Social innovation and development

Industry 5.0 encourages enterprises to actively participate in regional construction and social development while pursuing technological innovation. Enterprises can enhance the sustainable development capacity of the region by cooperating with local governments, non-profit organizations, and communities to carry out social welfare projects such as education support, infrastructure construction, environmental protection, etc. Industry 5.0 encourages enterprises to support local economic development while achieving global production, and promoting the economic and social sustainability of the community through localized production, hiring local employees, and using local resources.

Industry 5.0 can promote diversity and equality in the workplace through the transformation of technology and organizational culture. For example, the promotion of remote work and flexible work arrangements enables people of different genders, ages, and backgrounds to better participate in economic activities, reducing inequality between genders, regions, and races. For example, Microsoft is committed to creating a diverse and inclusive work environment to promote cooperation and innovation among employees from different backgrounds. Microsoft actively promotes gender equality, racial equality, and diverse culture in the process of recruitment, promotion, and training, and ensures that every employee has equal development opportunities by setting up internal support networks and resource groups (Figure 13.5) [11].

Industry 5.0 can also help promote the common progress of all social classes by making all social groups equally able to benefit from technology through education and technology popularization, narrowing the digital divide, avoiding the aggravation of inequality caused by technological progress, and promoting the common progress of all social classes.

For example, Schneider Electric conducts energy education projects around the world, especially in developing countries, to help residents understand and use sustainable energy technologies. Accessible online and free of charge worldwide, Schneider Electric University directly addresses the data center industry's skills gap, helping industry stakeholders to upskill and stay up to date on the latest technology, sustainability and energy efficiency initiatives impacting the sector. These projects not only improve the quality of life of residents but also promote the economic development and sustainability of the community [12].

Social sustainability is also reflected in encouraging cross-industry and cross-field cooperation and solving major challenges facing society, such as poverty,

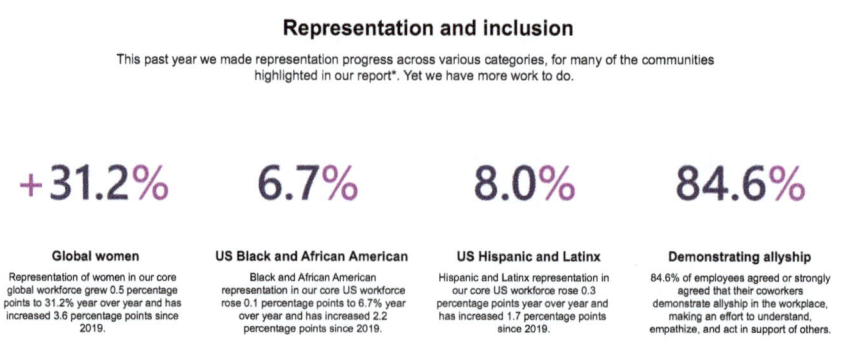

Representation and inclusion

This past year we made representation progress across various categories, for many of the communities highlighted in our report*. Yet we have more work to do.

+31.2%	6.7%	8.0%	84.6%
Global women	**US Black and African American**	**US Hispanic and Latinx**	**Demonstrating allyship**
Representation of women in our core global workforce grew 0.5 percentage points to 31.2% year over year and has increased 3.6 percentage points since 2019.	Black and African American representation in our core US workforce rose 0.1 percentage points to 6.7% year over year and has increased 2.2 percentage points since 2019.	Hispanic and Latinx representation in our core US workforce rose 0.3 percentage points year over year and has increased 1.7 percentage points since 2019.	84.6% of employees agreed or strongly agreed that their coworkers demonstrate allyship in the workplace, making an effort to understand, empathize, and act in support of others.

Figure 13.5 Microsoft global diversity and inclusion

health, education, and environmental issues, through collaborative innovation among public, private, and social sectors. Such cooperation can drive society to achieve sustainable development goals faster and more widely.

13.1.2.3　Environmental sustainability

In Industry 5.0, environmental sustainability refers to the full use of advanced technologies and innovative means by enterprises and society in the process of development to minimize negative impacts on the environment, protect natural resources, and support the health and stability of the ecosystem. Industry 5.0 emphasizes the coordinated development of technological progress and environmental protection and is committed to achieving long-term environmental sustainability.

Low-carbon manufacturing

In the context of continued global warming, it is urgent to control carbon emissions. Industry 5.0 enables enterprises to reduce greenhouse gas emissions such as carbon dioxide generated during the production process through technical means. Enterprises can reduce their dependence on fossil fuels by improving process flows, improving energy efficiency, and introducing clean and renewable energy (such as solar energy, wind energy, and biomass energy), as well as adopting carbon capture and storage technology to reduce the carbon footprint of the manufacturing industry, reduce environmental pollution during energy use, gradually achieve carbon neutrality goals, and reduce the impact on climate change [13].

Resource conservation and recycling

Industry 5.0 optimizes production processes and reduces resource consumption and waste through intelligent manufacturing technologies (such as the Internet of Things, artificial intelligence, and big data analysis). By precisely controlling the use of energy and raw materials in the production process, Industry 5.0 can significantly improve resource utilization and efficiency, thereby reducing the environmental burden. Industry 5.0 vigorously promotes the circular economy, extends the product life cycle through innovation in the design and production process, and reduces the demand for new resources through recycling, reuse, and remanufacturing. Companies take into account the recyclability and sustainability of products when designing products. Companies conduct environmental impact assessments on all links in the supply chain to ensure that every step of raw material procurement, production process, and product distribution meets environmental standards. Through supply chain transparency, companies can track and optimize environmental performance in the supply chain and reduce negative impacts on the environment. Through the circular economy and closed-loop supply chain management, Industry 5.0 reuses or recycles waste in production, reduces waste emissions, and recycles materials into new production processes, minimizing the consumption of natural resources and promoting industrial sustainable development [14].

Green consumption, green technology, and green transformation
Industry 5.0 needs to consider environmental impact at the product design stage and reduce the environmental impact of products throughout their life cycle by selecting environmentally friendly materials, reducing energy consumption, and improving product recyclability. This design concept not only helps to reduce the environmental footprint in production but also guides consumers to choose more environmentally friendly products. Industry 5.0 promotes sustainable consumption behavior through technological innovation and consumer education. The transparent information provided by enterprises enables consumers to understand the environmental impact of products so that they can make more environmentally friendly purchasing decisions and promote the market to develop in a sustainable direction.

Industry 5.0 uses the Internet of Things and sensor technology to monitor the impact on the environment (such as emissions, pollution, and resource use) during production and operation in real-time. This data can help companies identify and solve environmental problems on time and prevent pollution incidents.

Industry 5.0 is also promoting the development and application of green technologies, such as advanced wastewater treatment technologies, air purification systems, and resource recovery technologies. These technological innovations can not only reduce the environmental impact of the production process but also provide new solutions for environmental protection and restoration [15].

Industry 5.0 encourages enterprises to cooperate with governments, non-profit organizations, and academia to jointly develop and promote technologies and standards that contribute to environmental sustainability and promote the green transformation of the entire industry and society.

Ecological protection and environmental regulations
The environmental sustainability of Industry 5.0 is to protect the ecosystem during industrial development and avoid damage to the natural environment and biological habitats by reducing pollutant emissions and the use of harmful substances. When selecting factory sites and raw materials, enterprises will consider the impact on local ecosystems and take measures to reduce interference with the natural environment.

In Industry 5.0, enterprises must comply with various environmental laws and standards and actively support governments and international organizations in formulating and implementing stricter environmental protection policies. However, Industry 5.0 also encourages enterprises to participate in the formulation and evaluation of environmental policies, and help governments formulate more effective environmental protection measures by providing technical and data support. Enterprises are the main body of Industry 5.0. Enterprises can participate in the formulation and evaluation of environmental policies, which can not only improve

Figure 13.6   The importance of environmental regulations

their participation and execution but also set an example for environmental protection in the industry.

Environmental sustainability in Industry 5.0 emphasizes the balance between technological progress and environmental protection. Through innovation, intelligence, and collaboration, it promotes the green transformation of industrial development methods, ensures that economic development is coordinated with the protection of earth resources and the health of the ecosystem, and achieves long-term environmental sustainability (Figure 13.6) [16].

The sustainability of Industry 5.0 emphasizes the transformation of industrial production into an eco-friendly model that can meet current needs without compromising the needs of future generations through systematic thinking and overall optimization. This sustainability is not only at the environmental level but also a comprehensive transformation at the social and economic levels.

13.1.3   How does Industry 5.0 promote carbon emission implementation

The key technologies in modern manufacturing include the industrial Internet of Things, robotics, artificial intelligence, big data, cloud, network security, new materials, and material manufacturing, as well as modeling, simulation, visualization, and immersion. Manufacturing plants supported by these technologies have advantages such as high productivity, accuracy, safety, high quality, intelligence, and repeatability. The manufacturing industry also relies on these technologies to continuously make innovations and breakthroughs. However, the production process of manufacturing enterprises has the characteristics of high consumption and high emissions. Industry 5.0 uses technology and optimizes and innovates new technologies to achieve the transformation, upgrading and sustainable development of the manufacturing industry. Industry 5.0 is to accelerate the optimization and upgrading of the industrial structure, deeply implement green manufacturing, coordinate development and green and low-carbon transformation, and vigorously promote industrial energy conservation and carbon reduction. Industry 5.0 promotes the green development of the manufacturing industry, which can effectively reduce carbon emissions and achieve a virtuous cycle of economic growth and environmental protection.

13.1.3.1   Digitalization affects carbon emissions

Digitalization can optimize the energy utilization efficiency and resource allocation efficiency of manufacturing enterprises, enhance R&D and innovation capabilities, improve production processes, and transform management and operation models. At the same time, it can make the production and operation process more "automatic" and "precise", accurately control material input, reduce waste and consumption for production and manufacturing, greatly reduce the cost of green transformation, and gradually become a new driving force for the green transformation of manufacturing enterprises [17].

Existing research shows that digital technology plays an important role in helping the world cope with climate change. Digital technology can provide networked, digital, and intelligent technical means for the green development of the economy and society, enable the construction of a clean, low-carbon, safe, and efficient energy system, help industrial upgrading and structural optimization, promote green changes in production and lifestyle, and promote the reduction of overall social energy consumption. China's "1+N" policy system for carbon peak and carbon neutrality proposes to promote the deep integration of emerging technologies such as big data, artificial intelligence, and 5G with green and low-carbon industries; and promote the integrated development of digitalization, intelligence, and greening in the industrial field (Figure 13.7).

An important aspect of the impact of digitalization on the environment is the empowerment effect, that is, the effect produced by the use of digital technology in

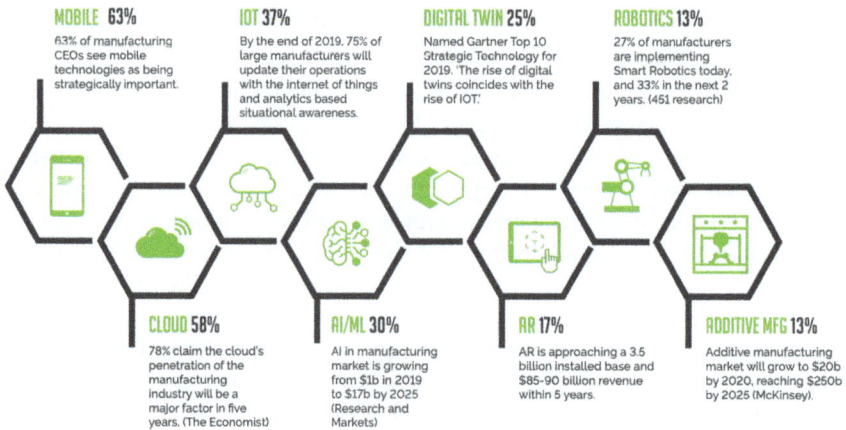

MOBILE 63%
63% of manufacturing CEOs see mobile technologies as being strategically important.

IOT 37%
By the end of 2019, 75% of large manufacturers will update their operations with the internet of things and analytics based situational awareness.

DIGITAL TWIN 25%
Named Gartner Top 10 Strategic Technology for 2019. 'The rise of digital twins coincides with the rise of IOT.'

ROBOTICS 13%
27% of manufacturers are implementing Smart Robotics today, and 33% in the next 2 years. (451 research)

CLOUD 58%
78% claim the cloud's penetration of the manufacturing industry will be a major factor in five years. (The Economist)

AI/ML 30%
AI in manufacturing market is growing from $1b in 2019 to $17b by 2025 (Research and Markets)

AR 17%
AR is approaching a 3.5 billion installed base and $85-90 billion revenue within 5 years.

ADDITIVE MFG 13%
Additive manufacturing market will grow to $20b by 2020, reaching $250b by 2025 (McKinsey).

Figure 13.7   Eight tools for digital technology

economic and social activities. On the one hand, digital technology can reduce unnecessary activities, and on the other hand, it can benefit the environment by optimizing and dematerializing economic activities. The empowerment process includes five links: green and low-carbon-related information acquisition, transmission, storage, processing, and standardization. Its basic logic can be summarized as "connection-mining-optimization, control-efficiency enhancement", and its action mechanism is to change the way of value creation, improve the efficiency of value creation, expand the carrier of value creation, and enhance the ability to obtain value. Specifically, digital technology enables users to participate in the value-creation process from research and development to production in various forms, changing the way enterprises create value; digital technology uses data logically to strengthen the control of production operations by enterprises and improve the efficiency of value creation; the new generation of information and communication technology realizes the integration and flow of data in the industrial chain, promotes the professional division of labor among enterprises, forms a value network, and expands the carrier of value creation; digital technology weakens the boundaries of the industry, spawns new business models such as "cross-border", and enhances the value acquisition ability of enterprises [18].

Digital technology implements technical improvements and optimized configurations for traditional industries, leads process and service innovation, and has great potential to support low-carbon development. In terms of carbon emission management, it can promote efficient carbon management and carbon emission tracking and monitoring. The application of new-generation information technology in the field of traditional energy consumption promotes the clean transformation of its energy structure and directly reduces carbon emissions by improving energy efficiency, reducing environmental impact, and recycling resources (Figure 13.8).

Digital technology will play an important role in carbon emissions, carbon removal and carbon management. Carbon neutrality mainly includes carbon emissions and carbon removal. If carbon emissions and removal are equal, carbon neutrality can be achieved, and carbon management is always accompanied in this process. Carbon emissions include energy supply and consumption, and energy supply includes traditional energy and clean energy. For traditional energy, digital technology improves energy supply efficiency and reduces environmental damage. For clean energy, it solves the two major problems of clean energy consumption and stability. The energy consumption side includes industry, construction, transportation and life. Digital technology enables industrial intelligent green manufacturing and energy management, enables the entire life cycle of buildings to reduce energy consumption, promote transportation, and improve transportation organization efficiency. In terms of life, digital technology enables smart medical care, education, culture and tourism, finance, etc. In terms of carbon removal, digital technology improves the efficiency of ecological carbon sequestration. In terms of carbon management, carbon accounting monitoring, carbon trading, and carbon finance are also inseparable from digital technology. Digital technology helps build a clean, low-carbon, safe, and efficient energy system and accelerates the realization of green changes in production and lifestyles (Figure 13.9).

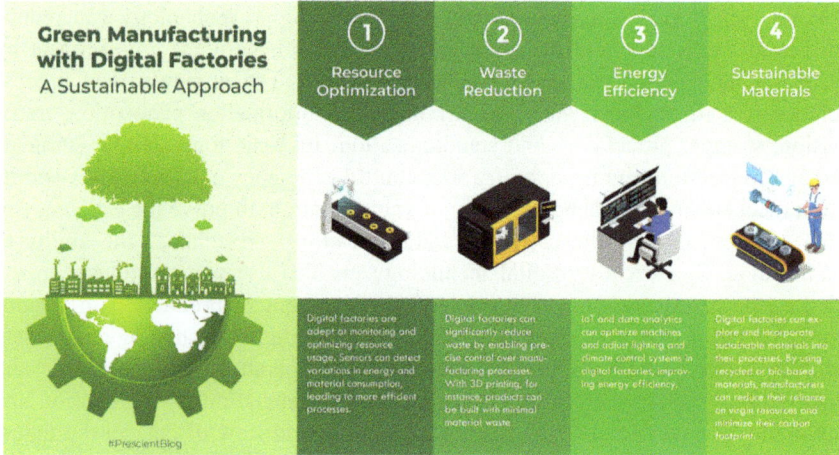

Figure 13.8   Digital technology enables green transformation of factories and manufacturing

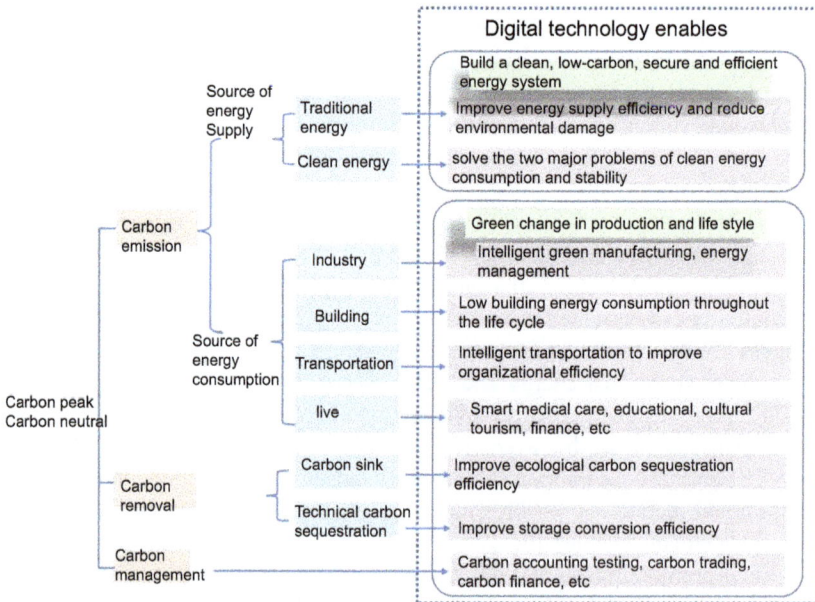

Source: China Academy of Information and Communications Technology

Figure 13.9   The main ways in which digital technology can help achieve carbon peak and carbon neutrality

13.1.3.2   Impact of AI and big data on carbon emissions

Artificial intelligence is a new technical science that studies and develops theories, methods, technologies and application systems for simulating, extending and expanding human intelligence. As a strategic technology for the new round of scientific and technological revolution and industrial transformation, AI is accelerating its integration with various fields of the economy and society and is gradually becoming a powerful means to reduce carbon emissions and achieve carbon neutrality.

AI helps reduce its carbon emissions

With the continuous acceleration of the deployment of AI infrastructure such as data centers, and the continuous expansion of the development and application of AI algorithm models, the consumption of energy such as electricity is constantly increasing. By applying AI technologies such as deep learning algorithms and promoting the development of AI open source platforms, the energy consumption of AI can be effectively reduced, and low-carbon operation of data centers and model development can be achieved [19].

Huawei uses "AI+Cloud" to help data centers save energy and reduce consumption

Huawei's AI-based icooling data center energy efficiency optimization solution realizes full-link intelligent management of temperature control without changing the hardware and product configuration of the data center. It generates AI energy-saving algorithms through machine deep learning and combines expert experience to achieve a 15%-18% reduction in PUE under the same conditions. iPower, based on AI, realizes visual intelligent monitoring of the entire power supply and distribution chain, as well as early warning of the health of key components of core equipment, transforming passive operation and maintenance into active prevention, ensuring the safety and reliability of the data center. iManager improves operational efficiency and benefits by solidifying tools, operation and maintenance processes, and experience into tools, achieving optimal utilization of data center resources (Figure 13.10).

Artificial intelligence helps monitor corporate carbon emissions

Through close-range perception by smart sensors, long-range observation by satellite remote sensing, and real-time monitoring and identification of emissions such as smoke, heat, and nitrogen dioxide by thermal infrared sensors, the use of artificial intelligence monitoring platforms for comprehensive analysis and processing of emission data to form visual results can achieve tracking, learning, and simulation of carbon footprints, effectively monitor the emissions of enterprises or industries, and optimize carbon emission activities.

US WattTime, a non-profit innovation organization in the United States, thermal infrared sensor technology monitors carbon emissions and uses thermal infrared sensors and other sensor monitoring equipment, combined with satellite imaging and artificial intelligence technology, to use satellite networks to monitor the carbon emissions and emission levels of all large power plants around the world and publish

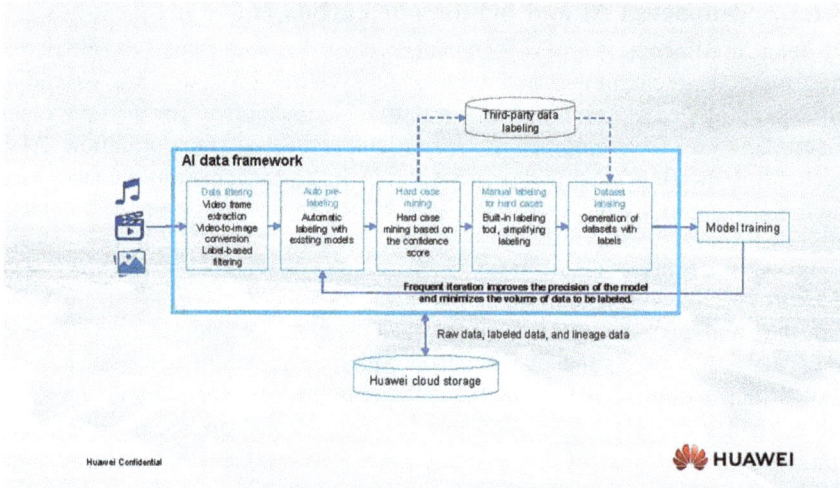

Figure 13.10   Model Arts, a one-stop development platform for AI developers

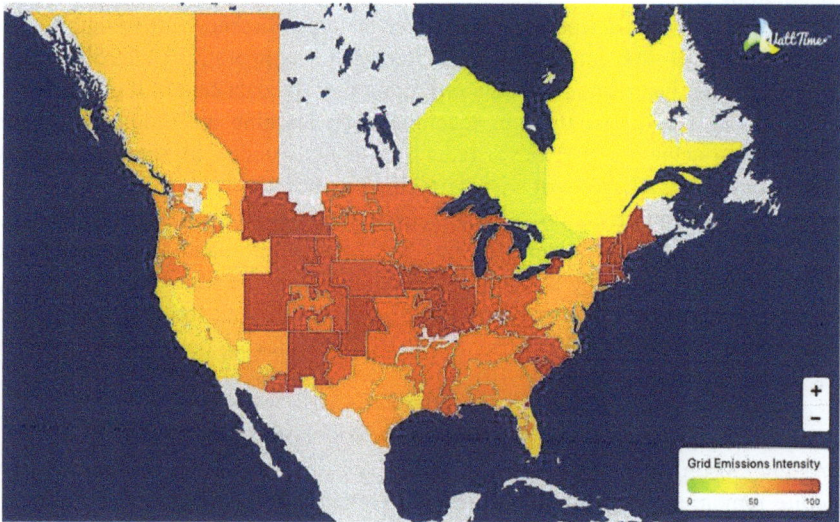

Figure 13.11   Carbon emission intensity in the United States in mid-to-late July

the monitoring data to the public. These sensors can work at different wavelengths, use artificial intelligence technology, and process the collected data through various algorithms to form visual images to detect whether there are signs of emissions, visually identify emissions, and better analyze carbon emissions. The system is open to the public and can solve the problem of falsification of power plant emission data supervision and promote fair emission reduction (Figure 13.11) [20].

Artificial intelligence and big data help the industry reduce carbon emissions

Industrial manufacturing can use AI and big data to monitor and analyze each link in real-time from design, production, assembly, transportation, etc., to make corresponding adjustments and optimizations. In the product design stage, through big data analysis, companies can obtain and analyze a large amount of user feedback, market demand, and historical data to optimize product design. AI can predict market trends and help design teams make more market-oriented and environmentally friendly design choices at an early stage, such as choosing more sustainable materials or designing more recyclable products. AI combined with big data can also create a virtual model of the product (digital twin technology) and conduct simulation tests in a virtual environment. This can significantly reduce the number of physical prototypes, and reduce the consumption of raw materials and energy use in the manufacturing process, thereby reducing carbon emissions.

In the production process, AI can first analyze material performance data, supply chain information, and environmental impact to help companies choose more environmentally friendly and sustainable materials. AI can also identify and recommend low-carbon materials or alternative materials, which not only help to improve the environmental performance of products and extend product life but also reduce the carbon footprint of products during their life cycle [21].

AI can analyze production data and identify the most energy-saving and efficient manufacturing processes. By optimizing production parameters or scheduling, such as temperature, pressure, and speed, unnecessary energy consumption and equipment idling in the production process can be reduced, scrap rate and energy loss can be reduced, and overall energy efficiency can be improved.

Big data and AI can monitor production lines and equipment in real-time, automatically adjust manufacturing processes to adapt to environmental changes (such as fluctuations in the power supply or changes in raw material quality), reduce carbon emissions caused by unstable processes, perform predictive maintenance on machinery and equipment, and reduce energy waste caused by equipment failures, thereby improving energy utilization (Figure 13.12).

During transportation, AI can combine big data to analyze real-time traffic conditions, weather conditions, vehicle performance, and other factors to automatically generate the optimal transportation route, reducing transportation time and avoiding unnecessary detours and waiting. AI can automatically plan the best loading method based on the volume, weight, and load capacity of the cargo and ensure that the loading rate of each vehicle is maximized, reducing unnecessary transportation times, and thereby reducing overall carbon emissions [22].

Big data and AI can monitor changes in transportation demand in real-time and dynamically adjust transportation plans. For example, when demand changes, the system can automatically rearrange routes or vehicles to ensure the best use of transportation resources and avoid vehicles running empty.

Most importantly, AI can integrate data from all links of the supply chain, from raw material procurement to final delivery, to achieve optimized management of

Applied Industrialized AI – Manufacturing

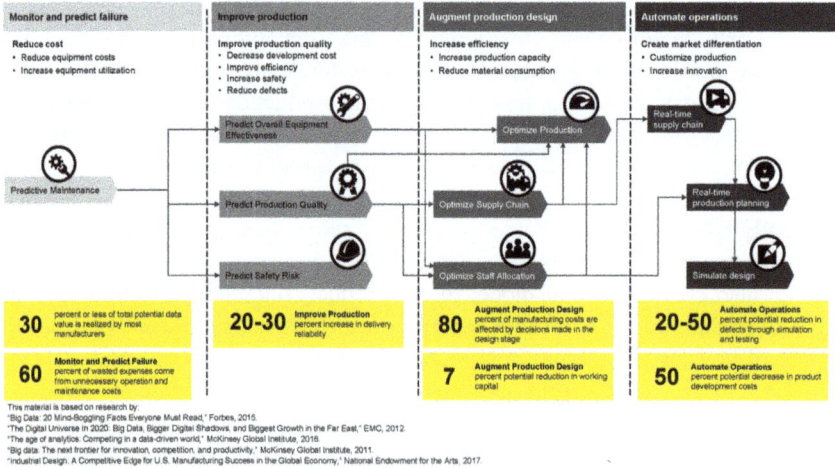

Monitor and predict failure	**Improve production**	**Augment production design**	**Automate operations**
Reduce cost	Improve production quality	Increase efficiency	Create market differentiation
• Reduce equipment costs	• Decrease development cost	• Increase production capacity	• Customize production
• Increase equipment utilization	• Improve efficiency	• Reduce material consumption	• Increase innovation
	• Increase safety		
	• Reduce defects		

30	percent or less of total potential data values is realized by most manufacturers	**20-30**	**Improve Production** percent increase in delivery reliability	**80**	**Augment Production Design** percent of manufacturing costs are affected by decisions made in the design stage	**20-50**	**Automate Operations** percent potential reduction in defects through simulation and testing
60	**Monitor and Predict Failure** percent of wasted expenses come from unnecessary operation and maintenance costs			**7**	**Augment Production Design** percent potential reduction in working capital	**50**	**Automate Operations** percent potential decrease in product development costs

This material is based on research by:
"Big Data: 20 Mind-Boggling Facts Everyone Must Read," Forbes, 2015.
"The Digital Universe in 2020: Big Data, Bigger Digital Shadows, and Biggest Growth in the Far East," EMC, 2012.
"The age of analytics: Competing in a data-driven world," McKinsey Global Institute, 2016.
"Big data: The next frontier for innovation, competition, and productivity," McKinsey Global Institute, 2011.
"Industrial Design: A Competitive Edge for U.S. Manufacturing Success in the Global Economy," National Endowment for the Arts, 2017.

Figure 13.12   Artificial intelligence in manufacturing

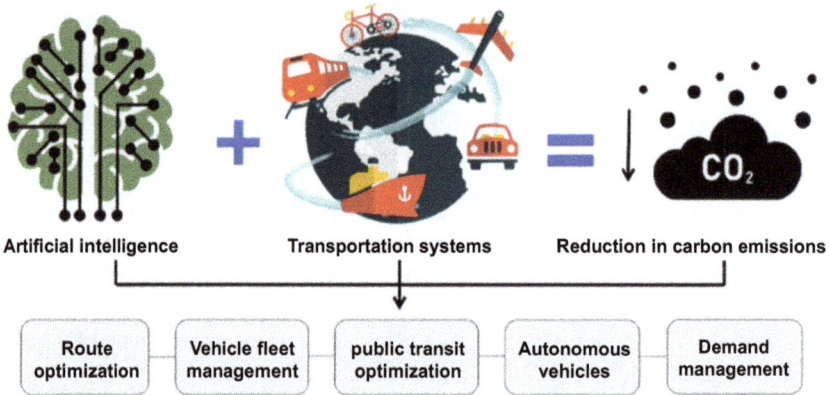

Figure 13.13   Artificial intelligence helps the transport industry reduce emissions

the entire process. By reducing inventory backlogs and optimizing production and distribution rhythms, AI helps companies reduce additional transportation demand and energy consumption caused by poor inventory management.

Finally, AI can combine big data to accurately track the carbon emissions of each transportation task and generate a detailed carbon footprint report. Companies can use this data for analysis, identify high-emission links and take corresponding emission reduction measures. This will more accurately predict the carbon emissions of different transportation plans, help companies choose the most environmentally friendly transportation method, and adjust strategies in real-time during transportation to ensure the lowest carbon emissions (Figure 13.13).

AI and big data can not only help companies effectively reduce carbon emissions but also drive the entire industry in a more sustainable direction. The application of these technologies enables companies to actively respond to climate change and environmental challenges while achieving economic benefits.

13.2   Summary

Industry 5.0 is a new generation of industrial concepts, which aims to create a more harmonious industrial ecosystem through the combination of intelligent technology, personalized manufacturing and sustainable development [23]. It not only promotes technological progress but also pays attention to the symbiosis of humans and the environment, and closely combines the improvement of productivity with social progress. The sustainability of Industry 5.0 is multidimensional and multifaceted. It combines the economy, environment, and society to truly achieve sustainable development. Industry 5.0 covers a wide range and involves a wide range of technologies and contents, but the fundamental purpose of Industry 5.0 is to promote the comprehensive upgrading of the industrial system and achieve a more efficient, flexible, and sustainable production method. The environmental sustainability of Industry 5.0 is exploring the balanced relationship between industry and the environment, using industrial technology to empower the industry, allowing the industry to reduce environmental pollution and carbon emissions, and comprehensively promoting the green transformation and upgrading of the industry from the industrial level, reducing damage to the environment, better management and reducing carbon emissions, and also allowing enterprises to get more opportunities and development from it, assume more social responsibilities, give back to society, and protect the environment.

References

[1]   Industry 5.0: Towards a sustainable, human-centric and resilient European industry. (n.d.). https://Op.Europa.Eu/En/Publication-Detail/-/Publication/468a892a-5097-11eb-B59f-01aa75ed71a1/ (accessed August 3, 2024).

[2]   Translation of the European Commission's "Industry 5.0" Series of Reports. (n.d.). https://www.Aii-Alliance.Org/Index/C185/N2918.html (accessed August 3, 2024).

[3]   Zhuang, C. (2021, August 3). The connotation, system architecture and enabling technologies of Industry 5.0. https://M.e-Works.Net.Cn/Articles/Article149005.htm (accessed August 3, 2024).

[4]   What the sustainable development goals (SDGs) mean for business. (n.d.). SDG Essentials for Business. https://sdgessentials.org/zh/what-the-sdgs-mean-for-business.html (accessed August 4, 2024).

[5]   Ivanov, D. (2022). The Industry 5.0 framework: Viability-based integration of the resilience, sustainability, and human-centricity perspectives. *International Journal of Production Research*, 61(5), 1683–1695. https://doi.org/10.1080/00207543.2022.2118892 (accessed August 3, 2024).

[6]   Krantz, T., and Jonker, A. (eds). (2023, November 30). What is sustainable development? IBM. https://www.ibm.com/cn-zh/topics/sustainability (accessed August 4, 2024).

[7]   Towards a smart supply chain for Industry 5.0. (2022, March 18). CeMAT ASIA PTC ASIA. https://www.cemat-asia.com/news_/shownews.php?lang=cn&id=4544 (accessed August 4, 2024).

[8]   A better IKEA catalogue. (n.d.). Except integrated sustainability. http://www.except.nl/en/projects/558-een-betere-ikea-catalogus (accessed August 4, 2024).

[9]   Nescafé outlines extensive plan to help make coffee farming more sustainable. (2022, October 7). https://www.vendingmarketwatch.com/coffee-service/news/21283162/nescafe-outlines-extensive-plan-to-help-make-coffee-farming-more-sustainable (accessed August 4, 2024).

[10]   Social sustainability. (n.d.). United Nations Globalcompact. https://cn.unglobalcompact.org/SocialSustainableDevelopment.html (accessed August 6, 2024).

[11]   Global diversity and inclusion. (n.d.). Microsoft. https://www.microsoft.com/en-us/diversity/default (accessed August 6, 2024).

[12]   Schneider electric creates professional education platform to address the data center talent shortage. (2022, July 12). Schneider Electric. https://www.se.com/ww/en/about-us/newsroom/news/press-releases/schneider-electric-creates-professional-education-platform-to-address-the-data-center-talent-shortage-62cbd1bd31d8d70be01792a4 (accessed August 9, 2024).

[13]   Zero carbon + green + smart, manufacturing factories "Paint" low-carbon and environmentally friendly colors. (2022, March 29). Intelligent Manufacturing Network. https://ecep.ofweek.com/2022-03/ART-93010-8420-30555441.html (accessed August 9, 2024).

[14]   Green development: Saving, recycling and efficient use of resources. (2021, April 30). Xinhua News Agency. https://www.gov.cn/xinwen/2021-04/30/content_5604152.htm#2 (accessed August 9, 2024).

[15]   Digitalization promotes green development of manufacturing. (2024, January 3). Tsinghua University Internet Industry Research Institute. https://www.iii.tsinghua.edu.cn/info/1165/3951.htm (accessed August 9, 2024).

[16]   The impact of ecological environmental safety on the international operation of enterprises and its response. (2024, April 8). https://www.kwm.com/cn/zh/insights/latest-thinking/the-impact-of-ecological-and-environmental-safety-on-corporate-international-operations.html (accessed August 10, 2024).

[17]   Zhou, J., and Liu, W. (2024). Carbon reduction effects of digital technology transformation: evidence from the listed manufacturing firms in China. *Technological Forecasting and Social Change*, 198, 122999. https://doi.org/https://doi.org/10.1016/j.techfore.2023.122999.

[18]   Digital carbon neutrality white paper. (2021, December). China Academy of Information and Communications Technology. http://www.caict.ac.cn/kxyj/qwfb/bps/202112/P020211220632111694171.pdf (accessed August 10, 2024).

[19]   Vatin, N.I., Negi, G.S., Yellanki, S., Mohan, C., and Singla, N. (2024). Sustainability measures: An experimental analysis of AI and Big Data insights in Industry 5.0. *BIO Web of Conferences*, 86, 01072.

[20]   Climate TRACE: using satellites and machine learning to pinpoint global emissions. (n.d.). Greentechmedia. https://www.greentechmedia.com/articles/read/climatetrace-using-satellites-and-machine-learning-to-track-global-greenhouse-gas-emissions. (accessed August 10, 2024)

[21]   Fei, X., Jiaqi, L., and Yamei, F. (2022). The effect of artificial intelligence technology on carbon emissions. *Science & Technology Progress and Policy*, 39(24), 1–9. https://doi.org/10.6049/kjjbydc.2022030149.

[22]   Aheleroff, S., Huang, H., Xun, X., and Zhong, R.Y. (2022). Toward sustainability and resilience with Industry 4.0 and Industry 5.0. *Frontiers in Manufacturing Technology*, 2. https://doi.org/10.3389/fmtec.2022.951643.

[23]   FengTang, Z., Leong, W.Y. (2024). Carbon emission under Industry 5.0. *Journal of Innovation and Technology*, 21, 239–259.

Chapter 14

BIM risk evaluation for comprehensive management corridor projects

Feng Yang[1,2] and Wai Yie Leong[2]

In the era of Industry 5.0, China's urbanization process is accelerating, and the development of urban comprehensive management corridors is rapidly progressing. The application of building information modeling (BIM) technology in urban comprehensive management corridors is an important way to achieve smart manufacturing and sustainable development. However, the application of BIM technology in urban comprehensive management corridor projects faces many risks, which is a major obstacle to the digital development of cities. This study is based on the analytic hierarchy process (AHP) to establish an index system of risk factors, calculate the relevant weights, and identify the risk factors for investors and constructors. Then, the fuzzy comprehensive evaluation method is used to assess the risks of actual application cases, obtain the risk categories and risk levels of applying BIM in urban comprehensive management corridor projects, and provide corresponding risk reduction suggestions.

The research aims to provide a reference for the application of BIM technology in urban comprehensive management corridor projects, promote the digital transformation of the construction industry, and facilitate the sustainable development of smart cities.

14.1 Introduction

In the era of Industry 5.0, also known as the era of smart manufacturing, it is an extension and development of Industry 4.0, which emphasizes a higher level of intelligence and automation, as well as the deep integration and application of artificial intelligence (AI) in the production process. In this context, urban comprehensive management corridors, as an important part of urban infrastructure, their intelligence and informationization level of construction and management directly affect the sustainable development and operational efficiency of the city.

[1]Department of Transportation Engineering, Fujian Forestry Vocational and Technical College, China
[2]Faculty of Engineering and Quantity Surveying, INTI International University, Malaysia

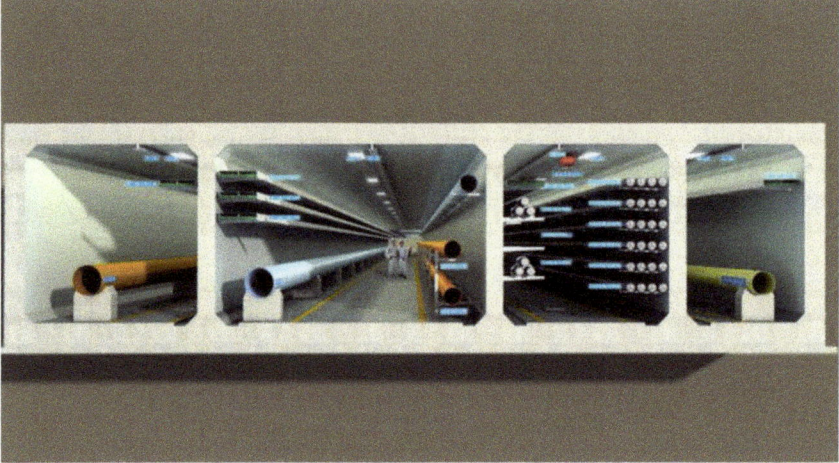

Figure 14.1   Urban integrated pipe corridor plan

In recent years, China has strongly supported the construction of urban comprehensive management corridors, and the investment in municipal infrastructure has been continuously increasing. Comprehensive management corridors, with their energy-saving and low-carbon characteristics, have begun to be constructed in more cities (Figure 14.1).

By combining comprehensive management corridors with BIM technology, the digitalization and intelligence of their entire life cycle, including design, construction, and operation and maintenance, can be improved. Through the latest multi-terminal management equipment, they can be better operated and maintained, keeping them in good condition, which lays a broad foundation for the intelligence of municipal facilities.

However, the application of BIM technology in comprehensive management corridor projects faces many risks, which is an adverse factor hindering the application of BIM technology.

14.2   Characteristics of urban comprehensive management corridor project risks and analysis of BIM application

14.2.1   Characteristics of urban comprehensive management corridor projects and the main risks they face

China's development of BIM technology started relatively late. However, in recent years, it has developed rapidly and has gradually become widespread in the construction and transportation industries. Despite this progress, the application of BIM in integrated corridor projects, which have experienced rapid growth in recent years, faces significant challenges. These challenges include a shortage of high-level BIM professionals, the lack of uniform BIM model standards, fragmented software development leading to insufficient information interoperability,

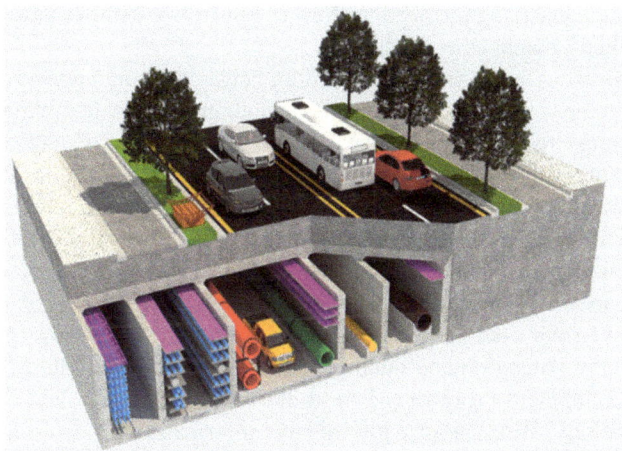

Figure 14.2   The model of an integrated urban pipeline corridor

the complexity of managing multiple systems and disciplines involved in integrated corridor projects, the difficulty of managing operations, and the large investment required for software development. The model of an integrated urban pipeline corridor is shown in Figure 14.2.

14.2.2   Application status of BIM technology in municipal comprehensive pipeline corridor projects

Li analyzed literature data from 2004 to 2015 and concluded that the application of BIM is mainly concentrated in eight areas: collaboration, innovation, stakeholders, visualization, implementation, culture, framework, and operation and maintenance. They observed a trend of BIM evolving from technology to application and from single-domain to cross-domain integration [1].

Darko *et al.* (2020) studied the application of AI in the Architecture, Engineering, and Construction (AEC) industry. After thoroughly analyzing 41,827 related literature records, they found that genetic algorithms, neural networks, fuzzy logic, fuzzy sets, and machine learning are the most commonly used AI methods in AEC. They also identified future research directions, including the application of robotic automation and convolutional neural networks to solve AEC problems [2].

Sacks *et al.* explored the use of digital information tools in the construction industry and discussed the future application of construction technologies that rely on digital information. The paper identified new challenges and proposed a series of research topics that have the potential to unlock a range of future applications, many of which will utilize AI [3].

Deng reviewed the evolution of smart building representations in the AEC-FM industry, particularly the development from BIM to digital twins. The article proposed a five-level classification system to categorize existing research and noted that most studies have yet to fully realize the concept of digital twins. The study concluded by emphasizing the need for further research to achieve advanced digital twins for building management [4].

In the process of urban integrated corridor construction and operation, there are deficiencies in the regulatory mechanisms, including inadequate archiving of pipeline data, incomplete quality assurance systems, and imperfect regulatory agencies, which increase the difficulty of construction and operation. Additionally, the low level of information management, including a focus on construction over maintenance, lack of data standards, and insufficient awareness of scientific planning, further complicates maintenance. The absence of a shared underground pipeline platform also makes it difficult to realize various information-based functions [5].

Yang and Leong used the PESTEL analysis framework to show that BIM can enhance the informatization and intelligence of municipal utility tunnels and has a promising development prospect [6]. With the rapid application and development of BIM technology, BIM can provide valuable information for decision-making in various management aspects, such as sustainable design, schedule planning, cost estimation, construction monitoring, and building performance analysis. BIM-centered project delivery has demonstrated excellent performance in digital management, leading to its widespread adoption, which has driven the high degree of automation, intelligence, and reliability in the AEC industry in the digital age.

14.3   Construction of a BIM risk evaluation index system

14.3.1   Construction of an index system based on the analytic hierarchy process

Based on the characteristics of municipal engineering project construction and referencing the research of relevant experts and scholars, this study analyzes the participating units of the project from the perspective of government investment and the construction side. The evaluation indicators were adjusted using the *Delphi method*. According to the characteristics of municipal engineering project construction, the index system was divided into five primary risk indicators: technical risk, management risk, economic risk, personnel risk, and environmental risk.

The primary indicators were further subdivided into 19 secondary indicators, forming the evaluation system as shown in Table 14.1.

14.3.2   Determining index weights

14.3.2.1   Constructing the judgment matrix and calculating weights

The construction of the judgment matrix is crucial for applying the BIM risk index system. The content of this study and the survey questionnaire are based on the 1–9 scale method, which defines the judgment matrix as shown in Table 14.2.

1.   Calculation of first-level indicator weights.

A total of 37 valid questionnaires were obtained from the AHP survey (those that passed the consistency check were considered as final valid

Table 14.1   *Risk index system for the application of BIM in municipal integrated corridor projects*

No.	Risk factor category	Primary indicators
1	Technical risk, a_1	Data information integrity, a_{11}
2		Geometric accuracy of models, a_{12}
3		Applicability of data interaction a_{13}
4		Stability of hardware and software systems, a_{14}
5		Reliability of simulation analysis, a_{15}
6	Management risk, a_2	Capability of management team, a_{21}
7		Reasonableness of responsibility allocation, a_{22}
8		Enterprise incentive positivity, a_{23}
9		Information sharing coordination, a_{24}
10	Economic risk, a_3	Contract economic risk, a_{31}
11		Intellectual property risk, a_{32}
12		Effectiveness of cost control, a_{33}
13		Reasonableness of return on investment, a_{34}
14	Personnel risk, a_4	Technical skill level, a_{41}
15		Learning and innovation ability, a_{42}
16		Job stability, a_{43}
17	Environmental risk, a_5	Extent of BIM software application, a_{51}
18		Degree of BIM software standardization, a_{52}
19		Policy support, a_{53}

questionnaires). Based on the survey results, the original data were collected and organized to obtain the expert scoring table (mean values) as shown in Table 14.3.

2. Calculation of weights using AHP results:

Based on the expert scoring in Table 14.3, a judgment matrix was constructed for the five risk categories: technical risk, management risk, economic risk, personnel risk, and environmental risk. Using the AHP, the eigenvectors, weight values, and the maximum eigenvalue were calculated, as shown in Table 14.4.

To verify the reasonableness of the obtained results, the maximum eigenvalue of the judgment matrix is calculated as:

$$\lambda_{max} = 5.355$$

Next, the consistency index (CI) value is calculated using the maximum eigenvalue:

$$CI = \frac{\lambda_{max} - n}{n - 1} = \frac{5.355 - n}{n - 1} = 0.089 \tag{14.1}$$

Using the *AHP method*, it is necessary to perform a consistency test. The CI value has been calculated, and the RI values are shown in Table 14.5.

$$CR = \frac{CI}{RI} = \frac{0.089}{1.12} = 0.0794 < 0.1 \tag{14.2}$$

Table 14.2   The 1–9 scale method and its interpretation

a_{ij}	Comparison between two indicators	Interpretation
1	Equally important	Indicator i is equally important as indicator j
3	Slightly more important	Indicator i is slightly more important than indicator j
5	Significantly more important	Indicator i is significantly more important than indicator j
7	Strongly more important	Indicator i is strongly more important than indicator j
9	Extremely more important	Indicator i is extremely more important than indicator j
2, 4, 6, 8	Intermediate values	Represents a value between two adjacent judgments
1/3	Slightly less important	Indicator i is slightly less important than indicator j
1/5	Significantly less important	Indicator i is significantly less important than indicator j
1/7	Strongly less important	Indicator i is strongly less important than indicator j
1/9	Extremely less important	Indicator i is extremely less important than indicator j
1/2, 1/4, 1/6, 1/8	Intermediate values	Represents a value between two adjacent judgments

Table 14.3   Expert scoring table

	Technical risk	Management risk	Economic risk	Personnel risk	Environmental risk
Technical risk	1	0.553	0.57	1.223	1.207
Management risk	1.807	1	0.522	1.631	0.717
Economic risk	1.756	1.917	1	0.832	0.532
Personnel risk	0.818	0.613	1.202	1	0.361
Environmental risk	0.829	1.395	1.88	2.767	1

In general, the smaller the CR value, the better the consistency of the judgment matrix. Typically, if the *CR* value is less than 0.1, the judgment matrix is considered to meet the consistency test. If the CR value is greater than 0.1, it indicates that the matrix lacks consistency. For the fifth-order judgment matrix in this study, the CI value is 0.089, and the RI value from the table is 1.120. The calculated CR value is 0.0794, which is less than 0.1, indicating that the judgment matrix satisfies the consistency test and the weights are consistent.

Table 14.4   Expert scoring table

	Eigenvector	Weight value (%)	Maximum Eigenvalue	CI value
Technical risk	0.853	17.051	5.355	0.089
Management risk	0.981	19.619		
Economic risk	1.077	21.539		
Personnel risk	0.705	14.097		
Environmental risk	1.385	27.695		

Eigenvector: W = (0.853, 0.981, 1.077, 0.705, 1.385)
The weights of each indicator are as follows:
a_1 = 17.051%, a_2 = 19.619%, a_3 = 21.539%, a_4 = 14.097%, a_5 = 27.695%.

Table 14.5   Random consistency index (RI) table

Order N	1	2	3	4	5	6	7	8	9	10	11	12
RI	0	0	0.52	0.89	1.12	1.26	1.36	1.41	1.46	1.49	1.52	1.54

From Table 14.5 above, the random consistency RI value for N = 5, RI = 1.12.

The weight calculation results from the AHP using the multiplicative method show that the weights are as follows: technical risk: 17.051%, management risk: 19.619%, economic risk: 21.539%, personnel risk: 14.097%, environmental risk: 27.695%.

Using the same method, separate judgment matrices were established for the secondary risk indicators within each of the following categories: technical risk, management risk, economic risk, personnel risk, and environmental risk. The eigenvectors, maximum eigenvalues were calculated, and consistency tests were performed. The results are presented in Tables 14.6, 14.7, 14.8, 14.9, and 14.10, respectively.

Through these calculations, the weights of the various indicators and the weights of the sub-indicators are obtained, which allows the calculation of the weight of the sub-indicators relative to the overall goal, as shown in Table 14.11.

From this, we can see that among the primary indicators, environmental risk, economic risk, and management risk are the most critical, followed by technical risk and personnel risk. The weight analysis of these risk factors suggests that in the application of BIM in municipal comprehensive utility corridor projects, the most significant factors are policy support guidance (19.9%), investment return risk (10.73%), job stability risk (9.41%), and information sharing coordination (8.92%). Additionally, the environmental risk category carries a relatively high proportion.

Table 14.6   Weight analysis of secondary indicators (technical risks)

Technical risks	Data information integrity	Model geometric accuracy	Data interaction suitability	Hardware and software system stability	Simulation analysis reliability
Data information integrity	1	0.194	0.296	0.173	0.159
Model geometric accuracy	5.143	1	1.105	0.813	0.538
Data interaction suitability	3.375	0.905	1	0.496	0.483
Hardware and software system stability	5.786	1.23	2.016	1	0.483
Simulation analysis reliability	6.286	1.857	2.071	2.071	1

λ_{max} = 5.06, CI = 0.015, CR = 0.014<0.1

Table 14.7   Weight analysis of secondary indicators (management risks)

Management risks	Management team capability	Reasonableness of responsibility distribution	Enterprise incentive positivity	Information sharing coordination	Weight (%)
Management team capability	1	0.311	0.194	0.163	6.172
Reasonableness of responsibility distribution	3.214	1	0.378	0.311	15.366
Enterprise incentive positivity	5.143	2.643	1	0.636	33.002
Information sharing coordination	6.143	3.214	1.571	1	45.459

λ_{max} = 4.042, CI = 0.014, CR = 0.016<0.1

14.3.3   Risk evaluation model based on fuzzy comprehensive evaluation

This chapter primarily adopts the fuzzy comprehensive evaluation method for the comprehensive evaluation of BIM technology application in utility tunnels. The comprehensive evaluation serves as the basis for determining whether the owner, construction unit, and BIM industry workers will adopt BIM technology.

Table 14.8   Weight analysis of secondary indicators (economic risks)

Economic risks	Contract economic risks	Intellectual property risks	Cost control effectiveness	Reasonableness of investment return	Weight (%)
Contract economic risks	1	1.541	0.226	0.175	8.505
Intellectual property risks	0.649	1	0.136	0.137	5.746
Cost control effectiveness	4.429	7.357	1	0.583	35.913
Reasonableness of investment return	5.714	7.286	1.714	1	49.836

λ_{max} = 4.027, CI = 0.009, CR = 0.01<0.1

Table 14.9   Weight analysis of secondary indicators (personnel risks)

Personnel risks	Technical skill level	Learning and innovation capability	Job stability	Weight (%)
Technical skill level	1	0.341	0.184	10.033
Learning and innovation capability	2.933	1	0.269	23.213
Job stability	5.429	3.714	1	66.754

λ_{max} = 3.054, CI = 0.027, CR = 0.052<0.1

Table 14.10   Weight analysis of secondary indicators (environmental risks)

Environmental risks	Extent of BIM software application	Degree of BIM software standardization	Policy supportiveness	Weight (%)
Extent of BIM software application	1	1.579	0.265	17.718
Degree of BIM software standardization	0.633	1	0.134	10.427
Policy supportiveness	3.778	7.444	1	71.855

λ_{max} = 3.005, CI = 0.003, CR = 0.005<0.1

The process of fuzzy comprehensive evaluation in this chapter is as follows: First, the indicator factors of the evaluation object and the set of evaluation levels are obtained. The weight and membership vector are calculated to form a fuzzy judgment matrix. Finally, the matrix and indicators are comprehensively computed

Table 14.11   BIM risk indicator system for municipal comprehensive utility corridor projects

No.	Risk factor category	Weight	Primary indicators	Weight
1	Technical risk, a_1	0.1705	Data information integrity, a_{11}	0.0077
2			Geometric accuracy of models, a_{12}	0.0335
3			Applicability of data interaction, a_{13}	0.0264
4			Stability of hardware and software systems, a_{14}	0.0415
5			Reliability of simulation analysis, a_{15}	0.0613
6	Management risk, a_2	0.1962	Capability of management team, a_{21}	0.0121
7			Reasonableness of responsibility allocation, a_{22}	0.0301
8			Enterprise incentive positivity, a_{23}	0.0647
9			Information sharing coordination, a_{24}	0.0892
10	Economic risk, a_3	0.2154	Contract economic risk, a_{31}	0.0183
11			Intellectual property risk, a_{32}	0.0124
12			Effectiveness of cost control, a_{33}	0.0774
13			Reasonableness of return on investment, a_{34}	0.1073
14	Personnel risk, a_4	0.141	Technical skill level, a_{41}	0.0141
15			Learning and innovation ability, a_{42}	0.0327
16			Job stability, a_{43}	0.0941
17	Environmental risk, a_5	0.277	Extent of BIM software application, a_{51}	0.0491
18			Degree of BIM software standardization, a_{52}	0.0289
19		0.1705	Policy support, a_{53}	0.199

(using the $M\,(.,\,+)$ operator for research) to achieve a comprehensive benefit evaluation for the project. The specific process is as follows:

1. Establish the indicator set (U):
    This includes m indicators. In this chapter, the object of evaluation is the comprehensive benefit of applying BIM technology to prefabricated box

girders, and the previous analysis identified five main aspects, including ten specific indicators used as evaluation indicators.

Based on the above information, the number of indicators mmm and the indicator set can be obtained as follows:

$$M = 10, U = u_1, u_2 \ldots \ldots u_m$$

2. Determine the evaluation set (V):

The evaluation object is assessed, and in the construction of the beam yard using BIM technology, the evaluation indicators can be divided into five levels: high risk, relatively high risk, average risk, relatively low risk, and very low risk. Therefore, the set is obtained as follows:

$V = \{$high risk, relatively high risk, average risk, relatively low risk, very low risk$\}$

3. Establish the judgment matrix:

For the single indicators $u_i (i = 1, 2, 3 \ldots m)$ in the indicator set, a single-indicator judgment is made to determine the membership degree r_{ij} of the evaluation level $V=\{v_1, v_2, \ldots, v_n\}$ for indicator u. Thus, a single-indicator evaluation set is obtained as follows:

$$r_i = \{r_{i1}, r_{i2}, \ldots, r_{in}\}$$

Finally, the evaluation of all mmm single indicators is combined to form the overall evaluation matrix R, where matrix R represents the fuzzy relationship between the indicator set U and the evaluation set V.

$$R = \left(r_{ij} \right)_{m \times n} \tag{14.3}$$

$$R_i = \begin{bmatrix} r_{11} & r_{11} & \cdots & r_{1n} \\ r_{21} & r_{22} & \cdots & r_{2n} \\ \cdots & \cdots & \cdots & \cdots \\ r_{m1} & r_{m2} & \cdots & r_{mn} \end{bmatrix} \tag{14.4}$$

4. Determining the weight vector of indicators:

After obtaining the fuzzy matrix, the impact of the weight of each evaluation indicator on the final evaluation result varies. Through calculation, a fuzzy subset A of U can be derived, which becomes the weight allocation set:

$$A = \{a_1, a_2 \ldots a_m\}, \quad a_i > 0, \sum a_i = 1 \tag{14.5}$$

The weights significantly influence the evaluation results. Typically, the weights of the indicators in an evaluation system are determined by authoritative experts or experienced participants according to relevant rules. This chapter utilizes the AHP and fuzzy evaluation method to conduct a comprehensive evaluation of BIM application benefits. The weight coefficients of the 19 indicators obtained earlier serve as the weight allocation set:

$$A = \left\{ \begin{array}{l} 0.0077, 0.0335, 0.0264, 0.0415, 0.0613, 0.0121, 0.0301, 0.0647, 0.0892, 0.0183, \\ 0.0124, 0.0774, 0.1073, 0.0141, 0.0327, 0.0941, 0.0491, 0.0289, 0.1990 \end{array} \right\}$$

5. Selection of a comprehensive evaluation algorithm

   At this point, there are two fuzzy sets: one is the fuzzy set $U \times V$, which is the fuzzy matrix R, and the other is the set of indicator importance degrees A within U. The fuzzy matrix R reflects the membership degree of evaluation indicators in various aspects. By performing the convolution operation between the weight vector A and the fuzzy matrix R, the fuzzy comprehensive evaluation result vector can be obtained, which in turn reflects the membership degrees of the fuzzy subsets for each grade of BIM benefits.

   In this chapter, the vector can represent the comprehensive membership degree of the evaluation subject. Based on the principle of maximum membership degree, the risk evaluation grade can be determined after calculation.

$$B = A^{o}R = a_1, a_2 \ldots a_m{}^{o} \begin{bmatrix} r_{11} & r_{11} & \cdots & r_{1n} \\ r_{21} & r_{22} & \cdots & r_{2n} \\ \cdots & \cdots & \cdots & \cdots \\ r_{m1} & r_{m2} & \cdots & r_{mn} \end{bmatrix} = b_1, b_2, b_3, b_4 \quad (14.6)$$

   In this chapter, the vector can represent the comprehensive membership degree of the evaluation subject. Based on the principle of maximum membership degree, the risk evaluation grade can be determined after calculation.

14.4   Case application

This chapter selects four comprehensive utility tunnel projects in the Southeast region as examples for study. The pipelines in the tunnels include electricity, communication, water supply and drainage, and gas. A panel of 12 experts was invited, including representatives from the owner, design, and construction parties, to conduct a questionnaire survey for the risk assessment related to the design and construction processes of the utility tunnels. The weights determined through the AHP were based on the survey results. The expert questionnaire is shown in Table 14.12.

Based on the survey results, the membership functions of each indicator were normalized. The ratio of the benefit situation for each indicator was calculated, as shown in Table 14.13.

Table 14.12   BIM risk indicator system for municipal comprehensive utility corridor projects

Indicator level	Overall weight	High risk	Relatively high risk	Moderate risk	Relatively low risk	Very low risk
Data information integrity, a_{11}	0.0077	1	2	2	4	3
Geometric accuracy of models, a_{12}	0.0335	1	4	4	2	1
Applicability of data interaction, a_{13}	0.0264	3	2	2	4	1
Stability of hardware and software systems, a_{14}	0.0415	0	2	4	3	3
Reliability of simulation analysis, a_{15}	0.0613	4	2	3	2	1
Capability of management team, a_{21}	0.0121	1	2	4	3	2
Reasonableness of responsibility allocation, a_{22}	0.0301	0	2	3	4	3
Enterprise incentive positivity, a_{23}	0.0647	4	2	2	2	2
Information sharing coordination, a_{24}	0.0892	2	4	3	2	1
Contract economic risk, a_{31}	0.0183	3	4	3	1	1
Intellectual property risk, a_{32}	0.0124	2	4	3	2	1
Effectiveness of cost control, a_{33}	0.0774	3	4	2	2	1
Reasonableness of return on investment, a_{34}	0.1073	3	4	2	2	1
Technical skill level, a_{41}	0.0141	2	2	4	3	1
Learning and innovation ability, a_{42}	0.0327	3	4	2	2	1
Job stability, a_{43}	0.0941	4	4	2	1	1
Extent of BIM software application, a_{51}	0.0491	3	2	2	3	2
Degree of BIM software standardization, a_{52}	0.0289	2	4	3	2	1
Policy support, a_{53}	0.199	2	2	4	3	1

Table 14.13   BIM risk indicator system for municipal comprehensive utility corridor projects

Indicator level	Overall weight	High risk	Relatively high risk	Moderate risk	Relatively low risk	Very low risk
Data information integrity, a_{11}	0.0077	0.0833	0.1667	0.1667	0.3333	0.2500
Geometric accuracy of models, a_{12}	0.0335	0.0833	0.3333	0.3333	0.1667	0.0833
Applicability of data interaction, a_{13}	0.0264	0.2500	0.1667	0.1667	0.3333	0.0833
Stability of hardware and software systems, a_{14}	0.0415	0.0000	0.1667	0.3333	0.2500	0.2500
Reliability of simulation analysis, a_{15}	0.0613	0.3333	0.1667	0.2500	0.1667	0.0833
Capability of management team, a_{21}	0.0121	0.0833	0.1667	0.3333	0.2500	0.1667
Reasonableness of responsibility allocation, a_{22}	0.0301	0.0000	0.1667	0.2500	0.3333	0.2500
Enterprise incentive positivity, a_{23}	0.0647	0.3333	0.1667	0.1667	0.1667	0.1667
Information sharing coordination, a_{24}	0.0892	0.1667	0.3333	0.2500	0.1667	0.0833
Contract economic risk, a_{31}	0.0183	0.2500	0.3333	0.2500	0.0833	0.0833
Intellectual property risk, a_{32}	0.0124	0.1667	0.3333	0.2500	0.1667	0.0833
Effectiveness of cost control, a_{33}	0.0774	0.2500	0.3333	0.1667	0.1667	0.0833
Reasonableness of return on investment, a_{34}	0.1073	0.2500	0.3333	0.1667	0.1667	0.0833
Technical skill level, a_{41}	0.0141	0.1667	0.1667	0.3333	0.2500	0.0833
Learning and innovation ability, a_{42}	0.0327	0.2500	0.3333	0.1667	0.1667	0.0833
Job stability, a_{43}	0.0941	0.3333	0.3333	0.1667	0.0833	0.0833
Extent of BIM software application, a_{51}	0.0491	0.2500	0.1667	0.1667	0.2500	0.1667
Degree of BIM software standardization, a_{52}	0.0289	0.1667	0.3333	0.2500	0.1667	0.0833
Policy support, a_{53}	0.199	0.1667	0.1667	0.3333	0.2500	0.0833

A fuzzy matrix R for the comprehensive evaluation of BIM technology application benefits in this project can be obtained as follows:

$$R = \begin{bmatrix}
0.0833 & 0.1667 & 0.1667 & 0.3333 & 0.2500 \\
0.0833 & 0.3333 & 0.3333 & 0.1667 & 0.0833 \\
0.2500 & 0.1667 & 0.1667 & 0.3333 & 0.0833 \\
0.0000 & 0.1667 & 0.3333 & 0.2500 & 0.2500 \\
0.3333 & 0.1667 & 0.2500 & 0.1667 & 0.0833 \\
0.0833 & 0.1667 & 0.3333 & 0.2500 & 0.1667 \\
0.0000 & 0.1667 & 0.2500 & 0.3333 & 0.2500 \\
0.3333 & 0.1667 & 0.1667 & 0.1667 & 0.1667 \\
0.1667 & 0.3333 & 0.2500 & 0.1667 & 0.0833 \\
0.2500 & 0.3333 & 0.2500 & 0.0833 & 0.0833 \\
0.1667 & 0.3333 & 0.2500 & 0.1667 & 0.0833 \\
0.2500 & 0.3333 & 0.1667 & 0.1667 & 0.0833 \\
0.2500 & 0.3333 & 0.1667 & 0.1667 & 0.0833 \\
0.1667 & 0.1667 & 0.3333 & 0.2500 & 0.0833 \\
0.2500 & 0.3333 & 0.1667 & 0.1667 & 0.0833 \\
0.3333 & 0.3333 & 0.1667 & 0.0833 & 0.0833 \\
0.2500 & 0.1667 & 0.1667 & 0.2500 & 0.1667 \\
0.1667 & 0.3333 & 0.2500 & 0.1667 & 0.0833 \\
0.1667 & 0.1667 & 0.3333 & 0.2500 & 0.0833
\end{bmatrix}$$

Using the indicator weight set calculated by the AHP as follows:

$$A = \left\{ \begin{array}{l} 0.0077, 0.0335, 0.0264, 0.0415, 0.0613, 0.0121, 0.0301, 0.0647, 0.0892, 0.0183, \\ 0.0124, 0.0774, 0.1073, 0.0141, 0.0327, 0.0941, 0.0491, 0.0289, 0.1990 \end{array} \right\}$$

The fuzzy evaluation set B is calculated as:

$$B = A^{o}R = a_1, a_2 \ldots a_m{}^{o} \begin{bmatrix}
r_{11} & r_{11} & \cdots & r_{1n} \\
r_{21} & r_{22} & \cdots & r_{2n} \\
\cdots & \cdots & \cdots & \cdots \\
r_{m1} & r_{m2} & \cdots & r_{mn}
\end{bmatrix}$$

$$= (0.2129\,0.2490\,0.2367\,0.1943\,0.1071) \tag{14.7}$$

Based on the maximum membership degree principle, the value 0.2490 corresponds to a relatively high risk, which suggests that, overall, the application of BIM technology in comprehensive utility tunnel construction is still considered to carry a significant level of risk.

14.5 Conclusion and outlook

14.5.1 Research conclusions

This chapter analyzes the application risks considered by various parties in the use of BIM technology in comprehensive utility tunnels. Based on a comparison of

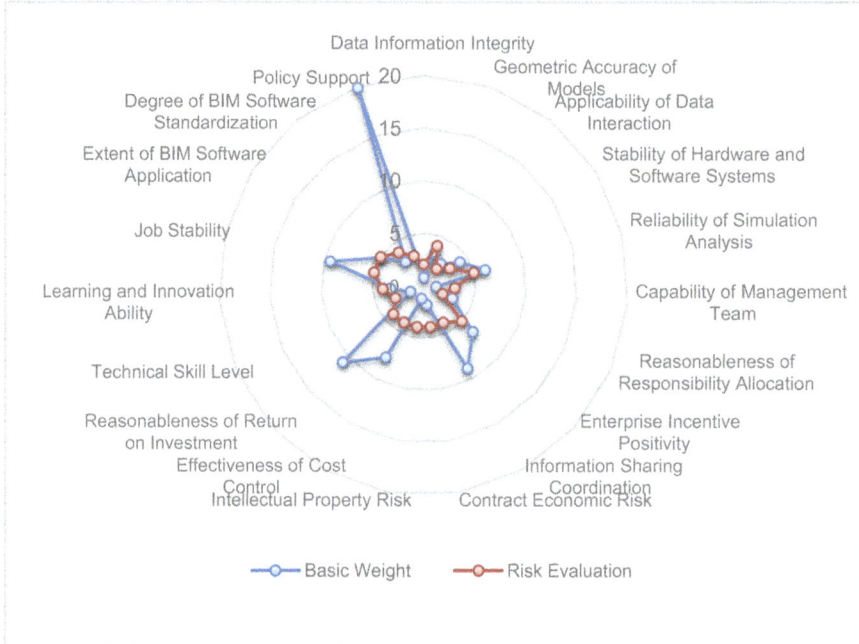

Figure 14.3   Risk diagram for the application of BIM technology in comprehensive utility tunnels

risks, the concerns and potential risks faced by different aspects were identified. Subsequently, through a real case study, project participants provided insights into the actual risks encountered during the application process. As a result, a risk diagram for the application of BIM technology in comprehensive utility tunnels was developed, as shown in Figure 14.3.

Based on Figure 14.3, the risks and recommendations for introducing BIM technology into comprehensive utility tunnels can be summarized as follows:

1. Risks faced by investors and construction managers:

    The general risks include technical risks, management risks, economic risks, personnel risks, and environmental risks. The primary concerns are policy support (19.9%), investment return risk (10.73%), job stability risk (9.41%), and information sharing coordination (8.92%), with environmental risks also being significant. Therefore, from the perspective of investors and construction managers, policy guidance is crucial for increasing the application of BIM in comprehensive utility tunnels. Relevant policies should be formulated based on the actual situation to improve the application rate. The relatively high investment return risk highlights the need to balance the benefits that BIM technology brings to the comprehensive utility tunnel projects with reducing the costs associated with BIM, including human, material, and financial resources, to improve the return on investment and reduce investment risks.

The risks associated with job stability and information sharing coordination suggest the need for a stable BIM technical team, and the collaborative effort between the company, team, and technical personnel to manage the risks related to information sharing and coordination.

2. Risk evaluation from actual participants:

In actual projects, participants identified significant risks in enterprise incentive positivity, job stability, and the reliability of simulation analysis. Therefore, during the implementation of BIM, the software implementation team needs to strengthen the training of technical personnel to improve their technical level and stabilize the technical team, ensuring the stability of technical positions and thereby reducing the risks associated with BIM technology application in projects.

3. Differences between risk forecasts by investors/construction managers and risk evaluations by participants:

There is a noticeable difference in policy support and investment return rationality between risk forecasts made by investors/construction managers and actual risk evaluations by participants. The pre-evaluated risks suggest that companies are concerned about policy continuity, but in specific projects, policies tend to be stable, leading to a significant difference in policy support risks, which also affects investment return risks. Companies can adjust relevant measures based on comparisons between pre-evaluated risks and actual risk evaluations to reduce the actual risks associated with BIM application.

14.5.2 Future research directions

This study assesses the pre-evaluated risks of applying BIM technology in comprehensive utility tunnels from the perspectives of investors and construction managers, highlighting the risk landscape of BIM application in these projects. Through the AHP, various risks were categorized, and their weights were determined. Subsequently, evaluations of several typical projects were conducted, revealing that the application of BIM technology in comprehensive utility tunnels still faces risks, particularly concerning policy support, which is a significant concern for investors and construction managers and exceeds the actual risks encountered. Therefore, the stability and continuity of policies have a substantial impact on the promotion and application of BIM technology.

Although this chapter categorizes the various risks, it does not conduct specific quantitative calculations for each risk. Future research could quantify the impact mechanisms of various risks and incorporate them into the analysis. Additionally, comprehensive utility tunnels face numerous risks in their operation and management. Future research could collect data on the application of BIM throughout the entire lifecycle of comprehensive utility tunnels to obtain more comprehensive risk assessment results.

References

[1] Li, X., Wu, P., Shen, G. Q., Wang, X., and Teng, Y. (2017). Mapping the knowledge domains of building information modeling (BIM): A bibliometric

approach. *Automation in Construction*, 84, 195–206. https://doi.org/10.1016/j.autcon.2017.09.011.

[2] Darko, A., Chan, A. P. C., Adabre, M. A., Edwards, D. J., Hosseini, M. R., and Ameyaw, E. E. (2020). Artificial intelligence in the AEC industry: Scientometric analysis and visualization of research activities. *Automation in Construction*, 112, 103081. https://doi.org/10.1016/j.autcon.2020.103081.

[3] Sacks, R., Girolami, M., and Brilakis, I. (2020). Building information modelling, artificial intelligence and construction tech. *Developments in the Built Environment*, 4, 100011. https://doi.org/10.1016/j.dibe.2020.100011.

[4] Deng, M., Menassa, C. C., and Kamat, V. R. (2021). From BIM to digital twins: A systematic review of the evolution of intelligent building representations in the AEC-FM industry. *Journal of Information Technology in Construction*, 26, 58–83. https://doi.org/10.36680/j.itcon.2021.005.

[5] Jiang, K. (2022). Discussion on the integrated technology development of urban underground pipeline surveying and mapping. *Structure*, 7(6). https://doi.org/10.26789/jzjg.v7i6.1582.

[6] Yang, F., Leong, W. Y. (2024). Application and challenges of BIM technology in China's integrated utility tunnels. *Journal of Innovation and Technology*, 41, 1–13.

Chapter 15

AI technology drives film industry into the 5.0 era

Mingda Gao[1] and Wai Yie Leong[1]

The film industry is currently undergoing a transformation, driven by the advent of Industry 5.0 and the rise of artificial intelligence (AI) technologies. This chapter explores the evolution of the film industry from its early stages to the present 5.0 era, paralleling the advancements seen in broader industrial revolutions, from Industry 1.0 to 5.0. In this context, the film industry is leveraging AI and artificial intelligence generated content (AIGC) to revolutionize various aspects of film production, including script writing, visual effects, 3D modeling, and even music composition. This chapter examines the impact of AI on the film industry, discusses key technologies enabling this transformation, and evaluates AI's potential to both complement human creativity and redefine traditional filmmaking processes. Film Industry 5.0 represents a hybrid model, where human intelligence and AI-driven innovation coexist, reshaping the future of cinematic storytelling.

15.1 Introduction

15.1.1 Problem statement

In 2015, the concept of Industry 5.0 was introduced and today this revolution has realized its effects on production [1]. All walks of life have begun to discuss how the Industrial 5.0 era should change, and the development of AI technology is seen as a signal of the arrival of a new era. Industry 5.0 allows manufacturers to produce and deliver products and services based on the specific needs and wants of customers. This industrial revolution allows industries to design appropriate production processes using AI to be able to personalize products and services for customers. This revolution actually improves manufacturing and automation processes [1]. The concept of Industry 5.0 originated from Industry 4.0. Industry 4.0 was publicly introduced in 2011 at the Hannover Fair [2,3]. Industry 4.0 is a technological revolution that includes cloud computing, augmented reality (AR), virtual reality (VR), additive manufacturing (3D printing), smart manufacturing, advanced

[1]Faculty of Engineering and Quantity Surveying, INTI International University, Malaysia

Industry 5.0

... promotes talents, diversity
and empowerment

... is agile and resilient with flexible
and adaptable technologies

... leads action on sustainability
and respects planetary boundaries

Figure 15.1   Core values of Industry 5.0 [3]

robotics, edge computing, data analytics, cyber-physical systems, cyber security solutions, smart supply chain management, image processing, ML-based object detection technology, nanotechnology, quantum computing, cognitive computing, digital twins, natural language processing, and advanced analytics, and every other thing that is relevant to automation in industries, businesses, and other sectors to reduce cost, time and labor all are driven by Industry 4.0 which is eventually giving rise to another industrial revolution, Industry 5.0 [4]. Industry 5.0 is a new concept, and its definition is not particularly clear. In 2021, the European Commission formally called for the Fifth Industrial Revolution (Industry 5.0), after discussions among participants from research and technology organizations as well as funding agencies across Europe in two virtual workshops organized by Directorate "Prosperity" of Directorate-General for Research and Innovation, on July 2 and 9, 2020, by the formal release of the document titled "Industry 5.0: Towards a Sustainable, Human-centric, and Resilient European Industry" on January 4, 2021 [5]. According to previous research, Industry 5.0 is centered on three core values human-centricity, sustainability, and resilience (Figure 15.1).

The human-centric approach puts core human needs and interests at the heart of the production process, shifting from technology-driven progress to a thoroughly human-centric and society-centric approach. As a result, industry workers will develop new roles as a shift of value from considering workers as "cost" to "investment." Technology is to serve people and societies, meaning that technology used in manufacturing is adaptive to the needs and diversity of industry workers 6].

Sustainable means industry needs to develop circular process, respect planetary boundaries recycling nature resources to lead a circle economy with better resource efficiency and effectiveness.

Resilient refers to the need to develop a higher degree of robustness in industrial production, arming it better against disruptions and ensuring it can provide and support critical infrastructure in times of crisis. The future industry needs to be resilient enough to swiftly navigate the (geo-)political shifts and natural emergencies [5].

The film industry is a film and television production industry system consisting of all products and services required for the production of film and television works, including equipment, technical services and manufacturers required for pre-production, post-production, and finished product production. The biggest difference between film and other industries lies in its cultural communication attributes. The products it produces are virtual audio-visual products, unlike traditional industries. In the early days, there was no perfect industrial system in the film industry. However, with the continuous improvement of the commercial value of films and the strengthening of commodity attributes, the emergence of genre films has also made film production tend to be industrialized, with industry production standards and processes, and formulaic and templated production ideas. Film production has gradually evolved into the film industry. The progress and revolution of industry often begin with the renewal and iteration of technology, and the same is true for the film industry. The continuous development of AI technology, especially the emergence of generative AI, has also changed the pattern of the film industry, and the film industry has begun to enter the 5.0 era. What changes have taken place from Film Industry 4.0 to 5.0? Has the only way of filming changed fundamentally due to the emergence of AI? What is the difference between Film Industry 5.0 and Industry 5.0 in a broad sense? These are the issues that this chapter wants to discuss.

15.1.2   Literature review

At the beginning of the existence of industry, all goods were made by human hands. With the advancement of technology, machines gradually replaced manual labor, and industrial revolutions began again and again. The same is true for the film industry. The advancement of technology has led to the renewal of film expression and the change of film production methods. As a result, the industry's demand for talents has also changed.

15.1.2.1   From Industry 1.0 to 5.0

Industry 1.0 steam and mechanization
During the early 18th century, mechanization led to an eightfold increase in production in the spinning industry. Steam as the main feature of this revolution in large industries led to the development and improvement of productivity [1]. During this stage, steam power replaced manpower and became the mainstream of industrial production. Following the use of steam power, improvements were made in steam ships and steam locomotives and caused another great change in this period because this technology reduced time for people and goods to travel long distances [7].

Industry 2.0 use of electricity
The electricity was used as main source of power from 19th century. It's easier to use as steam and water. This feature made this power supply usable in special

machines and different. Meanwhile management tools also made significant progress in this period and improved the performance and efficiency of industries [1]. In this period, mass production and Assembly lines were created and people like Frederick Taylor studied the jobs and productivity of workers, and as a result, the concepts and principles of agile and pure production were introduced, and increasing the quality and quantity of production was included in the producers' program [7].

Industry 3.0 electronics and computer technology
In the middle of 20th century, manufacturing automation systems were developed in factories. These technologies reduce the challenge of employees in various tasks and the concept of mass production [7].

Industry 4.0 information and communication technology
Production systems in Industry 3.0 were equipped with computer technology. In Industry 4.0 the connection of these computer systems through the network was expanded and the Internet platform created a wide connection between physical and cyber systems, and as a result, intelligent factories were created. In intelligent factories, production systems, components and people are connected through the network [7]. The theme of Industry 4.0 is "smart manufacturing for the future," which was a German initiative [1]. The purpose of this revolution is to use emerging technologies to significantly improve productivity, so many emerging technologies have also emerged in the process of Industry 4.0 maturing. The most important of which are the internet, the Internet of Things (IoT), robotics and AI, big data and cloud computing, VR and AR, smart factories, smart logistics and environmental intelligence [8].

Industry 5.0
For Industry 5.0, there are two approaches and perspectives. The first is the cooperation of robots and humans. In this perspective, humans and robots have close communication and cooperation at the required times and places. In this cooperation and interaction, human beings focus on activities that require creativity and has a supervisory role, and other tasks are the responsibility of robots. Another approach of Industry 5.0 is bio economics, which deals with the correct and smooth use of biological resources. The approach strikes a balance between the elements of the environment, industry and economics, and sustainability is considered [8] (Figure 15.2).

15.1.2.2   Film Industry1.0 to 5.0
Film Industry 1.0 silent films
At first, movies existed in the form of silent films, that is, movies could only show pictures and did not contain sound. In 1888, Thomas Edison proposed the concept of a kinetoscope, and his employee William Kennedy Dickson led a team to officially complete the development of the kinetoscope in 1893. Using the principle of persistence of vision, each image is temporarily "stopped" by an intermittent light source, so as to enable the viewer to save the detailed information of many different pictures, so that the illusion of continuous activity can be effectively produced,

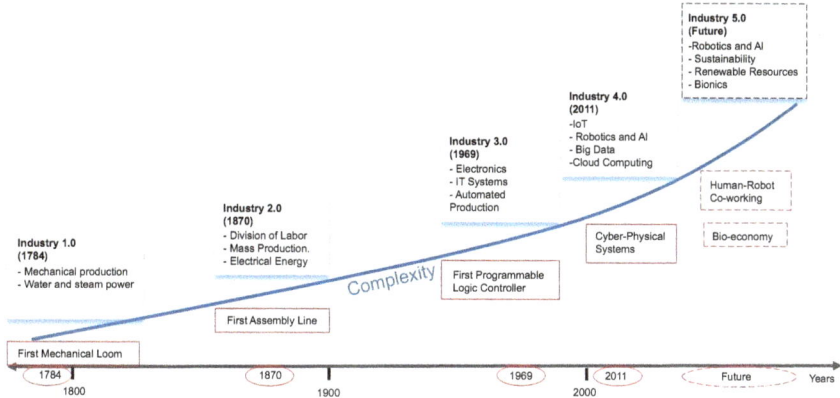

Figure 15.2   Industry revolution development [8]

using photosensitivity. The celluloid film serves as the base of the camera, and an electrically driven gear continuously rotates the film close to a magnifying lens and projects the image onto the curtain to create a movable picture. The film industry has officially entered the era of 1.0. Films in the 1.0 period of the film industry were black-and-white silent films shot with moving cameras using the principle of visual persistence. The narrative was mainly based on the performances of the actors and montage. In terms of film expression techniques, the Soviet film theorist Kuleshov proposed the **Kuleshov effect**: *which means that the mood of a film is not only reflected in the content of a single shot but is implied and expressed through the combination and connection of multiple pictures.* It constitutes the theoretical basis of the Soviet montage school.

Film Industry 2.0 talkies

The first major revolution in the film industry began with the advent of "talkies," or films with sound. The world's earliest talkies appeared in 1900, initially using wax discs to produce sound. During filming, a recording device captured sound waves and translated them into grooves on a wax disc; deeper grooves corresponded to stronger sound waves. As the film was projected, the wax disc would rotate in sync with the projector, emitting sound through an amplifier. However, this method often led to issues with synchronization between the characters' voices and their mouth movements.

A significant breakthrough came in the late 1920s with the development of synchronized sound technology, which greatly improved the alignment of sound with on-screen action. This milestone marked a revolutionary shift in cinema. In October 1927, "The Jazz Singer," the first feature-length film with synchronized dialogue, was released, redefining the audience's understanding of movies and sparking the global popularity of musicals as a commercial genre.

Many film professionals, especially actors, initially resisted sound films, believing the essence of cinema lay in expressive physical performances. Although sound added a new narrative dimension, it was seen as potentially diminishing the artistry of silent film acting. Despite this resistance, sound films ultimately became the new standard, ushering in what can be called the Film Industry 2.0 era.

Films in the Film Industry 2.0 period are sound films, which use synchronous sound technology to synchronize the actors' voices with the pictures, enriching the narrative dimension of the film. The main lines of background music (BGM) and theme songs enhance the entertainment of the film, and add new sound-related professions to the industrial system. It also changes the way actors perform. Unlike the exaggerated body performances of actors in the silent film era, actors in sound films need to perform in combination with lines, and their performances are more natural and real.

Film Industry 3.0 color films

The second major revolution in cinema came with the introduction of color. Although color films first appeared in the early 1900s, they initially relied on labor-intensive manual coloring techniques, where each frame was hand-painted. A notable example is the 1902 French film "A Trip to the Moon," which involved a studio of 200 women to hand-color the footage. A significant development occurred in 1906 when George Albert Smith introduced the Kinemacolor process, a two-color system using red and green filters rotating at twice the speed of the film to create the illusion of color. However, this method was limited, as it couldn't reproduce the full color spectrum, resulting in images that lacked the vibrancy of true-to-life colors.

The true breakthrough came in the 1930s with the advent of Technicolor, which revolutionized cinema by using dye-transfer technology and a three-color process that combined red, green, and blue to produce rich, vivid colors. This innovation not only enhanced visual storytelling but also became synonymous with the glamour of Hollywood's Golden Age, defining the aesthetic of countless classic films. This marked the beginning of the Film Industry 3.0 era.

Films in the 3.0 era of the film industry are color movies, using Technicolor technology to make movies appear as close to real colors as possible in real life. The emergence of color also provides a new position for the film industry, namely colorists. At the same time, the emergence of color also enriches the narrative dimension of movies, providing new expression methods for film directors, using color as a symbol, using different colors to represent different meanings, and also giving rise to different tonal styles, such as realism and magical realism.

Film Industry 4.0 CGI technology

It was not until the end of the 20th century that the emergence of computer-generated imaging (CGI) technology once again changed the landscape of the film industry. In the 1960s, computer graphics technology began to attract people's attention, but at that time, the images presented by computers were still relatively crude and difficult to apply to movies. Early computer graphics technology could only generate flat two-dimensional images. With the emergence of 3D graphics technology, people began to use computers to generate three-dimensional images. In the 1991 movie "Terminator 2" was released, the scene where the robot

"T-1000" appeared was stunning. This part was produced by Industrial Light and Magic, with 35 animators, computer scientists, technicians and artists, and it took ten completed within months. In addition to actual shooting, films had a richer creative space, which could present fantasy scenes and create realistic visual effects. Seamlessly blend live-action footage with computer-generated elements and push the boundaries of imagination. Now CGI technology has become an essential link in film production. The production cycle is shorter and the difficulty is lower. It has subverted the traditional visual era and changed the way of film production. It is Film Industry 4.0.

Film Industry 4.0 also provides new positions for the film industry, namely visual effects artists, who use CGI technology to produce composite images and visual effects for movies. During the Film Industry 4.0 era, many new technologies have emerged, such as the use of Steadicam stabilizers, the application of motion capture technology, and the improvement of video resolution and video frame rate, but these technologies have not fundamentally changed the way movies are made and presented. In essence, they are still upgrades of existing technologies and cannot be called a "film industry revolution."

Film Industry 5.0 artificial intelligence

We can find some rules from the previous reforms of the film industry, as follows:

The emergence of new technologies: Every update of the film industry must be accompanied by the emergence of new technologies, and these technologies can definitely change the way movies are presented to every extent and enrich the narrative dimension of movies.

The emergence of new positions: The emergence of new technologies is often accompanied by the emergence of new positions, the film production process changes, and the demand for talents will also change.

Changes in production processes: The evolution of the film industry must be accompanied by changes in the work process. New technologies will add new links to film production, and the way movies are made will also change with the emergence of new technologies.

The emergence of AI technology will also drive the film industry into a new revolution. Nowadays, in the process of film production, people have applied many AI technologies, such as AI motion capture, where real actors perform actions, and AI generates matching action models. For example, AI face-changing technology uses AI to replace the actor's face. AI can also very well assist in the generation of computer images and the production of special effects shots. Even AI can provide great help in the field of screenwriting. The 5.0 era of the film industry has arrived. From past experience, we can see that the emergence of new technologies must go through a stage from immaturity to maturity. Now we are in the stage of AI technology turning from raw to mature (Figure 15.3).

At this stage, new technologies will certainly be subject to various doubts. In fact, as long as human society continues to develop, the fear of new technologies will never stop. After all, AI is just a tool. Rather than being afraid of change, it is

Film Industry3.0
-color film
-Kinemacolor
-Technicolor

Film Industry4.0
-CGI film
-computer graphics

Film Industry1.0
-silent film
-persistence of vision

Film Industry2.0
-talkies
-wax disks
-audio feature film

AI Technology

Film Industry5.0
-AIGC
-VR/AR
-low cost

synchronized sound

CGI Technology

kinetoscope

Technicolor

1888   1927   1930s   1960s-21st century   Future

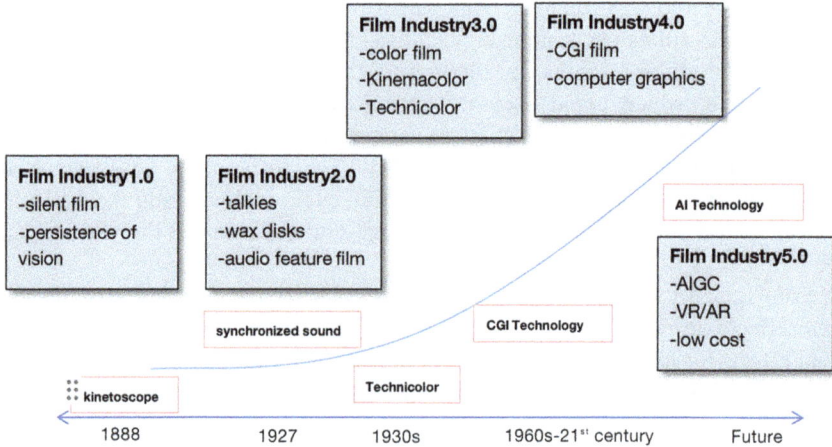

Figure 15.3   Film industry revolution development

better to adapt to change. Learning how to use this tool will be an essential ability for future film practitioners.

15.2   Film industry and artificial intelligence

Films are an essential component of our everyday pleasure. And the film industry, which generates billions of dollars in annual income, is one of the fastest-growing global revenue-generating sectors [9]. With the continuous expansion of the film market, film-related industries have also become very complete and have entered an assembly line industrial production model. The film industry includes a large number of jobs, and the division of labor between different jobs is clear. In recent years, the development of AI technology has also driven changes in the film industry, and the way movies are made is changing.

15.2.1   Film industry and occupation

The production and distribution of a film are mainly divided into four stages: pre-production, mid-production, post-production, and publicity and distribution. Each stage is divided into different work contents. An excellent film generally requires a team of hundreds or even thousands of people to produce. Among them, the director is the only role that needs to run through the entire film production. He needs to control the overall direction of the film, adjust the story plot, design the picture composition and shot connection of the film, control the film's set and art modeling, guide the actors' performances, guide the editor to edit the film, and discuss publicity and distribution strategies with the publicity and distribution team. Except for directors, other professions have their own different divisions of labor, see Table 15.1 for details.

Table 15.1   Film industry and occupations

Type of work	Duty
Director	Control the overall direction Participate in the work of all other professions
Screenwriter	Write the movie script including storyline and character dialogue
Performer	Play the role Be familiar with the lines
Camera man	Shoot the movie according to the instructions Adjust the lighting on the scene
Art	Design and arrange shooting scenes Design and make props Design actor costumes
Makeup artist	Makeup for actors Including daily makeup and special effects makeup
Sound engineer	Field recording
Editor	Editing movies Editing trailers
Effects artist	Creating visual effects
Foley artist	Foley
Promotion team	Promoting the film

According to Table 15.1, we can see that the film industry has different job divisions, and each team has its own responsibilities and at the same time, they form a whole, just like the operation of a large machine. The entire film production process can form a complete chain and have an industrialized system. Now with the emergence of new technologies, the way films are produced will also change.

15.2.2   Artificial intelligence and AIGC

AI encompasses machine-displayed intelligence that emulates human cognitive functions, such as problem-solving and decision-making. Combining machine learning and deep learning techniques, AI models are trained on extensive data sets to make intelligent decisions autonomously [4]. AI is expected to power 95% of customer interactions by 2025, with the market growing approximately 54% year-on-year, reaching $22.6 billion in size [10].

AIGC, which stands for artificial intelligence generated content, refers to a technology that leverages AI methods, such as generative adversarial networks (GANs) and large pre-trained models, to generate content based on learning and recognizing existing data. In November 2022, the American AI studio OpenAI released a multimodal large language model called Chatbot Generative Pre-Trained Transformer (ChatGPT), which brought AIGC into the spotlight. As an application of AI technology, AIGC is now being used in various scenarios in daily life, significantly improving work efficiency. The film industry has also seen several cases of AIGC applications.

15.2.3   Film industry combined with artificial intelligence

Actually, using AI to assist in film creation is not new; early AI applications focused on generating special effects and CGI. By 2016, AI started being used in scriptwriting. Initially, AI was viewed as a tool for automating repetitive tasks and was not considered creative. Since filmmaking heavily relies on human creativity and imagination, industry professionals did not feel significantly threatened. However, AI has now entered an era of "artificial intelligence gaining creativity," meaning that creative content production, once exclusive to humans, is becoming replaceable.

According to a survey conducted by the data website Statista on US film industry leaders, most respondents believe that AIGC will significantly impact the film industry, particularly in areas like 3D modeling, character and environment design, and voice generation and cloning (Figure 15.4).

With the continuous development of AI technology, more film-related industries have begun to connect with AI, and the AIGC-related market has also continued to expand. Data from Market Research shows that generative AI in product development market was valued at USD 71.9 million in 2023, and is expected to reach USD 1593.1 million in 2033 (Figure 15.5). The film industry can also use AI to complete more work, see Table 15.2 for details.

From Table 15.2, we can see that AI seems to be involved in all aspects of the film industry, and can even replace humans in film production. However, Table 15.2 only shows the ideal state of AI use. In fact, AI has not reached such a

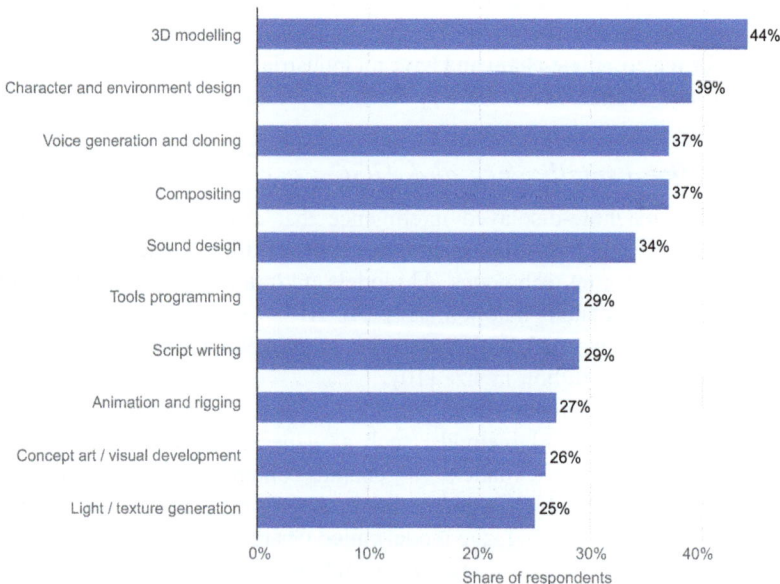

Figure 15.4   Relevant survey data from Statista

Generative Ai In Product Development Market
Size By Component 2023-2033(USD Million)

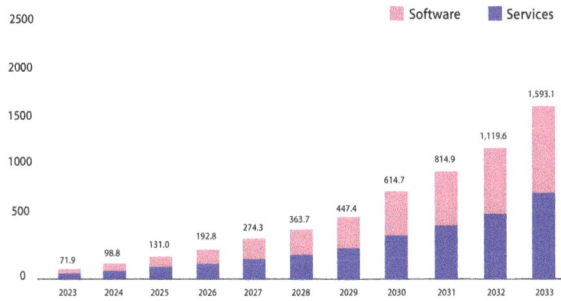

Figure 15.5   Data from market research

Table 15.2   Film industry using AI

Type of work	Column heads
Director	Using AI to modify scripts Using AI to generate storyboards Using AI to schedule filming
Screenwriter	Generate scripts Write dialogues
Performer	Using AI to generate virtual characters AI face-changing technology
Camera man	Using AI to generate some shots Using AI to design lighting
Art	Use AI to build scene references Use AI to design film and television props Use AI to design character costumes
Makeup artist	Design character makeup
Sound engineer	Using AI to generate character voices Using AI to create movie music
Editor	Using AI for video editing Add movie subtitles Edit movie trailers
Effects artist	Visual effects creation Using AI for motion capture AI character keying
Foley artist	Create film and television sound
Promotion team	Develop promotional strategies

ChatGPT ›

Scene 5: Debate - Day

Nick walks in the garden, discussing the family's business with Vito.

Nick:
"Dad, I really don't want this life. I have my own career in America."

Vito:
"Nick, some things you can't escape from. The family's responsibilities need someone to bear them."

Nick:
"But I don't want to be a part of the Mafia, I just want my own life."

Scene 6: Break Into Two - Night

Nick decides to help his father resolve the current crisis, but doesn't plan to stay involved long-term.

Nick:
"Okay, Dad, I'll help you with this trouble. But after this, I'm going back to America."

Vito:
"I understand, Nick. But I hope you see another side of our family."

Figure 15.6   Using ChatGPT to create a story similar to "The Godfather"

powerful level. According to experiments, AI at this stage still has certain problems in content generation.

During the experiment, the author tried AI content generation from the text level, video level and sound level. The specific results are as follows.

Text level: The main limitations of AI in script writing lie in its ability to convey emotions and themes effectively. While AI is highly capable in terms of technical screenwriting skills, it often falls short in the nuanced expression of feelings and thematic depth. A popular framework in drama theory is the "Save the Cat" beat sheet, which many successful works adhere to. Through interaction with ChatGPT, AI can grasp the mechanics of this framework and generate stories accordingly. For example, the author used the classic film "The Godfather" as a reference point, and after some adjustments and dialogue with the AI, ChatGPT produced a storyline resembling that of the original movie (Figure 15.6). However, when it comes to the depth of the plot and the quality of dialogue, the AI's output tends to be overly direct and lacking in subtlety. When given the theme of the breakdown of the American Dream, the AI simply incorporated the US setting into the characters' dialogue in a blunt manner, showing a deficiency in literary artistry.

Video level: On February 15, 2024, OpenAI officially launched the text-to-video model Sora, which aims to "simulate the world." Although this model has not yet been released for public use, its video content generation capabilities can be assessed from the demonstration videos shared by OpenAI. In these demos, Sora demonstrates the ability to create videos under 60 seconds with smooth camera transitions, lifelike character visuals, and basic editing techniques. The examples also showcased generated elements such as character movements, facial expressions, and intricate details like animal fur textures, which were quite impressive.

Sora's current capabilities suggest it could potentially replace certain types of live-action content, although it still falls short of fully substituting live footage. By

Figure 15.7   Video generated using Midjourney

comparison, other platforms like Midjourney or Stability AI face more challenges in video generation. Videos created from text descriptions on these platforms often resemble animated images, with minimal subject motion, limited to slight camera shifts or basic depth-of-field effects. Issues also persist in generating finer details, such as the articulation of fingers during character movements (Figure 15.7).

Audio level: In March 2024, Suno AI released its advanced music model, Suno V3, showcasing impressive music generation capabilities. Beyond creating BGM, Suno can also compose songs with both lyrics and vocals, suitable for use as theme songs in films. The model generates music from text descriptions provided by creators, and its output is often indistinguishable from that of human musicians. The author's own experiments reveal that Suno effectively captures the characteristics of various music styles, such as blues, jazz, concertos, and rock, and comprehends technical terms like C major and BPM.

Remarkably, Suno can also interpret and convey the emotional nuances in music, whether warmth, fear, tension, or sadness, with notable accuracy. While it excels at composing melodies and accurately reflecting a song's theme through its lyrics, the lyrical content tends to be somewhat straightforward and lacks the use of rhetorical devices or more nuanced poetic techniques.

In summary, while AI may never completely replace human input in film creation, in the film industry 5.0 AI will become an integral part of the filmmaking toolkit, enabling new forms of creative expression and expanding the possibilities of film. Film Industry 5.0 should be a hybrid of human intelligence and AI-driven innovation, with the two complementing each other to bring stories to life in ways we have only just begun to imagine.

15.3   Film Industry 5.0

The film industry is changing due to the rise of AI technology, and we are now at a critical juncture in the transition from Film Industry 4.0 to 5.0. In the future, in Film Industry 5.0, the responsibilities of different types of work will undergo various changes. For practitioners in the film industry, only by learning how to use new technologies can they avoid being eliminated in the wave of revolution.

15.3.1   Film Industry 4.0 to 5.0

Since the film industry began to apply CGI technology in the 1960s, the film industry has officially entered the 4.0 era. In the past few decades, new technologies have emerged and the way films are made has been evolving, but these technologies have not fundamentally changed the way films are made or presented. In the process of the film industry's innovation from 1.0 to 4.0, the working methods of many positions have not changed qualitatively, such as screenwriters are responsible for writing scripts and character dialogues, and camera man uses cameras to shoot films. In my opinion, the core of the film industry revolution lies in the change in the way films are presented, whether it is from silent to sound, black and white to color, real shooting to special effects, and the change of 5.0 is the addition of AI. In fact, the progress of movies in production over the years has been very significant. For example, the addition of the Steadicam stabilizer makes the shooting of long shots more convenient, the advancement of CGI technology makes the visual effects more and more realistic, the addition of Dolby sound effects makes the sound of movies more three-dimensional, and the improvement of picture resolution and frame rate gives the audience a better visual experience. However, the emergence of these new technologies did not bring about fundamental changes in film production and presentation. It was not until the emergence of AI that the 5.0 era truly arrived.

From Table 15.3, we can see that AI has become a key role in film production in Film Industry 5.0, and almost all positions can use AI for auxiliary work. In the past, films were regarded as the crystallization of human creativity and the output of creators' values. This has changed slightly in Film Industry 5.0. Humans will still be the leaders of film production, but AI will be more involved in the process of artistic creation. It is even difficult for us to distinguish which parts are created by AI and which parts are created by artists. Taking effects artists as an example, the creation of visual effects in the past required a large number of artists to participate in it, making a large number of visual reference graphic designs, action special effects design, and real-life performances for motion capture. Finally, effects artist created and synthesized visual effects. These pictures were all based on the imagination of the artists themselves. This also led to the emergence of a large number of CGI movies. It is difficult for people to gain a particularly novel experience of these effects. But in Film Industry 5.0, effects artists can describe the pictures they imagine to AI, and AI generates visual effects. Due to the difference between AI and humans in the underlying logic, the pictures generated by AI can break the limit of human imagination and let the audience gain a more novel visual experience. The emergence of AI has fundamentally changed the way movies are made. In the future, the requirements for film industry practitioners will also change. The use of AI will definitely become an important criterion for selecting talents. There should be AI-related positions in Film Industry 5.0, which will be mainly responsible for controlling the overall situation and providing technical guidance for the application of AI.

Table 15.3   Film Industry 4.0 to 5.0

Type of work	Film Industry 4.0	Film Industry 5.0
Director	Personally control the entire process of film production	Using AI to assist in creation Using big data technology to finalize the theme of the film
Screenwriter	Work with the director to write the script	Using AI to assist in creation Modifying the text content created by AI
Performer	Play the role with real people	Using virtual digital actors
Camera man	Shoot the film with real people	Using AI to generate videos Using AI to control cameras
Art	Design the art elements in the film	Using AI to design Modifying the content designed by AI Using 3D printing technology to make props
Makeup artist	Design the makeup of the characters in the film	AI provides makeup references
Sound engineer	On-site recording and sound production	AI assists in sound creation
Editor	Edit the film and trailer	Using AI to assist in editing
Effects artist	Use CGI technology to produce visual effects	Using AI to generate visual effects
Foley artist	Foxie	Using AI to design sound effects
Promotion team	Develop publicity strategy	Combining AI for publicity
Director of AI	Not to exist	Overall control of AI applications

15.3.2   Key technology empowerment

The development of key technologies has changed the way movies are made and promoted the arrival of Film Industry 5.0. The most critical one is the continuous progress of AIGC technology. The process of filmmaking itself is a kind of content output. The development of AIGC has changed the subject of content output and fundamentally affected the logic of filmmaking. In addition to AIGC, many other new technologies have also had a great impact on film production. See Table 15.4 for details.

15.3.3   Film Industry 5.0 application cases

Nowadays, there are actually many application cases of Film Industry 5.0. Creators from different countries have begun to use AI technology to assist their creation. Since film production is divided into front-stage work and behind-the-scenes work, many behind-the-scenes work is difficult to be directly reflected in the final

Table 15.4   Key technology empowerment

Type of work	Empowerment
AIGC	Use AI to generate the content needed for movies, including but not limited to scripts, storyboards, video content, music, character voices, visual effects, promotional posters, etc.
AI face-changing	Use AI face-changing technology to replace the face of the character.
AI editing	Use AI to assist in editing, including batch generation of trailers and editing of feature films.
AI virtual digital human	Virtual digital people will appear in the film as actors.
Cloud computer	Use cloud computer technology to achieve remote production of movies, and film production will not be restricted by time and place.
Virtual/augmented reality	Use VR/AR technology to enrich the narrative dimension of movies, enhance the realism of movies, and change the way movies are watched.
3D printing technology	Using 3D printing to create film and television props and visual references

presentation of the film, which also makes it difficult for people to perceive the application of AI such as AI writing scripts. Director Oscar Sharp and artist Ross Goodwin used AI with the given name Benjamin to generate a short film script named Sun Spring. They trained Benjamin with a recurrent neural network (RNN) together with long short-term memory (LSTM) by feeding it to science fiction movie scripts such as Star Trek, Truman's Worlds, and X-Men [11]. During the promotion of the film, some promotional materials were also generated by AI. Take "Go to your island," a Chinese animated film adapted from a novel of the same name on Jinjiang Literature website as an example, the official announcement poster released by ENLIGHT 100 PICTURES is AI-produced. ENLIGHT's team used ChatGPT, Midjourney, and Stable Diffusion, input tens of thousands of characters, generated thousands of posters with innovative styles, and finally presented a poster with fairy tale style and realistic details [12].

There are also more intuitive on-screen applications. For example, AI virtual digital human technology is used in the production of some short plays. In the short play "My Chinese Chic Boutique" produced in China in 2023, the digital human "Guoguo" played one of the characters "Ye Shiyi" and co-starred with real actors, giving the audience a very novel viewing experience (Figure 15.8).

Some content also relies on AI face-swapping technology to replace existing characters' faces, a technique that has seen some practical applications in the film industry. For example, in the movie Legend, AI face-swapping technology was used to recreate Jackie Chan's younger face on screen (Figure 15.9).

Figure 15.8   The role played by AI virtual digital human "Guoguo"

Figure 15.9   The real Jackie Chan and AI-generated younger Jackie Chan

15.4   Conclusion

15.4.1   Film Industry 5.0 workflow

The emergence of Film Industry 5.0 has changed the process of film production and added new roles to film production. The emergence of new technologies has

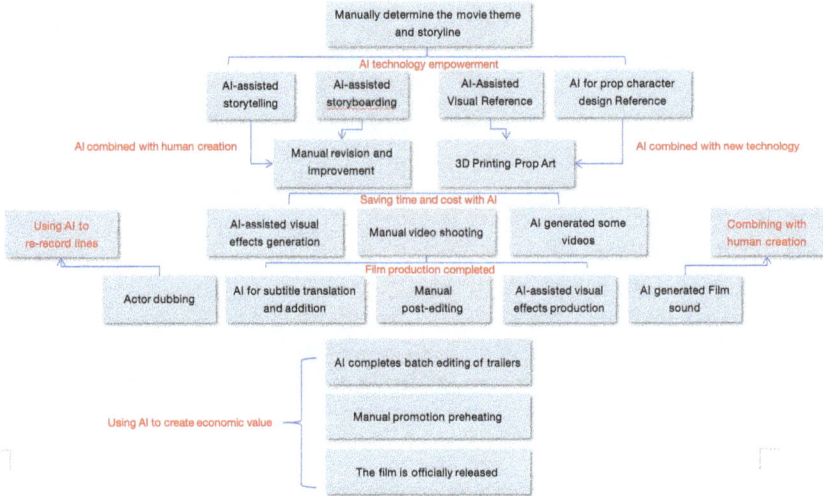

Figure 15.10   Film Industry 5.0 workflow

empowered the film industry, making the process of film production more convenient, saving time and cost, and truly achieving the goal of reducing costs and
increasing efficiency. From the workflow of Film Industry 5.0, we can see
(Figure 15.10) that AI is involved in the early, middle and late stages of film
production. AI can also be used to predict and analyze the revenue of the film, and
combined with big data technology, it can assist creators in creating more popular
films. A very critical point is the determination of the theme of the film. An
important feature of an excellent film is that it can arouse the emotional resonance
of the audience. This requires the creator to have rich emotions, delicate hearts and
observations of life, which is still difficult for the current AI. As 6G technology
continues to develop, collaborative machine learning may make AI more intelligent
[13], but understanding human emotions will still be a challenge for AI. Therefore,
in the workflow of Film Industry 5.0, humans still have to occupy a dominant
position. As the source of creativity and emotion, AI should be used as a new tool
and used properly. In the development of the film industry, it will always be
humans combined with AI, not AI replacing humans.

15.4.2   Composition of Film Industry 5.0

The emergence of new technologies often changes the existing system. These changes
are essential to make people's lives more convenient, improve work efficiency and
create greater value. For example, advanced forecasting, monitoring, and control of
energy generation are made possible by digital technologies [14]. 5G technology
enables traffic lights to collect real-time information on traffic patterns from cameras,
sensors, and drones placed around the smart city [15]. The IoT, big data analytics, AI,
and mobile applications promote sustainable development and enhance visitor
experiences [16]. Using the integration of IoT sensors in poultry farming for enabling

real-time monitoring, data-driven decision-making, and improved productivity [17]. For film creation, although films have their artistic attributes, the reason why the film industry can become an industrial system is essentially because of its commodity attributes, so the ultimate goal of the film industry must be to create greater economic value while providing entertainment for the public. The biggest change in Film Industry 5.0 compared to the past is to reduce costs and increase efficiency, improve production efficiency, and reduce production costs.

From the literature review, we can see that the progress of the film industry is inevitably accompanied by the emergence of new technologies, the emergence of new positions, and changes in production processes. The emergence of AI technology has changed the way movies are made. Practitioners no longer rely solely on human production but begin to use AI to assist in the generation of content. In this process, new positions will definitely be born, and these positions will definitely be closely related to AI. Therefore, Film Industry 5.0 can be defined as follows.

Film Industry 5.0: Using AI technology to optimize the film production process and provide audiences with a more novel technical visual experience. Fundamentally changing the way movies are made, a new form of the film industry created by AI combined with humans.

From Figure 15.11, we can see that Film Industry 5.0 is essentially a combination of human creation and AI technology. It uses AI technology to save labor, time and costs to create economic value, provide entertainment to the public with excellent standards, and output value, correctly guide public opinion, and gain greater social influence. Like Industry 5.0, Film Industry 5.0 also has its own core values. The core values of Film Industry 5.0 are mainly divided into three points.

The human-centric: No matter what era, the core of movies should be the output of human emotions. It can reflect social problems and also express emotions and arouse emotional resonance among the audience. Therefore, even if AI technology is applied to film production, it cannot affect the human-dominated nature of the film industry.

Cost reduction and efficiency improvement: As a cultural product, movies need to be profitable. Film Industry 5.0 uses AI technology to optimize the production process of movies, save costs and increase the profits of movies.

High-quality content: Movies need to provide audiences with a good audio-visual experience. No matter how the presentation and production methods of movies change, their content must be recognized by the audience and the market.

However, in this process, Film Industry 5.0 will also face some problems, such as the abuse of AI leading to the destruction of human creativity, or the copyright issues of AI-generated content, and the data security issues of AI technology [18]. The emergence of new technologies is always accompanied by doubts. As a commodity, movies must have good and bad products, and the market will naturally filter out those movies that abuse AI and are shoddy. In any case, we are at a critical juncture of change, and the era of Film Industry 5.0 has arrived.

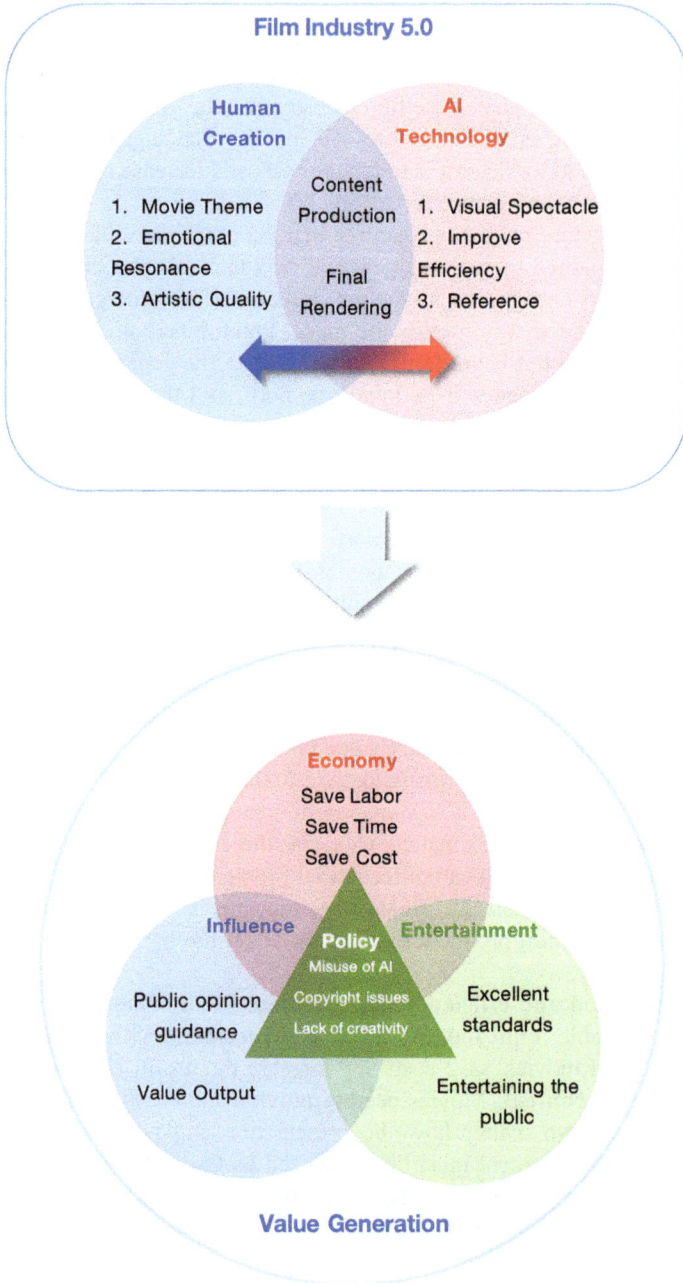

Figure 15.11   Composition of Film Industry 5.0

References

[1]   Pilevari, N. (2020). Industry revolutions development from Industry 1.0 to Industry 5.0 in manufacturing. *Journal of Industrial Strategic Management*, 5(2), 44.

[2]   Vogel-Heuser, B., Bauernhansl, T., and Hompel, M. (eds.) (2017). *Handbuch Industrie 4.0 Bd.4: Allgemeine Grundlagen*. Berlin, Heidelberg: Springer Vieweg.

[3]   Vogel-Heuser, B., and Hess, D. (2016). Guest editorial Industry 4.0–prerequisites and visions. *IEEE Transactions on Automation Science and Engineering*, 13(2), 411–413.

[4]   Rashid, A. B., and Kausik, A. K. (2024). AI revolutionizing industries worldwide: a comprehensive overview of its diverse applications. *Hybrid Advances*, 7, 100277.

[5]   Breque, M., De Nul, L., and Petridis, A. (2021). *Industry 5.0: Towards a Sustainable, Human-Centric and Resilient European Industry*. Luxembourg: European Commission, Directorate-General for Research and Innovation, 46.

[6]   Lu, Y., Adrados, J. S., Chand, S. S., and Wang, L. (2021). Humans are not machines—anthropocentric human–machine symbiosis for ultra-flexible smart manufacturing. *Engineering*, 7(6), 734–737.

[7]   George, A. S., and George, A. H. (2020). Industrial revolution 5.0: the transformation of the modern manufacturing process to enable man and machine to work hand in hand. *Journal of Seybold Report*, 15(9), 214–234.

[8]   Demir, K. A., Döven, G., and Sezen, B. (2019). Industry 5.0 and human–robot co-working. *Procedia Computer Science*, 158, 688–695.

[9]   Ni, Y., and Li, S. (2023). Decomposition-integration-based prediction study on the development trend of film industry. *Heliyon*, 9(11), e21211.

[10] Luu, V. (2024).10 remarkable artificial intelligence applications in 2024-bestarion. Available: https://bestarion.com/10-remarkable-artificial-intelligence-applications-in-2024/ (accessed 27 March 2024).

[11] Li, Y. (2022). Research on the application of artificial intelligence in the film industry. In *SHS Web of Conferences* (vol. 144, p. 03002). London: EDP Sciences.

[12] Zhao, M., Li, B., and Zhang, X. (2024). Application of AIGC in the design of film and television props. *Transactions on Computer Science and Intelligent Systems Research*, 3, 70–74.

[13] Leong, W. Y. (2024, May). Secure and efficient collaborative machine learning frameworks for 6G intelligent applications. In *2024 IEEE International Workshop on Radio Frequency and Antenna Technologies (iWRF&AT)* (pp. 324–328). Piscataway, NJ: IEEE.

[14] Leong, W. Y. (2023, August). Digital technology for ASEAN energy. In *2023 International Conference on Circuit Power and Computing Technologies (ICCPCT)* (pp. 1480–1486). Piscataway, NJ: IEEE.

[15] Leong, W. Y., and Kumar, R. (2023). 5G intelligent transportation systems for smart cities. In *Convergence of IoT, Blockchain, and Computational Intelligence in Smart Cities* (pp. 1–25). Boca Raton, FL: CRC Press.

[16] Leong, W. Y., Leong, Y. Z., and San Leong, W. Smart tourism in ASEAN: leveraging technology for sustainable development and enhanced visitor experiences. *International Journal of Social Sciences*, 4(3), 19–31.

[17] Leong, W. Y., Leong, Y. Z., and San Leong, W. (2024, July). Poultry precision: exploring the impact of IoT sensors on smart farming practices. In *2024 IEEE International Workshop on Electromagnetics: Applications and Student Innovation Competition (iWEM)* (pp. 1–2). Piscataway, NJ: IEEE.

[18] Leong, W. Y., Leong, Y. Z., and Leong, W.S. (2024, July). Strengthening security in computing. In *2024 IEEE Symposium on Wireless Technology & Applications (ISWTA)* (pp. 113–116). Piscataway, NJ: IEEE.

Chapter 16

Application and contribution of virtual reality technology in education

Guinong Peng[1] and Wai Yie Leong[1]

Virtual reality (VR) technology has been gaining traction across various sectors, with education being one of the primary fields experiencing significant advancements. By providing immersive and interactive environments, VR can enhance learning experiences and outcomes. This chapter delves into the applications of VR in education, discussing its benefits, challenges, and future prospects.

16.1 Introduction

Industry 5.0 (Industry 5.0) represents the latest stage of the industrial revolution, emphasizing the deep collaboration between human intelligence and intelligent machines, aiming to achieve more efficient and flexible production and service. VR technology plays a key role in this process, bringing a lot of innovation and change to Industry 5.0 through an immersive experience and interactive environment.

VR technology significantly enhances human–machine collaboration. Through the virtual training environment, employees can learn new skills in a safe virtual space to simulate the operation and troubleshooting of complex devices, thus reducing errors and accidents in practical operation. For example, manufacturing employees can simulate production line operations in a virtual environment, familiarize themselves with the equipment operation, and improve work efficiency and accuracy. This training method is not only safe but also can greatly reduce the training cost and time (Peng and Leong, 2024a).

VR technology improves production efficiency. The application of VR technology is particularly prominent in the process of product design, production planning, and maintenance. Enterprises can quickly iterate on product design through virtual prototype design and testing, reducing the cost and time of physical prototyping (Hamad and Jia, 2022). Virtual factory modeling helps companies optimize production processes, identify and eliminate bottlenecks, and reduce resource waste. For example, in automotive manufacturing, designers use VR

[1]Faculty of Engineering and Quantity Surveying, INTI International University, Malaysia

technology to quickly create and test vehicle models and perform virtual crash tests to accelerate the product development cycle (Zhang *et al.*, 2020).

Personalized manufacturing is an important feature of Industry 5.0, and VR technology has unique advantages in this field. Through VR, customers can experience product design personally, communicate and interact with designers intuitively, put forward suggestions for modification, and achieve highly personalized product customization (Paszkiewicz *et al.*, 2021). For example, furniture manufacturers use VR technology to allow customers to design and adjust furniture in a virtual environment, and see the final effect before production, so as to improve customer satisfaction and product competitiveness.

VR technology has promoted the development of Industry 5.0 and brought significant economic and social benefits to enterprises by enhancing human–machine collaboration, improving production efficiency, realizing personalized manufacturing, enhancing remote collaboration and maintenance, and improving staff training and safety management (Peng and Leong, 2024b). VR technology, as one of the core technologies of Industry 5.0, is changing the way we produce and live, and creating a new era of more intelligent and efficient industry (Fernandez, 2017).

16.1.1   Introduction to VR technology

With the rapid development of science and technology, VR technology has become an important force driving the global digital transformation (Häkkilä *et al.*, 2018). This field contains infinite potential and broad prospects (Gan *et al.*, 2023).

The concept of VR technology dates back to the 1960s, and the Sword of Damocles, developed by Ivan Suzeland, was seen as the earliest VR prototype (Hamad and Jia, 2022). Since then, with the continuous improvement of computer graphics, sensor technology, and real-time computing capabilities, VR technology has gradually moved from laboratory to practical application. In 1984, VPL's Jaron Lanier first proposed the term "virtual reality," marking a new stage of development of VR technology (Figure 16.1).

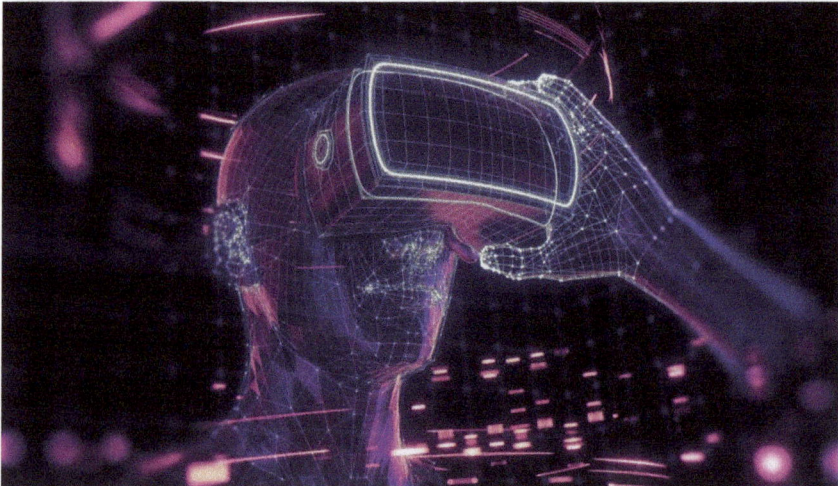

Figure 16.1   Virtual people with virtual glasses

Figure 16.2   SONY consumer-class VR virtual headset display

In the 21st century, with the rise of the concept of meta-universe, VR technology has ushered in unprecedented development opportunities. Globally, technology giants such as Meta, Apple and Huawei have increased their investment in the VR field, promoting the continuous evolution of underlying technologies such as near-eye display, rendering processing, and perceptual interaction. VR equipment has achieved significant improvement in picture quality, interaction, comfort, and other aspects, gradually from the professional field to the consumer market (Figure 16.2).

In terms of application field, VR technology has penetrated into games, education, medical care, industrial manufacturing, cultural tourism, and other industries. Through VR technology, users can experience the virtual environment personally, greatly enriching the entertainment methods, improving the learning efficiency, reducing the medical risks, and optimizing the industrial production process. Especially in the field of cultural tourism, VR technology has broken the geographical restrictions, allowing people to visit the global places of interest without traveling far, and feel the charm of different cultures.

While the impact of VR on games is often emphasized, the potential for transformational technology extends far beyond entertainment. From health care to real estate, recruitment and education, VR and extended reality (XR) could revolutionize a number of industries. For those unfamiliar with VR, it refers to a computer-generated environment, which the user can explore and interact with. The immersive nature of VR entices the brain to see these virtual experiences as real, thus creating a strong sense of presence.

16.1.2   The importance of VR technology in education

VR technology is increasingly important in the field of education (Peng and Leong, 2024c). With its unique immersion, interaction, and diversity, it has brought revolutionary changes to the education industry.

In terms of medical education, VR technology is revolutionizing medical education by providing an immersive training environment that simulates real medical scenarios (Gan *et al.*, 2023). For example, at the Children's Hospital of Los Angeles (CHLA), residents use VR to practice emergency treatment and surgery in a risk-free setting (Zhang *et al.*, 2020).

In addition, VR is used for detailed anatomical exploration and enables medical students to interact with 3D models of human organs. At Stanford University, VR is used to visualize and manipulate brain structures, providing a comprehensive understanding of the neuroanatomy. This approach can not only prepare students for real dissection but also speed up the learning process. As VR and AR technologies evolve and become affordable, their integration with medical training is expected to expand, promoting innovative educational practices and improving clinical skills (Figure 16.3).

Despite the enthusiasm for VR, supporters emphasize the need to exercise caution and the need to verify its effectiveness based on evidence. Concerns that incorporating new technology into the curriculum without removing existing

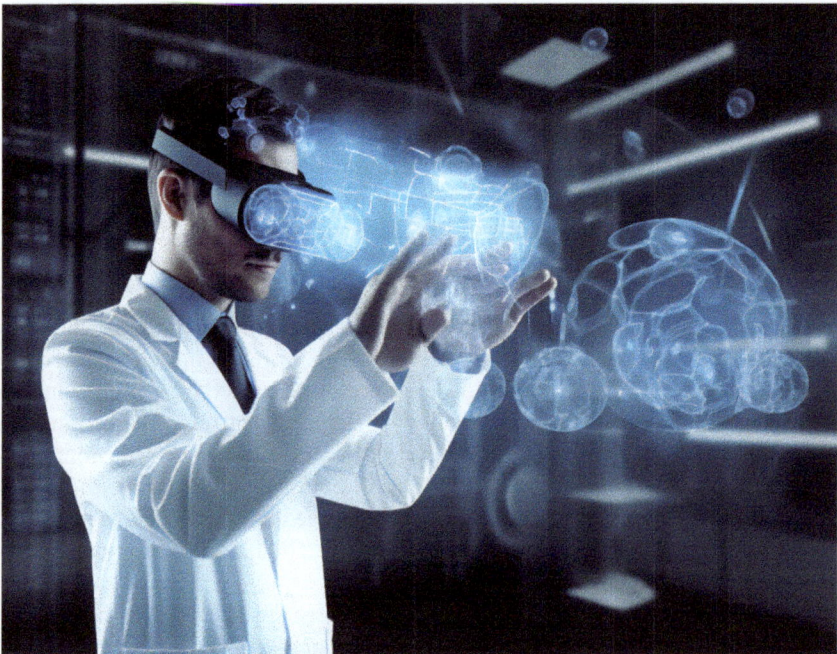

Figure 16.3   Virtual operating room

content may not always be beneficial. Although anecdotal evidence supports the benefits of VR, more comprehensive research is needed to determine its educational value compared to traditional training methods. Cross-institutional collaboration can help collect reliable data and facilitate informed decision-making for using VR in medical education. As technology advances and costs decline, its role in education may expand as long as VR implementation continues to prioritize student and patient needs and outcomes.

In the K-12 stage, VR technology provides a brand new learning platform for students. Through virtual environments, students can personally explore complex scientific phenomena, historical events, and geographical landscapes. For example, students can simulate chemical reactions, observe molecular structures, or experience major events in a virtual historical scene. This intuitive and three-dimensional learning method greatly improves students' interest and participation in learning and helps them to understand abstract concepts more deeply (Figure 16.4).

In higher education, VR technology is widely used in medicine, engineering, art, and other professional fields. In medical education, VR system can simulate real surgery scenarios, allow students to conduct operation exercises, improve practical operation ability, while avoiding risks in real surgery. In the field of engineering, VR technology can simulate complex mechanical structures and architectural scenarios, helping students to better understand and master professional knowledge. In art education, VR technology has become an innovative tool for students to create and experiment in virtual space (Figure 16.5).

For vocational education and skill training, VR technology is gradually becoming a traditional learning tool, and providing a safe and efficient practice platform. For example, in the fields of flight training, driving simulation, fire rescue, and emergency rescue, VR technology can simulate a variety of real scenes,

Figure 16.4   The K-12 stage VR learning scenario

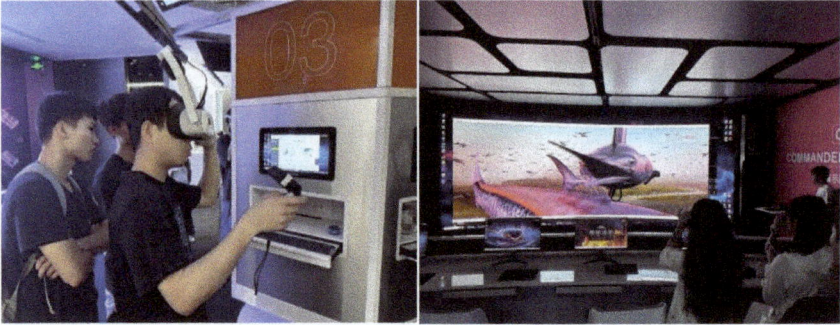

Figure 16.5   College students experience and learn virtual reality technology in a virtual reality technology company in China

Figure 16.6   Virtual reality technology is applied to fire simulation and emergency escape drill

allowing students to practice repeatedly in the virtual environment, so as to improve their skill level. This way not only reduces the cost of training but also reduces the risks in practical operation, which is more and more popular with students, and greatly improves students' interest and level of learning (Figure 16.6).

In terms of language learning, VR technology can simulate the real language environment and enable students to interact with virtual characters, thus improving their oral and listening abilities. At the same time, through VR technology, students can "stay" in different countries and regions, experience the local culture and life, and deepen their understanding and respect for different cultures. This immersive learning style is of great significance for improving the cross-cultural communication ability. Figures 16.7 and 16.8 show the virtual cartoon scene and the virtual real scene learning scene.

The research analysis in Figure 16.8 aims to assess the impact of the virtual learning environment on students' willingness to communicate (WTC), language anxiety, and self-perceived communication ability. Analyses used a paired t-test to compare the pre-and post-test scores for these three variables.1. * * Communication intention (WTC) * * − * * pretest * * *: $M = 2.87$, SD = 0.80− * * Posttest * *:

Figure 16.7   Language learning virtual learning scenario

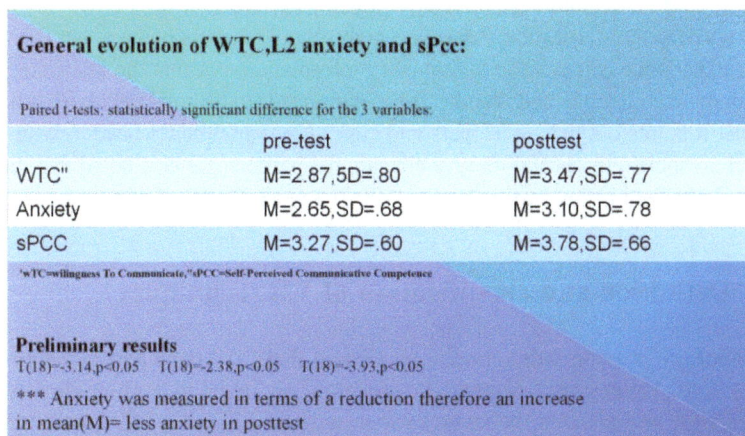

General evolution of WTC,L2 anxiety and sPcc:

Paired t-tests: statistically significant difference for the 3 variables:

	pre-test	posttest
WTC"	M=2.87,5D=.80	M=3.47,SD=.77
Anxiety	M=2.65,SD=.68	M=3.10,SD=.78
sPCC	M=3.27,SD=.60	M=3.78,SD=.66

'wTC=wilingness To Communicate,"sPCC=Self-Perceived Communicative Competence

Preliminary results
T(18)=-3.14,p<0.05   T(18)=-2.38,p<0.05   T(18)=-3.93,p<0.05

*** Anxiety was measured in terms of a reduction therefore an increase in mean(M)= less anxiety in posttest

Figure 16.8   Anxiety studies in language learning in virtual learning scenarios

M = 3.47, SD = 0.77 – * * Statistical significance * *: t (18) = −3.14, p <0.05 data showed that the WTC score was significantly increased from premeasurement to posttest. The mean score increased from 2.87 to 3.47, indicating a significant increase in students' willingness to participate in communication in the virtual environment. This statistically significant change ($p < 0.05$) indicates that the immersive nature of the VR setting had a positive impact on students' communication behavior.

Several studies have pointed out that the valuable value of VR language learning comes from its interactive nature. This approach goes beyond the simple sitting memory vocabulary, but allows learners to devote themselves to the learning process and realize the full activation and participation of the senses. VR language learning applications, as excellent auxiliary tools, are complementary to other language learning resources. For example, FluentU also advocates immersive teaching methods, but it guides learning through real native video content such as film clips and news materials.

VR technology offers many benefits for students with physical disabilities in special education settings. One of the major strengths is the creation of a barrier-free learning environment beyond physical limitations. For example, VR can take students to different locations and times to provide experiential learning without actual travel. These techniques also cater to a variety of learning styles and provide a multisensory experience, which is particularly beneficial for students who may struggle with adapting to traditional teaching methods. In addition, in addition to promoting engagement and motivation and making learning an exciting and interactive experience, VR can also work on some related diseases. This is also a very important role and experience.

The importance of VR technology in education is quite obvious. With its unique advantages, it has brought revolutionary changes to various education industries, improved students' interest and participation in learning, reduced training costs and risks, and promoted the balanced distribution and sharing of educational resources. With the continuous progress of technology and the continuous expansion of application fields, VR technology will play a more important role in the field of education. VR is growing in education, but limited funding makes it difficult for startups to gain a foothold. Although VR technology is sufficiently advanced to be a powerful educational tool, improving curriculum development is key to its effectiveness. The XR Association and IIT surveys show that most American high school teachers believe that the virtual learning experience can provide high-quality information, but that the associated costs may widen the equity gap and that the curriculum needs to be consistent with academic standards. It points out that developers and education communities need to work together to cross social classes and remove educational barriers. He also noted that the leap in digital light field technology is leading VR to a wearable-free future.

16.2   Definition and classification of VR techniques

VR technology is a computer simulation system that can create and experience the virtual world. It uses the computer to generate an interactive three-dimensional dynamic view and physical behavior system with multi-source information fusion, enabling users to immerse themselves in a digital space that feels like the real world. This technology is realized through computer graphics, sensor technology, and real-time computing, and integrates visual, auditory, tactile, and other sensory simulation, so that users can interact naturally with the virtual environment.

Specifically, VR technology covers the user's visual and auditory input through devices such as head-mounted displays, replacing them with computer-generated stereoscopic images and 3D sound effects, and combining handles, gloves, or other motion sensing devices to track the user's body movements, so as to interact with the virtual environment.

16.2.1   Top of a table VR

Use the keyboard, mouse, gamepad, and other computer peripherals to interact. The VR experience in this way is relatively simple. Users do not need to wear special devices but mainly observe the virtual environment through the computer screen, and interact with the traditional input devices.

16.2.2   Head-mounted VR

With a VR headset, it projects images and videos around the eyes to create a realistic virtual environment, while using devices such as a handle. Head-mounted VR devices can provide an immersive visual and auditory experience, making users feel immersive (Figure 16.9).

16.2.3   Wireless VR

Similar to the head-mounted VR, but without any cable or cable constraints, allowing users to move and explore more freely. Wireless VR technology solves the problem of cable limitations and improves user mobility and freedom.

16.2.4   Overall view VR

Similar to 360-degree panoramic photography, it will bring the user into a panoramic picture with a sense of space. Panoramic VR technology allows users to view virtual scenes by capturing or generating 360-degree images of the environment (Figure 16.10).

Figure 16.9   Head-mounted VR

16.2.5   Enhance augmented reality

Although it is not technically classified as VR technology, augmented reality (AR) is a technology that combines virtual elements with the real world, and it is also worth mentioning. By wearing smart glasses or headset, users can see the real world while also seeing the projection of virtual objects in the virtual object in the field of view, thus gaining an enhanced experience (Figures 16.11 and 16.12). This blended experience has been a focal point of research on improving accessibility in immersive environments, highlighting its potential to create more inclusive learning experiences (Dudley *et al.*, 2023).

Figure 16.10   The VR watches on the visual display

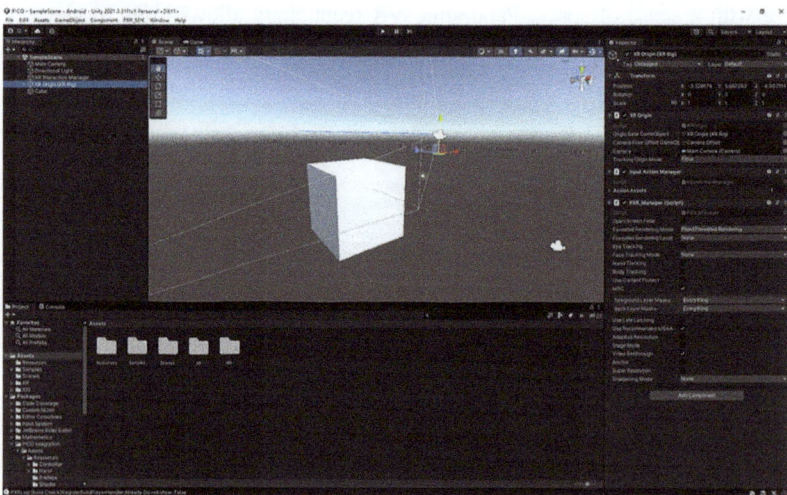

Figure 16.11   Unity3D Edit interface (from class design screenshots using Unity3D)

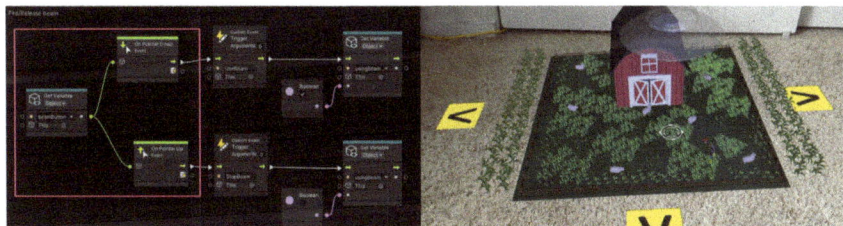

Figure 16.12   *Augmented reality production graphics (from using Unity3D design screenshots in class)*

Figure 16.13   *Mixed-reality-in-education*

16.2.6   Mixed reality

Mixed reality (MR) is the integration of virtual and real environments to create a new and comprehensive experience. Users using wearable devices or other interactive devices can interact with virtual objects that can interact with the real environment. Almost all educators still rely on videoconferencing software to perform their distance learning outcomes, and that won't change anytime soon. MR and other immersive solutions may provide an avenue as a new tool for teachers to prepare transition from traditional telecommunication solutions to improve student performance (Figure 16.13).

16.3   Basic components of VR technology (hardware and software)

16.3.1   Hardware part

VR headsets: Devices used to display VR scenes, usually including high-resolution display screens, sensors, lenses, and other components. These devices are able to cover the user's eyes, using different lenses and technologies to create realistic three-dimensional images that provide an immersive visual experience. Common VR headsets include Oculus Rift, PICO 4, and PlayStation VR. Specifically, VR technology covers the user's visual and auditory input through devices such as head-mounted displays, replacing them with computer-generated stereoscopic images and 3D sound effects, and combining handles, gloves, or other motion sensing devices to track the user's body movements, so as to interact with the virtual environment.

Other display devices, such as panoramic projection, VR glasses, can also provide VR display effect to some extent.

Controller: used to control characters or objects in a VR scene, usually including buttons, joysticks, touch pad, and other control elements. These devices allow users to interact with virtual environments, such as grabbing, moving virtual objects, or performing navigation and selection operations (Figure 16.14).

Tracking device: used to track user positions and movements in VR scenarios, usually including cameras and infrared sensors. These devices can capture the user's head, hands, and even the whole body movements in real time, ensuring that the interactive experience in the virtual environment is more natural and real (Figure 16.15).

Motion capture system: a high-level input device that can record the movements of actors in real time and apply them to virtual characters, further enhancing the reality of the interactive experience (Figure 16.16).

Calculating device: High-performance computer or game console: image, audio, and interactive content for processing VR scenarios. These devices require

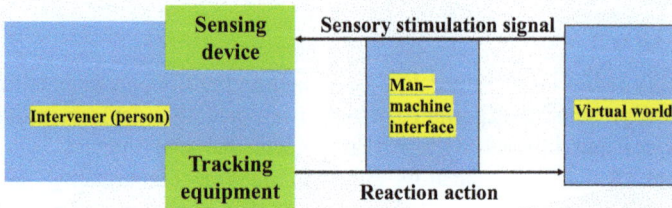

Figure 16.14   Composition of the virtual reality concept system

Figure 16.15   A schematic diagram of the motion-capture system

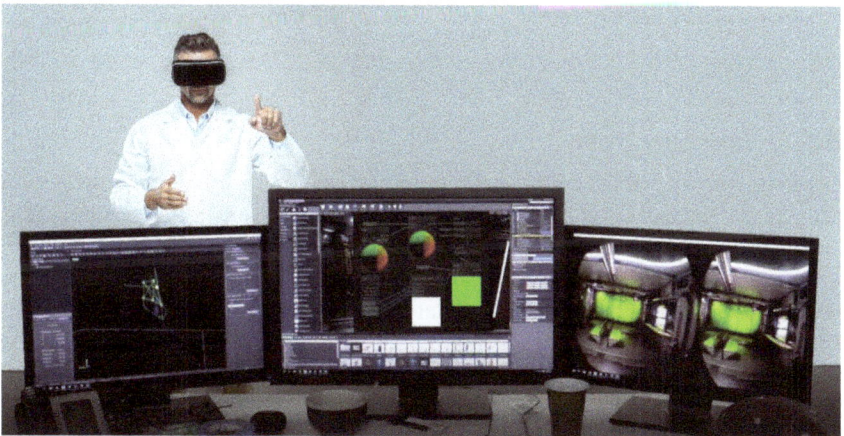

Figure 16.16   Action-capture system experiment

high performance and graphics processing capabilities to ensure a smooth running and high-quality display of VR applications.

Audio frequency apparatus: Headphone or speaker: used to provide sound and voice interaction in VR scenarios. These devices can enhance user immersion and make the virtual environment more realistic.

Other accessories: These include VR motion sensing kit and VR sports platform; these accessories enhance users' immersion and interactive experience in VR scenes (Figures 16.17 and 16.18).

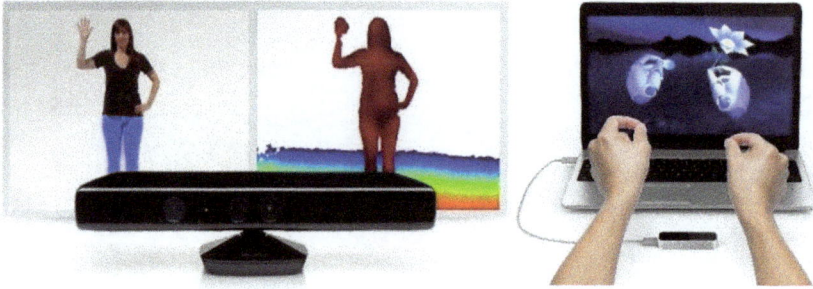

Figure 16.17   Kinect sensor and leap motion controller sensor

Figure 16.18   Monochrome egocentric articulated hand-tracking

16.3.2   Software part

Operating system: VR systems usually operate based on specific operating systems that may be second-developed on the basis of common operating systems (e.g., Windows, Android) to better adapt to the needs of VR applications (Figure 16.19).

Development engine: These include Unity and Unreal Engine; these engines provide rich functions and tools to facilitate developers to develop VR applications. They support 3D modeling, animation, special effects production, and interactive logic implementation.

Modeling tools: These include 3Ds Max and Maya that are used to create 3D models in a virtual scene. Mapping and texture making software: These include Photoshop and Substance Painter that are used to add maps and textures to 3D models to make them more realistic.

Animation and special effects software: These include After Effects and Houdini that are used to produce animation and special effects in VR content.

Interaction software: It is used to control the user interaction in a virtual environment. These software are able to capture user movements and gestures and translate them into instructions and feedback in a virtual environment. They are often used with devices such as sensors, cameras, or handles.

Content production and distribution platform: content production includes games, film and television, live broadcasting, social networking, and other fields. Distribution platforms are responsible for distributing the produced VR content to

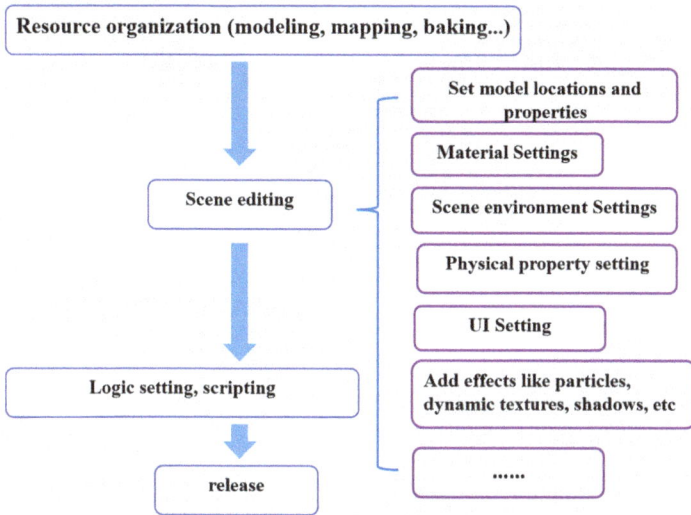

Figure 16.19   General process for developing a VR roaming display system based on the engine

users, such as the Steam platform is one of the earliest and most abundant VR content distribution platforms.

16.4   The development process of VR technology

The development process of VR technology has changed from the early embryonic stage to the productization and industrialization, showing its wonderful evolution from the concept to the wide application. In 1838, Charles Wheatstone proposed the concept of stereo vision, which laid the foundation of 3D display technology. Subsequently, Brewster David invented the lens-type stereoscope. In 1929, Edward Link designed the Link Trainer, which started the application of VR technology in flight simulation. In 1935, Stanley G Wynbaum conceived the holographic VR glasses in the novel, which inspired the later development. In the 1960s, Morton Helg introduced the first head-mounted display, and in 1968, Ivan Sutherland developed a computer graphics-powered helmet display, marking a major break-through in VR technology. In the 1980s, Jay Lanier proposed the "virtual reality" concept and developed a range of VR devices. In the 1990s, VR technology began to enter the consumer market. In 2012, the launch of Oculus Rift opened a new era of civilian VR devices, and in 2014, Google Cardboard further lowered the threshold of VR experience. 2016 is regarded as the first year of VR industrialization, and the release of high-end VR equipment has promoted the wide application in various fields. With the development of technologies such as 5G, the Internet of Things, and artificial intelligence, VR technology will continue to improve the user experience and play an important role in more areas (Figure 16.20).

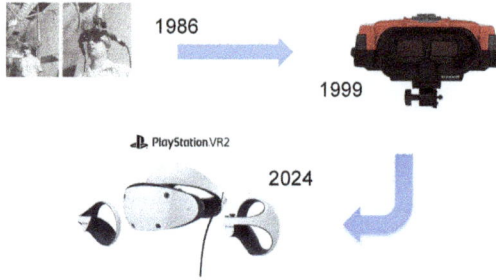

Figure 16.20   VR, the development of tools

16.5   Application characteristics of VR technology in teaching

16.5.1   Virtual real teaching

Virtual laboratory: VR technology can build a variety of virtual scenarios required for teaching such as chemistry laboratory and physics laboratory. This teaching method allows students to safely conduct experiments in a virtual environment, without having to worry about the risk of actual equipment and dangerous substances, thus improving the interactivity and interest of teaching. At the same time, VR laboratories can also simulate experimental scenarios that are difficult to achieve or costly in reality, such as nuclear reactions and cosmic exploration, to help students better understand experimental principles and phenomena.

Virtual scene experience: VR technology can also restore historical events, cultural scenes, or natural ecosystems, so that students can personally experience these scenes in a virtual environment, and enhance the immersion and intuitiveness of learning. For example, students can use VR technology to visit ancient Roman cities and experience ancient civilizations, or to enter ecosystem models to observe biodiversity and ecological cycles.

16.5.2   Interactive learning

Virtual mentors: VR technology can create virtual mentor roles that provide students with real-time feedback and guidance. Such virtual mentors can play an important role in language learning, skill training, and other fields, helping students to correct pronunciation, practice dialogue, and master the essentials of skills. By interacting with the virtual mentors, students can get a more personalized and targeted learning experience.

Exploratory learning: VR technology can create an open and free learning environment that encourages students to explore and learn on their own. Students can freely explore the unknown areas in the virtual environment, find answers and solve problems, so as to cultivate their independent learning ability and exploration spirit. Recent research on the benefits of immersive technologies, including AR and VR, has shown that these technologies are not only effective in supporting educational outcomes but also improve engagement and retention in subjects such as mathematics (Cevikbas *et al.*, 2023).

16.5.3 Distance education

Distance teaching: VR technology has realized a new mode of distance teaching. Students are not required to participate in online courses, lectures, and seminars through VR devices. This distance teaching method breaks the geographical restrictions and enables high-quality educational resources to cover a wider range of people.

Virtual campus: The virtual campus built through VR technology can provide students with an immersive campus experience. Students can visit the teaching buildings, libraries, sports fields, and other facilities in the virtual campus, and interact and communicate with the virtual students and teachers, so as to enhance their sense of identity and belonging to the campus culture.

One of the biggest drawbacks of working remotely is that employees are lazy or inefficient. In most cases, the opposite is true. According to a Harvard study, employees working from home can actually increase US labor productivity by more than 4%. Researchers also found that employees working from home actually take less sick time and are more comfortable in work environments, creating fewer errors and billions of dollars to the economy. Moreover, another study by Stanford University on a Chinese travel agency with 16,000 employees found that productivity increased by 13% as employees were more satisfied with their careers.

16.5.4 Professional skills training

Simulation training: VR technology has a wide range of applications in professional skills training, for example, in the fields of flight training, medical surgery, and fire drill, VR technology can simulate real scenes and operation processes, enabling students to repeatedly practice and master the essentials of skills in a safe environment. This simulation training method not only reduces the training cost and risk but also improves the training effect and efficiency. Studies have shown that VR-based training provides highly effective solutions across various industries, including engineering education (Soliman *et al.*, 2021), and can greatly enhance the learning process in technical fields.

16.5.5 Personalized learning

Customized learning paths: VR technology allows educators to create personalized learning paths for each student. By collecting information such as students' learning data and behavior patterns, educators can tailor their content and difficulty levels for students, thus improving the pertinence and effectiveness of learning. The current developments in immersive VR technology are shaping the future of education by offering more customized learning experiences, a trend likely to continue as immersive virtual reality becomes more prevalent (Wei and Yuan, 2023).

16.6   Cases of virtual technology application in education

16.6.1   Virtual classroom

Case study: A university uses a virtual classroom system to provide real-time online courses for distance learning students. With HD video and audio technology, students can attend classes at home or at other locations and interact with teachers and other students. This way breaks the limitation of time and space, and enables high-quality educational resources to cover a wider range of students (Figure 16.21).

16.6.2   Virtual laboratory

According to the Bureau of Labor Statistics, the number of STEM occupations will grow by nearly 11% between 2021 and 2031, while all other occupations will grow by less than 6%.

However, advanced laboratories with large amounts of practical STEM learning may be difficult to access and costly. Labster democratizes this process by providing a virtual laboratory environment for more than 20 course packages, including high school physics, nursing biological sciences, animal physiology, advanced biology, and engineering. These laboratories can be accessed through a web browser, without having to download or install other software, students can grow bacteria, track cell respiration during exercise, and even aim the mother for ultrasound, of course, this is virtual. Labster also owns UbiSim, a VR training platform for nurses that involves learners in risk-free clinical field scenes using VR headphones and controllers (Figure 16.22). Virtual reality technology has been widely used in distance education to enhance students' interaction and participation, especially in health education and shows its potential (Abbas *et al.*, 2023).

Figure 16.21   Virtual classroom scene

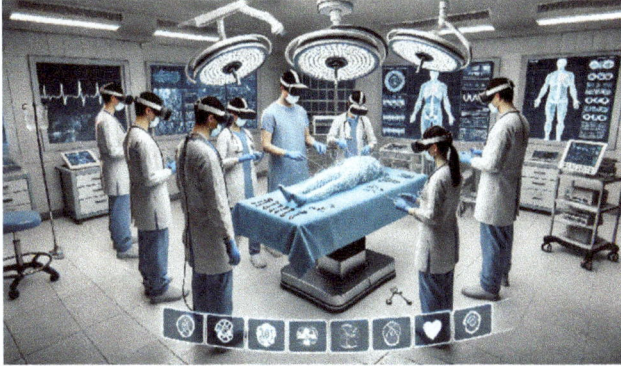

Figure 16.22   Virtual operating room scene display

16.6.3   Virtual training

Guangdong vocational colleges use VR technology to conduct automobile maintenance training. Students wear VR devices to simulate the real car repair process, and conduct fault diagnosis and maintenance operation. This virtual training method improves students' practical skills and safety awareness, while reducing training costs and risks. Virtual training can also provide personalized learning content and challenges according to students' learning progress and ability, and improve the training effect (Figure 16.23). The application of virtual laboratory technology enables students to simulate high-risk experiments, ensuring a safe and effective training process (Gagliardi *et al.*, 2023).

16.6.4   Virtual visit and travel

VR during travel is an innovative technology that provides users with an immersive travel experience through 360 degrees of comprehensive scene exploration. It allows users to not only display and flexibly explore destinations in detail but also create memorable interactions that help travel brands stand out. In addition, VR allows people facing travel restrictions to experience the joy of traveling and promote the sustainable development of tourism. As technology advances, VR travel is becoming more realistic, not only travel companies and hotels use VR to provide virtual visits, but also VR travel experiences designed for the elderly, VR flight experiences and landmark attractions are becoming increasingly popular, and these trends together drive the development and innovation of tourism (Figure 16.24).

16.6.5   The VR application cases in colleges and universities

VR technology is widely used in education and research, providing innovative teaching methods and rich learning experience, e.g., Stanford University through virtual human laboratory research psychology and communication, Harvard Medical School with virtual anatomy laboratory improve anatomy teaching effect, Massachusetts Institute of Technology in Architecture and Urban Planning for

Figure 16.23   Virtual surgery scenarios

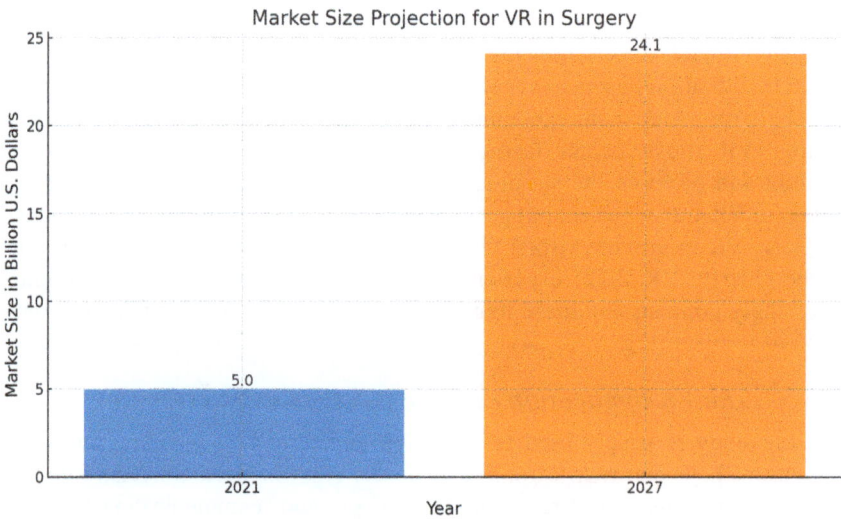

Figure 16.24   Market size projection for VR in surgery

Figure 16.25   University virtual reality teaching scene

virtual design simulation, Beijing University of Virtual History Museum enhances the immersion of history learning, the University of California, Berkeley's Virtual Chemistry Laboratory improves the security and repeatability of experimental teaching, the Chinese University of Hong Kong using VR development language learning application, the University of Queensland to create virtual archaeological site to provide real archaeological experience. These applications demonstrate the great potential of VR to improve students' skills, stimulate interest in learning and provide practical experience (Figure 16.25).

16.6.6   Application of VR in primary and secondary education

In primary and secondary education, the application of VR technology is gradually popularized, providing students with a more rich, intuitive and immersive learning experience.

A survey of the XR Association and Educational Technology International on the attitudes of more than 1,400 American high school teachers about XR technologies like VR showed that most believe virtual learning experiences can provide high-quality information. However, more than half also believe that the costs associated with these technologies may widen the equity gap. Ninety-four percent agreed that courses related to techniques such as VR need to be consistent with academic standards (Figure 16.26). The application of virtual reality in medical vocational education, such as in physician training in Jordan, has significantly improved the quality of the learning environment (Jawarneh *et al.*, 2023).

Guangzhou Foreign Language Middle School in China uses VR technology to conduct history teaching, allowing students to travel to the ancient Nanhai Temple

Figure 16.26   Learning scene of the high school virtual classroom

Figure 16.27   Virtual monuments scene roaming

through VR equipment to experience the historical changes and the fate of the characters (Figure 16.27).

VR technology has a wide range of application scenarios in primary and secondary education, covering experimental teaching, real teaching, personalized learning, social collaboration, and safety education. With the continuous development of technology and the popularization of its universal application, VR technology will bring more possibilities of innovation and change for primary and secondary school education. Through the immersive virtual reality environment, students can carry out various learning activities, such as scientific experiments and

Figure 16.28   VR applications in vocational education and training

historical re-enactment, in a simulated scene, thus obtaining a more profound learning experience (Abbas *et al.*, 2023).

16.6.7   VR applications in vocational education and training

The application of VR in vocational education and training is becoming increasingly widespread, revolutionizing the field of education. The following are the simulation scenarios of VR in vocational education and training (Figure 16.28).

16.7   Improve learning interest and participation, enhance practical operation ability, provide a safe learning environment, personalized learning experience

16.7.1   Application of VR in primary and secondary education to improve learning interest and participation

VR technology enables students to conduct immersive learning in them by creating highly simulated virtual environments. This novel learning method can greatly stimulate students' interest in learning, enabling them to more actively explore and try in the virtual environment. At the same time, VR technology can also provide a rich interactive experience, enabling students to maintain a high degree of participation in the learning process, thus improving the learning effect (Table 16.1).

Design and development of a traditional ceramic somatosensory interactive game made by Chinese college students. In view of the current offline ceramic museum exhibition vitality, developers through the use of field investigation, literature research, case study of SAN shek wan pottery technology, body feeling interactive

Table 16.1 *User experience map of the target user*

	Before experiencing VR	Wear a VR headset display	Start the VR experience	Follow the finger, and lead the operation	The VR experience is over	Share with each other
Emotion and thoughts	"VR high technology has no body, test, super, level want to try!"	"Finally, it's my turn. The VR headset is not as hard to wear as I expected."	"Good fierce, oneself, really on the body, in a virtual, quasi space inside!"	"Let me give it a try, the video introduction and interaction, the experience is very surprising."	"Oh! a little head sweating, but also, it is very interesting."	"Let Mei also experience, the interactive experience of the museum is worth it."
Feeling curve						
Concrete behavior	Tie up your shawl of hair	Adjust the tightness of the VR head display under the guidance of the staff and the staff	Try activity, hands, view, observe the virtual ring, virtual environment, hand synchronous interaction	Follow the game, guide to complete, virtual pottery	Under the guidance of the staff, take off the VR head display, look around the surrounding environment	Go to the small body, side, invite her to experience
Pain spot	Always feel that the VR head display wear will be uncomfortable, will tie the hair.	Although the tightness, the degree is more suitable, suitable, but, see the painting, the surface is still there, the point is not clear.	Sometimes, the virtual hand loses the sync.	Virtual hand grasp, take the detection, not very clever, sensitive, need, many attempts to succeed.	Take off the head to show that moment, I feel a little trance.	VR experience places are limited, but they will take a long time.

game, gesture recognition technology, discusses the significance and feasibility of ceramic museum exhibition, to SAN shek wan ceramic glaze as an example of body feeling interactive game development ideas, and the design and implementation of the game. In the development process of the game, Pico Neo3 is selected as the interactive device, Unity is used as the graphics engine, and the gesture interaction strategy is formulated based on the Pico SDK gesture recognition algorithm, so as to realize the function of users to glaze the virtual porcelain blank through their own gesture movements. The test results show that this game can not only bring real experience effects to users but also improve the users' attention and participation in Chinese traditional pottery culture, providing a new way for the inheritance and dissemination of Chinese traditional ceramic culture (Figures 16.29–16.31).

Run the game and observe the FPS data line chart during the game, as shown in Figure 16.32. In the screen recording state, the FPS value of the game running fluctuates only a little, stays at 90.0FPS for a long time, and reaches the refresh rate

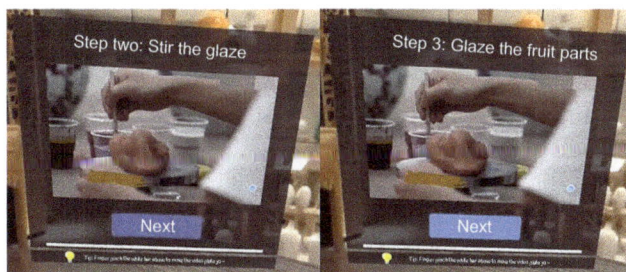

Figure 16.29   UI interfacial design

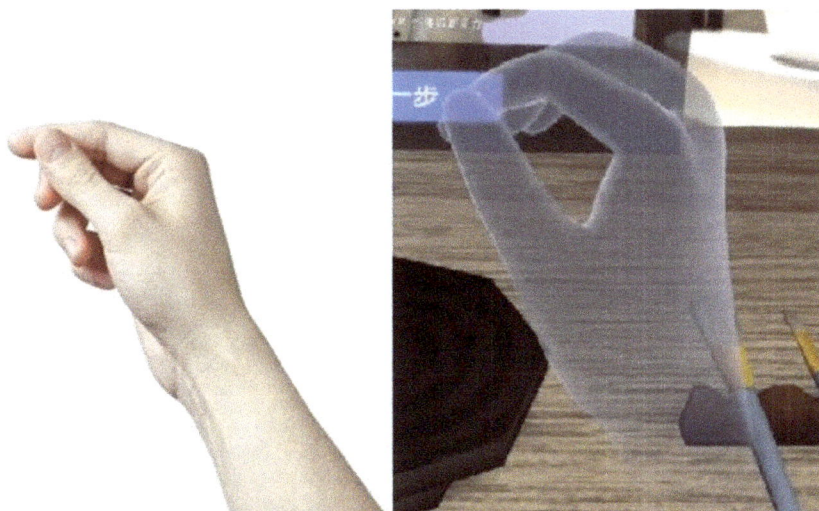

Figure 16.30   Real (left) and virtual (right) grasping item gestures contrast

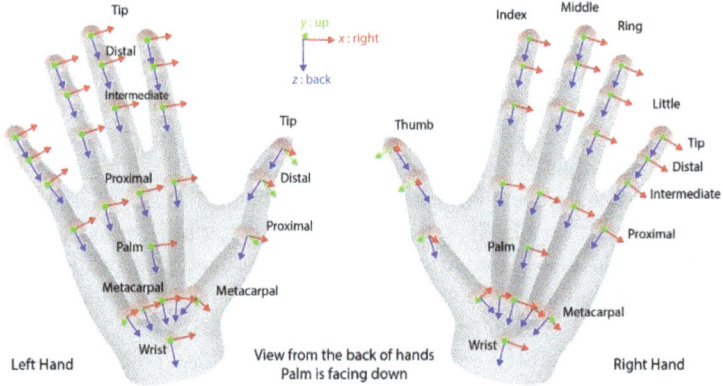

Figure 16.31   UI hand joint definition

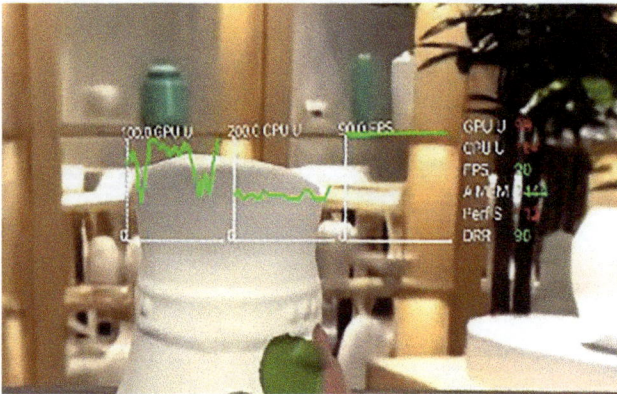

Figure 16.32   Data fold chart

Table 16.2   User test record

Number	01
Test content	VR visual adaptation
Test condition	The tester wore a VR headset
Testing procedure	The tester turns his body to observe the virtual environment
Expected result	The tester did not feel physical discomfort, such as dizziness and nausea
Actual result	No physical discomfort was seen in 90% of the participants

value set by VR header, which meets the frame requirements of VR experience. During the test process, it is difficult to feel the picture lag, and the operation experience is good (Table 16.2).

16.7.2   Enhancing the practical operation ability

In vocational education and training, practical operation ability is very important. VR technology simulates real career scenes and operation processes, enabling students to practice repeatedly in the virtual environment, so as to master practical operation skills. This way of practical operation can not only improve students' practical ability but also enable them to constantly find and solve problems in practice, and then improve their practical ability and innovative thinking.

16.7.3   Providing a safe learning environment

In some occupational fields, such as mechanical and electrical engineering, auto repair, and construction, there may be certain safety risks in the practical operation process. And VR technology can provide a safe learning environment, so that students can practice operations in the virtual environment, to avoid the dangers and accidents that may occur in the actual operation. This will not only ensure the safety of students but also enable them to play freely in a stress-free environment and better master skills.

16.7.4   Realizing the personalized learning experience

Each student has his or her own learning progress and ability level, and VR technology can provide a personalized learning experience based on these differences. Through VR devices, students can choose their own learning content and difficulty level for independent learning. At the same time, VR technology can also intelligently adjust the learning path and difficulty according to students' learning feedback and performance, so that each student can get the most suitable learning experience.

The application of VR technology in vocational education and training has significant advantages, which can greatly improve students' interest and participation in learning, enhance practical operation ability, provide a safe learning environment and realize personalized learning experience. Consuming any VR system radically progresses the students' performance. For this intention, it is clear that the VR application is critical in any educational institution (Al Farsi *et al.*, 2021). These advantages will bring more possibilities of innovation and change to the field of vocational education and training.

16.8   Contribution of VR technology to education

16.8.1   Providing an immersive learning experience

VR technology creates a three-dimensional, interactive virtual environment to immerse students in the virtual world related to the course content. This immersive experience greatly increases student engagement and interest, making the learning process more vivid and engaging. Students can carry out various learning activities in the virtual environment, such as scientific experiments, archaeological excavation, and language communication, as if they are in a real scene, so as to obtain a more profound learning experience (Figure 16.33).

16.8.2   Expanding the boundary of educational resources

VR technology has greatly expanded the boundaries of teaching resources. Teachers can use VR technology to provide students with rich teaching content and resources, including virtual simulation experiments, featured topics, and extended topics. These resources not only enrich the teaching content but also improve the interest and interactivity of teaching. At the same time, VR technology also breaks the limitation of time and space, enabling students to learn anytime and anywhere, and realizing the sharing and optimal allocation of educational resources.

16.8.3   Promoting educational equity

Under the traditional education mode, due to geographical and economic constraints, children in many areas cannot accept high-quality educational resources. However, through the application of VR technology, high-quality educational resources can be digitized and shared so that more children can have access to them. With VR equipment and Internet connections, whether in bustling urban areas or in remote rural areas, children can enjoy the same quality of educational resources. This helps to narrow the educational gap and promote the realization of educational equity. The contribution of VR technology to education is manifold. It provides not only an immersive learning experience and practical operation opportunities but also a safe learning environment, a personalized learning path, and rich educational resources. With the continuous development and

Figure 16.33   Student immersive learning scene

popularization of VR technology, it is believed that it will play a more important role in the future field of education.

16.9   Challenges facing VR technology in education

16.9.1   *Technology cost and equipment penetration rate*

In the past two years, the cost of VR equipment has been reduced relatively much. Hardware equipment in VR technology, such as VR glasses and handles, are now becoming more and more affordable. As of February 1, 2024, the mainstream VR devices on the market include HTC Vive Pro, Meta Quest, and Pico. The VR device used in this design is Pico Neo3, its basic configuration includes Qualcomm Snapdragon XR 2 platform, 6 DoF location tracking, single eye resolution 1832 * 1920, FOV 98, up to 120 Hz refresh rate, 5300 mAh large capacity battery, and support unlimited streaming Steam VR games and gesture recognition interaction. The Pico Neo3 package has a helmet and two wireless handles, eliminating the need for a tracker. The advantages of choosing Pico Neo3 as the target equipment are: no matching tracker, optical-based controller positioning tracking solution improves the development efficiency, occupies a certain share in the domestic market, abundant development resources; the equipment is affordable and easy to purchase channels. The first base version is priced at 2499 yuan, which is more cost-effective than other equipment. However, the cost of application in the education field is still high, which limits its popularization in the education field. For many schools and educational institutions, buying and maintaining these equipments is not a small financial burden. Penetration needs to increase: the penetration of VR devices in education is relatively low due to costs. This means that only a few students have access to VR technology to reach their full potential in teaching (Figure 16.34). Despite the great potential of virtual reality technology, its high equipment cost remains one of the barriers to widespread application in the field of education (Gagliardi *et al.*, 2023).

16.9.2   *Skills and quality requirements of teachers and students*

The application of VR technology requires teachers to have certain technical and operational knowledge, and can flexibly use VR technology for teaching. However, at present, many teachers have not received relevant training and lack the necessary technical support and teaching matching. Student quality requirements: students also need to have a certain computer operation ability and learning ability, in order to make better use of VR technology to learn. For some students, this may be a challenge.

16.9.3   *Content development and resource sharing*

Scenes, models and teaching resources in VR technology require professional software developers and designers to produce, with relatively few talents in this field. At the same time, as the knowledge of each subject area is constantly

VR Usage Adoption Rates In 2024

用于 ■ North America, ■ Europe, ■ Asia-Pacific, ■ Latin America, 以及 ■ Middle East & Africa

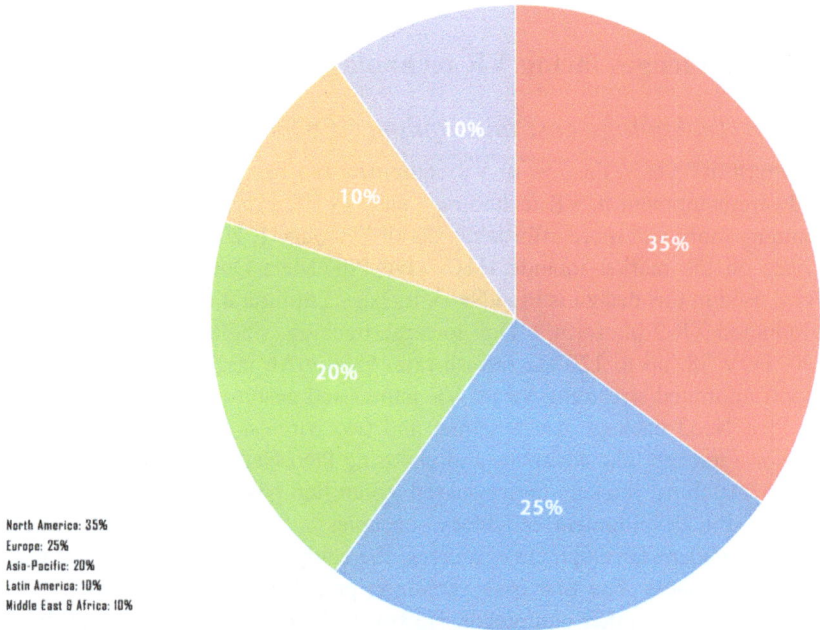

North America: 35%
Europe: 25%
Asia-Pacific: 20%
Latin America: 10%
Middle East & Africa: 10%

Figure 16.34   Virtual reality utilization rate in different regions in 2024

updated, we need to timely integrate the latest content into VR applications, which is also a challenge for content providers. Insufficient resource sharing: At present, the development and sharing of high-quality VR education resources are still insufficient, leading to the widening gap of educational resources between different regions and schools. This is not conducive to the realization of educational equity.

16.9.4   Health and safety issues

Long-term use of VR equipment may have some impact on students' vision and physical health. How to balance the advancement of technology and the health needs of students is a problem that educators need to seriously consider. Differences between virtual and reality: Although VR technology can provide realistic virtual scenes and objects, the differences from actual reality still exist. Excessive reliance on VR technology may lead to difficulties and in adaptation of students in practical operation, and affect the learning effect.

16.9.5   Market and application maturity

Although VR technology has great potential in education, its current market acceptance still needs to be improved. The cognition and application of VR technology in some schools and educational institutions are still in the initial stage. Application maturity: The application of VR technology in education is still in the stage of exploration and development, and a mature application mode and standard system has not yet been formed. This limits its generalization and application on a larger scale. VR technology faces many challenges in education, including technology costs, teachers' and students' skills, content development and resource sharing, health and safety issues, and market and application maturity. In order to overcome these challenges and fully realize the potential of VR technology in education, the joint efforts and cooperation of the government, schools, enterprises, and scientific research institutions are needed.

Although VR technology faces many challenges in education, with the continuous development of technology and the popularization of its application, its potential will be further tapped and played. In the future, with the reduction of cost and equipment popularization, more schools and educational institutions will be able to introduce VR technology for teaching; Meanwhile, with the strengthening of teacher training and curriculum design optimization, VR teaching will be more scientific, effective, and popular. In addition, with the continuous progress and innovation of technology, VR technology will be deeply integrated with other educational technologies such as AI and big data, bringing more possibilities for education. The challenges facing VR technology in education are multifaceted, but through continuous efforts and innovation, these challenges will gradually be overcome and solved. In the future, VR technology is expected to play a more important role in the field of education, providing students with a richer, more vivid and more efficient learning experience.

16.10 Future trends and applications of VR technology

Future VR technology will extend the full sensory experience to touch, smell, and taste, enabling users to feel more real sensory stimulation in the virtual world. At the same time, the boundary between AR and VR will become blurred, forming a seamlessly integrated MR, allowing the seamless integration of virtual and reality. In addition, advances in brain-computer interface technology will enable VR devices to connect directly to the brain for thought control and a more intuitive experience. Social virtual world will become an important part of people's life and social interaction, and the social activities in the virtual world will be more rich and diversified. Creators will also enjoy unlimited space for creation, breaking through the physical limitations of reality, and create unprecedented works.

VR technology will revolutionize traditional approaches in education and training, allowing students to experience historical scenarios, and doctors and pilots to improve their skills through virtual simulation. In medical and psychological treatment, VR will be used to treat psychological disorders and conduct surgical simulations to improve the quality and efficiency of medical services. Entertainment and tourism will welcome a whole new experience that people can play and explore in the virtual world. Remote work and collaboration will also become more efficient and convenient with VR technology, allowing employees to collaborate seamlessly in a virtual office. Architectural design and smart city planning will use VR for design and simulation to make the design more intuitive, anticipate problems and space for improvement. Finally, VR will also contribute to environmental protection and sustainable development, improving people's awareness of environmental protection and action by simulating and displaying environmental protection solutions.

References

Abbas, J. R., O'Connor, A., Ganapathy, E., *et al.* (2023). What is virtual reality? A healthcare-focused systematic review of definitions. *Health Policy and Technology, 12*(2), 100741.

Al Farsi, G., Yusof, A. B. M., Romli, A., *et al.* (2021). A review of virtual reality applications in an educational domain. *International Journal of Interactive Mobile Technologies, 15*(22), 99–110.

Cevikbas, M., Bulut, N., and Kaiser, G. (2023). Exploring the benefits and drawbacks of AR and VR technologies for learners of mathematics: Recent developments. *Systems, 11*(5), 244.

Dudley, J., Yin, L., Garaj, V., and Kristensson, P. O. (2023). Inclusive Immersion: a review of efforts to improve accessibility in virtual reality, augmented reality and the metaverse. *Virtual Reality, 27*(4), 2989–3020.

Fernandez, M. (2017). Augmented virtual reality: How to improve education systems. *Higher Learning Research Communications, 7*(1), 1–15.

Gagliardi, E., Bernardini, G., Quagliarini, E., Schumacher, M., and Calvaresi, D. (2023). Characterization and future perspectives of Virtual Reality Evacuation Drills for safe built environments: A Systematic Literature Review. *Safety Science, 163*, 106141.

Gan, W., Mok, T. N., Chen, J., *et al.* (2023). Researching the application of virtual reality in medical education: one-year follow-up of a randomized trial. *BMC Medical Education, 23*(1), 3.

Häkkilä, J., Colley, A., Väyrynen, J., and Yliharju, A. J. (2018). Introducing virtual reality technologies to design education. *In Seminar.net, 14*(1), 1–12.

Hamad, A., and Jia, B. (2022). How virtual reality technology has changed our lives: an overview of the current and potential applications and limitations. *International Journal of Environmental Research and Public Health, 19*(18), 11278.

Jawarneh, M., Alsharé, M., Dewi, D. A., Al Nasar, M., Almajed, R., and Ibrahim, A. (2023). The impact of virtual reality technology on Jordan's learning environment and medical informatics among physicians. *International Journal of Computer Games Technology, 2023*(1), 1678226.

Paszkiewicz, A., Salach, M., Dymora, P., Bolanowski, M., Budzik, G., and Kubiak, P. (2021). Methodology of implementing virtual reality in education for industry 4.0. *Sustainability, 13*(9), 5049.

Peng, G., and Leong, W. Y. (2024a). Brain wave response of style geometry in artistic creation in VR environments: an impact study on cognitive function and mental health of autistic children. In *2024 IEEE 8th International Conference on Signal and Image Processing Applications (ICSIPA)*. IEEE; pp. 1–6.

Peng, G., and Leong, W. Y. (2024b). Application of virtual reality technology combining sustainable development and history education. *Journal of Business and Social Sciences*. Retrieved from https://iuojs.intimal.edu.my/index.php/jobss/article/view/572.

Peng, G., and Leong, W. Y. (2024c). Virtual reality in waste management: evaluating its impact on community classification behavior. *Journal of Innovation and Technology*. Retrieved from https://iuojs.intimal.edu.my/index.php/joit/article/view/542.

Soliman, M., Pesyridis, A., Dalaymani-Zad, D., Gronfula, M., and Kourmpetis, M. (2021). The application of virtual reality in engineering education. *Applied Sciences*, *11*(6), 2879.

Wei, Z., and Yuan, M. (2023). Research on the current situation and future development trend of immersive virtual reality in the field of education. *Sustainability*, *15*(9), 7531.

Zhang, Q., Wang, K., and Zhou, S. (2020). Application and practice of VR virtual education platform in improving the quality and ability of college students. *IEEE Access*, *8*, 162830–162837.

Chapter 17

Integrating artificial intelligence into pedagogy: theoretical framework and application model of I-TPACK in intelligent education

Zhang Naixin[1,2] and Wai Yie Leong[3]

The era of Industry 5.0 signifies a profound integration of advanced technology with human creativity, with its core philosophy dedicated to fostering a symbiotic collaboration between technology and humanity. Education in the age of Industry 5.0 has embraced new transformations and opportunities. As a pivotal direction for the future of education, intelligent education aims to construct a more intelligent, personalized, and humanized educational ecosystem through the application of emerging technologies. This shift represents not only technological advancement but also a philosophical change, emphasizing the coexistence and cooperation between technology and humans.

Why should we study the changes in pedagogy brought about by technology? Hegel once stated that philosophy is its own history. By extension, we can understand that the philosophy of technology is the history of the philosophy of technology. The impacts of technology on pedagogy throughout history can provide insights and clues for our contemporary considerations of the influence of intelligent technology on teaching methods.

In the Industry 5.0 era, the influence of technology on teaching and learning activities has become more pronounced and complex. Virtual reality (VR), as a revolutionary technology, holds significant potential to reshape the educational landscape. Through the analysis of various case studies and research findings, it is evident that VR can enhance student engagement, knowledge retention, skill development, and inclusivity, creating a more dynamic and immersive educational environment (Leong *et al.*, 2023a).

Different technologies reflect different philosophies and pedagogies. For instance, computer-assisted instruction (CAI), based on behaviorist theory, focuses on repetitive practice and immediate feedback, while constructivist technologies such as VR emphasize immersive and contextual learning. Integrating VR into the

[1]Department of General Education, Chengdu Jincheng College, China
[2]Faculty of Education and Liberal Arts, INTI International University, Malaysia
[3]Faculty of Engineering and Quantity Surveying, INTI International University, Malaysia

intelligent educational environment can address the challenges and ethical considerations of implementing VR education, thus providing a comprehensive understanding of its strengths and weaknesses (Leong *et al.*, 2023a). Furthermore, EEG-based happiness recognition technology can also be integrated into the intelligent educational environment to monitor and enhance students' emotional well-being, creating a more supportive and responsive learning experience (Leong *et al.*, 2024a, 2024b). These technologies have their own advantages and disadvantages, and not all teaching activities are suitable for their application.

Therefore, we must not only review recent literature but also conduct a synthesis and summary of historical literature to gain a deeper understanding of the evolution and impact of technology on pedagogy.

This chapter provides a historical review of the underlying principles by which media technology influences pedagogy, offering a systematic synthesis of the deep logic behind the historical development of knowledge, Pedagogical Content Knowledge (PCK), and Technological Pedagogical Content Knowledge (TPACK). The chapter will conduct a comprehensive retrospective analysis of the changes brought to teaching by the introduction of technology into the educational field; the changes in teaching ushered in by the advent of intelligent technology in education; and the shifts in research on teacher professional development following the integration of technology into educational domains.

Through theoretical genealogy, this chapter endeavors to elucidate and dissect three pivotal inquiries: First, whether the evolution of technology has indeed precipitated changes in pedagogy? Should technological advancements fail to exert an influence on pedagogical methods, the significance of this study would be rendered moot; second, what is the ideal versus actual state of the educational environment constructed by intelligent technologies? Given the rapid flux in which intelligent technologies operate, discerning the current and prospective states of such technologies is immensely beneficial for comprehending how they impact the professional knowledge development of teachers; thirdly, what are the developmental trajectories and trends in the study of PCK integrated with technology? The confluence of diverse scholars' perspectives on this field of research forms the theoretical bedrock upon which this study constructs the concept of "Disciplinary Teaching Knowledge Integrated with Intelligent Technology."

17.1   The interactive mechanism and principles of technology and pedagogy in education

The field of educational technology originated in the 1940s, during World War II, when the extensive and efficient use of media technology in American military training established the legitimacy of this domain. Initially distinguished by visual and audiovisual education, educational technology emerged from the amalgamation of three research areas: educational media technology, instructional psychology, and educational systems methodology. As technology advanced, the research content and knowledge framework within this field underwent significant transformation. To accommodate these changes, the Association for Educational Communications and Technology (AECT) revised the definition and scope of educational technology in the years 1963, 1972, 1977, 1994, and 2004. To date, AECT has published five editions of the *Handbook of Research on Educational Communications and Technology*, providing crucial guidance and reference for educational technology researchers worldwide, and exerting a profound influence on educational technology research in the United States and internationally.

To cultivate a clearer, more precise, and profound comprehension of the interactions among the constituent elements of TPACK, it is imperative to delve into the historical trajectory of seminal research within the field of educational technology. This exploration aims to discern the intricate interplay between technology, disciplinary content, and pedagogical approaches and how this synergy transpires.

17.1.1 The influence of technological integration on pedagogy and learning outcomes

The advancement of educational technology has profoundly altered the landscape of teaching and learning. From early CAI to contemporary intelligent tutoring systems (ITS) and educational games, each technological innovation aims to meet specific pedagogical needs, offering learners a richer and more diverse educational experience. However, whether these technologies genuinely enhance learning outcomes remains a topic that warrants in-depth exploration (Anderson *et al.*, 1995).

Impact of technology on learning outcomes: The academic community holds divergent views on the role of technology in education. Some studies support the notion that technology can provide flexible learning environments and personalized learning paths, while others suggest that technological applications may have limitations, particularly in the absence of face-to-face interaction (Bernard *et al.*, 2009).

Interaction between technology and education: The debate between Clark and Kozma highlights differing perspectives on the role of technology in education. Clark posited that media does not directly affect learning (Clark, 1983), whereas Kozma argued that the application of new media can improve learning outcomes (Kozma, 1994). This debate underscores the importance of instructional design, rather than technology alone.

In summary, the development of educational technology provides a variety of tools and methods for teaching and learning, but its actual impact on learning outcomes is still a matter of debate. The use of technology should aim to promote the comprehensive development of students, rather than simply replacing traditional teaching methods. The academic community has different perspectives on the role of technology in education, emphasizing the importance of instructional design. At the same time, the diversity of education requires us to recognize the different needs of technology for different learners and disciplines, and the effectiveness of technology investment depends on its application in teaching practice.

Therefore, research and application of educational technology need to consider the characteristics of technology, instructional design, learner needs, and educational goals comprehensively, to achieve an organic integration of technology and education, and to promote innovation and development in education.

17.1.2 Correspondence patterns of technology and pedagogy

From the perspective of media ecology, technology not only serves as a vessel for information but also influences the encoding, transmission, and decoding of information, thereby affecting the outcomes of education. First is the non-neutrality of technology; as a medium of communication, it is not a neutral or transparent channel but rather dictates and shapes the encoding and transmission of information through its inherent material structure and symbolic forms. Second is the symbolic environment of media, where each medium constitutes a unique symbolic environment made up of specific codes and syntax, and the organization of the medium defines the nature of the information. Third is the impact of technology on culture, where there is a continuous and interdependent interactive relationship between technology and culture. Each technology is embedded with ideological biases and amplifies certain skills or tendencies, indicating that no technology is entirely neutral.

British educational technology scholar Laurie Lord has indicated that different technologies are closely associated with teaching methods. She proposes that teaching methods are the starting point for classifying instructional technologies and, based on five common teaching methods, categorizes instructional technologies into five media technologies: didactic teaching methods correspond to narrative media technologies, inquiry teaching methods correspond to interactive media technologies, discussion teaching methods correspond to communicative media technologies, exploratory teaching methods correspond to adaptive media technologies, and practical teaching methods correspond to productive media technologies (Laurillard, 2013), as shown in Table 17.1.

Canadian distance education scholar A. W. Bates believes that media should be selected and used based on teaching objectives and effectiveness. Each medium has its unique inherent laws, and there is no "super medium" that is suitable for all teaching objectives; all media have their strengths and weaknesses. Bates has

Table 17.1   Laurillard's matrix of technology and teaching methods

Media technology	Teaching method	Specific examples
Narrative media	Lecture	Print media, television, audio, video
Interactive media	Inquiry	Libraries, internet
Communicative media	Discussion	Forums, discussion groups, chat rooms
Adaptive media	Exploration	Laboratories, simulations
Productive media	Practice	Theses, products, models, constructions

proposed a decision outline for media selection and combination, providing a practical decision-making model for open and distance education practitioners. Singaporean scholar Chai Kin-Wah and colleagues found in a review study that future TPACK research should consider its high degree of contextuality, especially its specificity in technology, teaching, and subject matter. This indicates that the application of technology in different contexts needs to be specifically analyzed and optimized to match specific teaching needs (Chai *et al.*, 2013).

Therefore, the close association between technology and teaching methods, the need for different teaching methods to be supported by different technologies, and the existence of corresponding patterns between technology and teaching methods are evident. In terms of diversity and matching, in the selection and application of educational technology, it is necessary to choose the most suitable media and technology according to specific teaching objectives and contexts. Emphasizing non-neutrality and ideology, technology is not only a tool but also carries specific ideological biases, and the application of educational technology should consider its potential cultural and social impacts.

17.1.3   The transformative impact of technological advancements on education

Technological advancements have a profound and multidimensional impact on the field of education. First, technology has enhanced work efficiency, thereby promoting the development of social productivity and socio-economic progress, which directly impacts the improvement of quality of life. This is evident in the continuous growth of the American economy, where the information economy has played a crucial role. Particularly in the 1990s, the United States' "National Information Infrastructure" program, also known as the "Information Superhighway," aimed to improve work efficiency and quality of life, significantly boosting labor productivity.

In the field of education, a perspective on the "education revolution" suggests that there have been three major revolutions in the history of education: organized learning, the emergence of schools and universities, and the invention of printing and textbooks. Information technology has introduced the fourth revolution in education. Media technology has always been a major research area in educational technology; hence, many scholars define educational technology in terms of media technology. Educational technology is considered the convergence of three research branches: media in education, instructional psychology, and educational systems methodologies (Cohen *et al.*, 2000).

The American Association for Educational Communications and Technology (AECT) has defined educational technology approximately every decade, reflecting the changing role of technology in education. From early audio-visual communications and media to learning resources, programs, and devices, and finally to learning processes and technological processes and resources, these definitional changes highlight the increasing emphasis on technology integration and instructional strategies in educational technology.

In the *Handbook of Research on Educational Communications and Technology*, organized and published by AECT and involving representative experts and scholars in the field from across the United States, this trend is also reflected. To date, five editions of the handbook have been published. From the changes in the content composition of the handbook, as shown in Table 17.2, it can be observed that as specific technologies have emerged and developed, corresponding instructional content has increased. For example, with the development of mobile learning, flipped classrooms, and massive open online courses (MOOCs), the importance of instructional strategies has become increasingly prominent. Moreover, a significant characteristic of educational technology is the integration of subject teaching with information technology, emphasized in the handbook's focus on technology integration.

Reviewing the history of technological development, it is evident that there is an inseparable link between technological advancements and the development of educational theories (as shown in Table 17.3). From audio-visual technology in the 1940s to personal computers in the 1970s, and computer networks in the 1990s, each era's technological progress has been accompanied by innovations in educational theory. In the early 21st century, the widespread use of the internet spurred research in e-learning and computer-supported collaborative learning (CSCL), while the post-2010 era saw the rise of mobile learning and intelligent education driven by advancements in smartphones and artificial intelligence (AI) technologies (Alzahrani *et al.*, 2015).

In summary, technological advancements not only drive socio-economic development but also profoundly impact educational theories and practices. As a continually evolving field, educational technology's definitions and focus have changed with technological developments, reflecting ongoing exploration of technology integration and instructional innovation in education.

17.2   Current state of research in intelligent educational environments

In the current era, the extensive application of the internet, mobile internet, and AI technologies has transcended their traditional roles as providers of information and ideology. These technologies are now considered as constructors of social order. As scholars such as Harold Innis and John Durham Peters have pointed out, these technologies have become the new infrastructure of our time, defining a new type of social order distinct from the past (Innis, 1999, p. 19; Peters, 1999). New infrastructures like 5G, AI, and data centers are likened to "highways" leading to a fully digital society, while data is metaphorically referred to as the "new oil". This section examines the policies, initiatives, and the state of intelligent educational environments and related theories regarding the application of AI in education around the world.

Table 17.2 Thematic changes in the AECT handbook

Edition	Basics	Technology	Teaching strategies	Design and development	Research methods	Evaluation and assessment	Technology integration	Future directions
First edition	10	15	1	10	5	0	0	0
Second edition	11	12	5	8	5	0	0	0
Third edition	7	16	7	22	4	0	0	0
Fourth edition	10	12	13	13	8	9	5	4
Fifth edition	10	0	20	18	8	9	5	4

Table 17.3 Technological advances in EdTech paradigms

Phase	Period	Basic theory	Media technology	Learning center	Focus of teaching structure	Form
Educational Tech 1.0	1940–1990	Cognitivism, multimedia learning theory	Audio-visual media (broadcast TV, Film)	Knowledge center	Restructuring knowledge representation (symbol representation of multiple media)	Electric education
Educational Tech 2.0	1990–2010	Constructivism, learning environment theory	Computer and network media (computer, internet)	Learner Center, community of practice	Restructuring learning activities (structuring learner-centered learning environment and community)	Online education
Educational Tech 3.0	2010–now	System science, learning analytics theory	Mobile and intelligent media (mobile terminal, IoT, data center)	Evaluation center	Restructuring teacher-student relationship (data-driven teaching strategies, student-centered classroom)	Intelligent education

17.2.1   *Policies and initiatives for AI in education around the world*

Since the 1970s, the emergence of AI technology has been accompanied by explorations in educational applications. In 1970, Jaime Carbonell and others developed the SCHOLAR system, marking the beginning of intelligent computer-aided instruction. In 1973, Hartly and Sleeman proposed the basic framework for ICAI (Intelligent Computer-Aided Instruction) systems. In 1982, Sleeman and Brown formally defined ITS. In 1988, the first International Conference on Intelligent Tutoring Systems was held in Montreal, Canada, indicating that the research on ITS had entered the international academic community.

Since 2010, governments in Europe and the United States have invested significant resources in AI and big data infrastructure development and applications: In October 2016, the White House Office of Science and Technology Policy released "Preparing for the Future of Artificial Intelligence" and the "National Artificial Intelligence Research and Development Strategic Plan," proposing the integration of AI, data science, and related fields into the national education system. Stanford University released the report "Artificial Intelligence and Life in 2030" the same year, reflecting on the application of AI in education. In 2019, UNESCO released the report "Artificial Intelligence in Education: Challenges and Opportunities for Sustainable Development," focusing on the sustainable development of intelligent education. In February 2020, European Union released the "White Paper on Artificial Intelligence," emphasizing the cultivation of AI awareness among students and its integration with quality education. In September 2020, the EU released the "Digital Education Action Plan (2021–2027)," focusing on promoting the development of efficient digital education ecosystems and enhancing digital skills. In 2017, United Kingdom government released the report "Growing the Artificial Intelligence Industry in the UK," advocating for the widespread integration of data science and AI in education. In November 2017, Finland released "Finland's Age of Artificial Intelligence," proposing a multi-dimensional advancement of intelligent education and the construction of a coordinated and effective education system. In 2016, Japan formulated the "Education Informatization Acceleration Plan," proposing the concept of smart schools. In the 2019 "AI Strategy," Japan emphasized the cultivation of basic math, data science, and AI abilities for all citizens. In November 2019, Singapore released the "National AI Strategy," which includes the application of AI in the education sector.

In terms of big data infrastructure construction, Europe, the United States, and other countries have formulated corresponding policies. In 2004, a relatively complete framework for educational data infrastructure was proposed. In 2010, the federal government introduced the Federal Data Center Consolidation Initiative (FDCCI), promoting the large-scale, integrated, and green construction of data centers. In February 2020, European Union released the "European Data Strategy," planning to invest 4–6 billion euros in projects supporting the integration of EU data spaces and cloud infrastructure from 2021 to 2027. The National

Data of Australia Service Center established the Australian Scientific Data Discovery Platform Portal. Japan has initiated the construction of the AI society "Society 5.0".

China has also made rapid advancements in AI education and big data infrastructure construction. Since the release of the "New Generation Artificial Intelligence Development Plan" by the State Council in 2017, intelligent education has received comprehensive attention and rapid development. According to CNKI literature search results, since 2017, the number of Chinese papers on "artificial intelligence," "AI era," and "intelligent education" has increased significantly, accounting for about 70% of the total research.

Big Data Infrastructure Construction: In February 2022, China launched the "Eastern Data, Western Computing" project, constructing national computing power hub nodes and data center clusters, creating a multi-layered computing power facility system, demonstrating China's strategic determination in large-scale highly informationized infrastructure construction.

17.2.2 Knowledge graph analysis of intelligent education research

Intelligent education, as an emerging interdisciplinary field, has garnered significant academic attention since IBM incorporated it into the "Smarter Planet" initiative in 2008 (Castells, 2011). Numerous scholars have proposed definitions and research frameworks for intelligent education (Peters, 1999). However, the rapid advancements in AI technology have led to continuous evolution in the theoretical and practical domains, key issues, and trends of intelligent education, resulting in the absence of a universally accepted research paradigm in this field (Innis, 1999). This lack of a standardized paradigm provides room for academic exploration and innovation but also presents challenges in the development of research methodologies and theoretical frameworks. Therefore, conducting a comprehensive, extensive, and high-level bibliometric analysis of scholarly papers on intelligent education is essential to gain an overview of the development trajectory and future directions of this research field. Consequently, at the initial stage of this research (2023), we employed knowledge mapping analysis to visualize the current state of research in the international intelligent education domain from 2000 to 2023, and to analyze research frontiers and future trends.

17.2.2.1 Literature characteristics in the field of intelligent education research

This study employs a scientific knowledge mapping approach, focusing on the knowledge domain to illustrate the evolution and structural relationships within scientific knowledge. Utilizing CiteSpace, a free visualization tool for analyzing trends and patterns in scientific literature, we construct a network of interrelated knowledge based on co-citation, co-occurrence, and social network analysis theories (Table 17.4).

In this research, scientific knowledge mapping analysis is utilized to reveal the current state of national and institutional research in the field of intelligent education. As shown on the left side of Figure 17.1, the countries actively engaged in intelligent education research include China, the United States, South Korea, the United Kingdom, and Canada, with China, the United States, and South Korea holding central positions. The right side of Figure 17.1 presents the institutional research knowledge map, with representative research institutions identified by setting a threshold (threshold ≥ 2). The relevant information on countries and institutions is organized in Table 17.5.

Table 17.4   The search strategy

Type	Description
Database Source	Web of Science (WOS): SCI, SSCI, A&HCI, CPCI-S, CPCI-SSH, ESCI
Selection period	2000–2023
Screening strategy	Manual removal of documents not related to the topic
Search keywords	TI="smart education" OR "intelligent education" OR "smart learning" OR "intelligent learning"

National and Institutional.

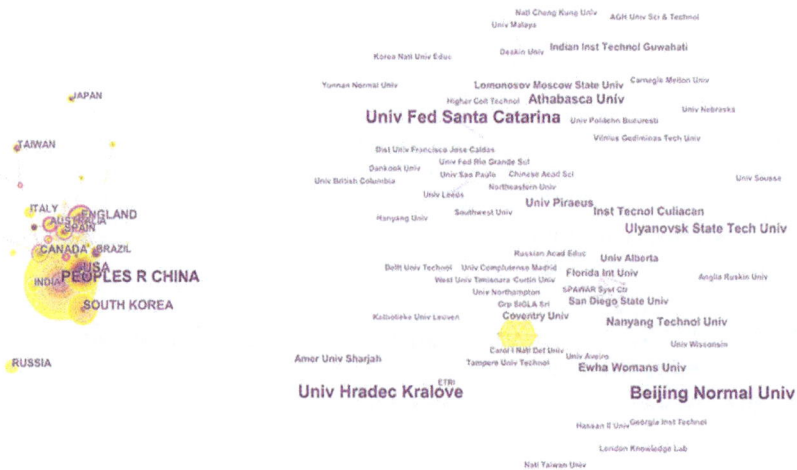

Figure 17.1   National and institutional knowledge maps

Table 17.5   Representative countries and institutions

No.	Country	Publications	Proportion	Institution	Publications	Proportion
1	China	116	17.05%	Beijing Normal University	11	1.6%
2	USA	61	8.97%	Federal University of Santa Catarina	10	1.5%
3	South Korea	46	6.76%	University of Hradec Králové	10	1.5%

Research Network: Author Publication and Co-citation Analysis.

Figure 17.2   Co-authorship knowledge map

In this study, which is tailored for academic journals or professional conferences, we employ quantitative analysis of co-citation among authors. This approach not only allows us to identify the core authors and research teams within the field but also reflects the academic relevance and influence of the authors. Utilizing the law proposed by Derek J. de Solla Price, a pioneer in scientometrics, we can calculate the publication volume of core authors (Levinson, 2004; Price, 1963) The co-authorship knowledge map, with a threshold of 15 or higher, is illustrated in Figure 17.2.

Table 17.6   Representative research of research teams

Teams	Research focus
Bajo team	Multimodal system supporting language learning for elderly or disabled people Intelligent system for automatic analysis of images and videos based on multimodal systems
Kinshuk team	Context-aware multi-document summarization method Rule-based visual analysis tool (Imran *et al.*, 2016) Learning analytics framework Intelligent tracking and forecasting system
Gomes team	Conflict resolution in virtual environments (Carneiro *et al.*, 2011)
Callaghan team	Enhancing reality, supporting intelligent learning environments in mixed reality (Alzahrani *et al.*, 2015) Support for learning in virtual collaborative learning environments based on social networks
Huang Ronghuai team	Theoretical model framework for intelligent learning (Liu *et al.*, 2016) Impact of spatial changes on students' learning experience (Du *et al.*, 2017) How to provide students with E-learning resources in intelligent learning environments (Fan *et al.*, 2015) Support for personalized meaningful learning in intelligent learning environments (Huang *et al.*, 2017) Intelligent campus weather monitoring system (Hu and Huang, 2015)

In this study, which is aimed at academic journals or professional conferences, the analysis reveals that the research teams with significant representation include those led by Javier Bajo, Dr. Kinshuk, Victor Callaghan, and Marco Gomes, as well as the team of Professor Huang Ronghua from Beijing Normal University. The representative research of these teams is organized in Table 17.6 as follows (Laurillard, 2013).

In Figure 17.3, the co-citation knowledge map of authors (threshold $\geq$ 8), it is evident from the collaboration among authors that the primary research teams in the international field of intelligent education are engaged in research on multi-agent systems, intelligent tracking and intervention systems, solutions related to intelligent learning environments and learning analytics systems, learning support within intelligent learning environments, learning spaces, and intelligent learning theories (D'Mello and Graesser, 2013).

The co-citation patterns reflect that influential authors are primarily investigating adaptive learning tutoring systems, affective computing, intelligent agents, theoretical frameworks, and practical explorations in intelligent education.

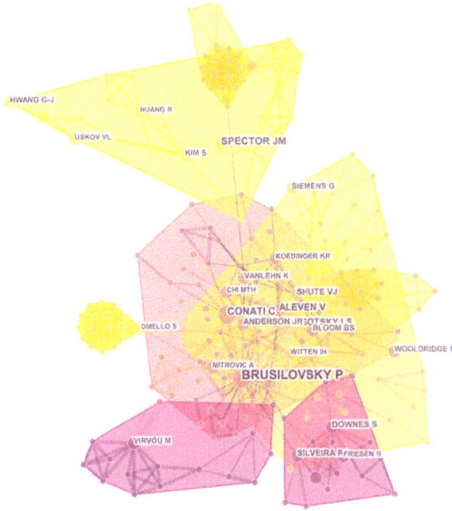

Figure 17.3   Co-citation knowledge map

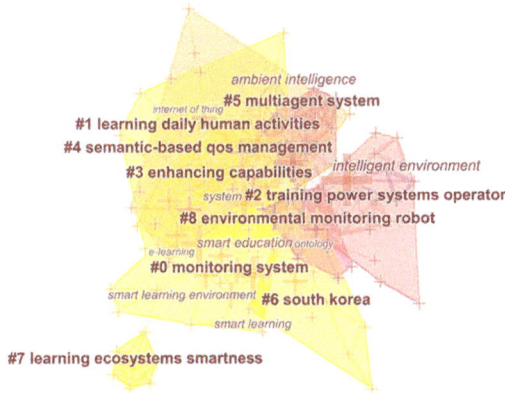

Figure 17.4   Keyword clustering knowledge map

17.2.2.2   Research hotspots and frontiers in the field of intelligent education

In this study, tailored for academic journals or professional conferences, keywords are recognized as one of the significant characteristic items of literature. By conducting co-occurrence cluster analysis on the keywords of international intelligent education research, the current research hotspots and significant topics within the field can be unearthed. The keyword clustering knowledge map is depicted in Figure 17.4. Through the analysis of frequency and centrality in the keyword clustering map, selecting key high-frequency terms enables the exploration of the research hotspots and frontiers in international intelligent education (Aleven *et al.*, 2009; Brusilovsky, 1996, 2001; Conati, 2002; Conati and Maclaren, 2009; Downes, 2007; Friesen, 2009; Silveira and Marques, 2011; Spector, 2014; Virvou and Katsionis, 2008).

The keyword clustering map reveals current hot topics in international intelligent education research. These include "monitoring system," "learning daily human activities," "training power systems operator," "enhancing capabilities," "semantic-based QoS management," "multiagent system," "learning ecosystem smartness," "environmental monitoring robot," and the country with a significant focus on intelligent education applications, "South Korea." By reviewing the relevant literature, representative studies of these hot topics have been identified and are presented in Table 17.7.

Table 17.7   Representative studies of current hot topics in intelligent education research

Hot topics	Representative studies
Monitoring systems	School bus tracking and monitoring system based on the Internet of Things Comprehensive teaching and learning support system Task-based context-aware intelligent interaction prediction system School bus tracking and monitoring system based on the Internet of Things
Human daily learning activities	Application of motion sensors to record human daily learning activities
Power systems staff training	Intelligent learning system for IBM power systems training based on SCORM learning objects
Enhanced capabilities	Video conferencing system to facilitate real-world interaction and family collaboration
Service quality management based on language definition	Intelligent service management platform based on semantic web calculations
Multimodal systems	Intelligent agents for personalized learning for the elderly Multimodal systems supporting language learning for the elderly or disabled people
Intelligent learning ecosystems	Intelligent monitoring and tracking of the learning ecosystem Research on the intelligent perception of behavior in mobile learning systems Intelligent context-aware systems for learning objects and mobile devices
Environmental monitoring robots	Robots for creating monitoring environments and assisting with curriculum construction

17.2.2.3   Trends in intelligent education research

Development of Components in Intelligent Educational Environments: In the domain of intelligent educational environment development, a representative perspective from international intelligent education theory is the ITS. There has been a general consensus among scholars since the 1980s regarding the components of an ITS. An ITS primarily consists of three parts: learning content, instructional strategies, and a mechanism for understanding the student's current knowledge state (Mahmoud *et al.*, 2013). These correspond to the domain knowledge module, pedagogical module, and learner module within the ITS, respectively. A representative achievement in this field is the Generalized Intelligent Framework for Tutoring (GIFT), published in 2012 by the Human Research and Engineering Directorate of the U.S. Army Research Laboratory (as shown in Figure 17.5) (Sottilare, 2017). Systems such as MindStar Books, AutoTutor, PAL3, and ALEKS, developed by American scholars, and the intelligent computer-assisted Chinese teaching system developed by Chinese scholars such as He Kegang are examples of such ITS.

In the field of smart classroom development, the Educause Learning Initiative in the United States has compiled extensive practices of smart classrooms and proposed various design schemes for smart classrooms tailored to different purposes such as lecturing, discussion, practice, and collaboration. Among these, the most representative is the TEAL (Technology-Enabled Active Learning) smart classroom created by the Massachusetts Institute of Technology (OEIT, 2011), as depicted in Figure 17.6. This classroom integrates lectures, simulations, and hands-on bench experiments, while also creating a rich collaborative learning experience (Dori and Belcher, 2005). This approach has served

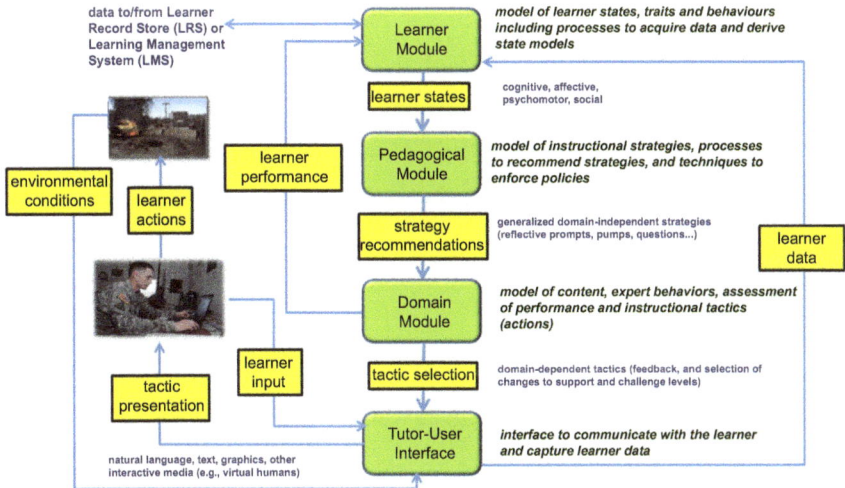

Figure 17.5   CITS composition and learner interaction (Sottilare, 2017)

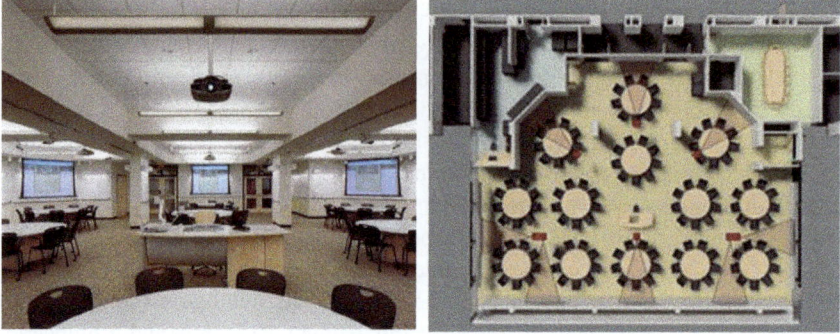

Figure 17.6   Smart classrooms advocated by MIT: The TEAL classroom design scheme (OEIT, 2011)

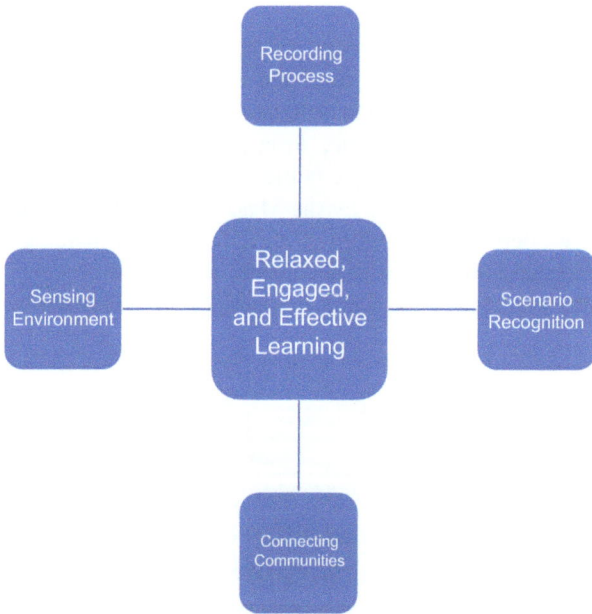

Figure 17.7   TRACE functional model of smart learning environments

as a template for smart classroom schemes widely adopted by universities worldwide, including those in China.

One of the representative perspectives within the intelligent education theories proposed by Chinese scholars is the TRACE3 smart learning environment theory put forward by Huang Ronghua and colleagues, as depicted in Figure 17.7. This study, synthesizing the representative views of scholars at home and abroad on the compositional elements of the environment, has summarized the constituents of a

Daily Learning Activities of Humans

Recording Process

Sensing Environment

Relaxed, Engaged, and Effective Learning

Scenario Recognition

1. Monitoring System
2. Environmental Monitoring Robot
3. Semantic-Based Service Quality Management

Connecting Communities

Enhanced Capabilities

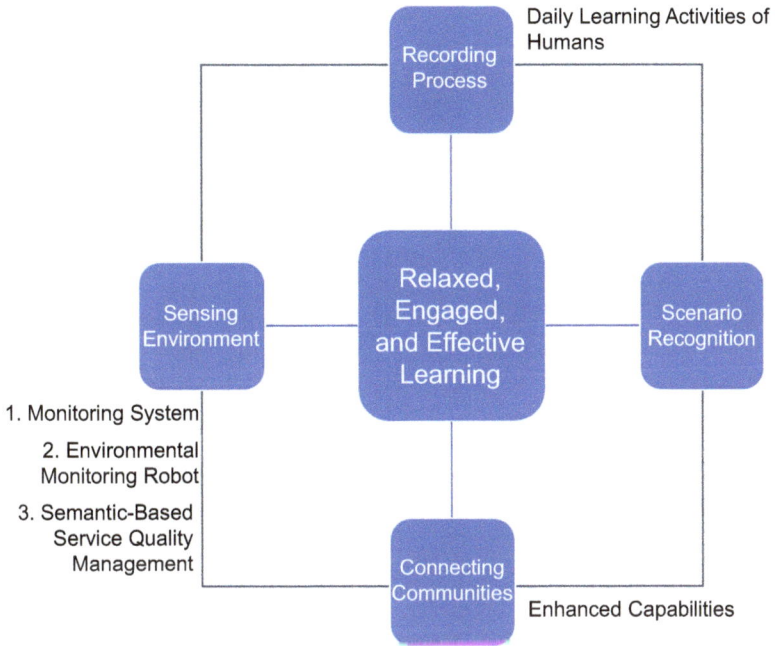

Figure 17.8   TRACE hot topic model of smart learning environments

smart learning environment, which encompass four major elements: learning resources, intelligent tools, learning communities, and teaching communities. Learners and teachers interact with the smart learning environment through the processes of learning and teaching, thus facilitating the occurrence of effective learning. Based on the TRACE smart learning environment functional model, the research hotspots in international intelligent education have been further organized, culminating in the TRACE smart learning environment hot topic model as shown in Figure 17.8 (Petrović *et al.*, 2022).

Currently, within the domain of intelligent educational environments, there is a dearth of research focused on the recording of learning processes, the recognition of learning contexts, and the bridging of learning communities. These three areas of study are poised to become future research trends.

Constructing an intelligent learning ecosystem

The rapid development of digital technologies has propelled the evolution and enhancement of educational software and platforms, bringing immense convenience to school education. However, the diversity, heterogeneity, and volatility of platforms, resources, and software have increasingly highlighted issues of digital divides and information silos. The intelligent learning ecosystem represents an upgrade and transformation of existing digital learning ecosystems, and

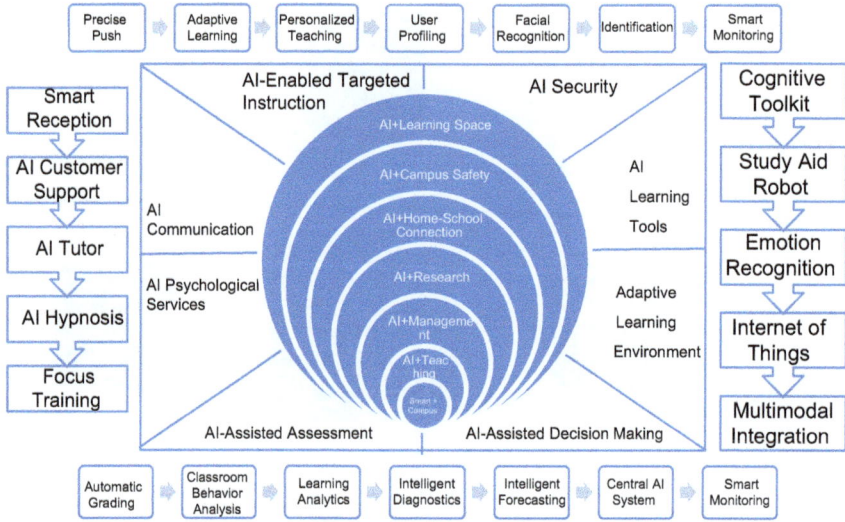

Figure 17.9   Eight typical artificial intelligence education application scenarios in "Smart+" campus

the international attention to intelligent learning ecosystems is showing a continuous upward trend.

In light of this trend, Chinese scholar Cao Xiaoming, based on the concept of educational informatization ecosystems, has summarized eight typical scenarios of AI applications in the "Smart+" campus context, as depicted in Figure 17.9.

Liu and Wang (2019), utilizing the top-level design methodology of information systems, have conducted a top-down approach to propose an overall architectural model for intelligent education. This model takes into account both the general system architecture and the technical architecture, as illustrated in Figure 17.10.

Research on the theories of intelligent education

Intelligent education, which emphasizes the support of smart media for learning, cannot be fully and effectively guided by traditional educational theories. Unlike traditional education, the development and transformation of intelligent education require theoretical frameworks that are still in need of breakthroughs. Currently, in international research on intelligent education, there is a tendency for technology and application to take precedence, and the formation and development of intelligent education theories and frameworks are in urgent need of advancement.

Chinese scholars have proposed numerous theories and practical frameworks regarding intelligent education and smart education. Professor Zhi-Ting Zhu has introduced a research framework for smart education, which mainly consists of four components: smart education, smart environment, smart pedagogy, and smart assessment. The smart environment encompasses digital environments such as smart terminals, smart classrooms, smart campuses, smart laboratories, maker spaces, and smart education clouds. Meanwhile, smart pedagogy refers to new

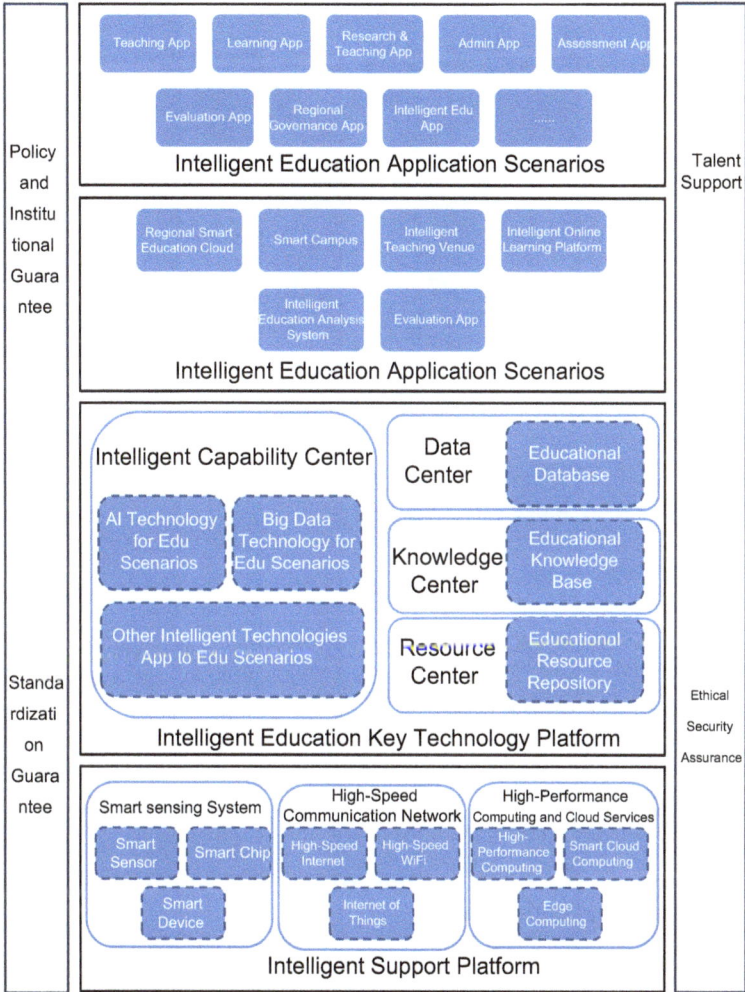

Figure 17.10 Intelligent education system architecture model intelligent education

types of teaching methods based on the smart environment, including differentiated instruction, personalized learning, collaborative learning, swarm intelligence learning, immersive learning, and ubiquitous learning (Gu *et al.*, 2021).

In the field of intelligent education pedagogy, Liu and Wang (2019) proposed in 2016 the "Three-Stage and Ten-Step Teaching Process" for smart classrooms, which is composed of three teaching stages and ten teaching steps, as shown in Figure 17.11. The "three stages" refer to the pre-class, in-class, and post-class segments that form a closed loop of classroom teaching, while the "ten steps" encompass activities such as student situation analysis, pre-class assessment, instructional design during the pre-class stage; situation creation, inquiry-based

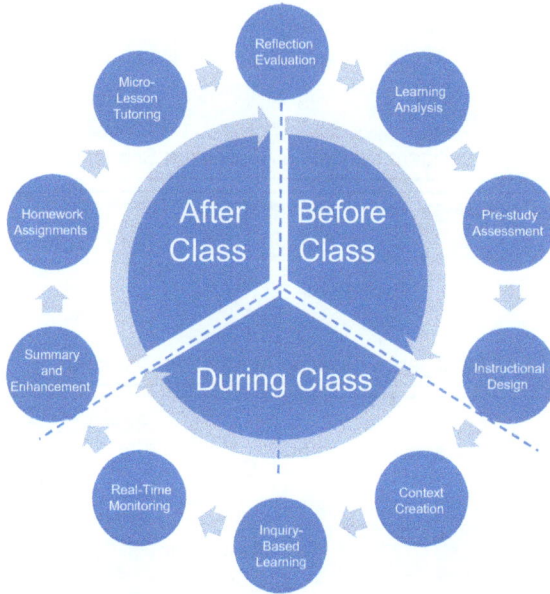

Figure 17.11   Smart education's three-phase ten-step instructional model

learning, real-time assessment, and summary enhancement during the in-class stage; and post-class homework, micro-lesson tutoring, and reflective evaluation in the post-class stage.

17.3   Review on the development of teacher knowledge in integrating technology

The contemporary teacher education reform that originated in the 1980s saw the United States as one of its initiators. In 1987, over 200 American scholars collaboratively authored the *Handbook of Research on Teacher Education*. The third edition of this work raised pivotal questions regarding the scope of subject matter and pedagogical knowledge that teachers should master during their training, how to balance these areas, and how to organize teacher training to meet the diverse individual needs of students while improving learning (Khong *et al.*, 2023; Koehler *et al.*, 2014).

Shulman's concept of PCK emerged as a representative response to these questions (Jin and Schmidt-Crawford, 2022). Subsequently, against the backdrop of the increasingly widespread and profound integration of information technology in education globally, Koehler and Mishra introduced the framework of TPACK in 2005, further enriching the understanding of teaching knowledge by integrating technology (Hennig and Liao, 2013; Karchmer-Klein and Konishi, 2023). This study aims to review the evolution of subject matter pedagogical knowledge and

TPACK, seeking to uncover changes, patterns, and trends in the recognition of teachers' professionalism.

17.3.1 The evolutionary development of pedagogical content knowledge research

In 1986, Professor Shulman from Stanford University proposed the foundational professional knowledge essential for effective teaching: PCK, which represents subject matter in forms that learners can grasp within specific contexts. The most significant contribution of this theoretical framework is its argument for the professionalism and irreplaceability of the teaching profession from the perspective of teachers' knowledge.

Today, the field has evolved, and researchers of PCK have convened two "PCK Summits" in 2012 and 2016, summarizing the "Consensus Model" (CM) and "Refined Consensus Model." (Berry *et al.*, 2015) discuss the significant contributions of these models to science education, emphasizing the ongoing development of PCK frameworks. Similarly, (Roy and Bairagya, 2019) highlight the repositioning of PCK within teachers' knowledge systems for teaching science. (Park, 2019) provide a detailed journey from Shulman's original notion of PCK to the Refined Consensus Model, illustrating the evolution of these concepts over time. This section attempts to provide a review and forecast of the consensus reached by researchers in the field of teacher education, the remaining disagreements, and future development trends by examining the evolution of PCK from 1985 to 2020 (Figure 17.12).

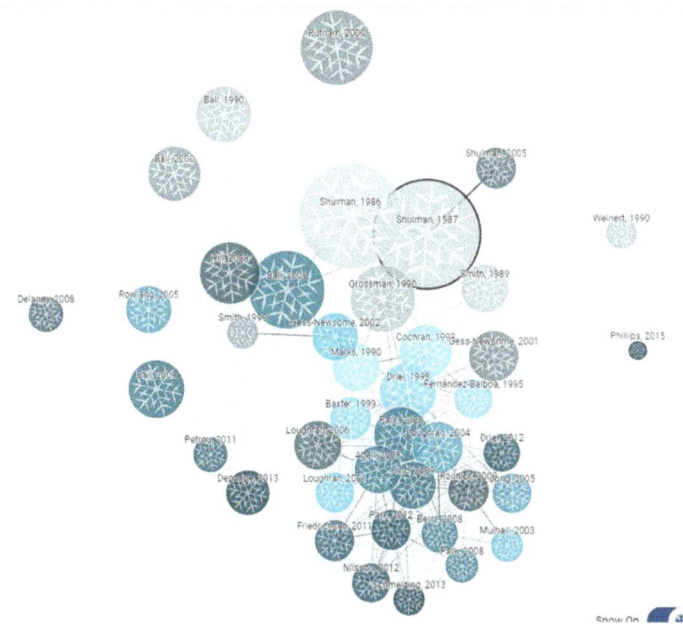

Figure 17.12   Connected papers knowledge graph based on Shulman's 1987 classic

17.3.1.1   PCK fundamentals: bridging a gap in teacher education research

Where lies the professionalism of teachers? In the early 1980s, Shulman, then a professor at Stanford University, proposed the addition of "pedagogical content knowledge" to teacher certification systems, leading to a new understanding that "teaching knowledge = subject matter knowledge + pedagogy knowledge + pedagogical content knowledge." He observed a lack of research in teacher education on how teachers transform their understanding of subject matter into teaching that students can comprehend (Ogren, 2005). He introduced the concept of PCK: "Within the domain of teaching subject matter, teachers provide the most effective expressions, the most powerful analogies, illustrations, examples, explanations, and demonstrations for the most frequently taught topics in their own subject areas, ways of expressing and representing topics so that they are understood by others. PCK also includes knowledge of whether learning certain topics is easy or difficult: understanding the preconceptions and prior knowledge that students of different ages and backgrounds bring to learning the most commonly taught topics and curriculum." He proposed that PCK includes the following components: (a) knowledge of the topics regularly taught in one's subject area; (b) understanding of the representation of these ideas; (c) knowledge of students' understanding of the topics, as shown in Figure 17.13.

In 1987, Shulman listed PCK as one of the seven knowledge bases of teachers (content knowledge, general pedagogy, curriculum knowledge, PCK, knowledge of learners and their characteristics, educational contexts, educational purposes), and he regarded PCK as the foundation of teaching knowledge and listed it as the preferred knowledge base. Tamir categorized the components of PCK into four: knowledge of students, curriculum knowledge, teaching knowledge (including teaching and management), and assessment knowledge. Grossman proposed that the four components of PCK are: (a) knowledge and beliefs about the purposes of

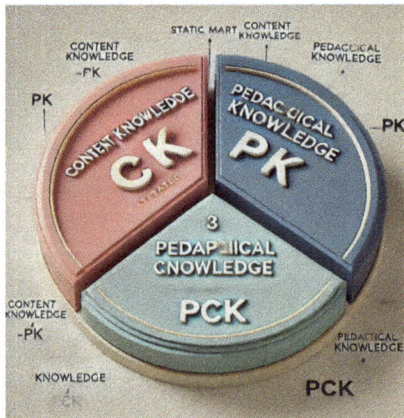

Figure 17.13   Static model of pedagogical content knowledge (PCK)

teaching subjects at different grade levels; (b) students' understanding, concepts, and misconceptions about a topic; (c) curriculum knowledge, including knowledge of curriculum materials available for teaching specific topics, and knowledge of the horizontal and vertical curriculum of a topic; and (d) "knowledge of teaching strategies and the performance of teaching specific topics." In addition to focusing on the affective domain of teaching purposes, this work also summarizes the interplay between teaching professional knowledge.

17.3.1.2   Dynamic models of PCK: PCKg and PCK spinning top models

Cochran *et al.* acknowledged the existence of PCK but redefined it as PCKg (pedagogical content knowing) through radical constructivism, distinguishing it from Shulman's static model of PCK. Cochran *et al.* (1993) describe PCKg as "teachers' integrative understanding of the four components of pedagogy, subject matter, student characteristics, and learning environments" (see Figure 17.14). This integrative process is ongoing and contingent upon teachers' backgrounds, experiences, and knowledge. Based on this understanding, teachers must develop their own PCKg; it cannot be acquired and transmitted as a body of knowledge. Furthermore, Geddis and Wood (1997) discuss the transformation of subject matter and the management of dilemmas in teacher education, highlighting the complexity and dynamic nature of PCKg.

Some scholars have proposed a complex, dynamically constructed perspective on PCK. Bishop and Denley further developed Shulman's view on PCK through the "Spinning Top" model (as depicted in Figure 17.15). Their model illustrates how PCK is dynamically generated from the other six knowledge bases of a teacher, which include Content Knowledge, Curriculum Knowledge, General Pedagogical Knowledge, Knowledge of Learners and Learning, Contextual Knowledge, and Knowledge of Values, Aims, and Purposes. Through this dynamic rotation process, these knowledge bases merge to form the teacher's unique PCK. The faster the rotation, the more the PCK appears white, signifying its greater impact on effective teaching and representing the dynamic construction process of PCK. This

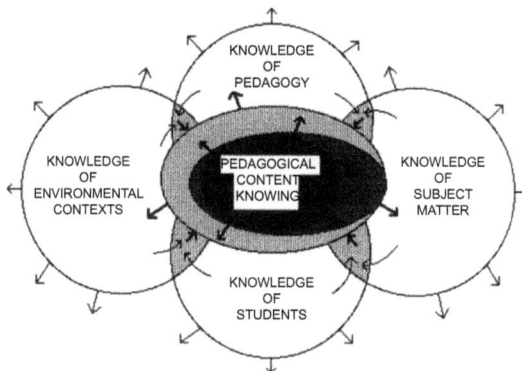

Figure 17.14   The PCKg model (Cochran et al., 1993)

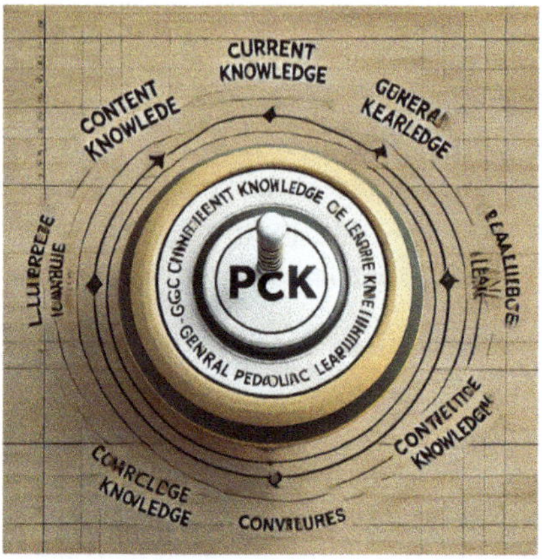

Figure 17.15   The Pie chart of PCK's spinning top model

model emphasizes that PCK is not static but continually evolves through the actual teaching practices of teachers. Teachers can integrate different knowledge bases during the teaching process to form effective teaching strategies and enhance teaching outcomes.

17.3.1.3   PCK knowledge in organizations: the consensus model and the refined consensus model of PCK

Smith and colleagues have emphasized the transformation between two forms of PCK, namely "Personal PCK" and "Normative PCK" (Normative or Collective PCK). The first PCK Research Summit was held in October 2012, where researchers collaboratively established the PCK CM in an attempt to reach a consensus on the differences and inconsistencies in PCK work. Within the PCK CM, student achievement is a key component as it shapes Teachers' Subject-Specific Pedagogical Knowledge (TSPK) and "Personal PCK." Teachers' reflection and re-evaluation of their own teaching is crucial for improving teaching quality (student achievement) through the reform and transformation of their Personal PCK, TSPK, and Teachers' Pedagogical Knowledge for Biology (TPKB) (Behling *et al.*, 2022; Park, 2019).

The second PCK Summit was convened in 2016, which led to the development of an improved version of the PCK CM, known as the "Refined Consensus Model" (RCM). The primary drawback of the PCK CM was its lack of detail regarding PCK and the absence of a precise visual representation of the components of PCK and their positions within the model, leading to much confusion in PCK research and unclear relationships among many PCK variants (Behling *et al.*, 2022) (Figure 17.16).

17.3.2   The evolution of TPACK research

Since the 1980s, the rapid development of digital technologies has significantly influenced the role of technology in teaching, drawing increasing attention from

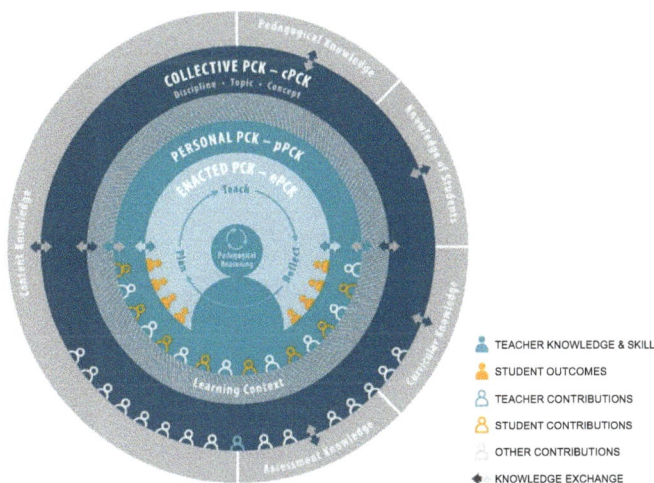

Figure 17.16   The refined consensus model of PCK (RCM)

scholars towards the integration of technological knowledge into the educational process. In 2005, Mishra and Koehler introduced the TPACK framework, which has become a crucial model for understanding how technology can be effectively integrated into teaching. Additionally, a comparative study on the development of PCK in the context of technological integration was conducted, while Lim *et al.* (2024) explored the TPACK growth of preservice teachers after technology integration courses in early childhood education.

17.3.2.1   TPACK core model: a new generation model of teacher knowledge

Mishra and Koehler from Michigan State University in the United States argue that the complexity of technology integration demands the development of new approaches to address this complexity. More importantly, there is a lack of theoretical foundations and conceptual frameworks to guide teachers in understanding the integration process within the field of educational technology. In 2005, Mishra and Koehler first introduced the concept of TPACK. The TPACK framework proposed by Mishra and Koehler is a complex, dynamic, and integrative structure of teacher knowledge, serving as the foundation for teachers to use technology for effective teaching. Lim *et al.* (2024) explored the TPACK growth of preservice teachers after technology integration courses in early childhood education. Additionally, preservice teacher cluster memberships were examined in an Edtech course, studying their TPACK development. Kim and Kim (2018) developed a TPACK scale for EFL teachers to promote 21st-century learning. Mishra and Koehler do not view TPCK as an extension of PCK but insist on an integration model.

Today, the TPACK framework has become one of the most influential teacher knowledge frameworks, with some referring to TPACK theory as the representative

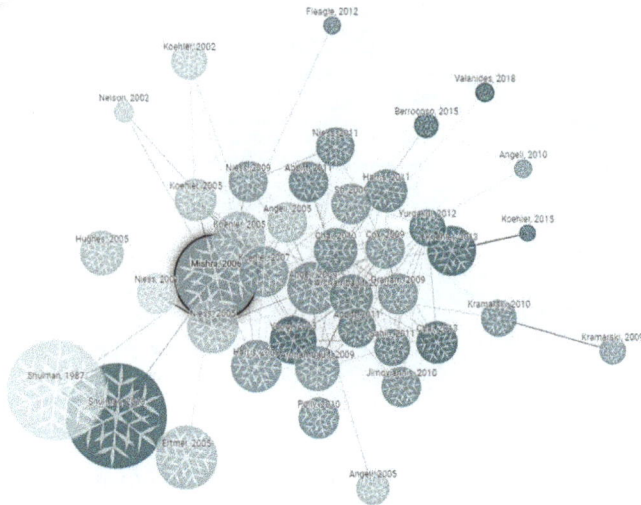

Figure 17.17   Connected papers knowledge graph constructed from Mishra and Koehler's (2006) classic paper

theory of the second generation of teacher professional knowledge following the first generation of PCK theory. In the latest research, Kim and Kim (2018) discussed the current status and future directions of TPACK in higher education faculty. Mishra points out that the future direction of the TPACK framework is to focus on the study of situational knowledge (XK) (Figure 17.17).

17.3.2.2   ITPACK dynamic model: ICT-TPCK transformation model

The ICT-TPCK transformation model, illustrated in Figure 17.18, posits that TPACK is a novel form of knowledge that emerges from the integration and transformation of information and communication technology, pedagogical knowledge, subject matter knowledge, knowledge of learners, and situational knowledge. critique the additive view of TPACK's components, arguing against the notion that an increase in any single element, such as technological knowledge, pedagogical knowledge, or content knowledge, would spontaneously lead to an overall increase in TPACK. They emphasize the need for a transformative approach to understanding TPACK. Furthermore, developed a TPACK scale to promote 21st-century learning among EFL teachers, while (Krauskopf *et al.*, 2015) explored the cognitive processes underlying TPCK. Otrel-Cass (2015) discussed the application of TPACK in networked inquiry learning in science, and (Ioannou and Angeli, 2015) highlighted its framework for integrating educational technology in computer science teaching.

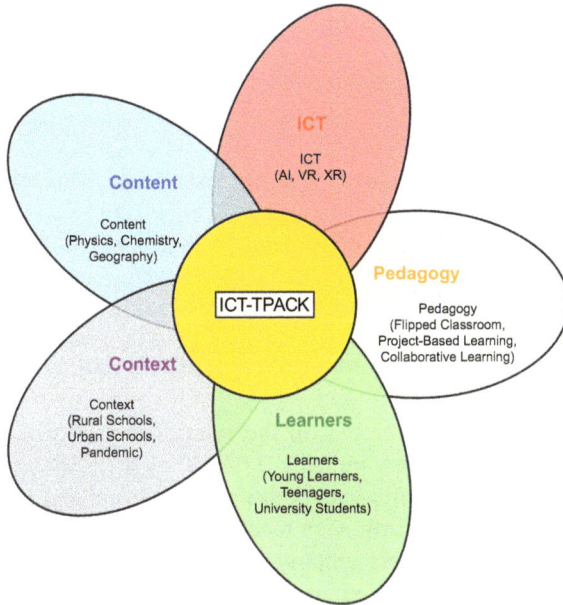

Figure 17.18   The ICT-TPCK knowledge framework

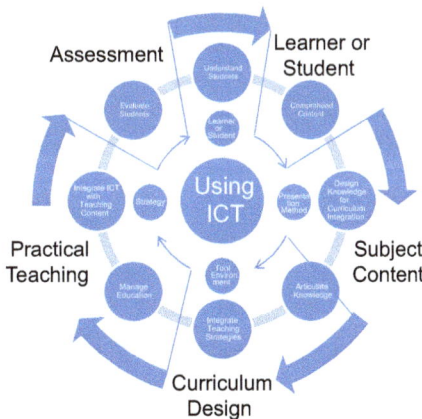

Figure 17.19   TPACK-Practical model (Yeh et al., 2014)

17.3.2.3   TPACK practical model: TPACK-Practical model

Studies have shown that teachers' teaching experience can demonstrate their proficiency in TPACK. Previous TPACK models have only discussed the content, components, and practical applications of TPACK, neglecting the impact of teaching experience on teachers' knowledge. Yeh *et al.* (2014) developed the TPACK-Practical framework model, as shown in Figure 17.19, which refers to the knowledge framework developed by teachers from practice with the help of information and communication technology. This framework is divided into five teaching domains and eight knowledge dimensions, highlighting the importance of practical experience for the development of teachers' knowledge. Additionally, McEwan and Bull (1991) discussed the pedagogic nature of subject matter knowledge, emphasizing the role of practical experience. Olofson *et al.* (2016) framed TPACK from a constructivist perspective, analyzing how teachers construct knowledge through experience. Jang and Tsai (2012) explored the use of interactive whiteboards by Taiwanese elementary teachers, further illustrating the importance of practical experience in developing TPACK.

17.3.2.4   ITPACK-in-Action model: the influence of social contexts

Bowers and Stephens have focused on the relationship between TPACK and teachers' beliefs and practices, viewing TPACK as a perspective for teacher development rather than a static knowledge base. Bowers and Stephens (2011) argue that different contexts influence teachers' knowledge and practices, and the type of knowledge teachers employ in teaching is contingent upon their practical experiences. At least four independent contextual factors have been identified that determine these practices. Building on this research, the TPACK-in-Action (TPACK-iA) model was proposed in 2014. Ling Koh *et al.* (2014) introduced this framework, which visualizes the interplay between TPACK and the four situational factors affecting teachers' ICT curriculum design: physical/technological, cultural/institutional, interpersonal, and personal (as shown in Figure 17.20).

The personal element refers to teachers' epistemological and pedagogical beliefs, which significantly influence their instructional decisions. The

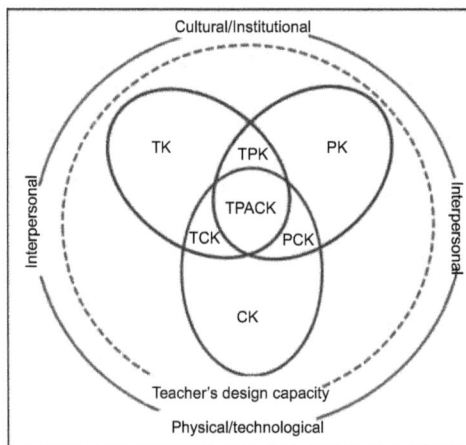

Figure 17.20   TPACK social context model (Ling Koh et al., 2014)

interpersonal element highlights the importance of relationships, as teaching design work is often carried out within teams, necessitating careful consideration of interpersonal dynamics. The cultural/institutional element views schools as sites of cultural reproduction, suggesting that these contexts have a profound impact on whether and how teachers use technology. Finally, the physical/technological supply element addresses the availability of resources and technology within schools, which directly affects teachers' decision-making processes.

17.3.2.5 Technology acceptance model: influences on teacher technology adoption

Why do teachers adopt technology in practice? Chin-Chung Tsai advocates that the TPACK framework can be further integrated with existing theoretical frameworks for research. The technology acceptance model (TAM) is one of the most widely used models in the field of educational technology research, as shown in Figure 17.21.

Initially, this model was developed to study the decisive factors in user acceptance of computers, and it has gradually evolved to primarily examine the degree of user acceptance of information systems. A substantial body of research has demonstrated that the TAM is effective in analyzing the factors influencing individual teachers' acceptance of information technology and in assessing whether teachers can accept information technology, providing a good explanation for teachers' technology adoption in the classroom (Hsu, 2016; Joo *et al.*, 2018). Researchers often employ a structural equation model that integrates TPACK with TAM to explore the relationships among various factors. Through testing in numerous subject and technology contexts, extensive research has found a close correlation between TPACK and technology acceptance (Mailizar *et al.*, 2021).

This model explains the mechanism of acceptance termination: technology use behavior is determined by behavioral intention, behavioral intention is determined by attitude and perceived usefulness, attitude is jointly determined by perceived usefulness and perceived ease of use, and external variables simultaneously affect

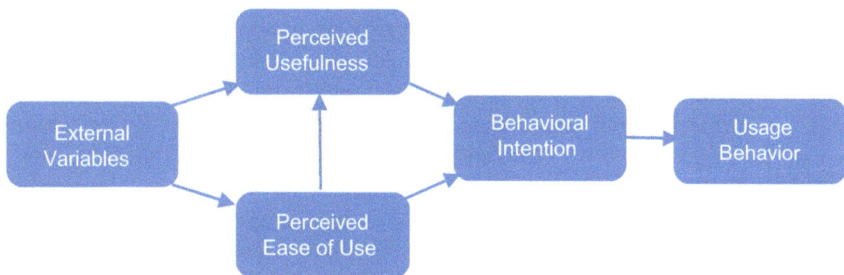

Figure 17.21   The technology acceptance model

perceived usefulness and perceived ease of use. It is evident that perceived usefulness is the primary determinant of people's use of computer systems, and ensuring perceived usefulness is the most critical issue that technology developers should focus on. Subsequently, Venkatesh and Davis further developed a more explanatory "Unified Theory of Acceptance and Use of Technology" (UTAUT) based on this.

17.4   Integration of Industry 5.0 and smart education

This chapter explores the relationship between technology and pedagogy, the current state of intelligent educational environments, and the evolution of subject-specific pedagogical knowledge through literature review and knowledge mapping analysis, aiming to summarize and construct a theoretical foundation that addresses the question, "Why should we study the changes in pedagogy brought about by technology?"

The era of Industry 5.0 signifies a deep integration of advanced technology with human creativity, with its core philosophy dedicated to fostering a symbiotic collaboration between technology and humans. It places greater emphasis on the harmonious fusion of human intelligence and creativity with cutting-edge technologies such as AI and robotics (Leong, 2024a). This cooperation is not only aimed at improving production efficiency and product quality but also focuses on the well-being, creativity, and sustainable development of human workers. The advent of this new era has transformed the landscape of manufacturing and profoundly influenced the development of the field of education.

In the context of Industry 5.0, education has welcomed new transformations and opportunities. Industry 5.0 is more than technological progress; it is a philosophical shift that emphasizes the coexistence and cooperation between technology and humans. In the age of Industry 5.0, machines and AI are no longer just tools but extensions of human creativity and skills. The application of this concept in the field of education signifies that intelligent education will enter a new phase, emphasizing the collaborative role of teachers and technology, and focusing on the personalized development and cultivation of creativity in students.

Intelligent education, as a key direction for the future of education, aims to build a more intelligent, personalized, and humanized educational ecosystem through the application of emerging technologies. The philosophy and technologies of Industry 5.0 provide strong support for intelligent education, enabling comprehensive upgrades in teaching methods, educational management, learning assessment, scientific research innovation, and educational services. By integrating technologies such as AI, big data, and ubiquitous computing into the educational system, intelligent education not only enhances teaching efficiency and learning outcomes but also promotes the comprehensive development and lifelong learning of students. Additionally, the integration of 3D printing technology in accelerating energy transition demonstrates how innovative manufacturing processes contribute to sustainable energy solutions (Leong *et al.*, 2023b). Incorporating sustainable development goals into educational modules, such as design thinking, can enhance students' understanding of global challenges and encourage them to propose innovative sustainable development solutions (Leong *et al.*, 2024c).

In the Industry 5.0 era, the impact of technology on teaching and learning activities has become more significant, complex, and diverse. The development of technology, especially the introduction of AI, VR, augmented reality (AR), and big data, has greatly changed traditional teaching and learning activities. These technologies not only provide new teaching tools and resources but also change the way teachers and students interact, making teaching more flexible and personalized (Leong *et al.*, 2023a). Specific applications of VR in education show that VR technology can be used to simulate complex scientific experiments, historical reenactments, and vocational training, providing students with immersive learning experiences that effectively improve learning outcomes and practical skills (Leong *et al.*, 2023c).

Different technologies reflect different philosophies and pedagogies. For example, CAI, based on behaviorist theory, focuses on repetitive practice and immediate feedback, while constructivist technologies such as VR emphasize immersive and contextual learning (Leong *et al.*, 2023a). These technologies have their own advantages and disadvantages, and not all teaching activities are suitable for their application.

In the context of Industry 5.0, the optimal matching of technology and pedagogy is particularly important. With the advancement of technology, educators need to understand the characteristics of each technology more deeply and combine them with appropriate pedagogies (Leong *et al.*, 2024c). For instance, AI can be used to design personalized learning paths, and VR can be used for practical operation and simulation environment teaching. Additionally, EEG technology can be integrated into intelligent education, providing real-time emotional feedback from students, enhancing the personalized learning experience. By recognizing students' emotional states, such as happiness, anxiety, or fatigue, educators can dynamically adjust teaching methods and content to improve student engagement and learning outcomes. These real-time data allow for the creation of a more responsive and supportive learning environment, closely monitoring and addressing students' emotional well-being.

Today, there is a growing recognition of the necessity to integrate specific technologies with specific pedagogies. For example, the application of AI technology in language learning, through intelligent voice assistants and adaptive learning systems, Leong can provide personalized practice and feedback; in science education, AR technology can help students intuitively understand complex scientific concepts and experimental processes.

The integration of emerging AI technologies and pedagogies is gradually opening the curtain on educational technology 3.0. This stage is characterized by highly personalized, intelligent, and interactive learning experiences. For example, ITS combined with AI technology can adjust teaching strategies and content in real-time based on students' learning behaviors and data, improving learning efficiency and effectiveness.

References

Aleven, V., McLaren, B.M., Sewall, J., and Koedinger, K.R. (2009). A new paradigm for intelligent tutoring systems: Example-tracing tutors. *International Journal of Artificial Intelligence in Education* 19, 105–154.

Alzahrani, A., Gardner, M., Callaghan, V., and Alrashidi, M. (2015). Towards measuring learning effectiveness considering presence, engagement and immersion in a mixed and augmented reality learning environment, in: *Workshop Proceedings of the 11th International Conference on Intelligent Environments*. Amsterdam: IOS Press, pp. 252–264. https://doi.org/10.3233/978-1-61499-530-2-252.

Anderson, J.R., Corbett, A.T., Koedinger, K.R., and Pelletier, Ray. (1995). Cognitive tutors: Lessons learned. *Journal of the Learning Sciences* 4, 167–207. https://doi.org/10.1207/s15327809jls0402_2.

Behling, F., Förtsch, C., and Neuhaus, B.J. (2022). The refined consensus model of pedagogical content knowledge (PCK): Detecting filters between the realms of PCK. *Education Sciences* 12, 592. https://doi.org/10.3390/educsci12090592.

Bernard, R.M., Abrami, P.C., Borokhovski, E., *et al.* (2009). A meta-analysis of three types of interaction treatments in distance education. *Review of Educational Research* 79, 1243–1289. https://doi.org/10.3102/0034654309333844.

Berry, A., Friedrichsen, P., and Loughran, J. (2015). *Re-Examining Pedagogical Content Knowledge in Science Education*. Milton Park: Routledge. https://doi.org/10.4324/9781315735665.

Bowers, J., and Stephens, B. (2011). Using technology to explore mathematical relationships: a framework for orienting mathematics courses for prospective teachers. *Journal of Mathematics Teacher Education*, 14(4), 285–304.

Carlson, J., Daehler, K.R., Alonzo, A.C., *et al.* (2019). The refined consensus model of pedagogical content knowledge in science education. In: Hume, A., Cooper, R., Borowski, A. (eds), *Repositioning Pedagogical Content Knowledge in Teachers' Knowledge for Teaching Science*. Singapore: Springer, pp. 77–94. https://doi.org/10.1007/978-981-13-5898-2_2.

Carneiro, D., Gomes, M., Novais, P., and Neves, J. (2011). Developing dynamic conflict resolution models based on the interpretation of personal conflict styles, in: Antunes, L., and Pinto, H.S. (eds), *Progress in Artificial Intelligence*. Berlin: Springer, pp. 44–58. https://doi.org/10.1007/978-3-642-24769-9_4.

Castells, M. (2011). *The Rise of the Network Society*. New York: Wiley.

Chai, C.S., Koh, J.H.L., and Tsai, C.-C. (2013). A review of technological pedagogical content knowledge. *Journal of Educational Technology & Society* 16, 31–51.

Clark, R.E. (1983). Reconsidering research on learning from media. *Review of Educational Research* 53, 445–459. https://doi.org/10.3102/00346543053004445.

Cochran, K.F., DeRuiter, J.A., and King, R.A. (1993). Pedagogical content knowing: An integrative model for teacher preparation. *Journal of Teacher Education* 44(4), 263–272. https://doi.org/10.1177/0022487193044004004.

Cohen, S.S., Zysman, J., and DeLong, B.J. (2000). *Tools for Thought: What is New and Important about the "E-conomy"?* Berkeley, CA: Berkeley Roundtable on the International Economy, University of California.

Conati, C. (2002). Probabilistic assessment of user's emotions in educational games. *Applied Artificial Intelligence* 16, 555–575. https://doi.org/10.1080/08839510290030390.

Conati, C., and Maclaren, H. (2009). Empirically building and evaluating a probabilistic model of user affect. *User Modeling and User-Adapted Interaction* 19, 267–303. https://doi.org/10.1007/s11257-009-9062-8.

D'Mello, S.K., and Graesser, A. (2013). AutoTutor and affective autotutor: Learning by talking with cognitively and emotionally intelligent computers that talk back. *ACM Transactions on Interactive Intelligent Systems* 2, 23:1–23:39. https://doi.org/10.1145/2395123.2395128.

Dori, Y.J., and Belcher, J. (2005). How does technology-enabled active learning affect undergraduate students' understanding of electromagnetism concepts? *Journal of the Learning Sciences* 14, 243–279. https://doi.org/10.1207/s15327809jls1402_3.

Du, J., Wang, X., Geng, M., and Huang, R. (2017). Manage learning space to improve learning experience: Case study in Beijing Normal University on classroom layout, in: *2017 IEEE 17th International Conference on Advanced Learning Technologies (ICALT)*. pp. 454–456. https://doi.org/10.1109/ICALT.2017.153.

Fan, L., Chang, T.-W., Huang, R., and Cheng, W. (2015). A framework of teaching and learning with e-textbooks in smart learning environment, in: *2015 IEEE 15th International Conference on Advanced Learning Technologies*. pp. 451–453. https://doi.org/10.1109/ICALT.2015.61.

Geddis, A.N., and Wood, E. (1997). Transforming subject matter and managing dilemmas: A case study in teacher education. *Teaching and Teacher Education* 13, 611–26.

Gu X., Du H., Peng H., and Zhu Z. (2021). The theoretical framework, development and future prospect of smart education. *Journal of East China Normal University (Educational Sciences Edition)* 39, 20. https://doi.org/10.16382/j.cnki.1000-5560.2021.08.002.

Hennig, C., and Liao, T.F. (2013). How to find an appropriate clustering for mixed-type variables with application to socio-economic stratification. *Journal of the Royal Statistical Society Series C: Applied Statistics* 62, 309–369. https://doi.org/10.1111/j.1467-9876.2012.01066.x.

Hsu, L. (2016). Examining EFL teachers' technological pedagogical content knowledge and the adoption of mobile-assisted language learning: A partial least square approach. *Computer Assisted Language Learning* 29, 1287–1297. https://doi.org/10.1080/09588221.2016.1278024.

Hu, Y., and Huang, R. (2015). Development of weather monitoring system based on Raspberry Pi for technology rich classroom, in: Chen, G., Kumar, V., Kinshuk, Huang, R., and Kong, S.C. (eds), *Emerging Issues in Smart Learning*. Berlin: Springer, pp. 123–129. https://doi.org/10.1007/978-3-662-44188-6_18.

Huang, R., Du, J., Chang, T., Spector, M., Zhang, Y., and Zhang, A. (2017). A conceptual framework for a smart learning engine, in: Popescu, E., Kinshuk, Khribi, M.K., *et al.* (eds), *Innovations in Smart Learning*. Singapore: Springer, pp. 69–73. https://doi.org/10.1007/978-981-10-2419-1_11.

Imran, H., Ballance, K., Da Silva, J.M.C., Kinshuk, and Graf, S. (2016). VAT-RUBARS: A visualization and analytical tool for a rule-based recommender system to support teachers in a learner-centered learning approach, in: Li, Y., Chang, M. M., Kravcik, *et al.* (eds), *State-of-the-Art and Future Directions of Smart Learning*. Singapore: Springer, pp. 31–38. https://doi.org/10.1007/978-981-287-868-7_4.

Innis, H.A. (1999). *The Bias of Communication*. Toronto: University of Toronto Press.

Ioannou, I., and Angeli, C. (2015). Technological pedagogical content knowledge as a framework for integrating educational technology in the teaching of computer science, in: Angeli, C., and Valanides, N. (eds), *Technological Pedagogical Content Knowledge: Exploring, Developing, and Assessing TPCK*. Boston, MA: Springer, pp. 225–237. https://doi.org/10.1007/978-1-4899-8080-9_11.

Jang, S.-J., and Tsai, M.-F. (2012). Exploring the TPACK of Taiwanese elementary mathematics and science teachers with respect to use of interactive white-boards. *Computers & Education* 59, 327–338. https://doi.org/10.1016/j.compedu.2012.02.003.

Jin, Y., and Schmidt-Crawford, D. (2022). Preservice teacher cluster memberships in an edtech course: A study of their TPACK development. *Computers and Education Open* 3, 100089. https://doi.org/10.1016/j.caeo.2022.100089.

Joo, Y.J., Park, S., and Lim, E. (2018). Factors influencing preservice teachers' intention to use technology: TPACK, teacher self-efficacy, and technology acceptance model. *Journal of Educational Technology & Society* 21, 48–59.

Karchmer-Klein, R., and Konishi, H. (2023). A mixed-methods study of novice teachers' technology integration: Do they leverage their TPACK knowledge once entering the profession? *Journal of Research on Technology in Education* 55, 490–506. https://doi.org/10.1080/15391523.2021.1976328.

Khong, H., Celik, I., Le, T.T.T., Lai, V.T.T., Nguyen, A., and Bui, H. (2023). Examining teachers' behavioural intention for online teaching after COVID-19 pandemic: A large-scale survey. *Education and Information Technologies* 28, 5999–6026. https://doi.org/10.1007/s10639-022-11417-6.

Kim, D., and Kim, W. (2018). TPACK of faculty in higher education: Current status and future directions. *Educational Technology International*, 19(1), 153–173.

Koehler, M.J., Mishra, P., Kereluik, K., Shin, T.S., and Graham, C.R. (2014). The technological pedagogical content knowledge framework, in: Spector, J.M., Merrill, M.D., Elen, J., and Bishop, M.J. (eds), *Handbook of Research on*

Educational Communications and Technology. New York: Springer, pp. 101–111. https://doi.org/10.1007/978-1-4614-3185-5_9.

Kozma, R.B. (1994). Will media influence learning? Reframing the debate. *ETR&D* 42, 7–19. https://doi.org/10.1007/BF02299087.

Krauskopf, K., Zahn, C., and Hesse, F.W. (2015). Cognitive processes underlying TPCK: Mental models, cognitive transformation, and meta-conceptual awareness, in: Angeli, C., and Valanides, N. (eds), *Technological Pedagogical Content Knowledge: Exploring, Developing, and Assessing TPCK*. Boston, MA: Springer, pp. 41–61. https://doi.org/10.1007/978-1-4899-8080-9_3.

Laurillard, D. (2013). *Rethinking University Teaching: A Conversational Framework for the Effective Use of Learning Technologies*, 2nd edn. London: Routledge, https://doi.org/10.4324/9781315012940.

Leong, W.Y., Leong, Y.Z., and Leong, W.S. (2023a). Virtual reality in education: Case studies and applications, in: *IET International Conference on Engineering Technologies and Applications (ICETA 2023)*, pp. 186–187. https://doi.org/10.1049/icp.2023.3332.

Leong, W.Y., Leong, Y.Z., and Leong, W.S. (2023b). Accelerating the energy transition via 3D printing, in: *2023 Asia Meeting on Environment and Electrical Engineering (EEE-AM)*. IEEE, pp. 1–5.

Leong, W.Y., Leong, Y.Z., and Leong, W.S. (2023c). Human-machine interaction in biomedical manufacturing, in: *2023 IEEE 5th Eurasia Conference on IOT, Communication and Engineering (ECICE)*, pp. 939–944. https://doi.org/10.1109/ECICE59523.2023.10383070.

Leong, W.Y., Leong, Y.Z., and Leong, W.S. (2024a). The impact of the Malaysia accreditation of prior experiential learning (APEL) programme. *EIET* 4, 8–19. https://doi.org/10.35745/eiet2024v04.02.0002.

Leong, W.Y., Leong, Y.Z., and Leong, W.S. (2024b). EEG-based recognition of happiness, in: *International Conference on Image Processing and Artificial Intelligence (ICIPAI 2024)*. Bellingham, WA: SPIE, pp. 447–453. https://doi.org/10.1117/12.3035200.

Leong, W.Y., Leong, Y.Z., and Leong, W.S. (2024c). Engaging SDGs agenda into a design thinking module. *Educational Innovations and Emerging Technologies* 4, 1–7.

Leong, W.Y. (2024). Secure and efficient collaborative machine learning frameworks for 6G intelligent applications, in: *2024 IEEE International Workshop on Radio Frequency and Antenna Technologies (iWRF&AT)*, pp. 324–328. https://doi.org/10.1109/iWRFAT61200.2024.10594448.

Levinson, M.H. (2004). Linked: The new science of networks. *ETC.: A Review of General Semantics* 61, 170–171.

Lim, B.Y., Lake, V.E., Beisly, A.H., and Ross-Lightfoot, R.K. (2024). Preservice teachers' TPACK growth after technology integration courses in early childhood education. *Early Education and Development* 35, 114–131. https://doi.org/10.1080/10409289.2023.2224219.

Ling Koh, J.H., Chai, C.S., and Tay, L.Y. (2014). TPACK-in-action: Unpacking the contextual influences of teachers' construction of technological pedagogical content knowledge (TPACK). *Computers & Education* 78, 20–29. https://doi.org/10.1016/j.compedu.2014.04.022.

Liu, B., and Wang, Y. (2019). Intelligent education: System framework, core technology platform construction and implementation strategy – CNKI [WWW Document]. https://www.cnki.net/KCMS/detail/detail.aspx?dbcode=CJFD&dbname=CJFDLAST2019&filename=ZDJY201910004&uniplatform=OVERSEA&v=HwP0KGtwwZHua3GG8D4L7lz2UmJLodafYJqkYlYW2dW8VERy3DAVXtKmpBHqgWZM (accessed 26 July 2024).

Liu, X., Huang, R., and Chang, T.-W. (2016). Design of theoretical model for smart learning, in: Li, Y., Chang, M.M, Kravcik, *et al.* (eds) *State-of-the-Art and Future Directions of Smart Learning.* Singapore: Springer, pp. 77–86. https://doi.org/10.1007/978-981-287-868-7_9.

Mahmoud, S., Lotfi, A., and Langensiepen, C. (2013). Behavioural pattern identification and prediction in intelligent environments. *Applied Soft Computing* 13, 1813–1822. https://doi.org/10.1016/j.asoc.2012.12.012.

Mailizar, M., Hidayat, M., and Al-Manthari, A. (2021). Examining the impact of mathematics teachers' TPACK on their acceptance of online professional development. *Journal of Digital Learning in Teacher Education* 37, 196–212. https://doi.org/10.1080/21532974.2021.1934613.

McEwan, H., and Bull, B. (1991). The pedagogic nature of subject matter knowledge. *American Educational Research Journal* 28, 316–334. https://doi.org/10.3102/00028312028002316.

OEIT (2011). TEAL – Technology enabled active learning [WWW Document]. iCampus. URL https://icampus.mit.edu/projects/teal/ (accessed 26 July 2024).

Ogren, C. (2005). *The American State Normal School: An Instrument of Great Good.* Berlin: Springer.

Otrel-Cass, K. (2015). Theorizing technological pedagogical content knowledge to support networked inquiry learning in science: Looking back and moving forward, in: Angeli, C., and Valanides, N. (eds), *Technological Pedagogical Content Knowledge: Exploring, Developing, and Assessing TPCK.* Boston, MA: Springer, pp. 193–207. https://doi.org/10.1007/978-1-4899-8080-9_9.

Park, S. (2019). Reconciliation between the refined consensus model of PCK and extant PCK models for advancing PCK research in science, in: Hume, A., Cooper, R., and Borowski, A. (eds), *Repositioning Pedagogical Content Knowledge in Teachers' Knowledge for Teaching Science.* Singapore: Springer Nature, pp. 119–130. https://doi.org/10.1007/978-981-13-5898-2_4.

Peters, J.D. (1999). *Speaking into the Air: A History of the Idea of Communication.* Chicago, IL: The University of Chicago Press.

Petrović, L., Stojanović, D., Mitrović, S., Barać, D., and Bogdanović, Z. (2022). Designing an extended smart classroom: An approach to game-based learning for IoT. *Computer Applications in Engineering Education* 30, 117–132. https://doi.org/10.1002/cae.22446.

Price, D.J. de S. (1963). *Little Science, Big Science.* New York: Columbia University Press.

Roy, S., and Bairagya, S. (2019). Conceptualisation of pedagogical content knowledge (PCK) of science from Shulman's notion to refined consensus model (RCM): A journey. *Education India Journal: A Quarterly Refereed Journal of Dialogues on Education*, 8, 9–53.

Sottilare, R.A., Brawner, K.W., Sinatra, A.M., and Johnston, J.H. (2017). An updated concept for a generalized intelligent framework for tutoring (GIFT). Retrieved from: https://gifttutoring.org/attachments/download/2076/Updated%20Concept%20for%20the%20Generalized%20Intelligent%20Framework%20for%20Tutoring_9%20May%202017.pdf.

Spector, J.M. (2014). Conceptualizing the emerging field of smart learning environments. *Smart Learning Environments* 1, 2. https://doi.org/10.1186/s40561-014-0002-7.

Yeh, Y.-F., Hsu, Y.-S., Wu, H.-K., Hwang, F.-K., and Lin, T.-C. (2014). Developing and validating technological pedagogical content knowledge-practical (TPACK-practical) through the Delphi survey technique. *British Journal of Educational Technology* 45, 707–722. https://doi.org/10.1111/bjet.12078.

Chapter 18

Industry 5.0 and personalised education: application of intelligent technology in the optimisation of vocational education resources and improvement of teaching methods

Hongli Zhang[1,2] and Wai Yie Leong[2]

With the ongoing development of global industry, Industry 5.0 [1] is emerging as a revolutionary stage following Industry 4.0 [2], emphasising not only intelligent manufacturing and automation but also human–machine collaboration. Its core technologies—IoT, AI, big data analytics, smart robotics, AR, VR, blockchain and cloud computing—significantly impact various sectors [3,4]. In education, these intelligent technologies offer new possibilities for personalised learning [5].

Personalised education, a student-centred model, focuses on individual differences and needs, tailoring teaching content and methods based on students' interests, abilities and progress to enhance engagement and outcomes. However, traditional personalised education faces challenges such as uneven resource distribution, complex management and data security issues [6]. Industry 5.0 technologies can support personalised education through real-time data analysis and intelligent systems, optimising resource allocation and improving teaching methods.

This study explores how Industry 5.0 technologies promote personalised education, focusing on AI and big data analytics. It examines these technologies' roles in enhancing education quality and efficiency through case studies and data analysis. Additionally, the chapter offers suggestions and strategies for future promotion and application of these technologies in education. The research aims to analyse the core technologies of Industry 5.0, their application in personalised education and how they can optimise resources and improve teaching methods, ultimately proposing actionable strategies for their implementation in the educational sector [7,8].

[1]Department of Information Engineering, Heilongjiang Institute of Construction Technology, China
[2]Faculty of Engineering and Quantity Surveying, INTI International University, Malaysia

18.1   Overview of Industry 5.0

18.1.1   Definition of Industry 5.0

Industry 5.0 is the next phase of the industrial revolution following Industry 4.0. Its core concept is the deep collaboration between humans and machines to achieve more efficient and flexible production and service models. Industry 5.0 emphasises not only intelligent manufacturing and automation but also the integration of human intelligence with machine intelligence, thereby enhancing production efficiency and product quality. The goals of Industry 5.0 include personalised customisation, flexible production and sustainable development. The concept of Industry 5.0 has not yet been widely standardised, and different institutions and experts may have varying definitions of it (Table 18.1) [9,10].

The transformation of industry from 1.0 to 5.0 is not only a history of technological advancement but also a reflection of the evolution of human thought and social structure (Figure 18.1).

Table 18.1   Definitions of Industry 5.0

Perspectives	Definitions
European Commission	In its industrial policy documents, the European Commission has put forward the vision of Industry 5.0, emphasising the integration of technological innovation with social needs, promoting green and digital transformation.
Academia	In many academic papers and research reports, Industry 5.0 is described as an era of human–machine collaboration, achieving more flexible and customised production by combining human creativity with the efficiency of intelligent systems.
Industrial organisations	Some industrial associations and organisations are also discussing the concept of Industry 5.0, usually emphasising the collaboration between intelligent technology and human labour, as well as the responsibility towards society and the environment.

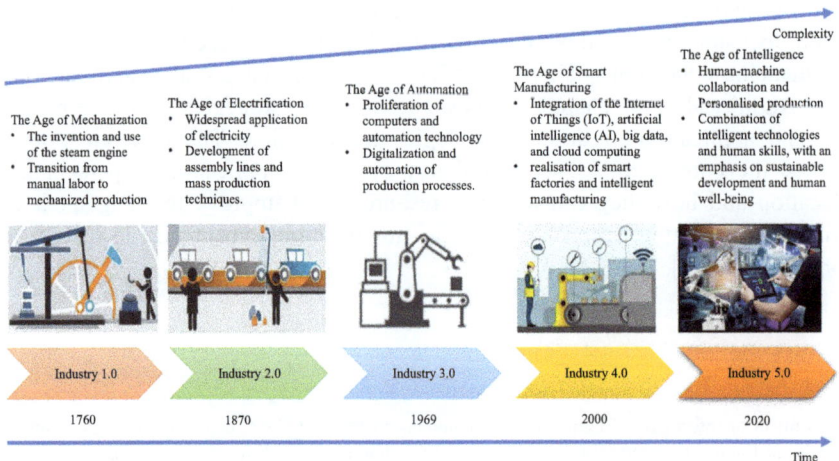

Figure 18.1   The evolution from Industry 1.0 to 5.0 [11]

Industry 1.0
The beginning of the industrial revolution marked the birth of mechanised production. Steam engines and other mechanical equipment made automation possible, laying the foundation for modern industry.

Industry 2.0
With the widespread use of electricity, industry entered the era of mass production. The introduction of assembly lines not only greatly increased production efficiency but also changed the organisation of labour.

Industry 3.0
The application of computers and automation technology ushered in a new era of digital production. During this phase, the production process became not only more precise and reliable but also significantly more efficient.

Industry 4.0
Characterised by the Internet of Things (IoT) and cloud computing, Industry 4.0 achieved the interconnection of devices and systems. Production during this period became not only smarter but also more flexible and adaptive.

Industry 5.0
The current phase emphasises human–machine collaboration and flexibility. The application of intelligent technologies enables humans and machines to cooperate more efficiently and allows the production process to be personalised and flexibly adjusted to meet changing market demands [12].

Industry 5.0 is not only a new stage in industrial production models but also symbolises a fundamentally different transformation optimise from previous revolutions. In this phase, human–machine collaboration and flexible production become the core features, integrating human creativity with machine efficiency to shape an intelligent and personalised production future (Table 18.2) [13].

Echoing the evolution of industrial development, the trajectory of education has also progressed from fixed models to more dynamic and personalised approaches. Scholars such as Ahmad and others [14] have summarised five significant milestones in the development of education (Figure 18.2).

Education 5.0 as the current frontier, is based on ethics and humanism, recognising the importance of value-driven education [15–17]. It emphasises ethical considerations and holistic development, integrating academic learning with moral and emotional growth. In Education 5.0, technology is a means to enhance value and make education more effective. Education 5.0 is not about reducing or increasing technology, but about making conscious, responsible choices without overlooking the bigger picture. Education 5.0 particularly focuses on aspects such as privacy, ethics, safety and technological concerns [18–20]. It prepares students for lifelong learning and lays the foundation for a broad range of skills far beyond the digital realm. Education 5.0 is not merely a shift in educational models; it represents a new understanding of learning, encompassing technology, ethics and the development of individuals and society. Within this framework, education is no longer confined to the transmission of knowledge but becomes an ecosystem for cultivating well-rounded individuals (Table 18.3) [21].

Table 18.2   Core concepts and key features of Industry 5.0

Core concepts	Key features
Human–machine collaboration	Intelligent machines and automated systems provide seamless integration and interaction, perfectly combining human creativity and flexible thinking with the precise execution of machines.
Flexible production	Personalisation and adaptability of the production process. Through intelligent production equipment and adjustable production lines, companies can quickly respond to changes in market demand, achieving small-scale, diversified production, thereby greatly enhancing efficiency and flexibility.
Digital technologies	Fully leveraging advancements in digital technologies such as the Internet of Things (IoT), cloud computing and big data analytics to digitise and intelligent the production process. The application of these advanced technologies not only optimises data collection and analysis but also provides real-time monitoring and precise forecasting, helping companies make more informed decisions.
Sustainable development	Emphasising the importance of ecological balance and advocating for the coordination between human development and the protection of natural resources. Industry 5.0 not only pursues economic benefits but also focuses on the long-term health of the environment and the well-being of society, striving to achieve harmonious coexistence between humans and nature.

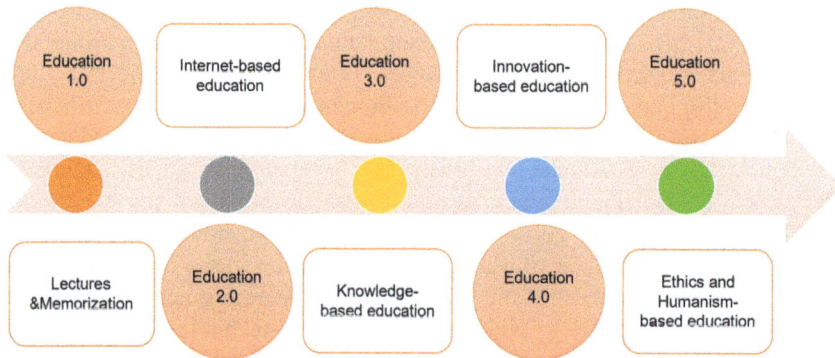

Figure 18.2   Five important milestones in the development of education

18.1.1.1   Literature review

In recent years, research on Industry 5.0 has been gradually increasing. Scholars generally believe that Industry 5.0 is a further development and enhancement of Industry 4.0, with its core focusing on driving transformations in production methods and service models through the application of intelligent technologies. The literature points out that the realisation of Industry 5.0 relies on the widespread

Table 18.3 Core concepts and key features of Education 5.0

Core concepts	Key features
All rounded person	The curriculum is designed to emphasise the balance of students' physical and mental development, including emotional intelligence, social skills and cultural and ethical awareness. Education is not only concerned with the cultivation of reason but also with the intrinsic value and emotional needs of students.
Personalised learning and intelligent assisted instruction	Through intelligent technologies such as AI teaching assistants and adaptive learning systems, students' learning journeys are personalised, ensuring that the content and pace of education match students' individual abilities and interests.
Sustainable development and ecological education	Integrate sustainability principles into the teaching philosophy, develop students' environmental awareness and social responsibility through project learning and community involvement, and encourage them to become active advocates for environmental protection.
Technology application and humanistic care	While promoting technological innovation, emphasis is placed on maintaining the concept of humanistic care to ensure that the application of technology promotes the overall well-being of students and the overall interests of society.
Social problem solvers	Students are encouraged to develop as active solvers of social problems through the cultivation of critical and innovative thinking that motivates them to come up with workable solutions to real-world problems.

application of advanced technologies such as the Internet of Things (IoT), artificial intelligence (AI) and big data analytics. These technologies provide a solid technical foundation for Industry 5.0 [22].

18.1.1.2 Differences between Industry 5.0 and Industry 4.0

Industry 4.0 emphasises the intelligent and automated production processes through IoT and automation technologies, while Industry 5.0 further focuses on the deep integration of human intelligence and machine intelligence. In Industry 5.0, the role of humans is not only to operate and monitor machines but also to collaborate with machines to accomplish complex tasks together. Additionally, Industry 5.0 places greater emphasis on personalised production and sustainable development, achieving on-demand production and efficient resource utilisation through flexible production lines and intelligent manufacturing systems (Table 18.4) [23–26].

Table 18.4   Comparison of Industry 4.0 and Industry 5.0

Comparison content	Industry 4.0	Industry 5.0
Development mode	Through the use of Digitalisation and artificial intelligence technologies, improve production efficiency.	Strengthen competitiveness and sustainability as the main driving forces for transformation.
Technical features	Focus on networking production activities as the main technical feature.	Balance various alternative technical governance models' different impacts.
Economic development concept	In terms of economic development concept, it aligns with traditional business model 'optimisation' standards, aiming to minimise costs and maximise profits, providing the greatest returns to shareholders.	Support human-centred technical methods, empower employees through the application of digital devices; emphasise sustainability and resilience, transition towards sustainable technology processes, expanding corporate responsibility throughout the value chain.
Social benefits	Relatively less involved in issues such as sustainable use of natural resources, climate change and social comprehensive development.	Highlight outcomes such as welfare, ecological benefits and comprehensive social development, and incorporate corresponding indicator systems.

18.1.1.3   Characteristics of Industry 5.0

1. **Human–machine collaboration**

   Industry 5.0 emphasises the collaboration between humans and machines, incorporating not only the application of robots and automation systems but also enhancing human workers' capabilities (Figure 18.3). By utilising collaborative robots, virtual reality (VR) and augmented reality (AR) technologies, workers can perform tasks more efficiently while avoiding dangerous and repetitive work. Robots can work alongside human workers to complete complex tasks, while VR and AR technologies are used for training and simulation, improving workers' skills and productivity [27].

2. **Sustainability**

   Industry 5.0 places a high priority on environmental sustainability, aiming to reduce resource waste and environmental impact during the production process. Specific measures include the use of smart materials, the promotion of circular economy practices and green technologies. Smart materials, such as renewable and recyclable materials, play a crucial role in Industry 5.0, ensuring the efficient use of resources and minimal environmental impact. Additionally, Industry 5.0 stresses the optimisation of production processes to reduce energy

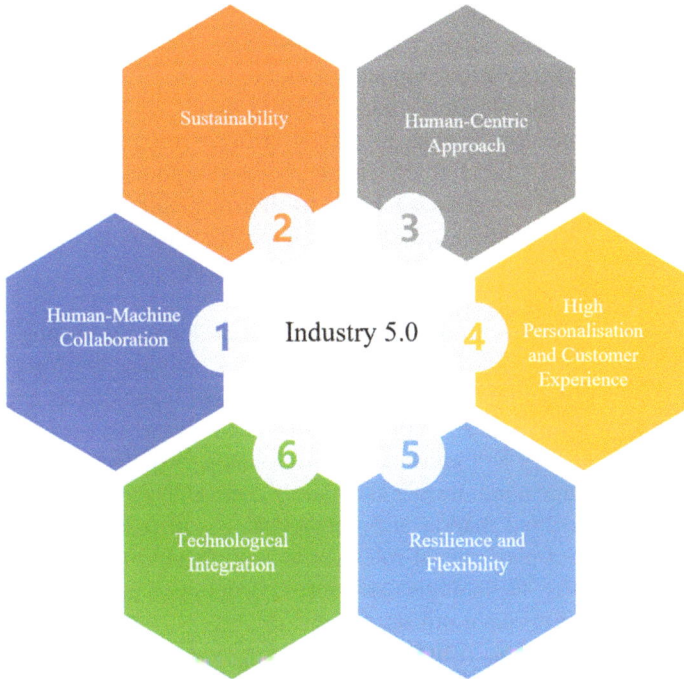

Figure 18.3   Characteristics of Industry 5.0

consumption and waste emissions, promoting leadership in environmental protection.

3. **Human-centric approach**

    Industry 5.0 centres on the well-being of workers in the production process, emphasising the enhancement of working environments and safety through technological support rather than merely replacing human workers. For example, advanced safety equipment and intelligent monitoring systems can continuously monitor workers' health and work conditions, preventing accidents and improving comfort and satisfaction at work.

4. **High personalisation and customer experience**

    Industry 5.0 is dedicated to providing highly personalised products and services to meet specific customer needs and expectations. By leveraging big data analytics, artificial intelligence and machine learning technologies, companies can gain deep insights into customer requirements and offer customised solutions. This personalisation not only increases customer satisfaction but also reduces costs through more efficient production techniques. For instance, personalised product design and custom production can meet unique customer demands while maintaining efficient production processes.

5. **Resilience and flexibility**

Industry 5.0 enhances supply chain transparency and operational flexibility, enabling companies to better respond to external shocks such as pandemics and natural disasters. By integrating and analysing supply chain data in real time, companies can quickly respond to market changes, optimise inventory management and reduce the risk of supply chain disruptions. Industry 5.0 also emphasises predictive maintenance and intelligent monitoring technologies to ensure efficient operation of production equipment, reducing faults and downtime and improving overall business resilience.

6. **Technological integration**

Industry 5.0 integrates multiple advanced technologies, including artificial intelligence, the Internet of Things (IoT), big data analytics, machine learning, smart robotics and virtualisation technologies. These technologies collectively drive the transformation and upgrading of industries. AI is used to optimise production processes and enhance decision-making efficiency, IoT connects devices for intelligent monitoring and management and big data analytics helps companies understand market and operational conditions. Machine learning and smart robotics play crucial roles in automation and precision production, while virtualisation technologies simulate and optimise production processes to ensure optimal resource utilisation.

Through these characteristics, Industry 5.0 represents not only technological advancement but also a redefinition of social and environmental responsibilities, promoting a more inclusive, sustainable and human-centric industrial future [28].

18.1.2   Core technologies of Industry 5.0

The realisation of Industry 5.0 is supported by a series of advanced technologies that collectively form its technological foundation, enabling the implementation of intelligent manufacturing and personalised education (Figure 18.4) [29].

18.1.2.1   Artificial intelligence (AI)

AI technologies, including machine learning, deep learning and natural language processing, can simulate human intelligence and perform complex data analysis and decision-making tasks. AI can be applied to personalised teaching systems, intelligent tutoring and assessment, automated management and predictive analytics [30].

18.1.2.2   Intelligent robotics

Intelligent robots have the ability to learn autonomously and adapt to their environment, capable of performing complex tasks such as operations on production lines and precision assembly. In education, intelligent robots can act as teaching assistants, providing personalised tutoring and participating in experimental and practical teaching.

18.1.2.3   Internet of things (IoT)

The IoT connects various devices and sensors, enabling real-time data collection, transmission and analysis, thus promoting intelligent manufacturing and

Figure 18.4   Core technologies of Industry 5.0

automation. In education, IoT can be used for smart campus management, student behaviour tracking and the creation of personalised learning environments [31].

18.1.2.4   Augmented reality and virtual reality

AR and VR technologies provide immersive experiences by blending virtual environments with the real world, enhancing users' perception and interaction capabilities. In education, AR and VR can be used for virtual laboratories, virtual training and interactive teaching, increasing students' interest and engagement in learning [32].

18.1.2.5   Blockchain

Blockchain technology, through its decentralised and distributed ledger, ensures data security, transparency and immutability. In education, blockchain can be used for academic certification, credit management and the sharing of educational resources, ensuring the security and reliability of data [33].

18.1.2.6   Big data analytics

Big data analytics processes and analyses large amounts of structured and unstructured data, revealing hidden patterns and trends to inform decision-making. In education, big data analytics can be used for collecting and analysing student data, improving curriculum design, teaching methods and resource allocation.

18.1.2.7   Cloud computing

Cloud computing provides computing resources and services via the Internet, supporting large-scale data storage and processing. In education, cloud computing can be used to build online learning platforms, educational resource libraries and data analysis systems, providing flexible and efficient computing resources [34].

18.1.2.8   5G communication technology

5G communication technology, with its high speed and low latency, supports real-time data transmission and remote control, significantly enhancing the performance of industrial IoT. It offers the capability for massive connectivity, allowing a large number of devices to interconnect, thus improving overall system efficiency. With its excellent network performance, 5G communication technology enables faster data exchange and more flexible automation control in industrial environments, laying the foundation for smart manufacturing and industrial automation [35].

18.1.2.9   Biometrics

Biometrics technology enhances identity verification and security in industrial environments through fingerprint, iris and facial recognition. These technologies effectively prevent unauthorised access and ensure the security of data and devices. By accurately identifying biological characteristics, biometrics technology provides reliable security in the industrial sector, helping to maintain workplace safety and information confidentiality.

Driven by the dual forces of Industry 5.0 and Education 5.0, cultivating students' core vocational skills has become increasingly important. Industry 5.0 emphasises human–machine collaboration, the integration of intelligent technologies and human creativity, while Education 5.0 focuses on personalised education and technology-enabled teaching models. By combining these cutting-edge fields, we can enhance students' vocational skills from an industrial perspective through the optimisation of educational resources and the improvement of teaching methods using intelligent technologies. Through precise personalised support, inter-disciplinary teaching methods, digital and intelligent technology training and the comprehensive development of soft skills, educators can foster well-rounded talents equipped with innovative thinking, technical skills and teamwork abilities. These students will be adept in future intelligent industrial environments, achieving both career success and personal value maximisation.

18.2   The needs and challenges of personalised education

18.2.1   Definition of personalised education

Personalised education is a student-centred educational model that tailors educational content and methods to the interests, abilities and learning progress of each student. Its core concept is to respect individual differences among students, fully harness their potential and enhance their learning outcomes and self-directed learning abilities [36].

18.2.1.1 Characteristics and goals of personalised education
Characteristics of personalised education
Student-centred approach: Personalised education emphasises a student-centred approach. Unlike traditional education models, which are often teacher-dominated with standardised content and methods, personalised education focuses on individual differences among students (Figure 18.5). It customises educational content and methods according to students' interests, abilities and learning progress. This personalised approach allows students to learn at their own pace, maximising their potential and improving their learning outcomes and engagement.

High flexibility: Personalised education is highly flexible. Unlike traditional classroom teaching, which is fixed in time and place, personalised education can adjust teaching strategies and content in real-time based on student feedback and learning progress. This flexibility helps students continuously adjust and optimise their learning methods and better meet the diverse learning needs and challenges of different students, ensuring that each student receives adequate attention and guidance.

Emphasis on teacher-student interaction: Personalised education values interaction between teachers and students. In traditional education models, interaction is often limited, making it difficult for teachers to fully understand each student's specific needs and learning conditions. In personalised education, teachers provide individualised tutoring and feedback to help students address learning issues, stimulate their interest and motivation in learning and improve teaching effectiveness [37].

Goals of personalised education
Improve learning outcomes: Through personalised teaching methods and content, students can learn effectively in environments tailored to their needs, enhancing their learning outcomes and engagement. Personalised education emphasises tailored instruction, designing personalised learning plans and activities based on students' unique characteristics and needs, helping them better master knowledge and skills.

Figure 18.5   Characteristics and goals of personalised education

Foster self-directed learning: Personalised education encourages students to take initiative in their learning, fostering their self-directed learning and problem-solving abilities. In personalised education, students are active participants rather than passive recipients of knowledge. They need to actively engage in the learning process, explore and solve problems independently and develop critical thinking and self-management skills.

Meet individual needs: Personalised education addresses individual differences among students, aiming to meet each student's learning needs and developmental goals. Each student has unique interests, abilities and learning styles. Personalised education provides learning opportunities and environments tailored to each student's development, helping them achieve their learning goals and potential [38].

Personalised education aims to enhance student learning outcomes, foster self-directed learning abilities and meet individual needs through student-centred teaching models, highly flexible teaching methods and interactive teaching processes. The following sections will explore the limitations of current educational models and the challenges faced by personalised education [39,40].

18.2.1.2   Limitations of current educational models

Current educational models predominantly rely on traditional classroom teaching, which has significant limitations. Traditional education typically uses standardised content and methods, making it difficult to cater to individual student needs. Each student has unique interests, abilities and learning styles, but standardised teaching content and pacing overlook these individual differences, preventing some students from fully realising their potential. Students who learn quickly may feel bored, while those who learn more slowly may struggle to keep up, leading to stress and frustration. This standardised approach restricts the development of individual differences and fails to achieve truly differentiated instruction.

Traditional classroom teaching is fixed in time and place, with limited flexibility to accommodate individual student differences. It primarily involves teacher-led lectures, with students passively receiving knowledge and lacking opportunities for interaction and engagement. Teachers find it challenging to understand each student's specific needs and learning conditions comprehensively, making it difficult to provide personalised guidance and support. Furthermore, the fixed schedule and location of traditional classrooms restrict students' flexibility in learning time and space. Students who need more time for review or preparation may find the fixed schedule insufficient for their needs, while those who require flexible learning arrangements may struggle to align the fixed schedule with their personal rhythms and habits.

Limited interaction between teachers and students in traditional educational models negatively impacts teaching effectiveness. Large class sizes make it difficult for teachers to address every student's needs, resulting in some students lacking guidance and support. Additionally, opportunities for students to ask questions and participate in discussions are limited, restricting their ability to express their viewpoints and inquiries fully. This lack of interaction is detrimental to fostering students' active learning and critical thinking skills, causing them to feel isolated and neglected, which affects their learning motivation and outcomes.

In summary, traditional educational models have evident limitations in standardisation, flexibility and interaction, making it challenging to meet the modern demands for personalised and diverse education.

18.2.2   Challenges of personalised education

18.2.2.1   Role changes for teachers and students

Personalised education requires changes in the traditional roles of both teachers and students.

Role of teachers: Teachers' roles need to undergo significant changes. In personalised education, teachers are not merely drivers of knowledge but also guides and facilitators of learning [41]. Teachers need higher professional skills and adaptability to meet diverse student needs. They must possess broad subject knowledge, educational psychology, teaching methods and technological skills. Teachers should design and adjust educational content and methods based on individual differences and provide personalised guidance and support. Additionally, they should proficiently use various educational technology tools, such as online teaching platforms and data analysis software, to optimise the teaching process and improve teaching effectiveness (Figure 18.6).

Role of students: Students need to assume greater responsibility for their learning, cultivating self-directed learning and self-management skills [42]. This is challenging for students accustomed to passively receiving knowledge. Students need to learn how to plan their studies, manage their time, seek resources and assistance, and engage in self-assessment and reflection. Personalised education emphasises student autonomy and initiative, encouraging them to explore and identify problems actively, fostering innovative thinking and problem-solving abilities. This approach not only helps students master knowledge better but also prepares them for lifelong learning, laying a solid foundation for their future development (Figure 18.7) [43].

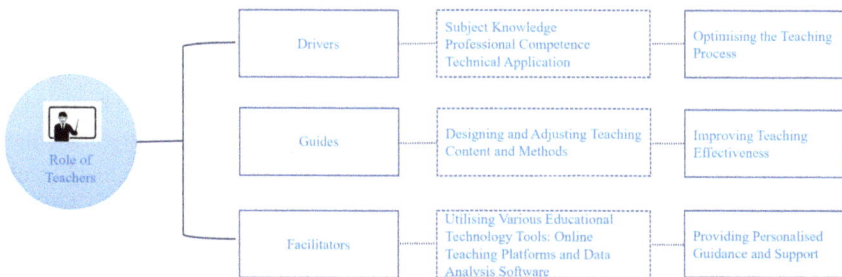

Figure 18.6   Role of teachers

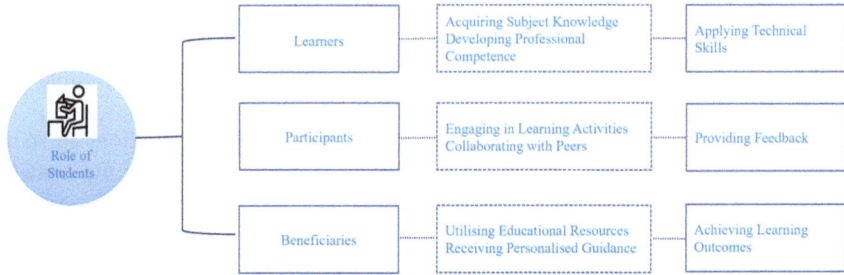

Figure 18.7　Role of students

18.2.2.2　Distribution and management of educational resources

Personalised education requires substantial educational resources, posing significant challenges in resource distribution and management [44].

Resource allocation: In traditional educational models, resource allocation is usually based on uniform standards, which cannot meet individual student needs. Personalised education requires resource allocation based on individual differences, including teacher time and effort, teaching materials, learning tools and technological support. For example, different students in the same class may need different textbooks, teaching methods and learning tools to meet their learning needs. Achieving reasonable resource allocation with limited resources is a major challenge for personalised education. Vocational education institutions need flexible resource allocation mechanisms to ensure every student receives the necessary resources [45].

Resource management: Personalised education involves a variety of resources, including teacher resources, teaching materials, technological equipment and learning platforms. Efficiently managing and coordinating these resources to ensure timely and appropriate provision to students is crucial for the success of personalised education. Vocational education institutions need advanced management tools and technologies, such as educational management systems and data analysis tools, to optimise resource allocation and utilisation. For instance, learning management systems can track students' learning progress and needs in real-time, dynamically adjusting resource allocation and teaching strategies. Additionally, institutions should strengthen cooperation with educational technology providers and other vocational education institutions to integrate resources and improve educational quality.

Teacher training and support: Personalised education places high demands on teachers, requiring deep subject knowledge and mastery of personalised teaching methods and technology tools. Vocational education institutions need to enhance teacher training and support to ensure they can meet the demands of personalised education [46]. Regular training courses can improve teachers' skills in data analysis, technology application and personalised teaching design. Institutions can also establish teacher collaboration and sharing platforms to promote experience exchange and resource sharing among teachers, collectively improving teaching standards.

Sustainable resource supply: Personalised education requires long-term resource investment, raising questions about ensuring sustained resource supply. Vocational education institutions need to plan budgets and resource allocation carefully to ensure the long-term implementation of personalised education. Institutions can also seek support from society, such as businesses and communities, to broaden resource acquisition channels. For example, partnerships with technology companies can introduce the latest educational technology equipment and tools, and collaboration with community organisations can provide more educational resources and volunteer support.

18.2.2.3   Data privacy and security issues

Personalised education relies heavily on data collection and analysis, involving substantial amounts of student personal information and learning data. Protecting the privacy and security of this data is one of the critical challenges faced by personalised education [47].

In personalised education, student learning data includes but is not limited to personal identification information, learning progress, learning habits, test scores and behaviour records. This data is extensively collected and used during the educational process to analyse students' learning conditions and provide personalised educational plans. However, the widespread use of student data also raises privacy protection risks. If data is misused or leaked, it can cause severe privacy violations for students. For example, if sensitive information such as students' grades and learning habits is leaked, it could be used for improper purposes, affecting students' personal rights. Therefore, vocational education institutions need to establish strict data privacy policies to ensure that the collection, storage and use of student data comply with relevant laws and regulations. Privacy protection measures include data anonymisation, access control and transparency in data usage, ensuring that only authorised personnel can access and use the data [48,49].

Data security involves preventing data loss, leakage and tampering during transmission and storage [50]. When implementing personalised education, vocational education institutions need to adopt advanced technical and management measures to ensure the security of student data. Data security measures include data encryption, network security protection and data backup. For instance, using encryption technology during data transmission can prevent data interception and tampering; using secure storage media and backup strategies during data storage can prevent data loss and damage. Additionally, vocational education institutions need to conduct regular security audits and vulnerability scans to identify and fix security vulnerabilities promptly, ensuring the security and reliability of data systems.

The use of student data in personalised education must comply with relevant laws and ethical standards. For example, vocational education institutions must obtain explicit consent from students or their guardians before collecting and using student data; when using data, they need to clearly inform students about the purpose and scope of data usage, ensuring transparency and compliance in data usage. Furthermore, vocational education institutions need to establish oversight mechanisms for data usage to ensure compliance with relevant policies and standards, preventing data misuse and unauthorised access.

Vocational education institutions need to strengthen data privacy and security training for teachers and relevant staff to enhance their awareness and skills in data protection. For example, regular data privacy protection training and security drills can help teachers and staff understand the basic principles and operational norms of data protection. Additionally, vocational education institutions can develop data management guidelines and operational manuals to clarify the responsibilities and processes of data protection, ensuring that data protection work is standardised and traceable.

Data privacy and security issues are significant challenges in personalised education, requiring vocational education institutions to adopt comprehensive measures in technology, management and education. By establishing strict data privacy policies, adopting advanced data security technologies, ensuring compliance in data usage and enhancing data protection education, vocational education institutions can effectively safeguard student data privacy and security, providing strong support for the smooth implementation of personalised education [51,52].

18.3 Application of intelligent technology in personalised education

18.3.1 Workflow and implementation of intelligent tutoring and personalised learning systems

18.3.1.1 Data collection and analysis

In personalised education, real-time data analysis is a crucial technology that enables a precise understanding of each student's learning status and needs by collecting and analysing their learning data. Student learning data includes, but is not limited to, attendance records, assignment completion, test scores, classroom participation and online learning behaviours. These data are gathered through various means such as online learning platforms, learning management systems (LMS), classroom interaction tools and educational applications [53].

Once collected, the data are subjected to in-depth analysis using big data analytics and machine learning algorithms to extract valuable information and patterns. For instance, by analysing students' assignment completion and test scores, it is possible to assess their knowledge mastery and learning outcomes. Similarly, by examining their online learning behaviours, one can gain insights into their study habits and interests. These analysis results assist teachers in promptly identifying the difficulties and issues students encounter in their learning process, providing targeted tutoring and support, thus enhancing the effectiveness of teaching.

18.3.1.2   Personalised learning path generation

In personalised education, generating a learning path tailored to each student is key to achieving personalised instruction. The generation of personalised learning paths relies on the construction of knowledge graphs and the application of behavioural analysis results. The knowledge graph module constructs a comprehensive knowledge graph by reflecting the relationships between subject knowledge points and mapping the students' knowledge mastery onto the graph. These knowledge graphs are constructed using knowledge graph construction tools and semantic analysis techniques.

Based on the knowledge graphs and behavioural analysis results, the personalised path generation module utilises recommendation algorithms and path planning algorithms to create personalised learning paths and learning suggestions for each student. For instance, for students struggling with specific knowledge points, the system can recommend relevant supplementary materials and practice questions to help them consolidate their knowledge. Conversely, for students demonstrating high comprehension, the system can suggest more complex and extensive learning resources to stimulate their interest and potential. By generating personalised learning paths, students can learn at their own optimal pace and manner, avoiding a 'one-size-fits-all' teaching method, thereby improving learning efficiency and outcomes.

18.3.1.3   Data-driven adjustment of teaching content and methods

Based on the analysis results of student learning data, teachers can dynamically adjust teaching content and methods to achieve data-driven personalised teaching. For example, if the analysis indicates that certain students are struggling with specific knowledge points, teachers can design supplementary teaching materials and exercises for these students to help them better understand and master the related knowledge. For students progressing quickly, teachers can provide more in-depth and expanded learning materials to stimulate their interest and potential [54].

Moreover, data-driven teaching methods can help optimise the classroom teaching process. By real-time monitoring of student participation and feedback, teachers can promptly adjust the teaching pace and interaction methods to enhance classroom interaction and engagement. For instance, if the majority of students in a class do not understand a particular knowledge point, the teacher can slow down the teaching pace, provide more explanations and examples, ensuring students can keep up with the teaching progress. Through data-driven adjustments, personalised education can better meet the individual needs of students, improving teaching effectiveness and learning experience.

18.3.1.4   Instant feedback and tutoring

In personalised education, instant feedback and tutoring are crucial for ensuring students can promptly resolve learning issues. The intelligent prompt module, utilising natural language processing technology and knowledge bases, can provide relevant prompts and explanations when students encounter learning difficulties, helping them understand and solve problems. For example, when students struggle with a specific question, the intelligent prompt module can offer problem-solving strategies and explanations of related knowledge points to help them overcome difficulties.

The exercise generation module creates personalised exercises based on students' learning progress and current level. These exercises not only target students' weak areas but also help consolidate their mastered knowledge. For instance, the system can generate more basic questions for students with lower knowledge mastery and more challenging questions for those with better mastery. Through personalised exercises, students can continually self-assess and improve, achieving better learning outcomes.

18.3.1.5   Interactive interface

The application of intelligent technology in personalised education also manifests in the design of interactive interfaces. The optimisation of student and teacher interfaces makes the display of personalised learning paths, the reception of instant tutoring and feedback more convenient and efficient.

The student interface focuses on user-friendliness and real-time updates, providing students with access to personalised learning paths, instant tutoring and feedback. Students can view their learning progress, receive personalised learning suggestions and exercises and see instant prompts and explanations through the interface. The simple and intuitive design of the interface enhances the student experience and motivates them to engage actively in their learning.

The teacher interface emphasises data visualisation and progress tracking, making it easier for teachers to view student reports, adjust teaching content and send feedback. Through the interface, teachers can access real-time data and analysis results for each student, identifying learning problems and weak areas to adjust teaching strategies accordingly. For example, teachers can use the interface to view the overall learning progress and comprehension of a class, promptly adjusting the teaching content and pace to ensure every student can keep up.

18.3.1.6   Technical support

Technical support is the cornerstone of intelligent education systems, providing a robust foundation for personalised education. The artificial intelligence engine supports the core technologies of the system, enabling machine learning and natural language processing and delivering intelligent analysis and recommendations. The implementation of these technical supports relies on deep learning frameworks and natural language processing tools.

Additionally, the data storage and management module ensure the secure storage and management of student learning data, utilising database management systems and data encryption technologies to guarantee data privacy and security. This not only protects the personal information of students but also ensures the accuracy and reliability of data analysis.

The application of intelligent technology in personalised education encompasses various aspects, from data collection and analysis, personalised learning

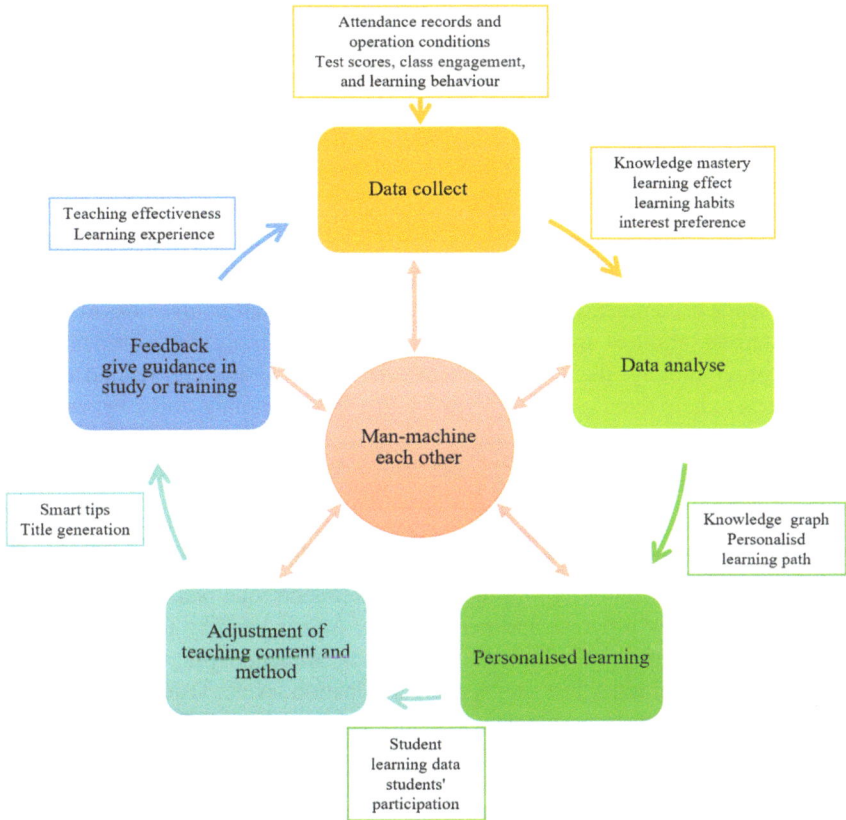

Figure 18.8   Man–machine interaction personalised learning process model

path generation and data-driven adjustment of teaching content and methods, to instant feedback and tutoring, interactive interfaces and technical support. These technological applications not only optimise vocational education resources but also improve teaching methods, providing students with a more efficient and personalised learning experience and advancing the development of personalised education (Figure 18.8) [55].

18.3.2   *Application of intelligent systems*

18.3.2.1   **Intelligent tutoring systems**

Intelligent tutoring systems are personalised learning tools based on artificial intelligence technology that provide instant learning guidance and feedback to students [56]. These systems analyse students' learning data and behaviours, identify their knowledge gaps and learning needs, and automatically generate personalised learning suggestions and practice questions. For example, if a student encounters difficulties while solving a math problem, the intelligent tutoring system can analyse the student's problem-solving steps, identify the issue and provide relevant hints and explanations to help the student gradually solve the problem. Intelligent tutoring systems not only improve learning efficiency but also reduce teachers' workload, allowing them to focus more on creative and interactive teaching activities [57].

18.3.2.2   Personalised learning path recommendation

Personalised learning path recommendation systems [58] analyse students' learning data and behaviours to customise personalised learning paths for each student. These systems recommend suitable learning resources and courses based on students' interests, abilities and learning progress. For example, for a student interested in programming, the system can recommend programming-related courses and projects to help them gradually improve their programming skills. Personalised learning path recommendation systems not only enhance students' learning efficiency but also motivate their interest and drive, making learning more targeted and effective [59,60].

18.3.2.3   Learning progress monitoring and feedback systems

Learning progress monitoring and feedback systems provide real-time monitoring of students' learning progress and performance, offering timely feedback to both teachers and students. These systems can track students' progress in courses, including completed tasks, test scores and participation levels. Teachers can use the system to understand each student's learning progress and difficulties, adjusting teaching strategies and providing support accordingly. Students can use the system to track their own learning progress and set learning goals and plans, enhancing self-management and learning capabilities. Learning progress monitoring and feedback systems enhance communication and interaction between teachers and students, increasing the transparency and effectiveness of teaching (Figure 18.9) [61,62].

18.3.3   Case studies

18.3.3.1   Practical application cases

Case 1: Chaoxing Learning App– personalised learning paths and intelligent tutoring systems

Chaoxing Learning App [63] is one of China's leading online education platforms, widely used in both higher education and vocational education. By leveraging big data analysis and artificial intelligence technology, Chaoxing Learning App provides vocational education students with personalised learning paths and intelligent tutoring, particularly in fields such as engineering technology, medical nursing and information technology, helping students enhance their skills and improve their employability.

Chaoxing Learning App uses big data analysis to comprehensively analyse students' learning behaviours, performance, interests and preferences, specifically targeting the needs of vocational education. The platform customises personalised

Intelligent Tutoring Systems	Personalised Learning Path Recommendation	Learning Progress Monitoring and Feedback Systems
Data Collection	Data Collection	Data Collection
Data Analysis	Data Analysis	Generate Feedback
Generate Personalised Learning Suggestions	Customise Learning Path	Adjust teaching strategies
Instant Tutoring & Feedback	Recommend Learning Resources	Set learning goals
Improve Learning Efficiency	Stimulate Learning interest	Enhance Communication & Interaction

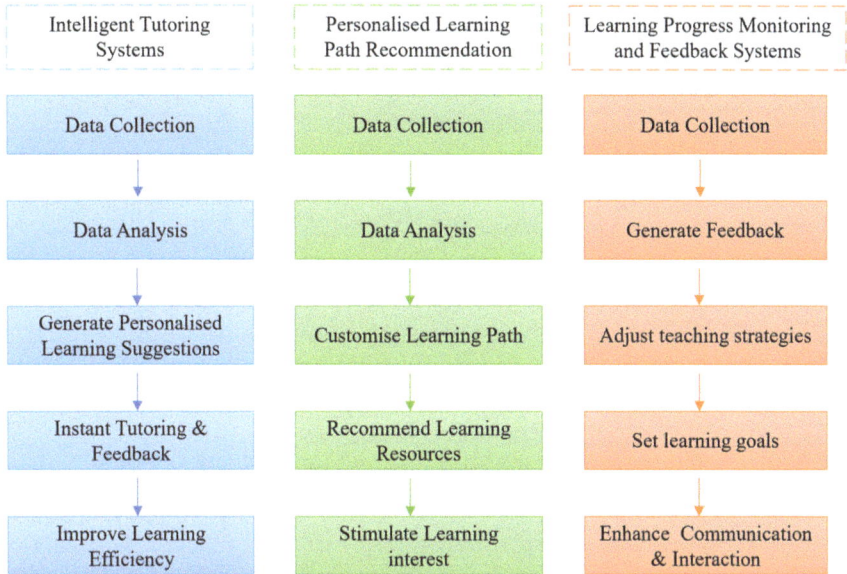

Figure 18.9   Process of intelligent tutoring system, personalised learning path recommendation system and learning progress monitoring and feedback system

learning paths based on students' foundational knowledge levels and career development goals. This includes recommending suitable vocational courses, learning resources and practical projects, helping students quickly master the required skills in areas such as electronics and electrical engineering, nursing operations and programming techniques.

The intelligent tutoring system of Chaoxing Learning App can monitor students' learning progress and performance in real-time, providing personalised tutoring suggestions based on their learning situations. For example, the system can detect if a student encounters difficulties while learning a particular skill and promptly recommend relevant supplementary materials and exercises. Additionally, the system automatically adjusts the learning plan based on students' progress, ensuring that they steadily improve their skills. The system particularly focuses on the combination of students' theoretical knowledge and practical skills, which is crucial for vocational education.

Through interactive teaching tools such as online discussions, real-time Q&A sessions and group collaborations, Chaoxing Learning App enhances students' learning experiences. The platform collects data on student interactions, analyses their engagement and learning outcomes and provides personalised feedback and guidance. Teachers can use this feedback to offer individualised tutoring to each student, helping them overcome learning challenges, particularly those related to vocational skills. The successful application of Chaoxing Learning App in

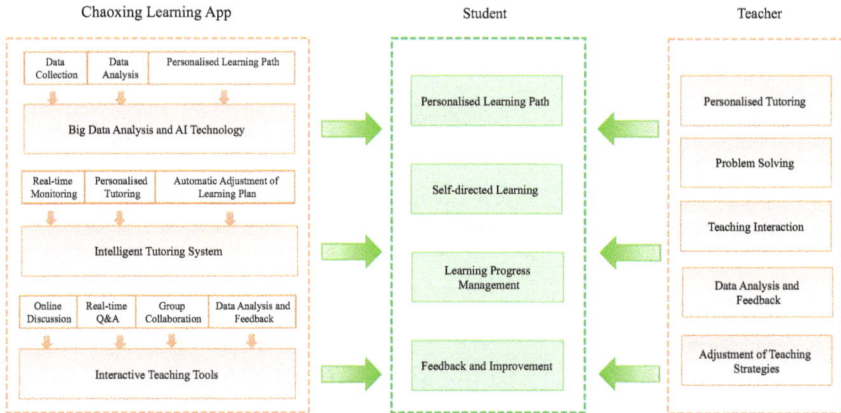

Figure 18.10　Application flowchart of Chaoxing learning app system in personalised learning

vocational education demonstrates the significant potential of big data analysis and AI technology in personalised learning paths and intelligent tutoring systems (Figure 18.10).

Case 2: XuetangX – Big data analysis and personalised learning platforms

XuetangX [64], developed by Tsinghua University, is a Chinese MOOC platform that provides online courses to a global audience. Any student with internet access can use the platform to study course videos online. XuetangX operates over 3000 high-quality courses, covering 13 major disciplines. XuetangX has been committed to developing both content and technology, promoting the sharing of educational resources and improving education quality. As of 31 March 2020, XuetangX had over 58 million users on its main site, with total course enrolments exceeding 160 million.

The platform is divided into an online learning system and a course management system. Students can freely select courses, watch lectures and participate in community discussions by registering and logging in. The system provides practice questions and scores based on their progress. Teachers can upload lecture videos, add teaching materials and exercises and use the big data analytics platform to view teaching feedback in real-time.

Students have shown significant improvements in academic performance and overall competence after using the platform. Teachers also report that intelligent technology helps them better understand students, optimising teaching strategies and improving teaching effectiveness. The success of XuetangX demonstrates the immense potential of big data analytics and intelligent systems in personalised education. Through precise data analysis and personalised learning path recommendations, student learning outcomes and engagement can be effectively enhanced, providing them with superior educational services (Figure 18.11).

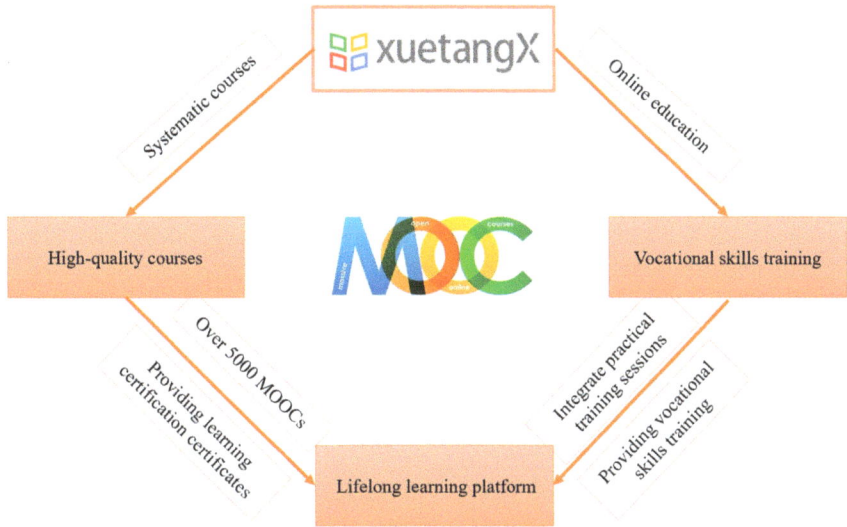

Figure 18.11   XuetangX learning platform overview

18.3.3.2   Successes and lessons learned

The study of the practical application cases of Shaoxing Learning App and XuetangX provides many successes and lessons learned.

Firstly, the application of intelligent technology needs to be closely integrated with teaching practices to ensure that the technological tools truly meet the teaching and learning needs of students. In both cases, personalised learning paths and intelligent tutoring systems not only help students better master knowledge but also provide strong support tools for teachers, enabling them to conduct teaching more efficiently.

Secondly, teachers need to receive adequate training in the use of intelligent technology, mastering the methods and techniques of using these technologies. Both Chaoxing Learning App and XuetangX place great emphasis on teacher training, ensuring that they can proficiently operate intelligent systems and apply them effectively in daily teaching. This not only enhances teachers' professional skills but also boosts their confidence and reliance on intelligent technology.

Additionally, when implementing intelligent technology, vocational education institutions need to establish comprehensive data management and security protection mechanisms to ensure the privacy and security of student data. In both cases, data privacy and security issues are given high priority. Vocational education institutions adopt advanced data encryption technologies and strict access control measures to ensure the security and privacy of student data.

Lastly, the application of intelligent technology requires continuous evaluation and improvement. Through constant feedback and adjustment, technological tools and teaching strategies can be optimised to ensure the effectiveness and sustainability of personalised education. Chaoxing Learning App and XuetangX regularly evaluate and improve their systems, continually optimising platform functions and teaching methods based on feedback from teachers and students, achieving significant results.

In summary, the cases of Chaoxing Learning App and XuetangX demonstrate the extensive application and remarkable effectiveness of intelligent technology in personalised education. Through personalised learning paths, intelligent tutoring systems and big data analysis, vocational education institutions can better meet individual student needs, improving teaching effectiveness and the student learning experience. As technology continues to develop and improve, intelligent technology will play an increasingly important role in personalised education, driving educational innovation and progress.

18.4   Optimising education resource allocation with intelligent technology

18.4.1   Current state of education resource allocation

There are significant imbalances in the current allocation of education resources, reflected in various dimensions such as geographic location, school type and subject areas [46,65].

18.4.1.1   Geographic imbalances

Geographic location is one of the critical factors influencing education resource allocation. Urban schools often receive more financial support, high-quality teachers and advanced equipment, while schools in rural and remote areas suffer from resource shortages, making it challenging to provide the same quality of education. This geographic disparity leads to unequal educational opportunities, with urban students generally enjoying better educational resources and learning environments, whereas rural students face resource shortages, outdated facilities and a lack of qualified teachers.

In many countries, the gap between urban and rural education resources is stark. Urban schools have modern teaching equipment, well-stocked libraries and diverse extracurricular activities, while rural schools often have basic infrastructure, limited teaching resources and poor learning conditions. Furthermore, urban schools can attract more high-quality teachers, whereas rural schools struggle to retain talented teachers due to their remote locations and harsh living conditions, resulting in weaker teaching staff.

18.4.1.2   Resource allocation differences between school types

There are also significant differences in resource allocation among different types of schools. Top vocational schools usually receive more funding and richer teaching resources, while regular vocational schools face resource shortages. This imbalance in resource allocation leads to substantial differences in educational quality and teaching effectiveness, affecting students' learning experiences and future prospects.

Top vocational schools often have robust financial support, enabling them to purchase advanced teaching equipment, offer diverse teaching activities and provide ample practical opportunities. These schools not only have advantages in hardware but also attract more excellent teachers and experts, offering high-quality teaching and guidance. In contrast, regular vocational schools, due to limited funds and resources, lag significantly in teaching equipment, curriculum setup and teacher quality, making it difficult to provide education services on par with top institutions.

18.4.1.3   Imbalances in resource allocation across subject areas

There are also imbalances in resource allocation across different subjects. Popular subjects tend to receive more attention and resources, while less popular subjects may struggle to develop due to a lack of support. For example, in vocational education, fields like information technology, engineering and healthcare often receive more teaching resources and funding due to high market demand, whereas traditional crafts and basic sciences receive relatively less due to uncertain employment prospects. This imbalance affects students' subject choices and future development.

Students may choose popular subjects due to their resource advantages, even if these subjects do not align with their interests and strengths. Additionally, the lack of resources in less popular subjects hinders their development and innovation, which is detrimental to the comprehensive and diverse development of education.

18.4.1.4   Uneven distribution of teaching staff

The uneven distribution of teaching staff is another significant issue in education resource allocation. Outstanding teachers tend to concentrate in top schools and urban areas, while regular schools and rural schools face shortages and variability in teacher quality. This imbalance directly impacts teaching quality and student learning outcomes.

In many regions, excellent teachers prefer urban schools and top institutions due to better salaries, working conditions and development opportunities. Rural and regular schools, due to poorer conditions, struggle to attract and retain high-quality teachers, adversely affecting their teaching quality. Furthermore, uneven opportunities for teacher training and professional development exacerbate the disparities in teaching staff. Teachers at top institutions usually have more opportunities for training and academic exchange, continuously improving their teaching levels and professional skills, while teachers at regular schools are limited by resources and opportunities, hindering their development.

18.4.1.5   Difficulty in accessing educational resources

Students' ability to access educational resources also varies due to various factors. Economically advantaged students can access more extracurricular tutoring, learning materials and practical opportunities, while economically disadvantaged students may struggle to obtain similar support due to financial pressures. Additionally, students' family backgrounds and social relationships influence their access to educational resources. For instance, students from highly educated families typically receive more learning guidance and resource support, while students from less educated families lack these advantages.

These differences in accessing educational resources further exacerbate educational inequality, affecting students' academic achievements and future

Table 18.5   Comparative analysis of educational resource allocation between urban and rural schools

Category	Description	Well-resourced urban schools	Resource-limited rural schools
Geographic location	The availability of financial support, quality teachers and advanced equipment varies significantly with location.	High financial support with access to quality teachers and modern equipment.	Limited resources, outdated facilities and a shortage of qualified teachers.
Type of school	Resource distribution varies among different types of vocational education institutions.	Top vocational schools with ample funding and resources for diverse and advanced educational activities.	Ordinary vocational schools face a shortage of resources, affecting the quality and range of education offered.
Field of study	Resource allocation differs among fields of study based on market demand and popularity.	Popular fields like IT, engineering and healthcare receive substantial support and resources.	Less popular or traditional fields suffer from limited resources, affecting development and innovation.
Teacher distribution	The quality and availability of teachers are unevenly distributed.	High-quality teachers prefer urban and top-tier schools due to better opportunities and conditions.	Rural and ordinary schools struggle to attract and retain skilled teachers due to less favourable conditions.
Access to resources	The ease of accessing educational resources is influenced by economic conditions and social background.	Students from better economic backgrounds have more access to extracurricular tutoring and materials.	Students facing economic challenges have limited support, which affects their educational opportunities and achievements.

development. To achieve educational equity, institutions need to reduce disparities in resource access through various means, ensuring that every student can enjoy high-quality educational resources fairly. The application of intelligent technology can effectively identify and address issues in resource allocation, optimising the distribution and utilisation of educational resources, and enhancing educational quality and student learning experiences (Table 18.5) [66,67].

18.4.2   Optimising resource allocation with big data analysis

18.4.2.1   Role of big data in education resource allocation

Big data technology plays a significant role in optimising education resource allocation. By collecting and analysing vast amounts of educational data, including

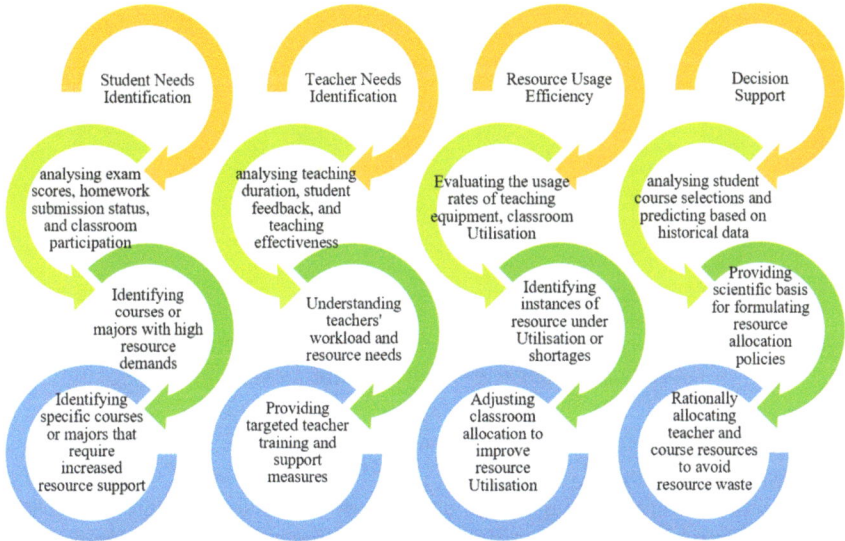

Figure 18.12   The role of big data in the allocation of educational resources

student learning data, teacher teaching data and school resource usage data, big data can reveal problems and deficiencies in resource allocation.

Big data analysis helps identify the needs of students and teachers. By analysing student learning data (e.g., exam scores, assignment submissions and classroom participation), administrators can determine which courses or subjects need more resources. For teachers, analysing teaching data (e.g., teaching hours, student feedback and teaching effectiveness) can reveal their workload and resource needs, enabling targeted support and training.

Big data provides detailed analyses of resource usage and demand. By analysing data on the usage rates of teaching equipment, classroom utilisation and library loan rates, school administrators can clearly understand actual resource usage, identifying instances of resource underuse or shortages. For example, if some classrooms are underutilised while others are consistently full, administrators can adjust classroom allocations to improve overall resource utilisation.

Additionally, big data provides a scientific basis for educational decision-making. By comprehensively analysing various educational data, administrators can formulate more reasonable resource allocation policies. For example, before a new term begins, administrators can predict and allocate resources based on student course selections and historical data, avoiding resource wastage and unmet student needs (Figure 18.12) [68,69].

18.4.2.2   Real-time monitoring and dynamic adjustment

One of the significant advantages of big data technology is its capability for real-time monitoring and dynamic adjustment. By establishing real-time data

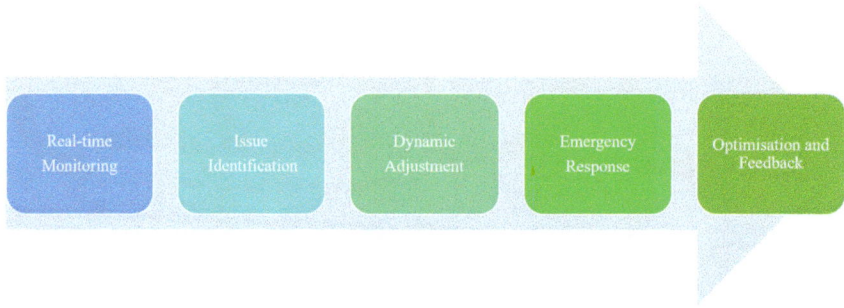

*Figure 18.13   Key steps in real-time monitoring and dynamic adjustment of
educational resources using big data technology*

monitoring systems, institutions can continuously monitor the usage of educational resources, promptly identifying problems in resource utilisation. For instance, real-time monitoring of the usage rates of classrooms and laboratories can ensure maximum resource utilisation; monitoring students' learning progress and needs in real-time allows timely adjustments in resource allocation (Figure 18.13).

Real-time monitoring enables dynamic adjustment of educational resources. Through big data analysis, systems can automatically identify changes in resource demand and make corresponding adjustments. For example, if the number of students in a particular major increase, the system can automatically allocate additional teaching resources for that major; if a course is found to be particularly challenging and students generally need more tutoring, the system can increase teacher resources and tutoring time for that course. This dynamic adjustment mechanism ensures that educational resources always match actual needs, enhancing the scientific and flexible allocation of resources.

Moreover, real-time data monitoring improves emergency response capabilities. For instance, if equipment in a classroom fails, the system can immediately notify relevant personnel for repairs and promptly adjust course schedules to ensure uninterrupted teaching activities. Through real-time monitoring and dynamic adjustment, educational resources can be efficiently managed and optimised, improving overall teaching quality and student learning experiences.

18.4.3   Intelligent systems enhance teaching efficiency

18.4.3.1   Automated management systems

Intelligent systems play a crucial role in enhancing teaching efficiency, with automated management systems being one of the key components. These systems can automate routine administrative tasks in schools, improving management efficiency and reducing human errors and delays. For example, tasks such as student attendance tracking, grade recording and teaching schedule arrangements can all be handled by automated systems. This not only alleviates the workload of teachers and administrative staff but also improves accuracy and efficiency.

Automated management systems also provide data support, aiding school administrators in making informed decisions. For instance, by analysing teaching data, the system can offer statistics on teacher workloads and optimisation suggestions for course schedules, helping administrators better plan and allocate teaching resources. The application of automated management systems promotes the intelligence and modernisation of school management, thereby enhancing overall teaching efficiency.

The Student Engagement and Attendance Tracking System (SEAtS) [70] is a smart attendance solution designed specifically for vocational education institutions. This system completes attendance tracking through student card swiping or QR code scanning, where each student is provided with a smart card containing personal information or can generate a unique QR code via a mobile application. Students simply swipe their cards or scan the QR code upon entering the classroom, and the system automatically records their attendance. Attendance data is uploaded in real-time to the school's central management system, which automatically generates attendance reports accessible to teachers and administrative staff at any time. The system also features an alert function for abnormal situations; for example, if a student misses multiple classes, the system automatically sends an alert to notify teachers and administrators to monitor the student's situation (Figure 18.14).

With the introduction of the SEAtS system, teachers no longer need to manually take attendance at the beginning of each class, as the system automatically records student attendance, significantly reducing time wastage and human error. Administrators can view and analyse attendance data in real-time through the system, allowing for more effective management and decision-making. For instance, if a student misses multiple classes, the system sends an alert,

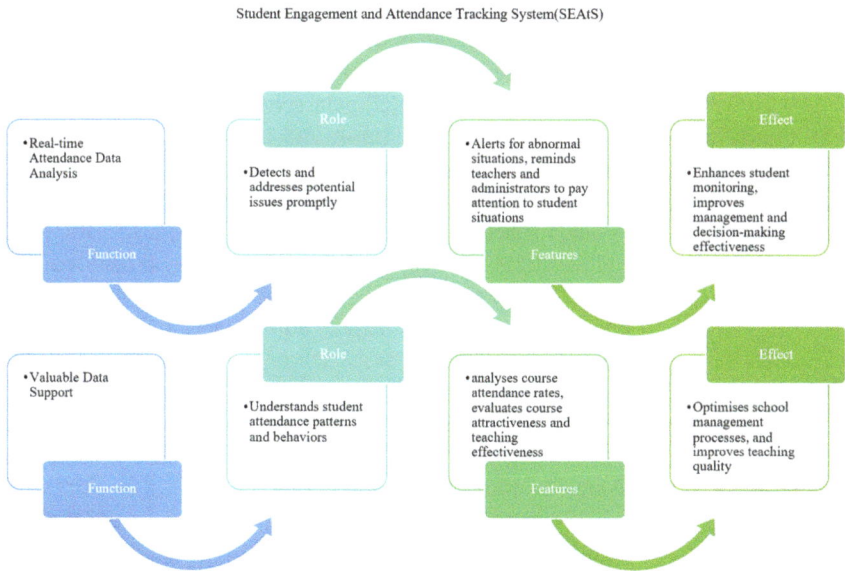

Figure 18.14   Student engagement and attendance tracking system

prompting teachers and administrators to check on the student's condition. This not only improves attendance efficiency but also better monitors students' learning status and attendance.

Furthermore, the SEAtS automated attendance system provides valuable data support for schools. By analysing attendance data, schools can understand student attendance patterns and behaviour, identifying and addressing potential issues promptly. For example, by analysing attendance rates for specific courses, schools can assess course attractiveness and teaching effectiveness, making necessary adjustments and improvements. Overall, the SEAtS system not only enhances attendance accuracy and efficiency but also optimises school management processes, improving overall teaching quality. This successful case demonstrates the immense potential and practical effects of intelligent technology in educational management, offering valuable experiences and references for other vocational education institutions.

18.4.3.2   Intelligent scheduling and resource allocation

Intelligent scheduling systems are another important application of intelligent technology in education. Traditional scheduling often requires a significant amount of manpower and time and is prone to conflicts and unreasonable arrangements. Intelligent scheduling systems use algorithms and big data analysis to automatically generate optimal course schedules, ensuring reasonable use of teachers' and students' time while avoiding schedule conflicts. These systems not only improve scheduling efficiency but also optimise resource allocation.

For instance, the system can allocate courses based on the usage of classrooms and laboratories, ensuring maximum resource utilisation. It can also assign teaching tasks based on teachers' expertise and workload, preventing overburdening teachers. Additionally, intelligent scheduling systems can flexibly adjust course schedules based on students' personalised needs, offering more choices and flexibility.

AscTimetables [71] is a widely used intelligent scheduling system in schools and universities worldwide, designed to optimise and manage course schedules. The system uses advanced algorithms and data analysis to consider factors such as classroom capacity, teacher availability and course requirements, automatically generating the optimal course schedule. Its main features include automatic scheduling, conflict detection, real-time updates, flexible adjustments and data analysis. The system can automatically detect and resolve scheduling conflicts, update scheduling information in real time, support manual adjustments and optimisation, and provide detailed data analysis reports to assist school management in decision-making and resource allocation.

AscTimetables intelligent scheduling system significantly improves scheduling efficiency and accuracy, reducing the time and workload associated with manual scheduling and maximising the utilisation of classroom and teacher resources. By using AscTimetables, schools can minimise human errors, optimise resource allocation and enhance overall teaching quality and management standards. The system has gained wide recognition globally, becoming the preferred tool for many vocational education institutions in managing course schedules.

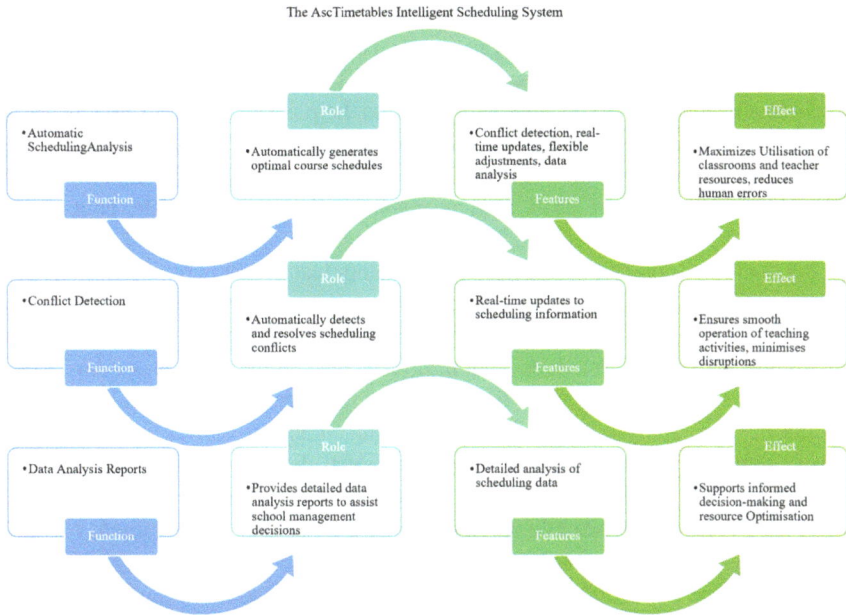

The AscTimetables Intelligent Scheduling System

| Function | Role | Features | Effect |

• Automatic SchedulingAnalysis

• Automatically generates optimal course schedules

• Conflict detection, real-time updates, flexible adjustments, data analysis

• Maximizes Utilisation of classrooms and teacher resources, reduces human errors

• Conflict Detection

• Automatically detects and resolves scheduling conflicts

• Real-time updates to scheduling information

• Ensures smooth operation of teaching activities, minimises disruptions

• Data Analysis Reports

• Provides detailed data analysis reports to assist school management decisions

• Detailed analysis of scheduling data

• Supports informed decision-making and resource Optimisation

Figure 18.15   The AscTimetables intelligent scheduling system

Moreover, the AscTimetables system excels in responding to unexpected events. When a teacher needs to take leave or laboratory equipment fails, the system can quickly respond by arranging a suitable substitute teacher or adjusting the usage schedule of the laboratory courses, ensuring the continuity and stability of teaching activities. This real-time response capability significantly enhances a school's ability to handle emergencies, ensuring the smooth conduct of teaching activities and setting a benchmark for modern vocational education institutions.

Through the application of big data analysis and intelligent systems, education resource allocation can be effectively optimised, improving teaching efficiency and resource utilisation. This not only helps address the issue of unequal education resources but also promotes the intelligent and modern development of vocational education, providing students with better educational services (Figure 18.15).

18.5   The advantages and future prospects of personalised education in the era of Industry 5.0

18.5.1   The advantages of personalised education in the era of Industry 5.0

In the era of Industry 5.0, personalised education demonstrates significant advantages, particularly in improving learning efficiency and optimising resource allocation. By leveraging advanced technologies such as big data, cloud computing and artificial intelligence, personalised education can tailor learning content and methods to each student's interests, abilities and learning styles, thereby enhancing learning outcomes and student engagement. Through intelligent systems and big data analysis, teachers can monitor student progress in real time, provide immediate feedback and adjust teaching strategies to help students overcome learning challenges. Additionally, personalised education utilises intelligent scheduling systems and learning management platforms to allocate educational resources more effectively, ensuring that each student receives the most suitable educational resources [72,73].

Personalised education excels in enhancing teacher-student interaction and data-driven decision-making. Personalised education platforms and tools facilitate interaction and communication between teachers and students, allowing teachers to better understand student needs and progress, providing targeted guidance and support. By leveraging big data and artificial intelligence, vocational education institutions can analyse student learning data, identify issues and areas for improvement and make more informed educational decisions to improve teaching quality. Moreover, personalised education can meet the needs of diverse students, including those with special educational needs, ensuring that each student receives a quality education in a suitable environment.

Personalised education promotes social and ecological value and strengthens the resilience of industrial and supply chains. By optimising educational resource allocation and improving educational efficiency, personalised education helps address imbalances in educational resources, promotes educational equity and fosters sustainable social and ecological development. Furthermore, personalised education integrates cutting-edge technologies such as intelligent teaching platforms, VR and AR into the teaching process, creating immersive and highly interactive learning environments, enriching the presentation of learning content and enhancing student perceptual experiences to improve learning outcomes. Education in the Industry 5.0 era emphasises interdisciplinary integration, encouraging students to break down academic barriers and engage in cross-disciplinary knowledge fusion, cultivating high-quality talents with international perspectives, innovative spirit and social responsibility, providing strong intellectual support and talent guarantees for sustainable societal development (Figure 18.16).

18.5.2   *Future prospects of personalised education*

In the era of Industry 5.0, personalised education demonstrates unprecedented potential and development prospects. As technology advances and educational philosophies continue to innovate, personalised education will gradually become a mainstream educational model.

1. **Technology-driven educational innovation**
   Deep Integration of Artificial Intelligence and Big Data: Artificial intelligence and big data will be more widely applied in the field of education. Through precise data analysis and intelligent algorithms, the comprehensive understanding of students' individual needs will be achieved. Educational systems

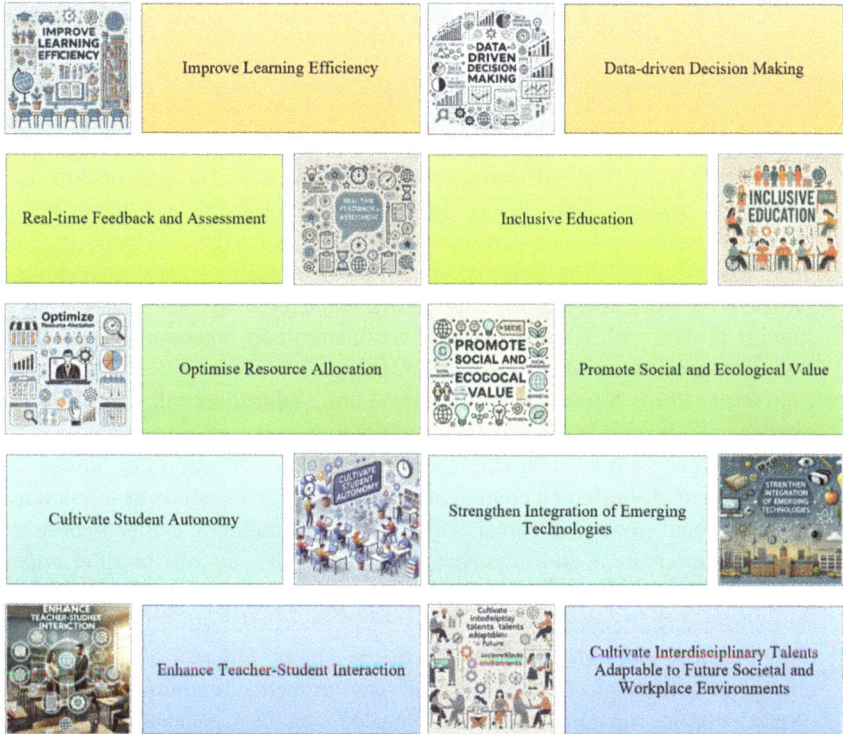

Figure 18.16   The advantages of personalised education in the era of Industry 5.0

will be able to adjust teaching content and methods in real-time, providing a more personalised and efficient learning experience. For example, intelligent tutor systems and adaptive learning platforms will become common teaching tools, helping students rapidly improve their learning outcomes under personalised guidance.

Application of Virtual Reality and Augmented Reality: VR and AR technologies will play significant roles in future education. These technologies can create immersive learning environments for students, allowing them to intuitively experience and explore complex concepts and knowledge. For instance, through virtual laboratories, students can conduct high-risk or high-cost experiments, enhancing their practical skills and depth of understanding. Through augmented reality, students can overlay digital information in real-world environments, increasing the fun and interactivity of learning.

2. **Educational equity and inclusion**

   Narrowing the Education Gap: Personalised education is expected to significantly narrow the education gap, especially in resource-poor areas. Through distance education and online learning platforms, students can overcome geographical limitations and access high-quality educational resources.

In the future, as technology becomes more widespread and infrastructure improves, more students from remote and impoverished areas will be able to enjoy the benefits of personalised education, thereby achieving educational equity.

Realisation of Diverse Learning Paths: Future education will pay more attention to individual differences and diverse needs. Through personalised education, students can choose different learning paths and development directions based on their interests and abilities. For example, some students may be more suited to academic research, while others may excel in vocational skills. With personalised learning plans, each student can find the most suitable growth path for themselves, maximising their potential.

3. **Deep integration of education and industry**

Cultivating Skills Needed for the Future: Future education will focus more on the deep integration with industries, cultivating various skills needed for the future. By cooperating with enterprises, vocational education institutions can design practically-oriented courses and projects, allowing students to learn and practice in real work environments. For instance, simulating company projects, enterprise internships and interdisciplinary collaboration will become important components of personalised education, helping students better adapt to future career challenges.

Realisation of Lifelong Learning: In the era of Industry 5.0, the rapidly changing technological and industrial environments require individuals to possess lifelong learning abilities. Future personalised education will not be limited to the school stage but will span entire careers. Through online learning platforms and intelligent learning assistants, individuals can learn and enhance their skills anytime, anywhere, keeping pace with developments in the era.

4. **Improvement of the educational ecosystem**

Sharing and Collaboration of Educational Resources: The future educational ecosystem will be more open and collaborative. Different vocational education institutions will achieve resource sharing and complementarity. By establishing educational resource libraries and online learning communities, teachers and students can easily access and share knowledge, improving the overall quality and efficiency of education.

Strengthening of Home-School Collaboration: Future personalised education will place greater emphasis on home-school collaboration. Through intelligent home-school communication platforms, parents and teachers can communicate in real-time about students' learning situations, jointly formulating and adjusting learning plans to support students' growth collaboratively.

In conclusion, personalised education in the era of Industry 5.0 has a broad future. With continuous technological advancements and the ongoing update of educational philosophies, personalised education will achieve widespread adoption and in-depth application, driving the innovation and upgrading of educational models. This will cultivate more high-quality talents who meet the needs of future society and industries. Through personalised education, we can achieve educational equity and social progress, welcoming a better future (Figure 18.17) [74,75].

18.5.3   Challenges and considerations

Personalised education holds great potential in the era of Industry 5.0, but it also faces numerous challenges. To fully realise the benefits of personalised education and provide a solid foundation for students' future development, it is essential to prepare adequately in terms of technology, ethics and support systems (Figure 18.18) [76].

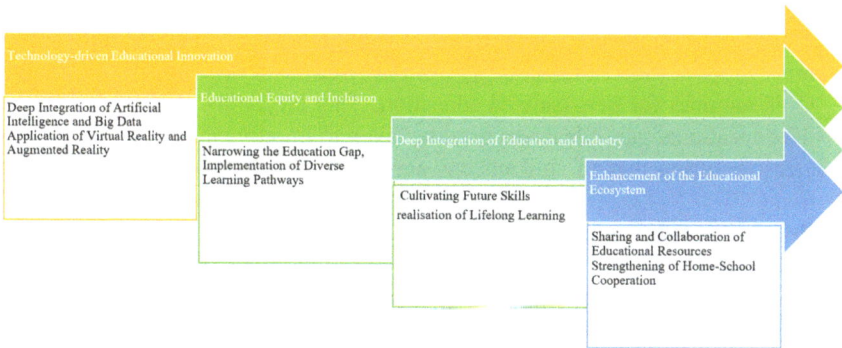

Figure 18.17   Future prospects of personalised education

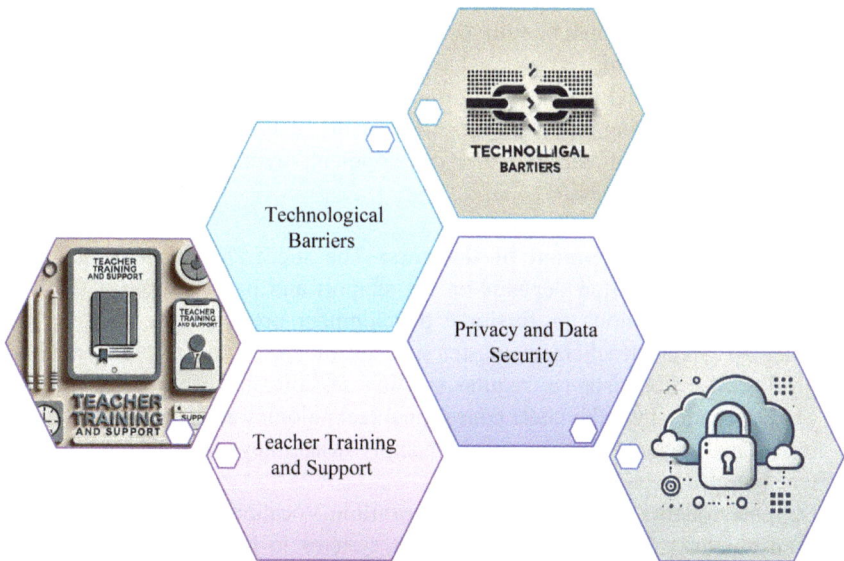

Figure 18.18   Challenges of personalised education

1. **Technological barriers**

   Difficulty in accessing advanced technology: Although personalised education has immense potential in the era of Industry 5.0, achieving this goal requires significant technological investment and resources. Many schools, especially those in economically underdeveloped areas, struggle to acquire advanced technological equipment and software. This technological barrier limits the widespread implementation of personalised education, leading to imbalances in the distribution of educational resources. Therefore, governments and vocational education institutions need to increase investment to ensure that all schools can receive the necessary technological support.

   Digital divide and infrastructure issues: The digital divide and infrastructure problems are another major challenge in realising personalised education. Schools in remote and rural areas may lack the necessary network and equipment support, preventing students from fully utilising the online resources and tools required for personalised education. This inequality further exacerbates the educational gap. Measures must be taken to improve infrastructure in these areas, ensuring that all students can equally benefit from personalised education [77].

2. **Privacy and data security**

   Protecting student data: Personalised education relies on a large amount of student data to customise teaching content and strategies. Therefore, protecting the security and privacy of this data is crucial. Vocational education institutions need to implement strict data protection measures to prevent misuse or leakage of student data. For example, encryption and anonymisation techniques can be used to ensure the security of data during transmission and storage.

   Ethical considerations in data usage: When collecting and using student data, ethical issues must be fully considered to ensure that data usage complies with moral standards. Vocational education institutions should transparently inform students and parents about the purpose and methods of data collection and usage, obtaining their consent. Additionally, data should be used primarily to promote student learning and development, avoiding any form of discrimination or bias [36,37].

3. **Teacher training and support**

   Professional development of teachers: The successful implementation of personalised education depends on the support and participation of teachers. Therefore, continuous professional development opportunities must be provided to enable teachers to master new technologies and teaching methods proficiently. For instance, regular training sessions and workshops can help teachers understand the latest educational technologies and teaching strategies, providing practical opportunities to apply what they have learned in their teaching practice.

   Support systems for effective implementation: Vocational education institutions need to establish comprehensive support systems to help teachers address the challenges encountered in implementing personalised education. This includes providing technical support, teaching resources and management systems. For example, dedicated technical support teams can be set up to offer technical assistance to teachers at any time; teaching resource libraries can be developed and shared to facilitate teachers' access and utilisation; feedback and evaluation mechanisms can be established to identify and solve problems promptly, ensuring the effective implementation of personalised education [78,79].

References

[1] Nahavandi, S. (2019). Industry 5.0—A human-centric solution. *Sustainability*, 11(16), 4371.

[2] Leong, W. Y., Chuah, J. H., and Tuan, T. B. (eds). (2020). *The Nine Pillars of Technologies for Industry 4.0*. Stevenage: Institution of Engineering and Technology.

[3] Leong, W. Y. (2022). *Human Machine Collaboration and Interaction for Smart Manufacturing: Automation, Robotics, Sensing, Artificial Intelligence, 5G, IoTs and Blockchain*. Stevenage: Institution of Engineering and Technology.

[4] Jain, V., Yie, L. W., and Teyarachakul, S. (2024). *Convergence of IoT, Blockchain, and Computational Intelligence in Smart Cities*. In R. Kumar (ed.). Boca Raton, FL: CRC Press.

[5] Özdemır, V., and Hekim, N. (2018). Birth of Industry 5.0: Making sense of big data with artificial intelligence, "the internet of things" and next-generation technology policy. *OMICS: A Journal of Integrative Biology*, 22 (1), 65–76. https://doi.org/10.1089/omi.2017.0194.

[6] Li, Y., Leong, W., and Zhang, H. (2024). YOLOv10-based real-time pedestrian detection for autonomous vehicles. In *2024 IEEE 8th International Conference on Signal and Image Processing Applications (ICSIPA)* (pp. 1–6). IEEE.

[7] Angelopoulos, A., Michailidis, E. T., Nomikos, N., *et al.* (2020). Tackling faults in the Industry 4.0 era—A survey of machine-learning solutions and key aspects. *Sensors*, 20(1), 109.

[8] Supriya, Y., Bhulakshmi, D., Bhattacharya, S., *et al.* (2024). Industry 5.0 in smart education: Concepts, applications, challenges, opportunities, and future directions. *IEEE Access*, 12, 81938–81967.

[9] Zhang, C., and Chen, Y. A. (2020). A review of research relevant to the emerging industry trends: Industry 4.0, IoT, blockchain, and business analytics. *Journal of Industrial Integration and Management*, 5(01), 165–180.

[10] Li, L. (2020). Education supply chain in the era of Industry 4.0. *Systems Research and Behavioural Science*, 37(4), 579–592.

[11] Li, Y., Zhang, H., and Liu, C. (2024). Teaching reform and practice of mechanical manufacturing and automation courses in the context of intelligent manufacturing. In *Proceedings of the 3rd International Conference on Educational Innovation and Multimedia Technology (EIMT 2024)*, Wuhan, China.

[12] Lu, Y. (2017). Industry 4.0: A survey on technologies, applications and open research issues. *Journal of Industrial Information Integration*, 6, 1–10.

[13] Aheleroff, S., Huang, H., Xu, X., and Zhong, R. Y. (2022). Toward sustainability and resilience with industry 4.0 and industry 5.0. *Frontiers in Manufacturing Technology*, 2, 951643.

[14] Ahmad, S., Umirzakova, S., Mujtaba, G., Amin, M. S., and Whangbo, T. (2023). Education 5.0: Requirements, Enabling Technologies, and Future Directions. *arXiv preprint* arXiv:2307.15846.

[15] Al-Emran, M., and Al-Sharafi, M. A. (2022). Revolutionizing education with industry 5.0: Challenges and future research agendas. *International Journal of Information Technology*, 6(3), 1–5.

[16] Zhang, H., Li, Y., and Liu, C. (2024). Application research on the construction of online teaching models in information-based environments for engineering colleges. In *Proceedings of the 3rd International Conference on Art Design and Digital Technology (ADDT 2024)*, Luoyang, China.

[17] Lantada, A. D. (2020). Engineering education 5.0: Continuously evolving engineering education. *International Journal of Engineering Education*, 36 (6), 1814–1832.

[18] Broo, D. G., Kaynak, O., and Sait, S. M. (2022). Rethinking engineering education at the age of industry 5.0. *Journal of Industrial Information Integration*, 25, 100311.

[19] Leong, W. Y., Leong, Y. Z., and Leong, W.S. (2024, July). Smart spectrum sensing and wireless coexistence technology for 6G networks. In *Fourth International Conference on Digital Signal and Computer Communications (DSCC 2024)* (vol. 13214, pp. 318–324). Bellingham, WA: SPIE.

[20] Ahmad, S., Umirzakova, S., Mujtaba, G., Amin, M. S., and Whangbo, T. (2023). Education 5.0: Requirements, Enabling Technologies, and Future Directions. *arXiv preprint* arXiv:2307.15846.

[21] Leong, W. Y., Leong, Y. Z., and Leong, W.S. (2024, July). Enhancing blockchain security. In *2024 IEEE Symposium on Wireless Technology & Applications (ISWTA)* (pp. 108–112). Piscataway, NJ: IEEE.

[22] Pereira, A. G., Lima, T. M., and Santos, F. C. (2020). Industry 4.0 and society 5.0: Opportunities and threats. *International Journal of Recent Technology and Engineering*, 8(5), 3305–3308.

[23] Alexa, L., Pîslaru, M., and Avasilcăi, S. (2022). From industry 4.0 to industry 5.0—An overview of European Union enterprises. In M. F. Silva and T. Gavrilă (eds), *Sustainability and Innovation in Manufacturing Enterprises: Indicators, Models and Assessment for Industry 5.0* (pp. 221–231). Cham: Springer.

[24] Leong, W. Y., Leong, Y. Z., and Leong, W.S. (2024, April). Nuclear technology in electronic communications. In *2024 IEEE 4th International Conference on Electronic Communications, Internet of Things and Big Data (ICEIB)* (pp. 684–689). Piscataway, NJ: IEEE.

[25] Souza, R., Ferenhof, H., and Forcellini, F. (2022). Industry 4.0 and industry 5.0 from the lean perspective. *International Journal of Management, Knowledge and Learning*, 11, 85–102.

[26] Alvarez-Aros, E. L., and Bernal-Torres, C. A. (2021). Technological competitiveness and emerging technologies in industry 4.0 and industry 5.0. *Anais da Academia Brasileira de Ciências*, 93, e20191290.

[27] Lu, Y., Zheng, H., Chand, S., *et al.* (2022). Outlook on human-centric manufacturing towards industry 5.0. *Journal of Manufacturing Systems*, 62, 612–627.

[28] Leng, J., Sha, W., Wang, B., *et al.* (2022). Industry 5.0: Prospect and retrospect. *Journal of Manufacturing Systems*, 65, 279–295.

[29] Deepa, N., Pham, Q. V., Nguyen, D. C., *et al.* (2022). A survey on blockchain for big data: Approaches, opportunities, and future directions. *Future Generation Computer Systems*, 131, 209–226.

[30] Maddikunta, P. K. R., Pham, Q. V., Prabadevi, B., *et al.* (2022). Industry 5.0: A survey on enabling technologies and potential applications. *Journal of Industrial Information Integration*, 26, 100257.

[31] Ghobakhloo, M., Iranmanesh, M., Tseng, M. L., Grybauskas, A., Stefanini, A., and Amran, A. (2023). Behind the definition of Industry 5.0: A systematic review of technologies, principles, components, and values. *Journal of Industrial and Production Engineering*, 40(6), 432–447.

[32] Tallat, R., Hawbani, A., Wang, X., *et al.* (2023). Navigating industry 5.0: A survey of key enabling technologies, trends, challenges, and opportunities. *IEEE Communications Surveys & Tutorials*, 26, 1080–1126.

[33] Zhang, P., Zhou, M., and Wang, X. (2020). An intelligent optimisation method for optimal virtual machine allocation in cloud data centers. *IEEE Transactions on Automation Science and Engineering*, 17(4), 1725–1735.

[34] Dai, W., Qiu, L., Wu, A., and Qiu, M. (2016). Cloud infrastructure resource allocation for big data applications. *IEEE Transactions on Big Data*, 4(3), 313–324.

[35] Leong, W. Y., and Kumar, R. (2023). 5G intelligent transportation systems for smart cities. In R. Kumar, V. Jain, W. Y. Leong, and S. Teyarachakul (eds), *Convergence of IoT, Blockchain, and Computational Intelligence in Smart Cities* (pp. 1–25). Boca Raton, FL: CRC Press.

[36] Bernacki, M. L., Greene, M. J., and Lobczowski, N. G. (2021). A systematic review of research on personalised learning: Personalised by whom, to what, how, and for what purpose(s)? *Educational Psychology Review*, 33(4), 1675–1715.

[37] Walkington, C., and Bernacki, M. L. (2020). Appraising research on personalised learning: Definitions, theoretical alignment, advancements, and future directions. *Journal of Research on Technology in Education*, 52(3), 235–252.

[38] Shemshack, A., and Spector, J. M. (2020). A systematic literature review of personalised learning terms. *Smart Learning Environments*, 7(1), 33.

[39]   Zhang, L., Basham, J. D., and Yang, S. (2020). Understanding the implementation of personalised learning: A research synthesis. *Educational Research Review*, 31, 100339.

[40]   Zmuda, A., Curtis, G., and Ullman, D. (2015). *Learning Personalised: The Evolution of the Contemporary Classroom*. New York: Wiley.

[41]   Bingham, A. J., Pane, J. F., Steiner, E. D., and Hamilton, L. S. (2018). Ahead of the curve: Implementation challenges in personalised learning school models. *Educational Policy*, 32(3), 454–489.

[42]   Mosier, A. D. (2018). *'Teachers' Challenges in Implementing Personalised Learning in Content Areas*, Doctoral dissertation, Walden University.

[43]   Patterson, C. (2015). Challenges faced by educators implementing programs that promote personalised learning. In M. Spector, D. Ifenthaler, D. Sampson, and P. Isaias (eds), *Competencies in Teaching, Learning and Educational Leadership in the Digital Age* (pp. 235–247). Cham: Springer.

[44]   Longo, F., Padovano, A., and Umbrella, S. (2020). Value-oriented and ethical technology engineering in industry 5.0: A human-centric perspective for the design of the factory of the future. *Applied Sciences*, 10(12), 4182.

[45]   Wan, P., Wang, X., Lin, Y., and Pang, G. (2021). A knowledge diffusion model in autonomous learning under multiple networks for personalised educational resource allocation. *IEEE Transactions on Learning Technologies*, 14(4), 430–444.

[46]   Pintilie, L. M., and Bedrule-Grigoruta, M. V. (2016). The effects of resource allocation on education system. In *ICERI2016 Proceedings* (pp. 8633–8643). Valencia: IATED.

[47]   Zhao, X. (2024). Optimisation of teaching methods and allocation of learning resources under the background of big data. *Journal of Computational Methods in Sciences and Engineering*, 24(2), 1025–1040.

[48]   Leong, W. Y., Leong, Y. Z., and Leong, W. S. (2024). Integrating SDGs education into a design thinking module. In *2024 IEEE 7th Eurasian Conference on Educational Innovation (ECEI 2024)* (pp. 26–28). IEEE.

[49]   Zhang, H. L., and Leong, W. Y. (2024). AI solutions for accessible education in underserved communities. *Journal of Innovation and Technology*, 2024 (11), 1–8.

[50]   Skitsko, V., and Osypova, O. (2022, September). Education 5.0 maturity index: Concept and prospects for development. In *International Conference on Electronic Governance with Emerging Technologies* (pp. 95–108). Cham: Springer Nature.

[51]   Leong, W. Y., Leong, Y. Z., and Leong, W.S. (2024). Engaging SDGs agenda into a design thinking module. *Educational Innovations and Emerging Technologies*, 4(2), 1–7.

[52]   Omoeva, C., Cunha, N. M., and Moussa, W. (2021). Measuring equity of education resource allocation: An output-based approach. *International Journal of Educational Development*, 87, 102492.

[53] Sun, Z., Anbarasan, M., and Praveen Kumar, D. J. C. I. (2021). Design of online intelligent English teaching platform based on artificial intelligence techniques. *Computational Intelligence*, 37(3), 1166–1180.

[54] Patel, V. L., and Dev, P. (2022). Intelligent systems in learning and education. In *Intelligent Systems in Medicine and Health: The Role of AI* (pp. 449–475). Cham: Springer International Publishing.

[55] Vinichenko, M. V., Melnichuk, A. V., and Karácsony, P. (2020). Technologies of improving the university efficiency by using artificial intelligence: Motivational aspect. *Entrepreneurship and Sustainability Issues*, 7 (4), 2696.

[56] Graesser, A. C., Conley, M. W., and Olney, A. (2012). Intelligent tutoring systems. In K. R. Harris, S. Graham, and T. Urdan (eds), *APA Educational Psychology Handbook: Vol. 3. Application to Learning and Teaching* (pp. 451–473). Washington, DC: American Psychological Association.

[57] Mousavinasab, E., Zarifsanaiey, N., Niakan Kalhori, S. R., Rakhshan, M., Keikha, L., and Ghazi Saeedi, M. (2021). Intelligent tutoring systems: A systematic review of characteristics, applications, and evaluation methods. *Interactive Learning Environments*, 29(1), 142–163.

[58] Mansouri, N., Soui, M., and Abed, M. (2023, September). Full personalised learning path recommendation: A literature review. *In International Conference on Advanced Intelligent Systems and Informatics* (pp. 185 195). Cham: Springer Nature.

[59] Lalitha, T. B., and Sreeja, P. S. (2020). Personalised self-directed learning recommendation system. *Procedia Computer Science*, 171, 583–592.

[60] Rahayu, N. W., Ferdiana, R., and Kusumawardani, S. S. (2023). A systematic review of learning path recommender systems. *Education and Information Technologies*, 28(6), 7437–7460.

[61] Barkham, M., Mellor-Clark, J., and Stiles, W. B. (2015). A CORE approach to progress monitoring and feedback: Enhancing evidence and improving practice. *Psychotherapy*, 52(4), 402.

[62] Marwan, S., Shabrina, P., Milliken, A., *et al.* (2021, November). Promoting students' progress-monitoring behaviour during block-based programming. In *Proceedings of the 21st Koli Calling International Conference on Computing Education Research* (pp. 1–10).

[63] Chaoxing. Retrieved July 15, 2024, from https://app.chaoxing.com/.

[64] Xuetang X. Retrieved July 15, 2024, from https://www.xuetangx.com/.

[65] Zhang, H., and Leong, W. Y. (2024). Industry 5.0 and education 5.0: Transforming vocational education through intelligent technology. *Journal of Innovation and Technology*, 2024(16), 1–9.

[66] Liefner, I. (2003). Funding, resource allocation, and performance in higher education systems. *Higher Education*, 46, 469–489.

[67] James, L., Pate, J., Leech, D., Martin, E., Brockmeier, L., and Dees, E. (2011). Resource allocation patterns and student achievement. *International Journal of Educational Leadership Preparation*, 6(4), n4.

[68] Della Sala, M. R., Knoeppel, R. C., and Marion, R. (2017). Modeling the effects of educational resources on student achievement: Implications for resource allocation policies. *Education and Urban Society*, 49(2), 180–202.

[69] Cheng, P., Ming, D., Man, X., and Dai, D. (2021). Optimised allocation of tennis teaching resources based on big data. *Journal of Physics: Conference Series* 1744(4), 042138.

[70] SEAtS Software. Student attendance management. Retrieved July 15, 2024, from https://www.seatssoftware.com/student-attendance-management/.

[71] ASCTimetables. Timetable software. Retrieved July 15, 2024, from https://www.asctimetables.com/.

[72] Leong, W. Y., Leong, Y. Z., and Leong, W.S. (2023, October). Smart manufacturing technology for environmental, social, and governance (ESG) sustainability. In *2023 IEEE 5th Eurasia Conference on IOT, Communication and Engineering (ECICE)* (pp. 1–6). Piscataway, NJ: IEEE.

[73] Rane, N., Choudhary, S., and Rane, J. (2024). Education 4.0 and 5.0: Integrating artificial intelligence (AI) for personalised and adaptive learning. *Journal of Artificial Intelligence and Robotics*, 1(1), 29–43.

[74] Yousuf, M., and Wahid, A. (2021, November). The role of artificial intelligence in education: Current trends and future prospects. In *2021 International Conference on Information Science and Communications Technologies (ICISCT)* (pp. 1–7). Piscataway, NJ: IEEE.

[75] Xiong, Z., Li, H., Liu, Z., *et al.* (2024). A review of data mining in personalised education: Current trends and future prospects. *Frontiers of Digital Education*, 1(1), 26–50.

[76] Alam, M., and Hasan, M. (2024). Applications and future prospects of artificial intelligence in education. *International Journal of Humanities & Social Science Studies*, 10, 209–218.

[77] Yu, S., and Lu, Y. (2021). Prospects and reflections: Looking into the future. In S. Yu and Y. Lu (eds), *An Introduction to Artificial Intelligence in Education* (pp. 189–198). Singapore: Springer.

[78] Walkington, C., and Bernacki, M. L. (2014). Motivating students by "personalizing" learning around individual interests: A consideration of theory, design, and implementation issues. *Motivational Interventions*, 18, 139–176.

[79] Qushem, U. B., Christopoulos, A., Oyelere, S. S., Ogata, H., and Laakso, M. J. (2021). Multimodal technologies in precision education: Providing new opportunities or adding more challenges? *Education Sciences*, 11(7), 338.

Chapter 19

Employment status analysis and response strategies of students majoring in mechanical manufacturing and automation in vocational colleges under the background of Industry 5.0

Yan Li[1,2] and Wai Yie Leong[2]

In the era of Industry 5.0, vocational education in mechanical manufacturing and automation has become crucial. This study examines key factors affecting students' employment prospects in these fields and offers strategies to address challenges. It highlights integrating theoretical and practical knowledge, aligning with industry needs and incorporating advanced technologies like intelligent manufacturing and robotics [1]. To meet evolving job market demands, the study advocates a dual-teacher model with academic and industry experts, emphasising the development of comprehensive skills such as innovation [2], teamwork and project management. Additionally, it stresses the importance of entrepreneurship education and industry training in enhancing employability and adaptability. The findings provide insights for educational institutions and enterprises to improve vocational education quality and meet modern manufacturing needs collaboratively. The research framework is shown in Figure 19.1.

19.1 Research background and significance

Industry 5.0 represents the latest phase of industrial development [2], emphasising not only the application of intelligent manufacturing and automation technologies but also focusing on social and environmental sustainability and human–machine collaboration [3]. This trend places new demands on production models, technology applications and the enhancement of quality and efficiency in manufacturing.

Vocational education in mechanical manufacturing and automation has become essential in this context. Vocational colleges are crucial in equipping students with modern manufacturing technologies and automation control skills, serving as a significant industry talent source. As Industry 5.0 progresses, the

[1]Department of Mechanical Manufacturing and Automation, Heilongjiang Institute of Construction Technology, China
[2]Faculty of Engineering and Quantity Surveying, INTI International University, Malaysia

Industry 5.0 Background: Analysis of Employment Situation for Vocational College Students in Mechanical Manufacturing and Automation

1 Research Background and Significance

1. Background
- 1.1.1 Definition of Industry 5.0
- 1.1.2 Transition from Industry 1.0 to Industry 5.0
- 1.1.3 Significant role of higher vocational education in industrial development

1.2 Purpose and Significance
- Understand market demand
- Evaluate current employment situation
- Identify skills gap
- Promote school-enterprise co-operation
- Push for educational reforms
- Guide career planning
- Address societal needs

2 Literature Review

2.1 Review: Research on Industry 5.0
- 2.1.1 Core concepts of Industry 5.0
- 2.1.2 Industry 5.0 - Human-centric Intelligent Manufacturing
- 2.1.3 Current State of Industry 5.0 Development Studies and in the EU

2.2 Development Stages of Higher Vocational Education
- 2.2.1 Early stage
- 2.2.2 Mobile stage
- 2.2.3 Modern stage

2.3 Impact of Industry 5.0 on Employment for Vocational College Students in Mechanical Manufacturing and Automation
- 2.3.1 Employment Rate Statistics for Vocational College Students in Mechanical Manufacturing and Automation under the Industry 5.0 Background
- 2.3.2 Employment Satisfaction Survey

3. Analysis of Employment Market Demand under the Background of Industry 5.0

3.1 Impact of Industry 5.0 on Manufacturing
- 3.1.1 Transition in production models
- 3.1.2 Expansion of technological applications
- 3.1.1 Quality and efficiency improvement

3.2 Changes in Employment Positions Brought by New Technologies in Mechanical Manufacturing and Automation
- 3.2.1 Job Demand in Traditional Mechanical Manufacturing Enterprises
- 3.2.2 Emerging Positions and Their Requirements
- 3.2.3 Analysis of Changes in Skill Requirements

4. Analysis of Employability for Vocational College Students in Mechanical Manufacturing and Automation

4.1 Skill Requirements for Emerging Positions in the Context of Industry 5.0
- 4.1.1 Insufficient integration of theory and practice
- 4.1.2 Inadequate mastery of emerging technologies
- 4.1.3 Lack of comprehensive abilities and soft skills

4.2 Analysis of the Differences Between the Employment Competencies of Vocational College Students in Mechanical Manufacturing and Automation and the Needs of Enterprises
- 4.2.1 Insufficient Mastery of Theory and Practice
- 4.2.2 Insufficient Mastery of Emerging Technologies
- 4.2.3 Insufficient Mastery of Emerging Technologies
- 4.2.4 Mismatch Between Enterprise-Specific Skill Requirements and Graduate Skills
- 4.2.5 Regional and Industry Differences

5. Research on Educational Reform for Vocational Colleges in Mechanical Manufacturing and Automation

5.1 Current Education and Teaching Reform Measures
- 5.1.1 Strengthening School-Enterprise Cooperation
- 5.1.2 Updating Course Content and Teaching Methods
- 5.1.3 Strengthening Training and Practical Teaching
- 5.1.4 Enhancing the Quality of the Teaching Staff
- 5.1.5 Improving Vocational Qualification Certification
- 5.1.6 Encouraging Comprehensive Skills Development

5.2 Practical Teaching Reforms in Smart Manufacturing College of Industry
- 5.2.1 Optimizing the Professional Curriculum Design for Mutual Development
- 5.2.2 Practice Teaching Management
- 5.2.3 Innovation and Entrepreneurship Education
- 5.2.4 Dual-Teacher Teaching
- 5.2.5 "Industry" Training and Certification
- 5.2.6 Employment System
- 5.2.7 Quality Assurance System

6. Conclusions and Prospects

6.1 Research Conclusions

6.2 Future Research Directions
- 6.2.1 Limitations of the research and improvement suggestions
- 6.2.2 The direction of future research

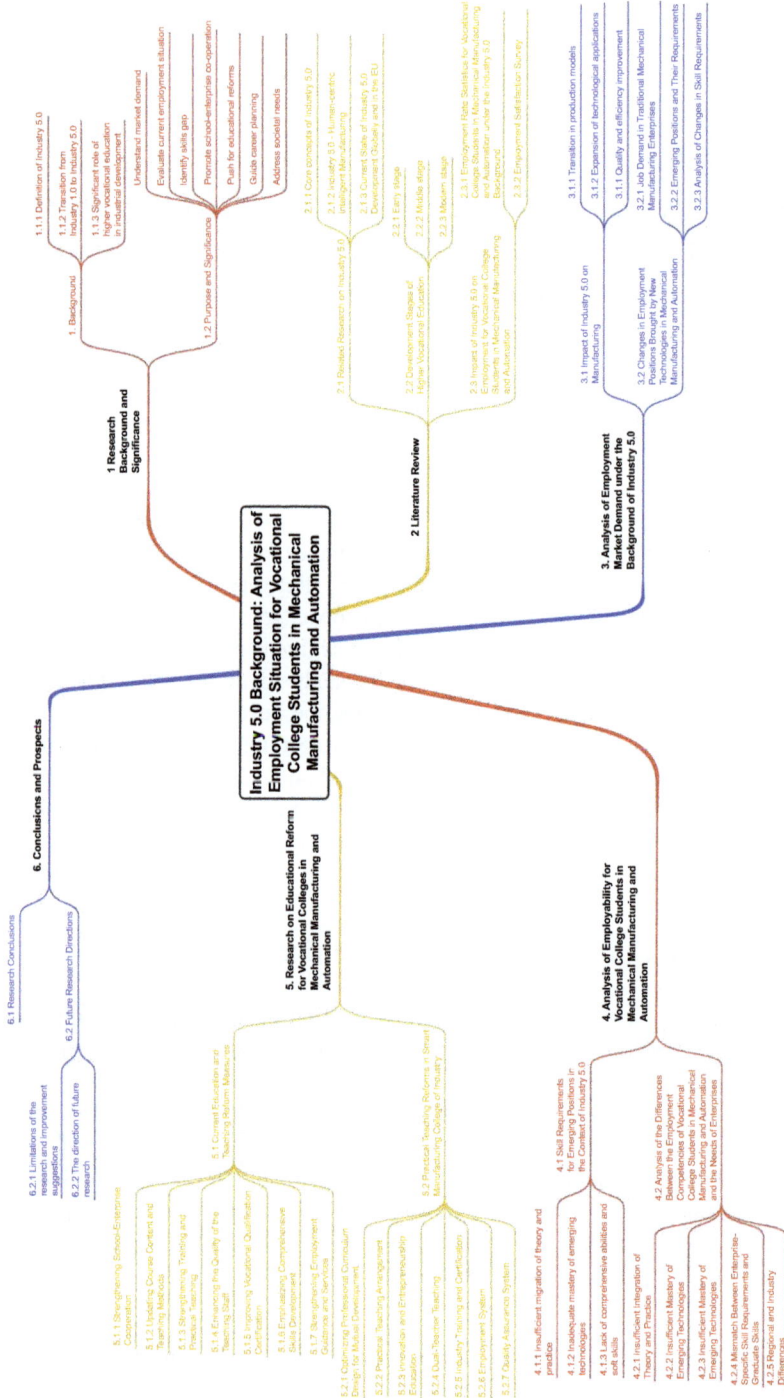

Figure 19.1   Research framework

changing demands of enterprises for talent present new challenges for vocational education institutions [4,5].

19.1.1   Research background

19.1.1.1   Definition of Industrial 5.0

Industry 5.0, introduced by the European Commission in 2021, marks the 'Fifth Industrial Revolution'. It represents a new industrial concept emphasising human–machine collaboration, sustainability and personalised production. By integrating human creativity with intelligent systems [6], it aims to comprehensively upgrade manufacturing, reshaping its future and profoundly influencing the global economy.

19.1.1.2   Definition of Industrial 5.0

From Industry 1.0 to Industry 5.0, the industrial sector has transformed mechanisation, electrification and automation into intelligence and human–machine collaboration. Each revolution has driven production efficiency and flexibility improvements, gradually focusing on environmental sustainability (Figure 19.2) [7].

19.1.1.3   Significant role of vocational education in industrial development

Vocational education plays a crucial role in industrial development in several aspects: enhancing practical skills, promoting technological innovation, improving employment opportunities, boosting enterprise competitiveness, facilitating regional economic development, and enhancing social recognition (Figure 19.3). These contributions underscore vocational education's pivotal role in driving industrial progress and societal development.

Vocational education plays a critical role in industrial development by training technical workers capable of operating and managing intelligent manufacturing equipment to meet enterprise demands [8]. These workers possess theoretical and practical skills, becoming the backbone of intelligent manufacturing enterprises. Vocational colleges closely collaborate with businesses, promptly updating curriculum content to ensure students master cutting-edge technologies, thereby driving the rapid application of smart manufacturing technologies [9].

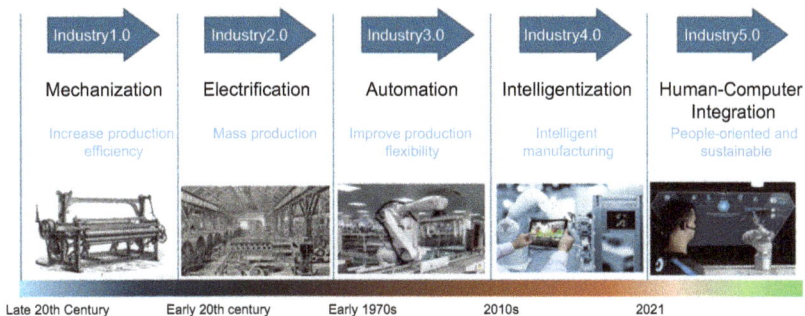

Figure 19.2   The transition from Industry 1.0 to Industry 5.0

Figure 19.3   The important role of vocational education in industrial development

In the context of industrial transformation and upgrading, vocational education is essential for cultivating the technical talents required for smart manufacturing, supporting the transition of traditional manufacturing to intelligent and digital industries and enhancing enterprise competitiveness [10]. Vocational education emphasises practical skills and job adaptability, enabling graduates to quickly integrate into enterprises, thus possessing strong employment competitiveness [11].

Additionally, vocational education promotes regional economic development by training local technical talents, optimising industrial structures and providing talent assurance for the sustainable development of local economies [12]. It also increases social cognition and emphasis on skills, driving social progress.

In summary, vocational education significantly promotes industrial development, technological innovation, industrial upgrading and regional economic growth.

19.1.2   Research purpose and significance

In the context of Industry 5.0, the employment situation for vocational college students majoring in Mechanical Manufacturing and Automation faces new opportunities and challenges [13]. To understand this phenomenon better, the research objectives of this paper include the following aspects (Figure 19.4).

Understand market demand: Analyse the demand for mechanical manufacturing and automation professionals in the intelligent manufacturing industry. Identify essential job roles and their requirements to enable vocational colleges to adjust their curriculum and teaching content, thereby training graduates who meet market needs [14].

Assess employment prospects: Evaluate the employment prospects for students in the vocational colleges' mechanical manufacturing and automation programmes by analysing employment data [15]. Helping students and parents better understand the employment situation in this field.

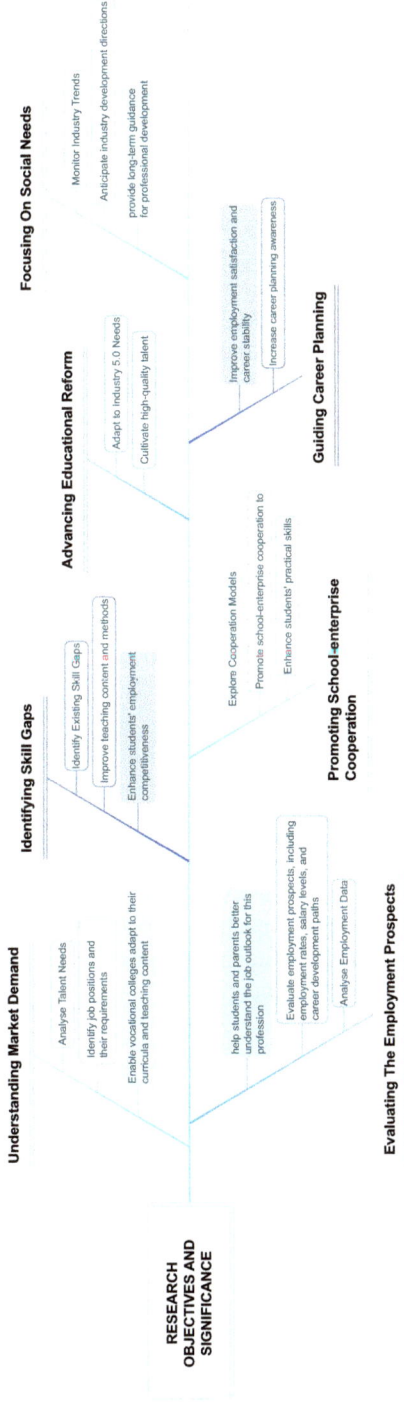

Figure 19.4 Research objectives and significance

Identify skill gaps: Investigate the specific skills required by employers for graduates and identify the skill gaps in the current teaching methods. Suggest improving teaching content and processes to enhance students' employability [16].

Promote school-enterprise cooperation: Explore models of close cooperation with enterprises through research, promoting school-enterprise collaboration [17]. Establish student internships and training bases to enhance students' practical skills and professional qualities.

Advance educational reform: Provide data support to vocational colleges to drive educational reform in mechanical manufacturing and automation programmes. Ensure the programmes better adapt to the development needs of intelligent manufacturing and Industry 5.0, cultivating high-quality talents with innovative abilities and comprehensive qualities [18].

Guide career planning: Offer career planning guidance to students, helping them understand the development trends and career paths in the intelligent manufacturing industry [19]. Enhance their career planning awareness and improve job satisfaction and career stability.

Address social needs: The research should focus on the current job market, future development trends and social needs. Anticipate the future direction of the mechanical manufacturing and automation profession to provide long-term guidance for its development [20].

19.2   Literature review

19.2.1   Related research on Industry 5.0

19.2.1.1   Core concepts of Industry 5.0

The core concepts of Industry 5.0 are built upon three fundamental pillars: 'human-centred', 'sustainability' and 'resilience' (Figure 19.5) [21]. These pillars not only define the goals and methods of Industry 5.0 but also point towards the future direction of manufacturing. Human–machine collaboration emphasises the close cooperation between humans and intelligent machines, leveraging their strengths to achieve more efficient and flexible production.

Figure 19.5   Core concepts of Industry 5.0

Human-centred: A new era of human–machine collaboration

The principle of 'human-centred' in Industry 5.0 aims to reshape the relationship between humans and machines [22], achieving harmonious collaboration between the two. This concept recognises that while robots, artificial intelligence and other digital technologies can significantly improve efficiency and precision, human creativity, intuition and sensitivity are equally indispensable [23]. By respecting and harnessing human value, we can enhance production quality and efficiency while creating a more humane working environment [3].

Sustainability: Towards green and sustainable production

Industry 5.0 focuses on environmental protection and resource conservation, striving to find the optimal balance between economic growth and ecological balance [24]. Strategies include adopting renewable energy, promoting recycling technologies and reducing carbon dioxide emissions and industrial waste through technological innovation [25]. These measures aim to minimise environmental impact, ensuring that our production activities leave a healthier planet for future generations.

Resilience: Enhancing the ability to respond to challenges

Resilience refers to the ability to quickly adapt and recover when facing unforeseen challenges, such as pandemics and natural disasters. This includes using advanced digital technologies and data analysis to predict potential risks and formulate effective response strategies. Manufacturing emphasises transparency and data sharing across the supply chain to improve system stability and responsiveness to unexpected events, enhancing the industry's competitiveness and sustainable development [26].

19.2.1.2   Industry 5.0 – Human-centred intelligent manufacturing

The first three industrial revolutions were driven by advanced technologies such as the steam engine, electric motor, and computer. Figure 19.6 illustrates the technological features that define Industry 4.0. These include advancements such as collaborative robots (cobots), artificial intelligence (AI), the Internet of Things (IoT), big data, augmented reality (AR) and virtual reality (VR), 3D printing, cloud computing, blockchain technology, and sustainable technologies. These interconnected technologies form the foundation of Industry 4.0, driving innovation, automation, and smarter industrial processes [27,28].

Industry 5.0 enhances Industry 4.0 technologies by strengthening human-robot collaboration. With Industry 5.0, the nine pillars of Industry 4.0 are expanded, placing human creativity and well-being at the forefront of the industry. It combines the speed and efficiency of machine technology with human intelligence and talent. The evolution from Industry 1.0 to Industry 5.0 represents a shift from artisanal workshops to human-centred intelligent manufacturing, as shown in Figure 19.7.

Industry 5.0 represents not only a pursuit of productivity and efficiency but also places greater emphasis on environmental protection and optimising the working environment for employees. Here are the significant benefits of implementing Industry 5.0 in manufacturing [29,30].

- Improved work efficiency and productivity: Through human–machine collaboration and digital twin technology, it simplifies manual labour, ensures

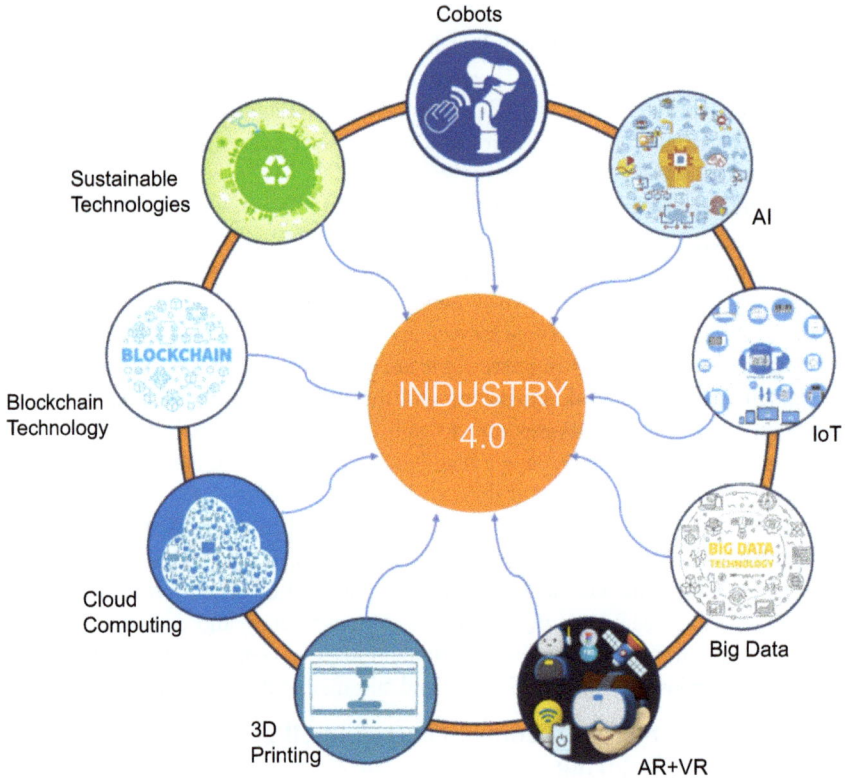

Figure 19.6   Technological features of Industry 4.0

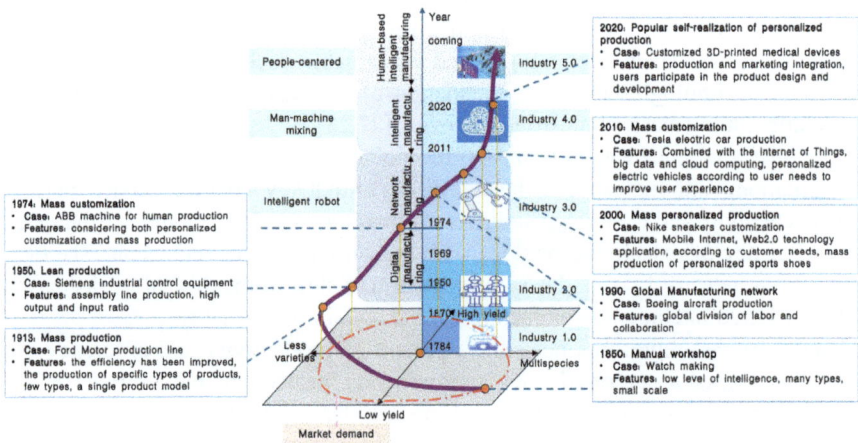

Figure 19.7   Development path from craft manufacturing to human-centred intelligent manufacturing: from Industry 1.0 to Industry 5.0

workplace safety, promotes customised production and enhances production efficiency and quality control [31].

- Comprehensive quality enhancement: Combining human creativity and robotic precision provides dual assurance for improving product quality. The deep integration of humans and machines increases production speed and accuracy, and real-time collected data is used for deep learning and prediction, significantly improving the accuracy of manufacturing processes. This advancement makes it possible to achieve defect-free production lines, fundamentally enhancing product quality [32].
- Positive contributions to the environment and society: Industry 5.0 emphasises the importance of sustainable development, contributing to environmental protection by promoting the circular economy and addressing climate change [33]. This includes using renewable energy, reducing waste, promoting reuse and reducing carbon dioxide emissions.

19.2.1.3   Current state of Industry 5.0 development in the EU

The EU is particularly advanced in Industry 5.0, promoting intelligent and personalised manufacturing [34,35] through initiatives like 'Industry 2025' and 'Horizon 2020.' On 20 March 2024, the European Commission released the second strategic plan for 'Horizon Europe' (2025−2027) [35,36], identifying key directions: green transition, digital transformation and a resilient, competitive Europe. The 'New European Bauhaus' initiative aims to achieve a green transition through inclusive, sustainable, artistic and cultural means.

With numerous intelligent manufacturing projects, EU countries lead in collaborative robots, AI, IoT and big data. Cross-national cooperation and joint R&D promote Industry 5.0 technologies across various industries [37]. Additionally, the EU enhances workforce skills through training programmes and educational projects to cultivate high-skilled talent.

In conclusion, the rapid development of Industry 5.0 in the EU underscores the importance of advanced manufacturing. Through policy support, technological innovation and enterprise transformation, Industry 5.0 is revolutionising traditional manufacturing towards intelligence, personalisation and sustainability, with the leading of the EU.

19.2.2   Development stages of vocational education

The development stages and status of vocational education vary by region and country and can be roughly divided into three stages.

Early stage: In the late 19th and early 20th centuries, with the development of the Industrial Revolution, the demand for skilled labour increased and many countries began establishing vocational colleges and training centres.

Mid-stage development: In the mid-20th century, vocational education developed in many countries. For example, the United States established a community college system offering two-year vocational and technical education. European countries like Germany developed the renowned 'dual system' vocational education, combining classroom learning with corporate internships.

Modern development: Entering the 21st century, vocational education has received global attention and development. Many countries have begun incorporating

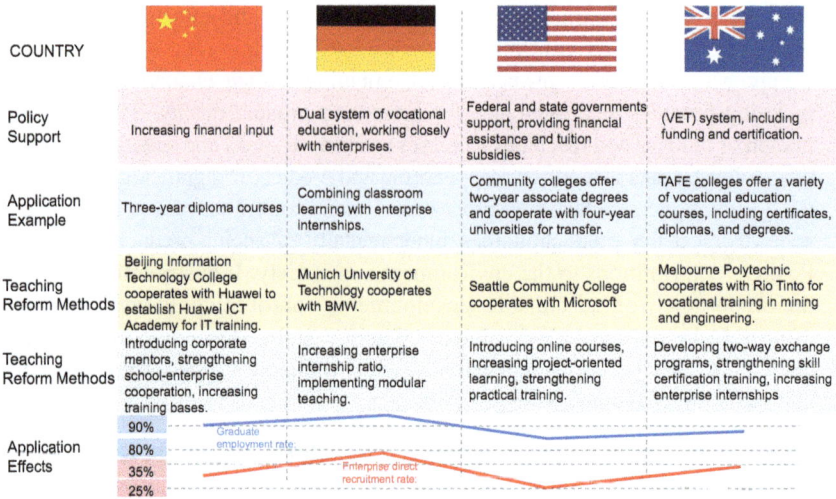

COUNTRY				
Policy Support	Increasing financial input	Dual system of vocational education, working closely with enterprises.	Federal and state governments support, providing financial assistance and tuition subsidies.	(VET) system, including funding and certification.
Application Example	Three-year diploma courses	Combining classroom learning with enterprise internships.	Community colleges offer two-year associate degrees and cooperate with four-year universities for transfer.	TAFE colleges offer a variety of vocational education courses, including certificates, diplomas, and degrees.
Teaching Reform Methods	Beijing Information Technology College cooperates with Huawei to establish Huawei ICT Academy for IT training.	Munich University of Technology cooperates with BMW.	Seattle Community College cooperates with Microsoft	Melbourne Polytechnic cooperates with Rio Tinto for vocational training in mining and engineering.
Teaching Reform Methods	Introducing corporate mentors, strengthening school-enterprise cooperation, increasing training bases.	Increasing enterprise internship ratio, implementing modular teaching.	Introducing online courses, increasing project-oriented learning, strengthening practical training.	Developing two-way exchange programs, strengthening skill certification training, increasing enterprise internships
Application Effects	90% — Graduate employment rate 80% 35% — Enterprise direct recruitment rate 25%			

Figure 19.8   Comparison of policy support, educational models, application examples, teaching reform methods and impact in various countries

vocational education into the higher education system, increasing the levels and types of vocational education, such as undergraduate and master's level vocational courses.

Currently, various countries support high-skilled talent training through policies and educational reforms to meet modern industry demands. Figure 19.8 compares countries' vocational education policies, education models, application examples, teaching reform methods and employment impacts.

19.2.3   Impact of Industry 5.0 on employment for vocational college students in mechanical manufacturing and automation

19.2.3.1   Employment rate statistics for vocational college students in mechanical manufacturing and automation under the Industry 5.0 background

Figure 19.9 shows the employment rates of vocational college graduates in mechanical manufacturing and automation over the past decade. From Figure 19.9, the following conclusions can be drawn.

Employment rate trends: From 2013 to 2023, the employment rate of vocational college graduates in mechanical manufacturing and automation showed an overall upward trend, with a slight decline in 2020 due to the COVID-19 pandemic but gradually recovering and continuing to rise after 2021 [38].

Gender differences: Overall, the employment rate of male graduates has consistently been higher than female graduates, but the gap has gradually narrowed.

Analysis of influencing factors: The upward trend in employment rates relates to economic recovery and industrial demand growth. Additionally, vocational education reform and market demand changes have driven this increase. Culturing

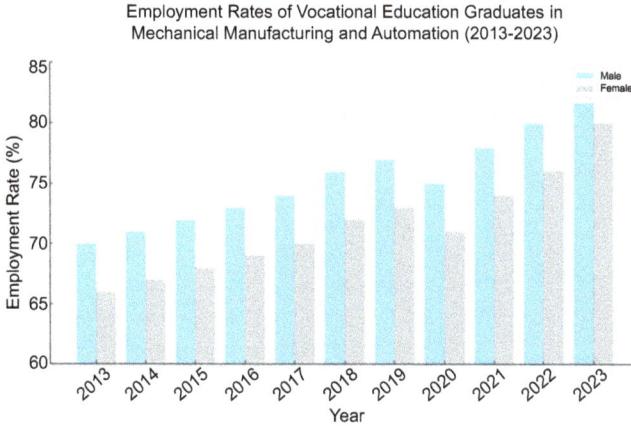

Figure 19.9   *Employment rates of vocational education graduates in mechanical manufacturing and automation (2013–23)*

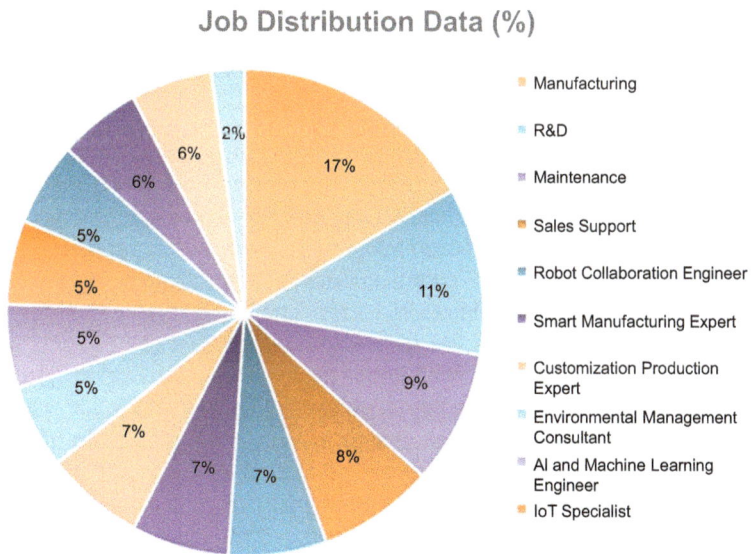

Figure 19.10   Job distribution

highly skilled talent and intelligent manufacturing development have enhanced employment opportunities for mechanical manufacturing and automation graduates.

    The increase in employment rates is due to economic recovery, industrial demand growth, vocational education reform and changes in market demand, which have led to more opportunities for mechanical manufacturing and automation graduates.

    As shown in Figure 19.10, the job demand for graduates includes traditional and emerging roles, with significant growth in emerging roles reflecting a strong

Employment Rate (%)

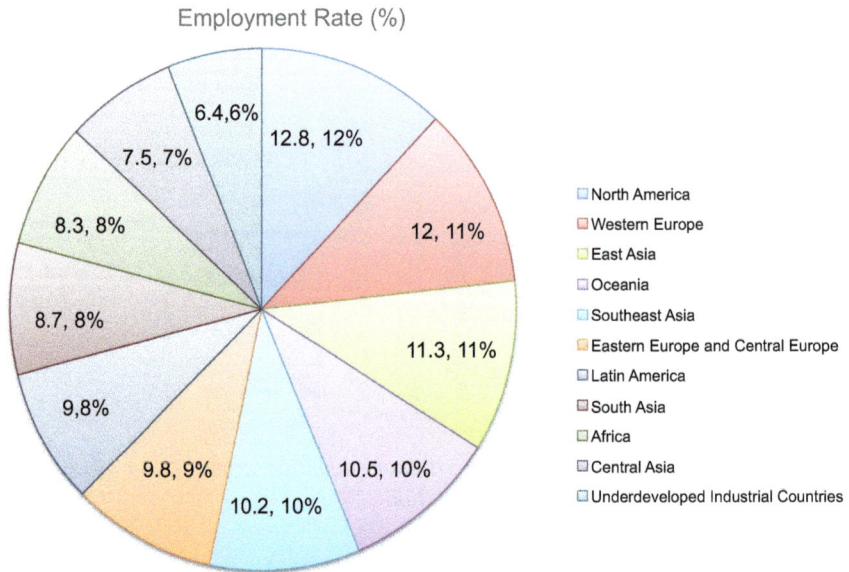

Figure 19.11   Employment region distribution analysis

demand for intelligent manufacturing and technological innovation. Vocational colleges must update their curricula and training programs to meet the changing market demands and enhance graduates' employability and career prospects.

As shown in Figure 19.11, vocational college graduates in mechanical manufacturing and automation show significant regional differences in employment situations. Developed countries have the highest employment rate, totalling 46.6%. Regions such as North America, Western Europe, East Asia and Oceania possess advanced industrial infrastructure, well-established vocational education systems and a high-skilled labour market, providing a broad employment market for graduates.

The employment situation for graduates of vocational colleges in mechanical manufacturing and automation varies with the level of industrial development. Developed countries have the highest employment rates due to their advanced industrial infrastructure and well-established vocational education systems. Emerging industrial countries follow, benefiting from rapid industrial development and high demand for skilled talent. Industrialising countries have weaker industrial bases, resulting in lower employment rates. Underdeveloped regions have the lowest employment rates due to low levels of industrialisation and underdeveloped vocational education systems. In summary, employment prospects are closely linked to each region's industrial base and economic development level.

19.2.3.2   Employment satisfaction survey

Figure 19.12 shows the employment satisfaction survey of mechanical manufacturing and automation graduates over the past five years.

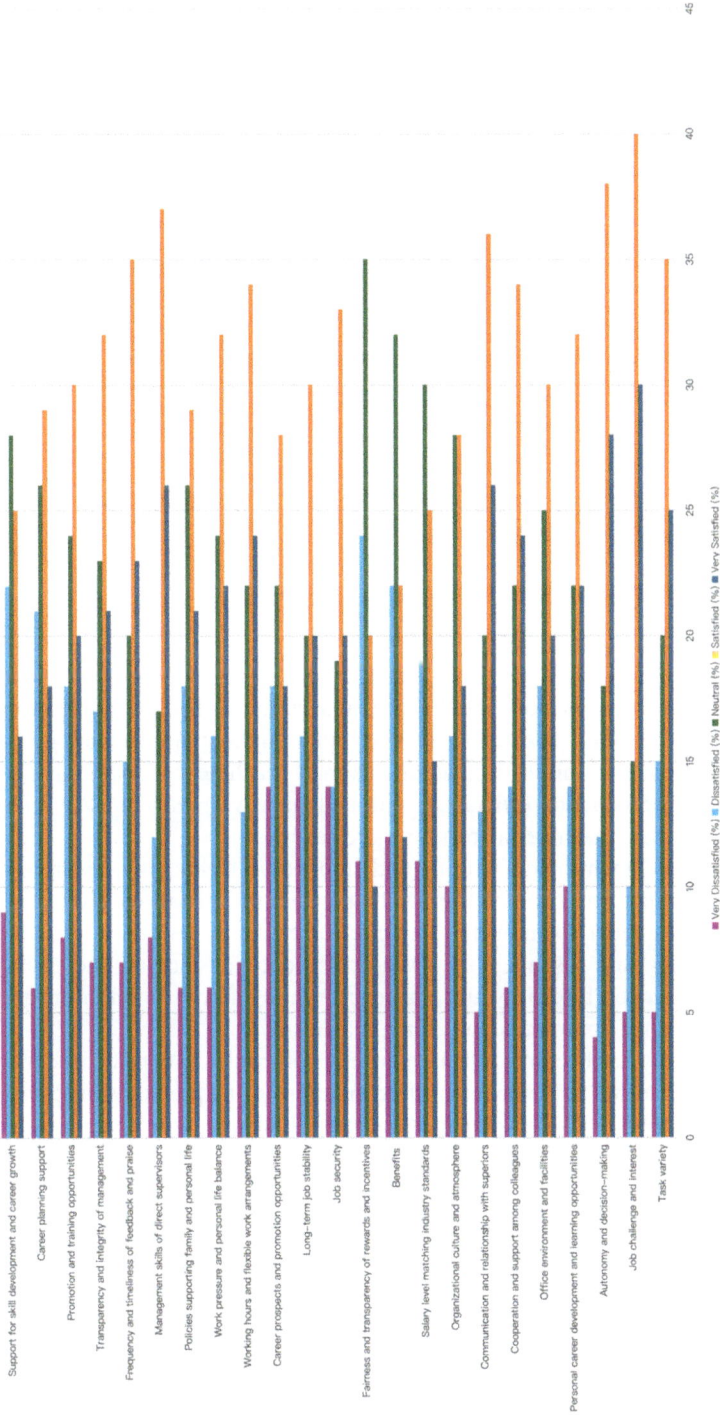

Figure 19.12 Survey categories and satisfaction levels

According to the employment satisfaction survey results, employees show different levels of job content, work environment, salary and benefits, job security and stability, work-life balance, management satisfaction and career development satisfaction [39,40].

Employees are satisfied with various tasks, challenges and autonomy but need more confidence in career development and learning opportunities. Regarding the work environment, collaboration among colleagues and communication with superiors are good, but the organisational culture needs improvement. Salary and benefits are significant dissatisfaction points, respectively. Job security and stability are decent, but satisfaction with career prospects and promotion opportunities could be higher. In management, employees are satisfied with their supervisors' skills, but the timeliness of feedback and transparency need improvement. While some areas have high satisfaction, there is a need to improve salary and benefits, organisational culture and career development support to enhance overall employee satisfaction.

19.3   Analysis of employment market demand under the background of Industry 5.0

19.3.1   Impact of Industry 5.0 on manufacturing

19.3.1.1   Transformation of production models

In the context of Industry 5.0, production models have undergone profound changes, emphasising personalisation, intelligence and sustainability. Manufacturing companies utilise flexible production systems and digital twin technology to achieve small-batch customised production. Human–machine collaboration significantly enhances production efficiency, with collaborative robots and AR technology making human–machine interaction more seamless.

The core of intelligent manufacturing and automation technology lies in IoT, big data analytics and cloud computing, leading to a fully digitalised production process. Companies adopt clean energy and circular economy models to promote green manufacturing and sustainability. Data-driven decision-making and operations improve production flexibility and efficiency, while real-time monitoring and predictive maintenance optimise supply chain management.

Network security has become critical, requiring companies to consider security from the design phase and implement real-time monitoring and protective measures to ensure system safety [41]. This demands continuous adaptation from companies to maintain competitiveness.

19.3.1.2   Expansion of technological applications

Industry 5.0 has significantly impacted manufacturing through the broad application of advanced technologies. AI and ML are used for real-time quality control and inspection, optimising production processes and reducing waste and rework [42]. IoT enables real-time monitoring and predictive maintenance of equipment and systems, reducing downtime. Cobots work alongside human workers to complete complex tasks, enhancing production flexibility. AR and VR technologies are used in employee training and production processes, providing real-time guidance and feedback. Digital twin technology creates digital models of physical assets, simulating and optimising production processes.

19.3.1.3 Quality and efficiency improvement

The integration of these technologies has significantly improved manufacturing quality and efficiency. AI and ML reduce downtime and resource waste, ensuring products meet the highest standards. IoT enhances equipment operational efficiency through real-time monitoring and predictive maintenance. Cobots improve production precision and speed while reducing labour consumption. AR and VR enhance employee training, reducing training time and costs. Digital twin technology increases the controllability and optimisation of production processes, supporting complete lifecycle management. These technologies have significantly improved manufacturing quality and efficiency, enhancing enterprises' competitiveness and market responsiveness.

19.3.2 Changes in job demand for mechanical manufacturing and automation due to technological transformation

From Industry 1.0 to Industry 5.0, the skill requirements for mechanical industry professionals have continuously evolved and advanced. During Industry 1.0, which focused on steam power and mechanical manufacturing, workers primarily performed basic tasks such as operating and maintaining machinery. Industry 2.0 introduced electricity and assembly line production, shifting job demands towards engineering design, production management and quality control. With the application of electronics and information technology, Industry 3.0 propelled the development of automated production, involving fields like automation engineering, robot maintenance and programming. Driven by the IoT, big data and artificial intelligence, Industry 4.0 expanded job roles to include smart manufacturing, data analysis and digital transformation consulting. Industry 5.0 emphasises human–machine collaboration, personalised production and sustainability [43]. The employment needs for mechanical engineering students have become more diverse, covering emerging roles such as human–machine collaboration systems engineer, sustainable manufacturing engineer and other new positions. These roles require mastery of interdisciplinary knowledge to meet the changing market demands.

Figure 19.13 illustrates the growth in demand for high-skill, high-tech positions under the background of Industry 5.0, reflecting the market's emphasis on emerging technologies and sustainable development.

The emergence of new fields like artificial intelligence, big data, IoT and blockchain has created demand for specialised roles and transformed traditional job positions. This has led to educational reforms, changes in office models and a focus on lifelong learning and career development [44,45].

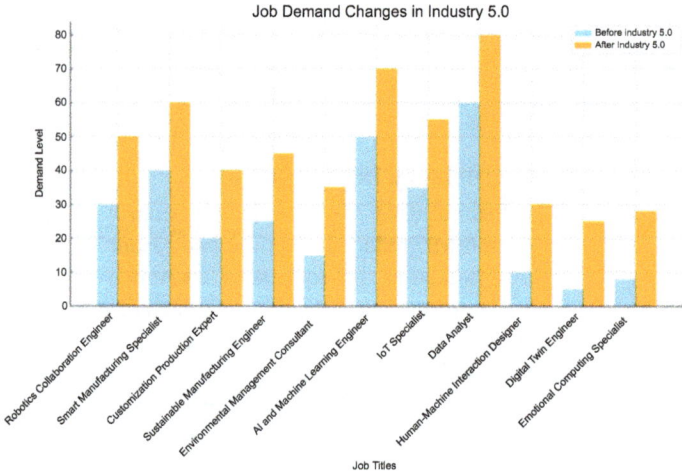

Figure 19.13   Job demand changes in Industry 5.0

Table 19.1   The job demand in traditional mechanical manufacturing

Position	Technical skills	Soft skills	Experience requirements
Mechanical engineer	Mechanical design, CAD software, manufacturing processes	Problem-solving	Mechanical design and manufacturing experience
Production line operator	Equipment operation, maintenance skills	Attention to detail	Production line operation experience, handling common equipment issues
Quality control inspector	Inspection tools, quality management systems	Patience, precision	Quality inspection experience
Purchasing agent	Supply chain management, purchasing strategies	Negotiation skills	Procurement and supply chain management experience
Maintenance Technician	Equipment repair, maintenance techniques	Troubleshooting abilities	Equipment repair experience

19.3.2.1   Job demand in traditional mechanical manufacturing enterprises

The job demand in traditional mechanical manufacturing is shown in Table 19.1.

19.3.2.2   Emerging positions and their requirements

Based on the analysis of the development trends of Industry 5.0 technology and the current manufacturing and related fields, combined with industry reports, corporate job postings, technology white papers released by tech companies and industry associations, and the views of experts and scholars published at conferences, seminars and online platforms, the emerging positions and their requirements in the context of Industry 5.0 have been summarised, as shown in Table 19.2.

19.3.2.3   Analysis of changes in skill requirements

Emerging roles emphasise intelligence, data-driven decision-making and inter-disciplinary skills compared to traditional manufacturing positions.

From Table 19.3, the emerging roles have more comprehensive technical and soft skills requirements.

Table 19.2   Emerging positions and their requirement

Position	Technical skills	Soft skills	Experience requirements
Human-robot collaboration engineer	Robot programming, AI, ML	Communication, teamwork	Automation system design, production line automation
Digital twin specialist	Digital Twin tech, 3D modelling, simulation tools	Data analysis, 3D modelling	Manufacturing or engineering experience, IoT knowledge
Predictive maintenance analyst	ML, data analysis tools	Equipment knowledge, maintenance strategy	Maintenance engineering, equipment management
Augmented reality developer	AR platforms, programming	Creative design, UI/UX design	Software or game development, industrial AR applications
Smart manufacturing system architect	Smart manufacturing systems, MES, ERP	System integration, process optimisation	Large-scale manufacturing system design, production management
Supply chain data analyst	Data analysis tools (Python, R, SQL)	Business understanding, KPI analysis	Supply chain or logistics management
Cybersecurity specialist	Cybersecurity tech, threat identification	Compliance knowledge, security strategy	Cybersecurity or information security experience

Table 19.3   Comparison of skill requirement changes between traditional mechanical manufacturing positions and emerging position

Aspect	New emerging positions	Traditional positions
Technical skills	AI, ML, IoT, AR, digital twin; data analysis, system integration	Mechanical design, CAD software, equipment operation and maintenance
Soft skills	Communication, teamwork, creative design, data analysis	Problem-solving, attention to detail, negotiation skills
Experience requirements	Experience in smart manufacturing, system integration, data analysis, IoT knowledge	Experience in mechanical design, production line operation, equipment maintenance

Enterprises must adjust their training and recruitment strategies to develop their employees' interdisciplinary capabilities and innovative thinking to adapt to new challenges and opportunities arising from technological changes. By understanding these changes in skill requirements, companies can more effectively plan their human resources to maintain a competitive edge in the future.

19.4   Analysis of employment competencies for students majoring in mechanical manufacturing and automation in vocational colleges

19.4.1   Skill requirements for emerging positions in the context of Industry 5.0

Table 19.4 highlights the essential skills for manufacturing professionals, emphasising technical expertise in CAD, automation and CNC technology, alongside critical soft skills and professional qualities such as communication, problem-solving, innovation, ethics, self-management and adaptability.

Table 19.4   Skill requirements under the background of Industry 5.0

Category	Specific analysis
Technical skills	Mechanical design and manufacturing: Proficient in using CAD software (such as AutoCAD, SolidWorks), understand basic mechanical manufacturing processes and procedures and be able to assemble and debug mechanical parts. Automation technology: Master PLC programming and applications, design and debug simple automated control systems, be familiar with the working principles and applications of standard sensors and actuators and understand basic electrical control principles. CNC technology: Able to operate CNC machines for processing, master CNC programming (such as G-code programming) and perform daily maintenance and troubleshooting of CNC equipment.
Soft skills	Communication and teamwork: Able to communicate effectively with team members, coordinate tasks, possess team spirit. Problem-solving skills: Ability to analyse and solve technical problems in the production process, quickly find solutions when encountering faults and ensure smooth production. Innovation ability: Possess certain innovative thinking, propose suggestions to improve production processes and procedures, focus on the application of new technologies and actively learn and practice new technologies.
Professional qualities	Professional ethics: Possess good professional ethics, comply with professional standards and corporate regulations and have a high sense of responsibility and dedication. Self-management ability: Able to arrange work time reasonably, improve work efficiency, possess self-learning and improvement abilities, and continuously enhance professional skills. Adaptability: Able to adapt to different work environments and job requirements, possess strong environmental adaptability and have good psychological qualities to cope with pressure and challenges.

19.4.2 Analysis of the differences between the employment competencies of vocational college students in mechanical manufacturing and automation and the needs of enterprises

There is a particular gap between the employment competencies of graduates from vocational colleges majoring in mechanical manufacturing and automation and the actual market demand. The specific differences are as follows.

19.4.2.1 Insufficient integration of theory and practice

Challenges: Vocational colleges often need to catch up in updating teaching equipment, causing a mismatch between school-taught skills and industry requirements. This results in graduates needing extra time for retraining when they enter the workforce.

Cause analysis: The delay in updating equipment and technology in vocational colleges is mainly due to funding and policy constraints. This creates a gap between educational content and real-world applications, as colleges cannot fully replicate enterprise environments and technologies.

Solution: To bridge this gap, colleges should strengthen cooperation with enterprises by sharing resources and jointly training students. Establishing joint laboratories and training centres and bringing industry lecturers to update course content regularly can help align education with industry needs

19.4.2.2 Insufficient mastery of emerging technologies

Challenges: Modern manufacturing is advancing towards intelligent manufacturing and Industry 5.0, increasing the demand for automation and digital technologies. However, vocational colleges must update curricula and equipment faster, leaving students behind in using new technologies.

Cause analysis: Updating curricula and purchasing equipment in vocational colleges involves lengthy approval and planning, hindering quick response to market changes. Limited teaching resources and faculty further restrict comprehensive coverage of emerging technologies, causing graduates' skills to lag behind market demand.

Solution: Colleges should expedite updating curricula and equipment to teach the latest technologies. They can invite industry experts for technical support and training materials. Additionally, faculty should pursue continuous professional development to stay current with industry trends and integrate them into teaching.

19.4.2.3 Lack of comprehensive skills and soft skills

Challenges: The curriculum and teaching overlook the development of students' essential skills, leading to graduates who need help to meet workplace expectations.

Cause analysis: The education system in vocational colleges emphasises technical training and needs more systematic training in soft skills [46]. Traditional teacher-centred teaching methods and tight course schedules limit students' autonomous learning and teamwork opportunities.

Solution: Colleges should incorporate courses that develop comprehensive abilities and soft skills, such as project management, innovative thinking, teamwork and communication skills. Colleges should encourage students to participate in real projects to enhance their comprehensive qualities. Simulating enterprise environments and teamwork tasks can help students practice and improve their comprehensive abilities in real-world scenarios.

19.4.2.4   Mismatch between enterprise-specific skill requirements and graduate skills

Challenges: Different companies have specific skill requirements, such as proficiency in certain software or the ability to operate specific equipment, which is challenging to cover thoroughly in the general education system of vocational colleges. This results in graduates often facing skill mismatches during job hunting, requiring targeted training after entering the corporation, thereby increasing the corporation's training costs and time.

Cause analysis: The educational content of vocational colleges is more general and cannot be customised to meet the specific needs of each corporation. Additionally, more communication between colleges and companies is necessary for colleges to promptly understand the exact skill needs of companies and adjust their curricula accordingly.

Solution: Colleges should collaborate with enterprises to tailor curricula and offer courses that align with specific skill needs. Regular surveys on enterprise requirements can help colleges promptly adjust educational content to meet market demands.

19.4.2.5   Regional and industry differences

Challenges: Talent demands vary across different regions and industries. Developed areas and high-end manufacturing industries require higher skill levels and academic qualifications. In contrast, some vocational college graduates may need more competitiveness in these areas and industries, failing to meet job requirements. On the other hand, graduates might need to concentrate more on positions with relatively lower skill requirements, leading to an oversupply [47].

Cause analysis: Differences in regional economic development levels and industrial structures lead to significant variations in skill requirements across regions and industries. Vocational colleges often need help to tailor talent training to the specific needs of each area and industry. Students' employment perceptions and regional choices also affect their employment matching.

Solution: Colleges should formulate targeted talent training programmes based on the needs of their region and primary industries. For example, they collaborate with local enterprises to understand the specific skill requirements of the area and industry and customise curricula and training programs accordingly. Colleges should also enhance employment guidance to help students understand employment prospects in different regions and industries, encouraging them to choose career paths based on market demands.

Vocational colleges should update their curricula and equipment, strengthen school-enterprise cooperation, provide customised training and adjust talent cultivation based on regional and industry needs. This will enhance the employability of graduates in mechanical manufacturing and automation, meet market demands and supply high-quality talent to enterprises and society.

19.5 Research on education and teaching reform in mechanical manufacturing and automation programmes in vocational colleges

19.5.1 Existing education and teaching reform measures

Vocational colleges have taken a series of measures in education and teaching reform in the mechanical manufacturing and automation programmes to address the gap between graduates' employment competencies and actual market demand [48]. The following details these reform measures and their implementation effects.

19.5.1.1 Strengthening school-enterprise cooperation

Vocational colleges establish long-term school-enterprise cooperation mechanisms to ensure course content aligns with industry needs. This cooperation includes jointly developing training programmes, inviting industry experts to participate in course design and teaching and regularly updating course content. Many colleges and enterprises co-establish laboratories and training centres to enhance students' practical skills, enabling them to adapt to the work environment quickly. Integrated education models, such as the 'order-based class,' address enterprise talent needs and provide stable employment opportunities for students.

19.5.1.2 Updating course content and teaching methods

Vocational colleges continuously update course content, introducing new fields such as intelligent manufacturing, Industry 4.0 and robotics technology. Project-driven teaching methods and blended learning models are adopted to improve students' comprehensive application abilities and innovation capabilities [49]. The combination of online and offline teaching modes utilises modern information technology to enhance students' hands-on skills and learning enthusiasm [50].

19.5.1.3 Strengthening training and practical teaching

High-level training bases are constructed to allow students to access advanced equipment and technology, improving their practical operational skills. Students are encouraged to intern at enterprises to understand production processes and job requirements, enhancing their employability [51].

19.5.1.4 Enhancing the quality of the teaching staff

Regularly organising industry training and technical development for teachers enhances their professional level and practical capabilities. Developing 'dual-qualified' teachers involves bringing in professionals with industry backgrounds and encouraging current teachers to obtain vocational qualification certificates, improving the overall quality of the faculty.

19.5.1.5   Improving vocational qualification certification

Incorporating vocational qualification certificate training into the curriculum ensures that students graduate with relevant certifications, boosting their employability. Integrating vocational qualification exam content into course instruction enhances students' practical skills and learning enthusiasm [52].

19.5.1.6   Perfecting vocational qualification certification

Increasing courses in project management, innovative thinking, teamwork and communication skills help cultivate students' comprehensive qualities. Organising extracurricular activities and professional competitions improves students' practical skills and innovative abilities.

19.5.1.7   Strengthening employment guidance and services

Employment guidance courses help students master career planning, resume writing and interview techniques, enhancing their employability. School-enterprise recruitment fairs and alums networks provide students with employment information and opportunities, strengthening employment support.

Vocational colleges should innovate teaching methods and enhance education quality through international cooperation to foster students' innovative and comprehensive abilities. Continuous educational reforms are crucial to cultivating high-quality technical talents, meeting market demands, promoting economic development and achieving seamless integration between education and the market.

19.5.2   *Teaching reform practises: the case of Haier smart home intelligent manufacturing industrial college*

The Industry Academy aims to cultivate high-quality technical talent and support regional economic development through industry-education integration. However, achieving full integration poses challenges due to potential conflicts between educational and corporate regulations.

Against the backdrop of Industry 4.0 and 5.0, vocational colleges across China have established intelligent manufacturing industry academies to achieve seamless integration between education and industry, cultivating highly skilled talent to meet the demands of the new era. Taking the Haier Smart Home Intelligent Manufacturing Industry Academy training model as an example, this approach emphasizes collaborative education mechanisms, comprehensive training arrangements, and a well-structured training system. By integrating resources from schools, enterprises, and students, it enhances students' professional skills while maintaining the authority of educational quality and certification (Figure 19.14). Through collaboration with enterprises to develop training programs and design demand-oriented curricula, these academies ensure that students graduate with the skills required for modern manufacturing (Figure 19.15).

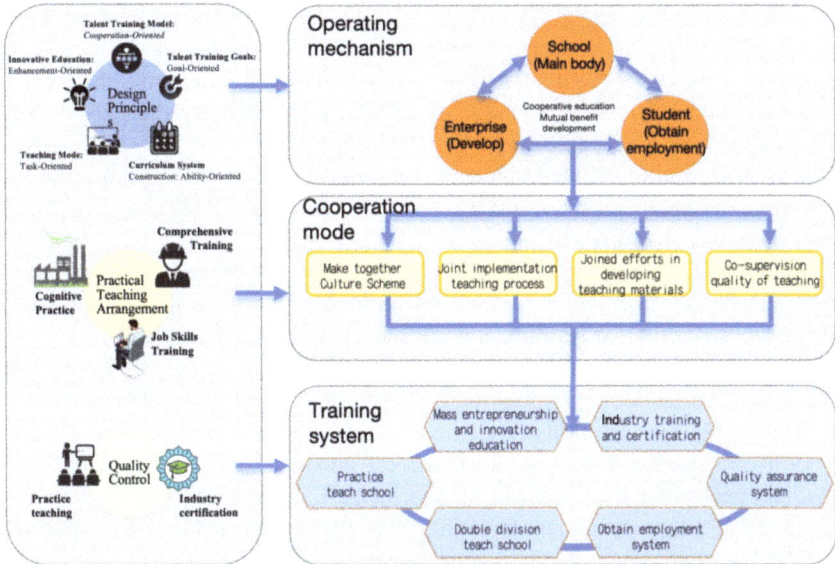

Figure 19.14   Training paradigm of Haier smart home intelligent manufacturing industrial college

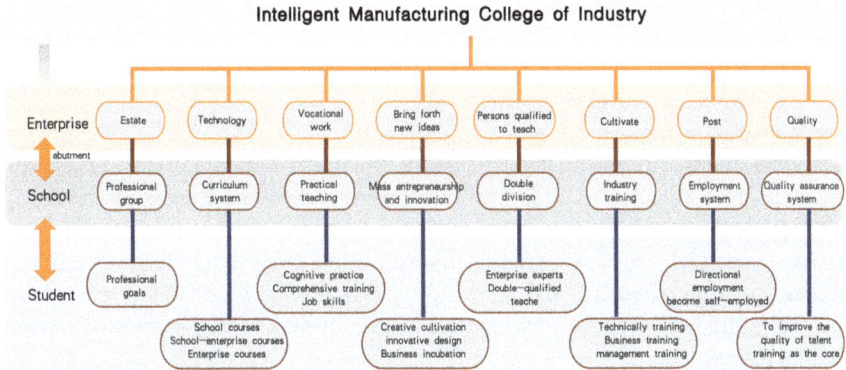

Figure 19.15   Training system of Haier Smart Home Intelligent Manufacturing Industrial College

19.5.2.1   Optimising professional curriculum design for mutual development

The design principles of the Industrial College's specialised curriculum include:

Talent training model: Cooperation-Oriented. The training model emphasises cooperation, fostering students' teamwork abilities and communication skills through collaboration and interdisciplinary cooperation.

Talent training goals: Goal-Oriented. Ensure students have clear career directions and development paths upon completing their studies [53].

Curriculum system construction: Ability-Oriented. The curriculum design focuses on cultivating students' practical abilities and enhancing their professional skills and application capabilities [54].

Teaching mode: Task-Oriented. The teaching method is task-oriented, developing students' problem-solving abilities and practical experience through actual tasks and case analysis.

Innovative education: Enhancement-Oriented. Innovative education aims to enhance students' innovation capabilities, stimulating their innovative thinking and abilities through creative learning activities and innovation projects.

The Mechanical Manufacturing and Automation programme curriculum consists of college, school-enterprise and enterprise courses to solidify theoretical foundations, master core skills and enhance comprehensive abilities.

College courses focus on fundamental theories and professional skills, including mechanical design, mechanical principles, material mechanics, automatic control, CNC technology, manufacturing processes, robotics and CAD/CAM.

School-enterprise courses integrate theory and practice through training and enterprise project research, allowing students to consolidate theoretical knowledge in practical operations and understand enterprise needs and project management.

Enterprise courses, conducted in enterprise training bases, emphasise skill enhancement and professional quality development, including advanced CNC technology, intelligent manufacturing systems training and classes on professional ethics and corporate culture, helping students quickly adapt to the enterprise environment.

Teaching adopts a dual-teacher model, with college teachers and enterprise experts instructing students. The use of project-driven and blended learning ensures theoretical depth and practical relevance. Regular evaluations of student learning outcomes are conducted to adjust course content and teaching methods promptly, ensuring continuous improvement in teaching quality.

In summary, this systematic, multi-level curriculum ensures that the Mechanical Manufacturing and Automation programme graduates possess high-level professional knowledge and practical skills to meet the demands of modern manufacturing.

19.5.2.2   Practical teaching arrangement

In the Intelligent Manufacturing Industrial College training model, practical teaching runs through the entire training process. Through cognitive practice, comprehensive training and job skills training, students can continuously improve their technical level and application ability through hands-on operations.

This model of practical teaching has the following characteristics:

Cognitive practise: Through visits to enterprises, laboratories and production workshops, students gain a preliminary understanding of professional knowledge and its application scenarios [55]. This stage enhances their perceptual knowledge of the theoretical concepts, laying a foundation for in-depth learning.

Comprehensive training: Conducted at on-campus training bases, this stage involves hands-on activities such as mechanical processing, CNC programming and robot operation [56]. A project-driven approach is used, allowing students to consolidate theoretical knowledge and develop operational skills, problem-solving abilities and teamwork spirit through completing real projects.

Job skills training: This advanced training stage occurs in an enterprise environment. Under the guidance of enterprise mentors, students engage in actual production operations and project practises, applying their acquired knowledge to practical work, thus enhancing their operational skills and professional quality while adapting to the enterprise work environment and culture.

Through these three stages of systematic practical teaching, students continuously improve their technical and application skills by combining theory and practice. Ultimately, upon graduation, they possess high-level professional knowledge and practical skills, meeting the demands of modern manufacturing.

19.5.2.3   Innovation and entrepreneurship education

In the Intelligent Manufacturing Industry Academy training model, dual-innovation education is crucial to cultivating students' innovative thinking and entrepreneurial abilities, thus enhancing their competitiveness. Dual-innovation education encompasses three stages: creativity cultivation, creative design and entrepreneurship incubation.

Creativity cultivation: Through educational activities and projects, students learn to propose innovative solutions by integrating AI, big data, IoT and other technologies, stimulating their creative consciousness.

Innovative design: Students transform initial ideas into feasible design plans, mastering skills in product design, process design and system integration. This stage also enhances their project management and teamwork abilities [57].

Entrepreneurial incubation: The academy collaborates with enterprises to provide resources and support, helping students turn design plans into actual entrepreneurial projects. This allows them to experience the entrepreneurial process and increase their chances of success.

Innovation and entrepreneurship education in the Intelligent Manufacturing Industrial College aims to enhance students' capabilities to meet Industry 5.0 demands. This education system includes three stages: idea cultivation, innovative design and entrepreneurial incubation. Students integrate AI, big data and IoT technologies to generate creative solutions, transform ideas into feasible designs and develop actual entrepreneurial projects with support from school-enterprise collaborations.

19.5.2.4   Dual-teacher teaching

The dual-teacher teaching model combines theoretical instruction by college teachers with practical guidance from enterprise experts, ensuring that students acquire solid theoretical knowledge and cutting-edge operational skills [58].

Curriculum design is collaboratively developed by colleges and enterprises, using a project-driven approach and covering foundational theories and professional skills. A multidimensional evaluation system assesses students through exams and project evaluations, with course content adjusted based on feedback. Colleges and enterprises jointly establish training bases, providing advanced equipment and natural production environments where students participate in internships under the guidance of enterprise experts. The dual-teacher teaching model enhances teaching quality, promotes school-enterprise cooperation and comprehensively improves students' employability.

19.5.2.5   Industry training and certification

In the context of Industry 5.0, vocational colleges design and implement industry training and certification systems through school-enterprise cooperation, combining technical, business and management training and utilising enterprise resources and online learning support to ensure students master the latest technologies and skills.

Training methods emphasise integrating theory and practice through classroom teaching, laboratory training and real enterprise cases, consolidating theoretical knowledge and enhancing practical skills. The project-driven teaching method, through real enterprise projects, improves students' ability to solve practical problems [59]. The blended teaching model, combining online and offline methods, utilises online platforms for learning resources while ensuring close integration of theory and practice during hands-on training. The certification system adopts the 1 +X certificate system, evaluating students' learning outcomes and skill proficiency through exams, project assessments and practical tests. In addition to completing their academic education (the '1' refers to the academic certificate), students can also obtain several vocational skill level certificates (the 'X' refers to the vocational skill level certificates). A multi-dimensional evaluation system, including theoretical knowledge tests, practical skill assessments and comprehensive quality evaluations, ensures continuous improvement in training quality and effectiveness.

In conclusion, this training and certification system, through clear training objectives, systematic training content, dual-teacher teaching models, project-driven and blended teaching methods, and a rigorous certification system, comprehensively enhances students' professional skills and competitiveness, ensuring they meet the needs of modern manufacturing [60].

19.5.2.6   Employment system

In the context of Industry 5.0, the school-enterprise cooperation employment system promotes not only student employment but also the quality of employment through two pathways: targeted jobs and entrepreneurship.

Targeted employment: Colleges and enterprises collaborate to create training plans and curricula that align with industry needs, ensuring students acquire relevant skills and knowledge. Long-term internships and exposure to enterprise culture provide practical experience, improving employment stability and alignment.

Enterprise experts offer career mentoring, guiding students in career planning and job searches.

Entrepreneurship: Colleges offer innovative courses to foster entrepreneurial thinking and skills. They help students develop ideas into projects through creative design and support from school-enterprise incubation bases, which provide space, funding and guidance. Colleges and enterprises join forces to invest in promising entrepreneurial projects.

Summary: This new model, through school-enterprise cooperation, offers diversified employment channels and dynamically adjusts curriculum content based on enterprise needs, ensuring students master the latest technologies and skills. Students acquire solid vocational abilities and competitiveness through vocational training and practical operations. Additionally, continuous career development support, including career planning, skills training and job recommendations, ensures that students continue to develop and advance in their careers, promoting talent cultivation and industry development in manufacturing.

19.5.2.7 Quality assurance system

The quality control system of the Industrial Academy is the core of the school-enterprise cooperative employment system, aiming to improve the quality of talent cultivation and ensure that students possess high-level professional knowledge and practical skills upon graduation.

Organisation of the quality control system: The college and enterprises jointly form a quality assurance committee to set standards, supervise teaching and evaluate outcomes. The committee includes college teachers, enterprise experts and industry advisors, ensuring comprehensive and professional quality control. The standards and evaluation metrics cover course content, teaching methods, practical operations and student performance, aligning with educational department requirements and enterprise needs.

Implementation of the quality control system: The college and enterprises collaborate to develop courses, dynamically adjusting content to ensure foresight and practicality. The dual-teacher teaching model enhances student capabilities through classroom instruction, laboratory training and enterprise internships. Students conduct practical operations and research under the guidance of enterprise mentors, with joint evaluations by both college and enterprise.

Effects of the quality control system implementation: Students receive comprehensive education and training through strict quality control, mastering solid theoretical knowledge and practical skills. Extensive monitoring and evaluation improve students' professional qualities and sensible abilities. Enterprises gain high-quality talent, reducing retraining time and increasing production efficiency and competitiveness. Continuous feedback and improvement mechanisms ensure the foresight and practicality of teaching quality [61].

This model improves talent development and enterprise satisfaction through rigorous standards and monitoring, ensuring sustainable school-enterprise collaboration and high-quality talent production.

19.6   Conclusion and prospect

19.6.1   Research conclusions

The advent of Industry 5.0 presents new challenges and opportunities for vocational students in mechanical manufacturing and automation. This paper analyses their employment status, identifies influencing factors and proposes strategies to enhance student quality and employability by integrating curriculum design, practical training, innovation and entrepreneurship education, dual-teacher instruction and industry training to achieve school-enterprise cooperation and industry-education integration.

1. **School-enterprise cooperation and curriculum design**: By integrating theory with practice and designing course content guided by industry needs, students are ensured to master the latest technologies and comprehensive qualities upon graduation, solidifying their theoretical foundation and enhancing core skills and abilities.
2. **Practical teaching**: Through cognitive practice, comprehensive training and job skills training, students continuously improve their technical level and application abilities through hands-on operations, mastering professional knowledge and cultivating vocational abilities and comprehensive qualities.
3. **Innovation and entrepreneurship education**: Systematic innovation and entrepreneurship education, including idea cultivation, innovative design and entrepreneurial incubation, cultivates students' creative thinking and entrepreneurial abilities, enabling them to meet the technical requirements of Industry 5.0 and play critical roles in the new industrial revolution.
4. **Dual-teacher teaching**: Combining theoretical teaching by college teachers with practical guidance from enterprise experts, this model ensures students master solid theoretical knowledge and cutting-edge operational skills. Students enhance their practical abilities and comprehensive qualities through project-driven teaching while completing real enterprise projects.
5. **Industry training**: The industry training system, including technical, business and management training, ensures students receive comprehensive vocational training during their studies. Through scientific organisation, comprehensive implementation and practical evaluation, students' overall qualities and vocational abilities are comprehensively improved, laying a solid foundation for future career development.

19.6.2   Research prospect

19.6.2.1   Limitations of the study and improvement suggestions

Although this study has achieved some results in exploring education and teaching reforms for mechanical manufacturing and automation programmes in vocational colleges under the backdrop of Industry 5.0, there are still certain limitations:

1. **Timeliness of data**: With the rapid development of technology, education and industry demands are constantly changing. The data and analysis in this

study are based on current information, requiring regular updates and re-evaluation in the future to ensure the timeliness and accuracy of the research results.

2. **Evaluation of specific implementation effects**: Although this study discusses teaching reform measures, their specific implementation effects need further exploration. Future research should conduct long-term evaluations by tracking graduates' employment status and enterprise feedback, providing more detailed data support.

19.6.2.2   Directions for future research

Interdisciplinary research: With the development of Industry 5.0, mechanical manufacturing and automation programmes require more multidisciplinary knowledge and skills. Future research can explore how to better integrate emerging technologies such as artificial intelligence, big data and the Internet of Things into the curriculum to cultivate versatile talents.

1. **Tracking student career development**: Establish a graduate career development tracking system to follow and evaluate students' career paths and achievements over the long term, providing data support and improvement suggestions to optimise talent training programmes further.
2. **Innovation in technology and teaching methods**: With the advancement of educational technology, new teaching methods and tools, such as the application of VR, augmented reality and AI in teaching, can be explored to enhance teaching effectiveness and student learning experiences.

In conclusion, by continuously deepening education and teaching reforms and promoting in-depth cooperation between vocational colleges and enterprises, more high-quality technical talents can be cultivated to meet the demands of Industry 5.0, making more significant contributions to social and economic development.

References

[1]   Wang, S., Meng, J., Xie, Y., Jiang, L., Ding, H., and Shao, X. (2023). Reference training system for intelligent manufacturing talent education: platform construction and curriculum development. *Journal of Intelligent Manufacturing*, 34(3), 1125–1164.

[2]   Leong, W. Y. (2022). *Human Machine Collaboration and Interaction for Smart Manufacturing: Automation, Robotics, Sensing, Artificial Intelligence, 5G, IoTs and Blockchain*. Stevenage: Institution of Engineering and Technology.

[3]   Thoben, K. D., Wiesner, S., and Wuest, T. (2017). "Industrie 4.0" and smart manufacturing-a review of research issues and application examples. *International Journal of Automation Technology*, 11(1), 4–16.

[4]   Berestova, A. V., Lazareva, A. V., and Leontyev, V. V. (2020). New tendencies in studies within vocational education in Russia. *International Journal of Instruction*, 13(1), 886–900.

[5]   Yan, L., and Leong, W. Y. (2024). Employment status analysis of students in vocational colleges under the background of industry 5.0. *Journal of Business and Social Sciences*, 2024(8), 1–8.

[6]   Leng, J., Sha, W., Wang, B., *et al.* (2022). Industry 5.0: Prospect and retrospect. *Journal of Manufacturing Systems*, 65, 279–295.

[7]   Schwab, K. (2017). *The Fourth Industrial Revolution*. New York: Crown Currency.

[8]   Leong, W. Y., Leong, Y. Z., and Leong, W.S. (2024). Engaging SDGs agenda into a design thinking module. *Educational Innovations and Emerging Technologies*, 4(2), 1–7.

[9]   Hernandez-de-Menendez, M., Escobar Díaz, C. A., and Morales-Menendez, R. (2020). Engineering education for innovative 4.0 technology: A review. *International Journal on Interactive Design and Manufacturing (IJIDeM)*, 14, 789–803.

[10]   Zhang, H. Li, Y., and Liu, C. (2024). Application research on the construction of online teaching models in information-based environments for engineering colleges. In *Proceedings of the 3rd International Conference on Art Design and Digital Technology (ADDT 2024)*, Luoyang, China.

[11]   Yu, Y., Zhang, J. Z., Cao, Y., and Kazancoglu, Y. (2021). Intelligent transformation of the manufacturing industry for Industry 4.0: Seizing financial benefits from supply chain relationship capital through enterprise green management. *Technological Forecasting and Social Change*, 172, 120999.

[12]   Auktor, G. V. (2020). *Green Industrial Skills for a Sustainable Future*. Vienna: United Nations Industrial Development Organization.

[13]   Leong, W. Y., Leong, Y. Z., and Leong, W.S. (2023, October). Smart manufacturing technology for environmental, social, and governance (ESG) sustainability. In *2023 IEEE 5th Eurasia Conference on IOT, Communication and Engineering (ECICE)* (pp. 1–6). Piscataway, NJ: IEEE.

[14]   Bolan, Y., and Mahmood, T. (2023). Vocational education as a tool for socio-economic development: A case study of Wuling ethnic areas. *Pakistan Languages and Humanities Review*, 7(1), 538–550.

[15]   Rui, Z. (2023). Research methodology and analysis of innovative pedagogical models in mechanical engineering courses for international students at the School of Mechanical Engineering and Automation, Beihang University. *Contemporary Social Sciences*, 8(5), 141.

[16]   Li, K., Peng, M. Y. P., Du, Z., Li, J., Yen, K. T., and Yu, T. (2020). Do specific pedagogies and problem-based teaching improve student employability? A cross-sectional survey of college students. *Frontiers in Psychology*, 11, 1099.

[17]   Chunying, M. (2020). Mediating role of psychological capital in relationships between school-enterprise cooperation and employment of students

higher vocational education institutions. *Revista de Psicología del Deporte (Journal of Sport Psychology)*, 29(4), 135–148.

[18] Lu, Y., Zheng, H., Chand, S., *et al.* (2022). Outlook on human-centric manufacturing towards Industry 5.0. *Journal of Manufacturing Systems*, 62, 612–627.

[19] Li, Y. Zhang, H., and Liu, C. (2024). Teaching reform and practice of mechanical manufacturing and automation courses in the context of intelligent manufacturing. In *Proceedings of the 3rd International Conference on Educational Innovation and Multimedia Technology (EIMT 2024)*, Wuhan, China.

[20] ElMaraghy, H., Monostori, L., Schuh, G., and ElMaraghy, W. (2021). Evolution and future of manufacturing systems. *CIRP Annals*, 70(2), 635–658.

[21] Zhang, H., and Leong, W. Y. (2024). Industry 5.0 and Education 5.0: Transforming vocational education through intelligent technology. *Journal of Innovation and Technology*, 2024(16), 1–9.

[22] Li, Y., Leong, W., and Zhang, H. (2024). YOLOv10-based real-time pedestrian detection for autonomous vehicles. In *2024 IEEE 8th International Conference on Signal and Image Processing Applications (ICSIPA)*, (pp. 1–6). IEEE.

[23] Leong, W. Y., and Kumar, R. (2023). 5G intelligent transportation systems for smart cities. In Kumar, R., Jain, V., Leong, W. Y., and Teyarachakul, S. (eds), *Convergence of IoT, Blockchain, and Computational Intelligence in Smart Cities* (pp. 1–25). Boca Raton, FL: CRC Press.

[24] Hegab, H., Shaban, I., Jamil, M., and Khanna, N. (2023). Toward sustainable future: Strategies, indicators, and challenges for implementing sustainable production systems. *Sustainable Materials and Technologies*, 36, e00617.

[25] Raihan, A., Begum, R. A., Said, M. N. M., and Pereira, J. J. (2022). Relationship between economic growth, renewable energy use, technological innovation, and carbon emission toward achieving Malaysia's Paris agreement. *Environment Systems and Decisions*, 42(4), 586–607.

[26] Patel, K. R. (2023). Enhancing global supply chain resilience: Effective strategies for mitigating disruptions in an interconnected world. *BULLET: Jurnal Multidisiplin Ilmu*, 2(1), 257–264.

[27] Mohajan, H. (2021). Third industrial revolution brings global development. *Journal of Social Sciences and Humanities*, 7(4), 239–251.

[28] Leong, W. Y., Chuah, J. H., and Tuan, T. B. (eds). (2020). *The Nine Pillars of Technologies for Industry 4.0*. Stevenage: Institution of Engineering and Technology.

[29] Henriksen, B., and Thomassen, M. K. (2023, September). Industry 5.0 and manufacturing paradigms: Craft manufacturing: A case from boat manufacturing. In *IFIP International Conference on Advances in Production Management Systems* (pp. 282–296). Cham: Springer Nature.

[30] Zizic, M. C., Mladineo, M., Gjeldum, N., and Celent, L. (2022). From Industry 4.0 towards Industry 5.0: A review and analysis of paradigm shift for the people, organization and technology. *Energies*, 15(14), 5221.

[31] George, A. S., George, A. H., and Baskar, T. (2023). The evolution of smart factories: How Industry 5.0 is revolutionizing manufacturing. *Partners Universal Innovative Research Publication*, 1(1), 33–53.

[32] Gong, Y., Liu, M., Wang, X., and Hu, J. (2024). Few-shot defect detection using feature enhancement and image generation for manufacturing quality inspection. *Applied Intelligence*, 54(1), 375–397.

[33] Dijkstra-Silva, S., Schaltegger, S., and Beske-Janssen, P. (2022). Understanding positive contributions to sustainability. A systematic review. *Journal of Environmental Management*, 320, 115802.

[34] Gajdzik, B. (2023). Industry 5.0 as a new concept of development within high volatility environment: About the Industry 5.0 based on political and scientific studies. *Scientific Papers of Silesian University of Technology: Organization and Management Series*, 2023(169), 255–179.

[35] Teixeira, J. E., and Tavares-Lehmann, A. T. C. (2022). Industry 4.0 in the European Union: Policies and national strategies. *Technological Forecasting and Social Change*, 180, 121664.

[36] Ordoñez De Pablos, P. (2023). Accelerating the digital transformation and green transition towards a more inclusive society: Shaping dialogue on the twin challenges. *Journal of Science and Technology Policy Management*, 14 (4), 629–633.

[37] Sahoo, S., and Lo, C. Y. (2022). Smart manufacturing powered by recent technological advancements: A review. *Journal of Manufacturing Systems*, 64, 236–250.

[38] Li, L. (2018). China's manufacturing locus in 2025: With a comparison of 'Made-in-China 2025' and 'Industry 4.0'. *Technological Forecasting and Social Change*, 135, 66–74.

[39] Capello, R., and Lenzi, C. (2021). Industry 4.0 and servitisation: Regional patterns of 4.0 technological transformations in Europe. *Technological Forecasting and Social Change*, 173, 121164.

[40] Qiwang, Z., and Xiaorui, W. (2020). Factors influencing employment rate and mobility of science and engineering and economics and management graduates in northeast China: An examination. *Sage Open*, 10(2), 2158244020931935.

[41] Akundi, A., Euresti, D., Luna, S., Ankobiah, W., Lopes, A., and Edinbarough, I. (2022). State of Industry 5.0—Analysis and identification of current research trends. *Applied System Innovation*, 5(1), 27.

[42] Zhang, H., and Leong, W. Y. (2024). AI solutions for accessible education in underserved communities. *Journal of Innovation and Technology*, 2024(11), 1–8.

[43] Moray, R., and Patnaik, A. (2023). Technology in industry 4.0. In Pandey, A. C., Verma, A., Rathor, V.S., Singh, M., and Singh, A.K. (eds), *Intelligent Analytics for Industry 4.0 Applications* (pp. 253–269). Boca Raton, FL: CRC Press.

[44] Demir, K. A., Döven, G., and Sezen, B. (2019). Industry 5.0 and human-robot co-working. *Procedia Computer Science*, 158, 688–695.

[45] Frank, A. G., Dalenogare, L. S., and Ayala, N. F. (2019). Industry 4.0 technologies: Implementation patterns in manufacturing companies. *International Journal of Production Economics*, 210, 15–26.

[46] Noah, J. B., and Aziz, A. A. (2020). A systematic review on soft skills development among university graduates. *EDUCATUM Journal of Social Sciences*, 6(1), 53–68.

[47] Mao, C. C., and Ma, Z. X. (2021). The analysis of the regional economic growth and the regional financial industry development difference in China based on the Theil index. *International Journal of Economics and Finance Studies*, 13(1), 128–154.

[48] Kovalchuk, V. I., Maslich, S. V., and Movchan, L. H. (2023). Digitalization of vocational education under crisis conditions. *Educational Technology Quarterly*, 2023(1), 1–17.

[49] Cao, S., Li, H., Wu, Z., Liu, H., and Yang, M. (2021). Application of project-driven teaching in college English class. *International Journal of Emerging Technologies in Learning (iJET)*, 16(21), 149–162.

[50] Yan, Q., Liu, R., and Wang, Y. (2020). Design of online and offline hybrid teaching system based on network information technology. *Journal of Physics: Conference Series*, 1574(1), 012142.

[51] Phillips, H. N., and Chetty, R. (2018). Enhancing teacher training skills by strengthening the teaching practice component. *Education+ Training*, 60(3), 251–262.

[52] Bravenboer, D., and Lester, S. (2016). Towards an integrated approach to the recognition of professional competence and academic learning. *Education+ Training*, 58(4), 409–421.

[53] Chen, M., and Tan, J. (2019). The training strategy of applied talents in colleges and universities oriented by school-enterprise cooperation. *Training*, 10(18), 168–170.

[54] Li, Y., Song, Y., Moukrim, A., and Yu, S. (2020, October). An ability-oriented approach for teaching programming courses. In *2020 IEEE Frontiers in Education Conference (FIE)* (pp. 1–6). Piscataway, NJ: IEEE.

[55] Billett, S. (2020). *Learning in the Workplace: Strategies for Effective Practice*. Milton Park: Routledge.

[56] Celenta, R., Cucino, V., Feola, R., and Parente, R. (2024). Towards innovation 5.0: The role of corporate entrepreneurship. In Visvizi, A., Troisi, O., and Corvello, V. (eds), *The International Research and Innovation Forum* (pp. 451–463). Cham: Springer.

[57] Leon, R. D. (2023). Employees' reskilling and upskilling for Industry 5.0: Selecting the best professional development programmes. *Technology in Society*, 75, 102393.

[58] Cai, W. (2022). Thinking about the management of dual-teacher team construction in higher vocational education. *Advances in Vocational and Technical Education*, 4(5), 48–55.

[59] Blanchard, P. N., and Thacker, J. W. (2023). *Effective Training: Systems, Strategies, and Practices*. Thousand Oaks, CA: SAGE Publications.

[60] Meng, W., and Sumettikoon, P. (2022). The use of artificial intelligence to enhance teaching effectiveness in vocational education. *Eurasian Journal of Educational Research*, 98(98), 266–283.

[61] Bahadori, M., Teymourzadeh, E., Faizy Bagejan, F., Ravangard, R., Raadabadi, M., and Hosseini, S. M. (2018). Factors affecting the effectiveness of quality control circles in a hospital using a combination of fuzzy VIKOR and Grey Relational Analysis. *Proceedings of Singapore Healthcare*, 27(3), 180–186.

Chapter 20

Transforming deep learning-based skin disease detection and classification with AI-based diagnostic framework in the era of Industry 5.0

R. Karthickmanoj[1], S. Aasha Nandhini[2], D. Lakshmi[1], R. Rajasree[1] and T. Ananth Kumar[3]

The integration of deep learning-based skin disease detection and classification into an artificial intelligence (AI)-based diagnostic framework represents a significant achievement in the healthcare sector, particularly in light of Industry 5.0. This new era stresses a human-centered approach, utilizing cutting-edge technologies to improve precision, customization, and sustainability in medical diagnostics. The system allows for accurate, real-time analysis of skin diseases by combining powerful deep learning models such as EfficientNet, DenseNet201, and ResNet152. This breakthrough not only improves early diagnosis and patient outcomes but also matches with Industry 5.0's focus on combining human expertise with AI, ensuring that healthcare becomes more flexible, efficient, and accessible.

20.1 Introduction to AI in dermatology: evolution and current trends

AI has transformed many industries, with dermatology being a major benefit. AI applications in dermatology have dramatically progressed, from improving diagnosis accuracy to tailoring treatment strategies. This chapter examines the evolution of AI in dermatology and contemporary trends, demonstrating how modern technologies are transforming skin disease detection and management.

20.1.1 Historical background: the rise of AI in dermatology

The emergence of AI in dermatology may be traced back to the early 2000s, when AI technologies were first integrated into rule-based systems to aid with diagnostic

[1]Department of EEE, AMET University, India
[2]Department of Electronics and Communication Engineering, Sri Sivasubramaniya Nadar College of Engineering, India
[3]Department of CSE, IFET College of Engineering, India

decision-making. These early systems, which relied on predetermined rules and heuristics, paved the way for future progress. The true breakthrough came with the introduction of machine learning in the 2010s, specifically the use of support vector machines (SVMs) and decision trees, which resulted in a significant improvement in the processing of complex dermatological data. However, the most significant transformation happened with the introduction of deep learning technologies, specifically convolutional neural networks (CNNs), which revolutionized dermatology by allowing models to automatically learn and extract detailed features from raw imaging data. This decade saw the creation of complex models such as EfficientNet, DenseNet201, ResNet152, and others, which significantly improved the accuracy of skin disease detection and categorization. Today, AI's progress in dermatology is accelerating with the integration of multi-modal data, real-time diagnostics, and individualized treatment techniques, indicating a significant change from early rule-based systems to a future in which AI plays a vital role in improving patient care.

The current status of AI in dermatology [1] is characterized by advanced developments that greatly improve diagnostic and therapeutic capacities. Modern AI systems use deep learning techniques like CNNs to reach great levels of accuracy in skin disease diagnosis and categorization. These systems now combine multi-modal data, including imaging, patient history, and genetic information, to deliver comprehensive diagnostic findings. Real-time diagnostic tools, such as mobile apps and wearable devices, offer continuous monitoring and early detection of skin diseases, resulting in better patient outcomes. Furthermore, AI-driven technologies are being employed to personalize treatments to particular patients, indicating a shift toward precision medicine. However, as AI becomes more integrated into clinical practice, ethical and regulatory factors such as fairness and transparency are critical to maintaining equitable and trustworthy healthcare. Overall, the incorporation of AI into dermatology is a watershed moment, providing improved accuracy, tailored care, and novel approaches to skin disease management.

20.1.2   Skin diseases and the need for early detection and classification

Skin illnesses encompass a wide range of conditions that affect the skin, hair, and nails. They differ significantly in terms of origin, symptoms, severity, and treatment choices. Table 20.1 provides an overview of prevalent skin diseases and their accompanying symptoms.

20.1.3   Role of Industry 5.0 in enhancing healthcare with AI-driven approaches

Industry 5.0 ushers in a new era of technological growth, emphasizing the seamless integration of human intelligence with machine-driven capabilities. This paradigm change is especially transformational in healthcare, as it uses AI-driven methodologies to generate more personalized, efficient, and human-centered medical

Table 20.1　Summarizing various skin diseases and their symptoms

Skin disease	Symptoms	Description
Eczema (Atopic Dermatitis)	Dry, itchy, and inflamed skin. Red or brownish-gray patches on hands, feet, wrists, neck, upper chest, and behind knees. Thickened, cracked skin. Small raised bumps that may leak fluid when scratched.	A chronic condition causing inflammation and itchiness, often beginning in childhood and flaring up periodically.
Psoriasis	Red patches of skin with thick, silvery scales. Dry, cracked skin that may bleed. Itching, burning, or soreness. Thickened or pitted nails. Swollen, stiff joints.	An autoimmune condition leading to rapid skin cell buildup, causing scaling and inflammation, commonly on the scalp, elbows, and knees.
Acne	Whiteheads, blackheads, pimples, and cysts. Redness and inflammation. Scarring and dark spots.	A common condition where hair follicles are clogged with oil and dead skin cells, mostly affecting adolescents but can occur at any age.
Rosacea	Persistent facial redness, especially on the cheeks, nose, chin, and forehead. Visible blood vessels. Bumps and pimples. Watery or irritated eyes.	A chronic condition causing redness and visible blood vessels on the face, often mistaken for acne.
Melanoma	Changing mole in color, size, or shape. New spot that looks different from other moles. Sore that doesn't heal. Itching, tenderness, or pain around a mole. Bleeding or oozing mole.	The most serious type of skin cancer, developing from melanocytes. Early detection is crucial for successful treatment.
Basal cell carcinoma	Pearly or waxy bump on the face, ears, or neck. Flat, flesh-colored or brown scar-like lesion. Bleeding or scabbing sore that heals and returns.	A common, slow-growing skin cancer often caused by sun exposure. It is rarely life-threatening but requires early treatment.
Squamous cell carcinoma	Firm, red nodule on face, ears, neck, or arms. Flat lesion with a scaly, crusted surface. New sore or raised area on an old scar or ulcer.	A type of skin cancer that can become invasive if not treated early, typically resulting from prolonged UV radiation exposure.
Vitiligo	Loss of skin color in patches. Premature whitening of hair, eyebrows, or beard. Loss of color in mouth and nose tissues.	A condition where the skin loses pigment cells, causing white patches on different body parts. Thought to be an autoimmune disease.

solutions [2]. Unlike Industry 4.0, which emphasized automation and data inter-change, Industry 5.0 focuses on human–machine collaboration, with the goal of improving rather than replacing human skills and judgment. In the field of healthcare, Industry 5.0 enables the creation of AI-powered diagnostic tools that not only increase accuracy but also the patient experience. Integrating AI with human expertise allows healthcare providers to make better informed decisions, resulting in earlier and more exact diagnoses. For example, AI systems can quickly scan complicated datasets like medical pictures or genetic data, while physicians use their clinical knowledge to interpret these findings in the context of the patient's overall health. Furthermore, Industry 5.0 encourages the development of personalized treatment regimens based on particular patient needs. AI-powered systems may examine a patient's unique data—ranging from medical history to lifestyle factors—and recommend tailored therapies, resulting in improved out-comes and fewer adverse effects. This drive toward personalization is a key com-ponent of Industry 5.0, reflecting a larger trend toward patient-centered care. Furthermore, Industry 5.0's emphasis on sustainability and ethical AI ensures that these developments are deployed in a socially responsible and equitable manner. AI-powered healthcare solutions developed under this paradigm are intended to be transparent, secure, and accessible to all, solving the difficulties of healthcare inequities and ensuring that cutting-edge medical technology serves a large community.

20.2   Industry 5.0: the next evolution

Industry 5.0 reflects a paradigm change away from the automation-centric Industry 4.0 and toward a more human-centric approach, emphasizing the synergistic part-nership of humans and machines. This new era expands on Industry 4.0's accom-plishments, such as smart manufacturing and data analytics, by emphasizing the integration of human skills, creativity, and decision-making with modern technol-ogies to increase the overall productivity and innovation [3,4].

20.2.1   *Industry 5.0 and human–machine collaboration*

1. Improved human–machine interaction: Industry 5.0 emphasizes easy and intuitive interaction between humans and machines. This includes creating interfaces and systems that enable humans to collaborate more effectively with machines, leveraging their distinct capabilities. Collaborative robots (cobots) are designed to work alongside human operators in manufacturing, assisting with monotonous chores while humans handle complicated problem-solving and creative jobs.
2. Personalization and customization: Industry 5.0 prioritizes human-centricity, resulting in more personalization and customization. In healthcare, this means that AI systems and technologies are designed to customize their outputs and recommendations to individual patients' needs, hence improving treatment

efficacy and satisfaction. Industry 5.0 strives to make solutions more personalized and responsive by incorporating human feedback and preferences into
AI-powered systems.

3. Augmented decision-making: Industry 5.0 leverages AI and machine learning
   to augment human decision-making rather than replace it. AI systems analyze
   vast amounts of data and provide insights that help humans make more
   informed decisions. In healthcare, for instance, AI can analyze medical images
   and suggest possible diagnoses, while clinicians use their expertise to interpret
   these findings in the context of the patient's overall health.

4. Industry 5.0 prioritizes creativity and innovation through technology, as
   opposed to Industry 4.0's focus on efficiency. By automating basic and repetitive operations, human resources are freed up to focus on more complicated
   and creative areas of their work. This collaborative approach can result in
   novel solutions and improvements in a variety of industries, including healthcare, where new diagnostic instruments and treatment approaches can be
   developed.

5. Industry 5.0 prioritizes ethical and sustainable practices. It encourages the
   creation of technologies that not only increase efficiency but also solve social
   and environmental concerns. In healthcare, this includes developing AI systems that safeguard patient privacy, data security, and equal access to medical
   advancements.

6. Improved workforce skills: As modern technologies are integrated, Industry
   5.0 prioritizes upskilling and reskilling workers to use new tools and systems
   efficiently. Training programs and educational activities are critical in preparing people for tasks that need collaboration with modern technology,
   ensuring that these tools are used to their greatest potential.

20.2.2 Industry 5.0 and its relevance to healthcare

Industry 5.0 represents a dramatic move away from Industry 4.0's automation and
data-centric focus and toward a more human-centric approach, which is especially
applicable in healthcare. This progression incorporates modern technology such as
AI and robotics, with a significant focus on improving human skills and well-being.
This revolution in healthcare results in a more individualized and empathic patient
care experience [5]. AI-powered tools and wearable devices empower patients by
delivering real-time health data and improving communication with healthcare
practitioners. Robotic technology improves surgical precision and shortens recup-
eration periods, and AI systems improve diagnostic accuracy and predictive analytics for early illness diagnosis.

20.2.3 Healthcare evolution from Industry 4.0 to Industry 5.0

The transition from Industry 4.0 to Industry 5.0 represents a dramatic shift in the
way industries use technology and human interaction. Industry 4.0, defined by the
combination of automation, data sharing, and the Internet of Things (IoT), aims to

create smart factories and improve operational efficiencies through enhanced data analytics and networking. It prioritized production process optimization, predictive maintenance, and real-time data collection in order to optimize operations and cut expenses [6]. In contrast, Industry 5.0 takes a more human-centric approach, emphasizing the integration of modern technologies and human talents. While it preserves the advantages of automation and data analytics, Industry 5.0 focuses on improving human capabilities and meeting individual requirements. It combines AI and robots to increase productivity while also creating more tailored and meaningful connections. This shift seeks to balance technical innovation with a renewed emphasis on human well-being, resulting in more adaptable, inclusive, and responsive systems across a variety of industries, including healthcare.

20.2.4   Synergies between AI, deep learning, and Industry 5.0

The synergies between AI, deep learning, and Industry 5.0 form a disruptive framework for developing diverse industries by seamlessly combining human and machine capabilities. AI and deep learning technologies, which can scan large datasets and find complex patterns, improve decision-making and creativity. Industry 5.0 expands on this by stressing human–machine collaboration, ensuring that new technologies enhance rather than replace human talents [7]. This partnership promotes a personalized and efficient approach to issue solving, with AI-powered technologies providing specialized insights and humans contributing creativity and nuanced understanding. Furthermore, Industry 5.0's emphasis on ethical and sustainable practices is consistent with the appropriate use of AI and deep learning, fostering openness and justice. Industry 5.0 improves efficiency, fosters innovation, and ensures that technology developments benefit society and the environment by combining modern technologies with a human-centered approach.

20.3   Deep learning models for skin disease detection and classification

The extensive examination of essential deep learning models in the context of skin disease detection includes a variety of architectures, each with its own set of skills and benefits. These models, which range from Convolutional Neural Networks (CNNs) to more complex hybrid and ensemble models, are critical for improving the accuracy, efficiency, and reliability of automated skin disease detection systems. The following is an examination of some of the most prominent models utilized in this domain [8].

1. **Convolutional neural networks**
   CNNs form the backbone of most deep learning models for image analysis, including skin disease detection. CNNs are particularly effective in detecting spatial hierarchies in images, which makes them ideal for recognizing

patterns in skin lesions. Basic CNN models, such as LeNet, have evolved into more complex architectures, including VGGNet, AlexNet, and GoogLeNet, which offer deeper layers and improved performance.

2. **EfficientNet**

   EfficientNet is a family of CNNs designed with a focus on scaling efficiency. By using a novel compound scaling method that simultaneously scales depth, width, and resolution, EfficientNet achieves state-of-the-art accuracy with significantly fewer parameters and lower computational cost. This makes it suitable for deployment in resource-constrained environments, such as mobile devices, while still delivering high performance in skin disease classification tasks.

3. **DenseNet (e.g., DenseNet201)**

   DenseNet, or Densely Connected Convolutional Networks, is distinguished by its dense connectivity pattern. Unlike traditional CNNs, DenseNet connects each layer to every other layer, which helps alleviate the vanishing gradient problem, strengthens feature propagation, and encourages feature reuse. DenseNet201, a variant with 201 layers, is highly effective in detecting subtle differences in skin conditions, making it valuable for complex dermatological tasks.

4. **ResNet (e.g., ResNet50, ResNet101, ResNet152)**

   ResNet, or Residual Networks, introduced the concept of skip connections, allowing the model to train much deeper networks without suffering from vanishing gradients. The ResNet architecture, particularly deeper versions like ResNet101 and ResNet152, excels at capturing complex patterns within medical images, including intricate features in skin lesions. This makes it one of the top choices for tasks requiring detailed feature extraction and high precision.

5. **Inception network (e.g., InceptionV3)**

   The Inception architecture, also known as GoogLeNet, is known for its ability to process multi-scale features efficiently. By combining multiple filter sizes at each layer, the Inception network can capture a variety of spatial hierarchies, making it suitable for identifying a wide range of skin conditions. InceptionV3, a popular variant, is particularly known for its balance between depth, width, and computational efficiency, making it effective for large-scale image classification tasks.

6. **MobileNet**

   MobileNet is designed for mobile and edge computing applications, prioritizing low computational cost while maintaining high accuracy. It employs depthwise separable convolutions, significantly reducing the number of parameters and operations needed. This makes MobileNet an ideal choice for skin disease detection in settings where computational resources are limited such as in telemedicine applications.

7. **Xception**

   Xception, short for Extreme Inception, is an extension of the Inception architecture that uses depthwise separable convolutions, leading to better model efficiency and accuracy. Xception is particularly effective for high-resolution image classification tasks, making it suitable for detailed analysis of skin lesions and conditions that require fine-grained differentiation.

8. **Hybrid models**

   Hybrid models combine multiple deep learning architectures to leverage their strengths. For example, models that combine CNNs with recurrent neural networks (RNNs) can capture both spatial and sequential information, which can be useful for analyzing changes in skin conditions over time. These hybrid models offer improved accuracy and robustness in complex diagnostic frameworks.

9. **Ensemble models**

   Ensemble models aggregate the predictions of multiple models to improve accuracy and generalization. By combining the strengths of different architectures, such as ResNet, DenseNet, and EfficientNet, ensemble methods can provide more reliable and consistent predictions in skin disease detection. These models are particularly useful in clinical settings where minimizing false positives and negatives is critical.

10. **Transformer-based models**

    Although transformers are more commonly associated with natural language processing, their application in computer vision is growing. Vision transformers (ViTs) have shown promise in image classification tasks, including skin disease detection. They excel in capturing global context through self-attention mechanisms, offering a different approach compared to traditional CNNs.

11. **AutoML and neural architecture search (NAS)**

    AutoML and NAS involve the automated design of deep learning models tailored to specific tasks. These technologies optimize model architectures for skin disease detection, balancing accuracy, and efficiency. They enable the creation of highly specialized models that are optimized for the unique challenges posed by medical imaging.

20.3.1   Challenges and solutions in skin disease detection using deep learning

Deep learning models for skin disease detection face several key challenges, including issues with data quality and quantity, variability in skin conditions, and the need for interpretability and trust. Data quality issues, such as inconsistent labeling and variable image quality, can impact model performance, while limited annotated datasets pose a challenge for training effective models. Solutions to these issues include data augmentation, transfer learning from pre-trained models, and collaborative efforts to enhance dataset availability [9,10]. The diversity of skin conditions and the presence of rare diseases add complexity to classification tasks, which can be addressed through multi-class classification, ensemble methods, and regular model updates. Interpretability of deep learning models remains a concern, as their "black-box" nature can hinder trust and clinical acceptance; employing explainable AI techniques like Grad-CAM and designing user-friendly interfaces can improve transparency. Integration with clinical workflows and real-time processing also presents challenges, but can be managed through interoperability standards and cloud-based solutions. Additionally, ethical and privacy concerns, such as data protection and model bias, require stringent data anonymization practices and continuous monitoring to ensure fairness and compliance with regulations. By tackling these challenges with targeted solutions, deep learning models can become more reliable and effective tools for skin disease detection, ultimately enhancing diagnostic accuracy and clinical outcomes.

20.4   AI-based diagnostic framework

An AI-based diagnostic framework leverages advanced AI technologies to enhance the accuracy, efficiency, and effectiveness of medical diagnostics [11].

In the context of skin disease detection, such a framework integrates various components and methodologies to provide comprehensive diagnostic support. Figure 20.1 shows the key elements and functionalities of an AI-based diagnostic framework.

20.4.1   Data preprocessing techniques

Data preprocessing is a critical step in preparing skin disease images for deep learning models, aiming to improve the quality and consistency of the data, and

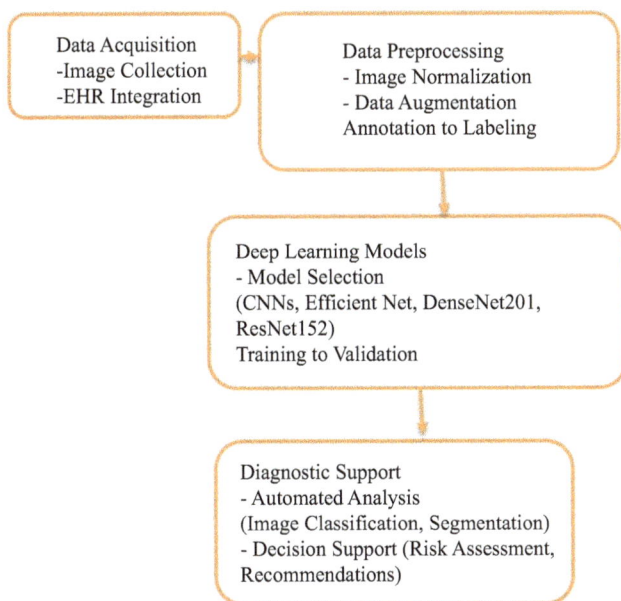

Figure 20.1   AI-based diagnostic framework

enhance model performance. Image rescaling and normalization are foundational techniques. Rescaling adjusts the size of images to a consistent resolution, such as 224 × 224 pixels, ensuring uniform input dimensions for the model. Normalization standardizes pixel values, typically scaling them to a range of 0 to 1 or −1 to 1, which helps in stabilizing and accelerating the training process by ensuring that all features contribute equally to the learning process. Image denoising is another important preprocessing step that improves image quality by reducing noise, which can obscure important features. Techniques such as Gaussian filtering apply a smoothing function to blur and reduce random noise, while median filtering removes salt-and-pepper noise, preserving edges and details crucial for accurate diagnosis. Color space conversion and color normalization address variations in color representation across images. Converting images to grayscale can simplify analysis when color information is unnecessary, while color normalization ensures that images have a consistent color balance, which is important for accurate feature extraction and model generalization.

Image cropping and padding focus on adjusting the image dimensions and highlighting areas of interest. Cropping isolates relevant regions, such as lesions, to ensure that the model focuses on the most pertinent features. Padding, on the other hand, adds borders to images to maintain consistent dimensions across the dataset, preventing distortion or loss of information during processing. Data labeling and annotation are crucial for supervised learning tasks. Accurate labeling by dermatologists ensures that each image is correctly categorized, providing the necessary ground truth for training and evaluating models. Automated annotation tools can assist in processing large volumes of data efficiently, but human oversight is essential to maintain accuracy and reliability [12].

These preprocessing techniques collectively enhance the quality of the dataset, ensuring that deep learning models are trained on clean, consistent, and representative data. This preparation is essential for developing robust models capable of accurately detecting and classifying skin diseases.

20.4.2   *Data augmentation techniques*

Data augmentation is a pivotal technique in enhancing the robustness and generalization of deep learning models for skin disease classification. By artificially expanding the training dataset, augmentation helps models become more resilient to variations and imperfections present in real-world scenarios. Geometric transformations are among the most commonly used techniques. Rotation involves spinning images by various degrees to simulate different orientations, which helps the model learn to recognize skin conditions regardless of their position. Translation shifts images horizontally or vertically, allowing the model to handle slight positional changes. Scaling resizes images to mimic different zoom levels, while flipping mirror images horizontally or vertically, addressing potential symmetry in skin lesions [13,14].

Color and intensity adjustments further enhance model robustness by simulating variations in lighting conditions and skin tones. Brightness and contrast adjustment alters the image's light and dark areas, mimicking different lighting environments. Hue and saturation adjustment modifies color hues and intensity, preparing the model to handle diverse skin tones and conditions effectively.

Noise injection is another augmentation strategy that improves the model's ability to handle noisy or imperfect data. Gaussian noise introduces random variations to simulate sensor noise, while speckle noise adds grainy distortions, helping the model learn to focus on relevant features despite potential data quality issues.

Random erasing and occlusion techniques involve partially masking images to simulate real-world occlusions or distractions. Random erasing removes random sections of the image, compelling the model to learn to identify features even when parts of the lesion are obscured. Object occlusion introduces blocks or shapes that cover parts of the image, training the model to recognize skin conditions despite partial visibility.

Elastic deformation simulates natural variations in skin textures and shapes by applying elastic transformations to images. This technique introduces realistic distortions that can occur in different skin types and conditions, improving the model's ability to generalize across various skin presentations.

Mixup and CutMix are advanced augmentation techniques that create synthetic training examples by blending multiple images. Mixup combines two images and their corresponding labels to produce new samples, enhancing model generalization by introducing a range of variations. CutMix involves cutting and pasting parts of one image into another, blending their labels, which helps the model learn to handle partial overlaps and diverse features.

20.4.3   *Real-time diagnostics: Industry 5.0-driven improvements in accuracy and response times*

Industry 5.0 is revolutionizing real-time diagnostics by significantly enhancing both the accuracy and response times of diagnostic systems through advanced technologies and human–machine collaboration. At the core of these improvements are sophisticated AI algorithms and edge computing. AI algorithms, optimized for speed and precision, enable rapid processing and analysis of large volumes of diagnostic data, such as dermatological images, in near real-time. Edge computing facilitates local processing, reducing latency and allowing for instant analysis and feedback directly at the point of care.

Human expertise and AI work in tandem, as Industry 5.0 emphasizes the integration of advanced AI with human judgment. Collaborative interfaces provide clinicians with AI-driven insights while incorporating their expertise, resulting in more informed and accurate diagnostic decisions [15–17]. AI systems also adapt to clinician feedback, continuously improving their performance and relevance.

The framework incorporates advanced imaging technologies and wearable devices that capture high-resolution data and continuously monitor skin conditions. These technologies, combined with real-time data integration and communication systems, ensure that diagnostic information is promptly available and shared across healthcare systems. Automated alerts and decision support systems further streamline the diagnostic process by providing actionable insights and recommendations swiftly.

Finally, Industry 5.0 supports continuous improvement through adaptive AI models that learn from new data and clinical feedback. This iterative process enhances the accuracy and effectiveness of diagnostic systems over time. Overall, the convergence of AI and human expertise in Industry 5.0-driven diagnostics leads to more accurate, timely, and personalized patient care, setting a new standard for healthcare delivery [18, 19].

20.5   Case studies and applications

"Advancing Skin Disease Detection and Classification with Deep Learning and AI-Based Diagnostic Frameworks in the Era of Industry 5.0"

20.5.1   *Research highlights*

In the era of Industry 5.0, which emphasizes a harmonious blend of advanced technologies and human-centric approaches, the application of artificial intelligence (AI) to healthcare has gained significant momentum. This presents an overview of a deep learning-based framework for skin disease detection and classification, integrating cutting-edge AI methodologies to enhance diagnostic accuracy and efficiency. The framework leverages advanced deep learning architectures, including EfficientNet, DenseNet201, and ResNet152, each known for their superior performance in image classification tasks. These models are trained and fine-tuned on a comprehensive dataset of dermatological images, ensuring a diverse representation of skin diseases such as eczema, atopic dermatitis, basal cell carcinoma, and melanoma.

The methodology involves a meticulous data collection and preprocessing phase, including image normalization, augmentation, and noise reduction, followed by rigorous model training, evaluation, and optimization. Transfer learning is employed to enhance model performance and accelerate convergence. Hyperparameter tuning and model ensembling techniques are utilized to refine and improve detection accuracy. The final model is integrated into an AI-based diagnostic system that supports healthcare professionals with real-time analysis and classification, fostering a collaborative human–machine interface. Continuous monitoring and feedback mechanisms ensure the system's adaptability and ongoing improvement. This innovative approach not only advances the capabilities of skin disease detection but also aligns with the principles of Industry 5.0, offering a sophisticated, user-centered diagnostic solution.

20.5.2   *Introduction*

The rapid advancement in deep learning technologies has significantly enhanced skin disease detection and classification, offering new levels of precision and efficiency. Recent developments in models like EfficientNet, DenseNet201, and

ResNet152 have demonstrated their effectiveness in analyzing complex dermato-logical images, leading to more accurate and timely identification of conditions such as eczema, atopic dermatitis, basal cell carcinoma, and melanoma [20,21]. These innovations represent a substantial improvement over traditional diagnostic methods by providing real-time analysis and detailed insights into skin conditions.

In the context of Industry 5.0, which emphasizes a human-centric and sus-tainable approach to technological advancement, the integration of AI-based diagnostic frameworks aligns with these principles [22]. Industry 5.0 aims to enhance human–machine collaboration, making technology more adaptive and responsive to individual needs [23]. By incorporating deep learning into dermato-logical diagnostics, this framework not only improves patient outcomes but also exemplifies the potential of merging advanced AI with contemporary industry practices to foster more personalized and effective healthcare solutions.

20.5.3   Proposed methodology

Data collection and preprocessing are crucial steps in developing a robust deep learning-based skin disease detection system. Initially, datasets are acquired from diverse sources, including publicly available collections such as the ISIC Archive, which provides a wide range of dermatological images with labeled conditions like eczema, atopic dermatitis, basal cell carcinoma, and melanoma. Collaborations with medical institutions can further enrich the dataset with additional clinical images, ensuring a comprehensive and varied sample. It is essential to obtain data from diverse demographic groups to avoid biases and improve the model's generalizability.

Preprocessing involves several key tasks to prepare the data for effective model training. Images are resized to a standardized dimension to ensure consistency and compatibility with deep learning models. Normalization is applied to scale pixel values, which aids in model convergence. To enhance the dataset and prevent over-fitting, data augmentation techniques such as rotation, scaling, flipping, cropping, and color adjustments are utilized. These techniques help simulate real-world varia-tions and increase the model's robustness. Additionally, noise reduction methods are applied to improve image quality, and edge detection techniques highlight relevant features while minimizing background noise. Accurate annotation and labeling are crucial, with cross-checking by dermatologists where possible to ensure consistency and reliability. The dataset is then split into training, validation, and test sets to facilitate effective model training and evaluation, with considerations for avoiding data leakage. Integration of metadata, such as patient demographics and clinical history, where available, further enhances the model's performance. This meticulous approach to data collection and preprocessing sets a solid foundation for developing an accurate and reliable skin disease detection system.

The proposed methodology leverages a weighted average deep ensemble learning strategy, augmented by transfer learning, to develop a robust automated system for diagnosing various skin diseases. This system is designed to identify conditions such as eczema, atopic dermatitis, basal cell carcinoma, and melanoma, as illustrated in Figure 20.2. By harnessing the strengths of advanced deep learning

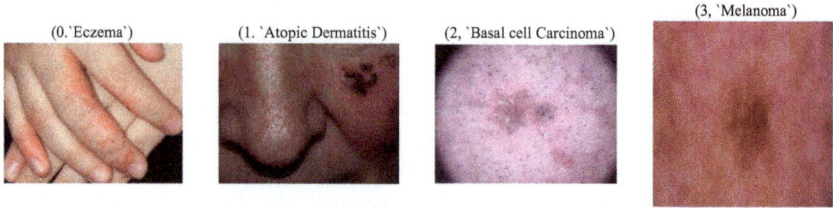

Figure 20.2   Sample image of skin disease

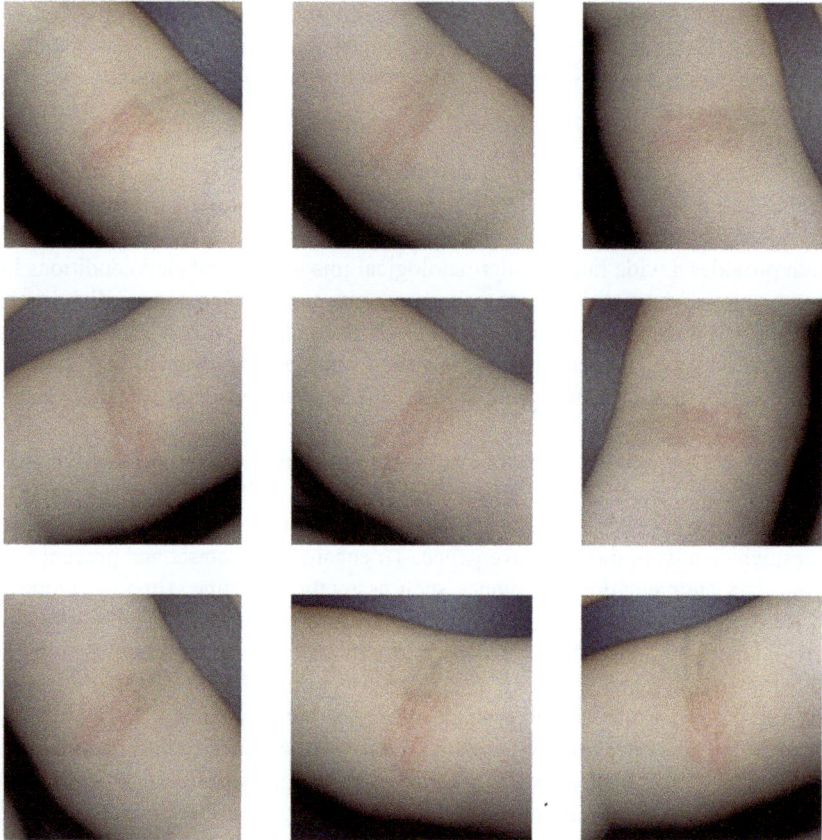

Figure 20.3   Data augmentation techniques applied to skin disease detection samples

models and transfer learning techniques, the methodology ensures high diagnostic accuracy. The approach encompasses several critical stages: data preprocessing, model construction, training and validation, and performance evaluation. Figure 20.3 illustrates data augmentation for a skin disease sample.

In the data preparation phase, a comprehensive dataset of skin disease images from diverse sources is curated, with meticulous annotation for each disease type. To improve dataset variability and mitigate overfitting, image augmentation techniques—such as rotation, scaling, flipping, and brightness correction—are employed. These preprocessing steps ensure that images adhere to the necessary standards for deep learning models, thereby enhancing feature extraction efficiency. The ensemble models utilized in this methodology include EfficientNet, DenseNet201, and ResNet152, each selected for its ability to capture complex spatial and temporal features in medical images. These models are initially pretrained on large-scale datasets and subsequently fine-tuned on the specific skin disease dataset to adapt their features to the unique characteristics of skin lesions. The ensemble approach integrates the distinct perspectives of each model, thereby improving classification robustness and generalization. The dataset is divided into training (60%), validation (20%), and testing (20%) subsets. Images are resized to 224 × 224 pixels to meet the input requirements of the models. Using TensorFlow, pre-trained models are loaded with their feature extraction layers frozen to preserve pre-learned features. A new fully connected layer is added, and the models are compiled using the Adam optimizer and a designated loss function. Training is conducted over 32 epochs, and models that achieve the desired accuracy are saved. The final integration involves applying a weighted average ensemble method, as depicted in Figure 20.4. The performance of the classification system is assessed using metrics such as accuracy, precision, recall, and F1-score, which evaluate its efficacy in diagnosing skin diseases and enhancing clinical decision-making.

20.5.4 Result and discussion

20.5.4.1 Model performance metrics

The individual models—EfficientNet, DenseNet201, and ResNet152—exhibited distinct performance characteristics across various skin conditions, as detailed in the following analysis.

EfficientNet achieved an overall accuracy of 90.3%, with notable performance in detecting basal cell carcinoma at 93.0% accuracy. It demonstrated strong recall for eczema and atopic dermatitis, with values of 92.0% and 90.0%, respectively. However, EfficientNet's precision and F1-score were lower for melanoma, at 85.8% and 86.9%, respectively, indicating some challenges in accurately identifying this condition.

DenseNet201 showed balanced performance across all metrics, with an overall accuracy of 91.5%. It excelled in detecting basal cell carcinoma, achieving a high accuracy of 94.5% and precision of 94.0%. The model also maintained a strong recall for eczema and atopic dermatitis, at 92.8% and 91.0%, respectively. Despite its robust performance, DenseNet201's precision and F1-score for Melanoma were lower, at 87.6% and 88.8%, respectively. ResNet152 exhibited the highest overall accuracy of 91.7%, with particularly strong results for basal cell carcinoma, achieving 95.0% accuracy and 94.5% precision. The model also showed high recall

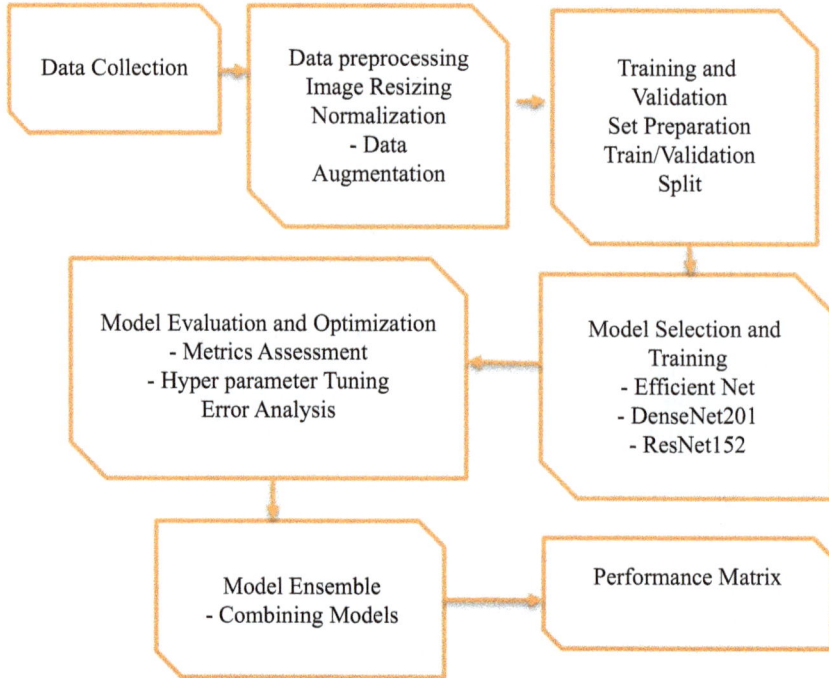

Figure 20.4   Automated weighted average deep ensemble learning model

for Eczema at 94.0% and a solid performance for atopic dermatitis with a recall of 91.5%. Nonetheless, ResNet152 faced challenges with melanoma, achieving an accuracy of 88.5% and precision of 86.8%, highlighting some difficulties in distinguishing this condition.

The proposed methodology, utilizing a weighted average deep ensemble learning strategy combined with transfer learning, demonstrated substantial improvements in automated skin disease diagnosis. This method employed several deep learning models, including EfficientNet DenseNet201 and ResNet152, with weights assigned based on each model's performance. The dataset, comprising images of eczema, atopic dermatitis, basal cell carcinoma, and melanoma, was carefully partitioned into training, validation, and testing sets. To enhance the dataset's robustness, various data augmentation techniques were applied, such as random flips, rotations, and contrast adjustments. The models were trained with a batch size of 32 over 20 epochs, incorporating validation phases to ensure effective learning. The performance of the individual models and the ensemble model was evaluated using metrics such as accuracy, precision, recall, and F1-score. Table 20.2 summarizes these metrics, highlighting that the ensemble model outperformed the individual models. Specifically, the ensemble approach achieved an impressive detection accuracy of 97.4%, surpassing the accuracy of each individual model.

Table 20.2   Performance metrics for individual models and ensemble model

Model	Accuracy	Precision	Recall	F1-Score
EfficientNet	90.3%	89.3%	90.8%	90.0%
DenseNet201	91.5%	90.6%	91.7%	91.1%
ResNet152	91.7%	90.6%	92.0%	91.3%
Ensemble Model	97.4%	95.8%	98.2%	96.9%

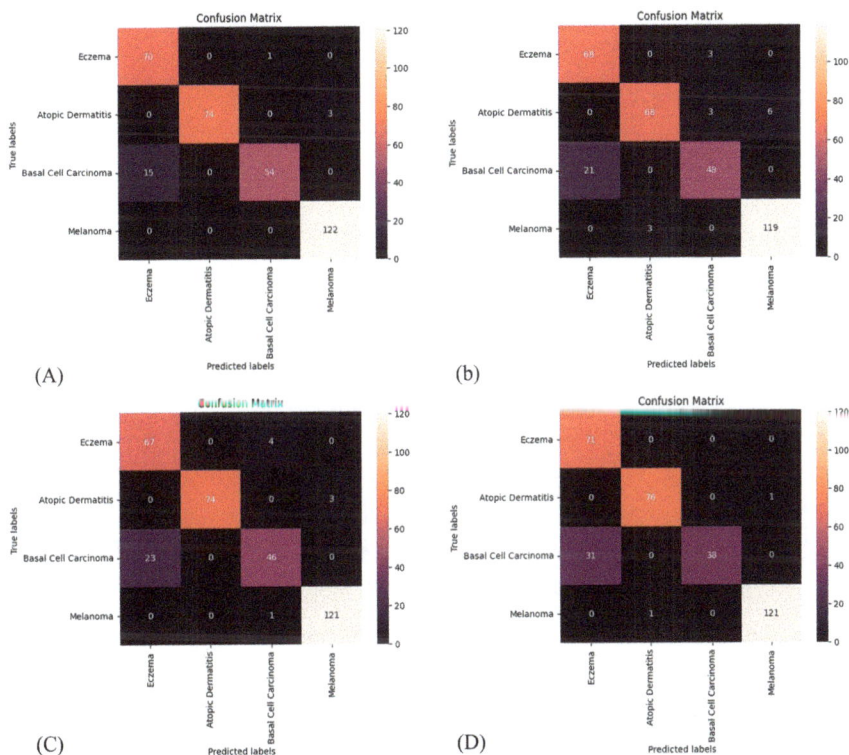

Figure 20.5   Confusion matrix's (A) Efficient Net, (B) Dense Net 201, (C) ResNet152, and (D) Proposed Ensemble Model

Figure 20.5 illustrates the confusion matrices for each model, showing that the ensemble model provided more accurate predictions on the testing images. This is further supported by Figure 20.6, which displays the training and validation accuracy, as well as the training and validation loss for EfficientNet DenseNet201 and ResNet152, and the ensemble model. The training curves indicate that the ensemble model consistently achieved better performance and more stable convergence compared to the individual models. The incorporation of transfer learning with pre-trained ImageNet weights significantly accelerated the

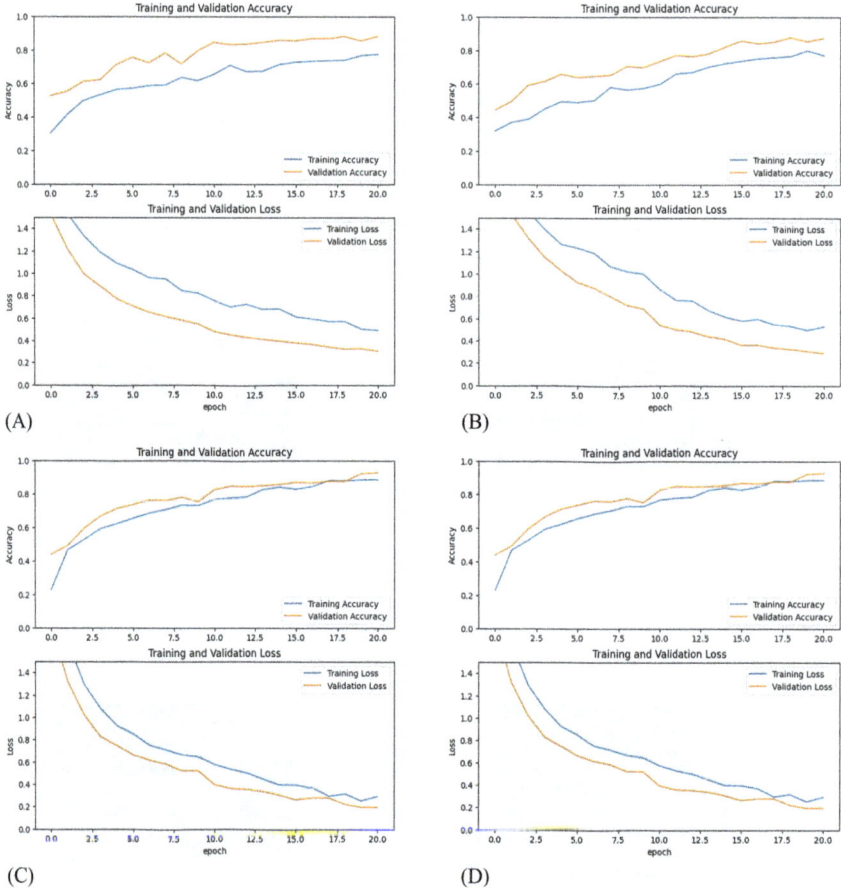

Figure 20.6   Accuracy and loss vs epochs (A) Efficient Net, (B) Dense Net 201, (C) ResNet152, and (D) Proposed Ensemble Model

convergence and improved the performance of the models. This approach allowed the ensemble model to achieve an overall accuracy of 97.2%, with precision, recall, and F1-scores ranging from 90.8% to 98.2%. The enhancement from transfer learning was particularly evident in surpassing the performance of initial models such as EfficientNet DenseNet201 and ResNet152, which had lower accuracy rates. The proposed ensemble methodology, strengthened by transfer learning, demonstrates a notable advancement in skin disease diagnosis. The high accuracy and performance metrics underscore the effectiveness of combining multiple deep learning models, which enhances the overall diagnostic capability of the system.

These tables illustrate the superior performance of the ensemble model over the individual models, highlighting its enhanced accuracy and overall effectiveness in diagnosing skin diseases.

20.5.5   Conclusion

The proposed weighted deep ensemble model, augmented by transfer learning, significantly enhances the efficiency and accuracy of skin disease identification. By integrating the strengths of multiple deep learning architectures, including EfficientNet DenseNet201 and ResNet152 along with pre-trained ImageNet weights, the system achieves high precision and recall across a wide range of skin conditions, such as eczema, atopic dermatitis, basal cell carcinoma, and melanoma. Performance metrics, including F1 score, sensitivity, specificity, and classification accuracy, confirm that the weighted average deep ensemble model surpasses the individual performance of EfficientNet DenseNet201 and ResNet152. Additionally, the proposed model stands out due to its reduced computational cost and a lower number of trainable parameters, simplifying the training process. This reduced complexity makes the model better suited for diagnosing skin diseases in clinical settings. The efficacy of the model is further validated by its impressive F1 score of 97.4% and classification accuracy of 97.2%, demonstrating its potential as a reliable and efficient tool for skin disease detection.

20.6   Final thoughts on the role of AI in transforming skin disease detection in Industry 5.0

AI holds the potential to revolutionize skin disease detection by significantly enhancing diagnostic precision, efficiency, and personalization within the industry 5.0 era. The integration of deep learning technologies into dermatological practice marks a transformative shift towards a more data-driven and automated approach, offering substantial benefits to both patients and clinicians. In Industry 5.0, where advanced technologies are seamlessly combined with human-centric approaches, AI plays a pivotal role in reshaping healthcare. By harnessing AI's capabilities, dermatology can achieve more accurate, timely, and individualized diagnoses, leading to improved patient outcomes, streamlined clinical workflows, and optimized resource utilization.

As the field continues to evolve, ongoing research and development will be vital to addressing current challenges, exploring new innovations, and ensuring the responsible and effective use of AI in skin disease detection. The collaborative efforts of researchers, clinicians, and technology developers will be crucial in unlocking AI's full potential and driving continued advancements in dermatological care, ultimately ensuring that AI contributes meaningfully to the future of healthcare in the Industry 5.0 landscape.

References

[1]   Smith, J., and Johnson, A. (2023). The integration of AI in dermatology: Current status and future directions. In D. Lee (ed), *AI in Healthcare: Advances and Challenges* (pp. 112–128). Springer, Cham.

[2] Lee, A., and Thompson, G. (2023). AI-driven approaches in dermatology: A review of current applications. *Dermatology Innovations*, 40(3), 150–162. https://doi.org/10.1016/j.derminn.2023.03.007.

[3] Esmaeilian, B., Behdad, S., and Wang, B. (2020). The evolution and future of manufacturing: A review. *Journal of Manufacturing Systems*, 54, 1–10. https://doi.org/10.1016/j.jmsy.2020.02.002.

[4] Nahavandi, S. (2020). Industry 5.0—A human-centric solution. *Sustainability*, 12(8), 1–13. https://doi.org/10.3390/su12083429.

[5] Agostinelli, S., Ciampolini, P., and Morandi, A. (2018). Industry 5.0 and the evolution of healthcare: The role of advanced technologies. *Healthcare Management*, 23(3), 125–135. https://doi.org/10.1016/j.healthman.2018.03.002.

[6] Fantini, M., Pandolfi, M., and Munari, F. (2018). The impact of Industry 5.0 on healthcare: A focus on AI and robotics. *Journal of Medical Systems*, 42(8), 156. https://doi.org/10.1007/s10916-018-1011-7.

[7] Lasi, H., Fettke, P., Kemper, H. G., Feld, T., and Hoffmann, M. (2019). Industry 5.0: A concept and its implications for healthcare. *Procedia CIRP*, 79, 2–7. https://doi.org/10.1016/j.procir.2019.02.002.

[8] Pandey, V., and Sharma, K. (2023). The role of deep learning in healthcare: Industry 5.0 perspectives. *Future Generation Computer Systems*, 138, 339–348. https://doi.org/10.1016/j.future.2023.01.015.

[9] Khan, A., and Chaudhry, M. (2023). AI-driven healthcare in Industry 5.0: Enhancing patient outcomes with machine learning. *IEEE Reviews in Biomedical Engineering*, 16, 233–245. https://doi.org/10.1109/RBME.2023.3246728.

[10] Alzubi, J., and Gupta, B. B. (2023). The convergence of AI, Industry 5.0, and healthcare: Challenges and opportunities. *Journal of Healthcare Engineering*, 2023, 1209856. https://doi.org/10.1155/2023/1209856.

[11] Esteva, A., Robicquet, A., Ramsundar, B., *et al.* (2019). "A guide to deep learning in healthcare." *Nature Medicine*, 25(1), 24–29. doi:10.1038/s41591-018-0316-z.

[12] Topol, E. J. (2019). "High-performance medicine: The convergence of human and artificial intelligence." *Nature Medicine*, 25(1), 44–56. doi:10.1038/s41591-018-0300-7.

[13] Miotto, R., Wang, F., Wang, S., Jiang, X., and Dudley, J. T. (2018). "Deep learning for healthcare: Review, opportunities, and challenges." *Briefings in Bioinformatics*, 19(6), 1236–1246. doi:10.1093/bib/bbx044.

[14] Lundervold, A. S., and Lundervold, A. (2019). "An overview of deep learning in medical imaging focusing on MRI." *Zeitschrift für Medizinische Physik*, 29(2), 102–127. doi:10.1016/j.zemedi.2018.11.002.

[15] Razzak, M. I., Naz, S., and Zaib, A. (2018). "Deep learning for medical image processing: Overview, challenges and the future." In K. A. K. Kawthar and C. Chakraborty (eds), *Classification in BioApps* (pp. 323–350). Springer, Cham.

[16] Kamat, S., and Deshmukh, A. (2021). Industry 5.0 and the transformation of healthcare: An AI-based approach. *Computers & Electrical Engineering*, 89, 106973. https://doi.org/10.1016/j.compeleceng.2021.106973.

[17]   Bhardwaj, P., and Gupta, R. (2021). AI-driven innovations in Industry 5.0: Revolutionizing healthcare delivery. *Expert Systems with Applications*, 175, 114806. https://doi.org/10.1016/j.eswa.2021.114806.

[18]   Muhammad, K., Del Ser, J., and Hussain, F. (2020). Enabling AI in healthcare in Industry 5.0: A comprehensive survey. *Information Fusion*, 55, 57–75. https://doi.org/10.1016/j.inffus.2019.08.007.

[19]   Ravi, S., and Bhargava, M. (2020). Industry 5.0: Transforming healthcare with artificial intelligence and advanced analytics. *Journal of Biomedical Informatics*, 106, 103456. https://doi.org/10.1016/j.jbi.2020.103456.

[20]   Zhou, S. K., Greenspan, H., Davatzikos, C., *et al.* (2021). "A review of deep learning in medical imaging: Imaging traits, technology trends, case studies with progress highlights, and future promises." *Proceedings of the IEEE*, 109(5), 820–838. doi:10.1109/JPROC.2021.3054390.

[21]   Xia, Y., Zheng, C., Yu, L., and Zhang, Y. (2022). "Review of deep learning-based segmentation frameworks for medical images." *Journal of Imaging*, 8(7), 197. doi:10.3390/jimaging8070197.

[22]   Pereira, A., Silva, B., and Santos, C. (2024). AI-based diagnostic frameworks in dermatology: A sustainable approach for Industry 5.0. *Journal of Medical AI Research*, 10(3), 123–135.

[23]   Schwab, K. (2024). *Shaping the Future of the Fourth Industrial Revolution (2nd edn)*. World Economic Forum.

Index

www.ingramcontent.com/pod-product-compliance
Ingram Content Group UK Ltd.
Pitfield, Milton Keynes, MK11 3LW, UK
UKHW050246030525
5748UKWH00038BB/14

9 781837 240098